T0205466

Birkhäuser

Advances in Mathematical Fluid Mechanics

Lecture Notes in Mathematical Fluid Mechanics

Editor-in-Chief
Giovanni P Galdi
University of Pittsburgh, Pittsburgh, PA, USA

Series Editors
Didier Bresch
Université Savoie-Mont Blanc, Le Bourget du Lac, France

Volker John
Weierstrass Institute, Berlin, Germany

Matthias Hieber
Technische Universität Darmstadt, Darmstadt, Germany

Igor Kukavica
University of Southern California, Los Angles, CA, USA

James Robinson
University of Warwick, Coventry, UK

Yoshihiro Shibata
Waseda University, Tokyo, Japan

Lecture Notes in Mathematical Fluid Mechanics as a subseries of "Advances in Mathematical Fluid Mechanics" is a forum for the publication of high quality monothematic work as well lectures on a new field or presentations of a new angle on the mathematical theory of fluid mechanics, with special regards to the Navier-Stokes equations and other significant viscous and inviscid fluid models.

In particular, mathematical aspects of computational methods and of applications to science and engineering are welcome as an important part of the theory as well as works in related areas of mathematics that have a direct bearing on fluid mechanics.

More information about this subseries at http://www.springer.com/series/15480

I. V. Denisova • V. A. Solonnikov

Motion of a Drop in an Incompressible Fluid

Birkhäuser

I. V. Denisova
Institute of Problems
of Mechanical Engineering
Russian Academy of Sciences
St. Petersburg, Russia

V. A. Solonnikov
St. Petersburg Department
of Steklov Math. Institute
Russian Academy of Sciences
St. Petersburg, Russia

Translation from Russian language edition: Dvizhenie kapli v neszhimaemoy zhidkosti: monografiya by I. V. Denisova and V. A. Solonnikov, © Izdatel'stvo lan' 2020, Izdatel'stvo lan'. All Rights Reserved.

ISSN 2297-0320 ISSN 2297-0339 (electronic)
Advances in Mathematical Fluid Mechanics
ISSN 2510-1374 ISSN 2510-1382 (electronic)
Lecture Notes in Mathematical Fluid Mechanics
ISBN 978-3-030-70052-2 ISBN 978-3-030-70053-9 (eBook)
https://doi.org/10.1007/978-3-030-70053-9

Mathematics Subject Classification: 35Q30, 76D05, 35R35

© The Editor(s) (if applicable) and The Author(s), under exclusive license to Springer Nature Switzerland AG 2021
This work is subject to copyright. All rights are reserved by the Publisher, whether the whole or part of the material is concerned, specifically the rights of translation, reprinting, reuse of illustrations, recitation, broadcasting, reproduction on microfilms or in any other physical way, and transmission or information storage and retrieval, electronic adaptation, computer software, or by similar or dissimilar methodology now known or hereafter developed.
The use of general descriptive names, registered names, trademarks, service marks, etc. in this publication does not imply, even in the absence of a specific statement, that such names are exempt from the relevant protective laws and regulations and therefore free for general use.
The publisher, the authors, and the editors are safe to assume that the advice and information in this book are believed to be true and accurate at the date of publication. Neither the publisher nor the authors or the editors give a warranty, expressed or implied, with respect to the material contained herein or for any errors or omissions that may have been made. The publisher remains neutral with regard to jurisdictional claims in published maps and institutional affiliations.

This book is published under the imprint Birkhäuser, www.birkhauser-science.com by the registered company Springer Nature Switzerland AG
The registered company address is: Gewerbestrasse 11, 6330 Cham, Switzerland

Contents

Chapter 1
Introduction

Abstract A literature review, problem statement, and definition of spaces are given. Free boundary problems for the Navier–Stokes system are set. Boundary conditions on an unknown interface are discussed. The transformation from the Euler coordinates to the Lagrangian ones is carried out which leads to a problem in a known domain but with nonlinear coefficients. This system is one of the main problems studied in the monograph the content of which is described at the end of the chapter.

The problem on the unsteady motion of two viscous incompressible immiscible liquids with an unknown closed interface belongs to the class of free boundary problems. The theory of these problems for the Navier–Stokes equations has in their development only about four decades, although their statement goes back to the classical works of the nineteenth century. Most authors working in this direction consider stationary problems. This also applies to the problem of the motion of a finite volume of one fluid in another.

This problem was first studied in the works of J. Hadamard [38] and W. Rybczynski [70]. They obtained an analytical solution to the Stokes system corresponding to the axially symmetric fall of a spherical drop at a constant speed. Stationary motion of two liquids with a free closed surface of their interface was investigated by V. Y. Rivkind [65–67] and J. Bemelmans [8]. In work [67] based on a priori estimates, the solution of the Navier–Stokes system was justified by approximate methods by which the motion and shape of a droplet in a liquid medium [65, 66] were calculated. Assuming fluid density is small, J. Bemelmans established the existence of a unique solution to the nonlinear problem of a drop falling with a constant speed.

Unsteady problems on the motion of fluids with free boundaries are more difficult to research. They are less studied. The general formulation of these problems is as follows (see [63]): find the region $\Omega_t \subset \mathbb{R}^n (n = 2, 3)$ occupied fluid with viscosity $\nu > 0$ and density $\rho > 0$ at the moment time $t > 0$; find

© The Author(s), under exclusive license to Springer Nature Switzerland AG 2021
I. V. Denisova, V. A. Solonnikov, *Motion of a Drop in an Incompressible Fluid*, Advances in Mathematical Fluid Mechanics,
https://doi.org/10.1007/978-3-030-70053-9_1

the vector field $\boldsymbol{v} = (v_1, \cdot, v_n)$ and pressure function p satisfying the initial-boundary value problem for the Navier–Stokes equations:

$$\mathcal{D}_t \boldsymbol{v} - \nu \Delta \boldsymbol{v} + (\boldsymbol{v} \cdot \nabla)\boldsymbol{v} + \frac{1}{\rho}\nabla p = \boldsymbol{f}, \ \nabla \cdot \boldsymbol{v} = 0 \ \text{ in } \Omega_t, \ t > 0, \quad (1.0.1)$$

$$\boldsymbol{v}|_{t=0} = \boldsymbol{v}_0(x), \ x \in \Omega_0, \ \boldsymbol{v}|_{x\in\Sigma} = 0, \ \mathbb{T}\boldsymbol{n}|_{\Gamma_t} = \sigma H \boldsymbol{n}. \quad (1.0.2)$$

Here $\mathcal{D}_t = \partial/\partial t$, $\nabla = (\partial/\partial x_1, \partial/\partial x_2, \partial/\partial x_3)$, Σ is a solid boundary, $\Gamma_t = \partial\Omega_t \backslash \Sigma$ is the free surface, $\boldsymbol{f}$ is the field of external forces, $\boldsymbol{v}_0$ is the vector field of initial velocity in the given region Ω_0, $\mathbb{T}$ is the stress tensor with the elements

$$T_{ik} = -\delta_i^k p + \mu(\partial v_i/\partial x_k + \partial v_k/\partial x_i), \qquad i, k \leqslant n;$$

$\boldsymbol{n}$ is the outward normal vector to Γ_t at the point x, $\sigma \geqslant 0$ is surface tension coefficient, and $H(x,t)/(n-1)$ is the mean curvature of Γ_t ($H < 0$ at the points of convexity of Γ_t to the outside of the fluid); the dot denotes the scalar product in $\mathbb{R}^n$.

In addition, to exclude mass loss across the free surface Γ_t, it is assumed to be always composed by the same liquid particles. It is analytically written like this: for $t > 0$, Γ_t consists of such points $x(\xi, t)$ whose corresponding radius vector $\boldsymbol{x}(\xi, t)$ is a solution to the Cauchy problem

$$\mathcal{D}_t \boldsymbol{x} = \boldsymbol{v}(x(t), t), \ \ \boldsymbol{x}|_{t=0} = \boldsymbol{\xi}, \quad \xi \in \Gamma_0, \ \ t > 0, \quad (1.0.3)$$

where $\Gamma_0 = \partial\Omega_0 \backslash \Sigma$ is the surface given at the initial moment. On a solid surface, the non-slip condition is usually set.

Unsteady motion of a heavy viscous fluid over a solid bottom Σ was studied in the works of J. Beale [6, 7] and G. Allain [4]. In the three-dimensional case, local-in-time unique solvability was proved for the problem (1.0.1)–(1.0.3) with $\sigma = 0$ in Sobolev spaces [6], and, moreover, for $\sigma > 0$ and small initial data, global-in-time solvability was done also in these spaces [7]. Without limiting the size of the initial data, G. Allain established a local existence theorem for this problem with $\sigma > 0$, $n = 2$ [4].

The most advanced problem is that on the motion of an isolated liquid mass. In this case, the surface Σ is absent, and the free boundary Γ is closed. Thus, V. O. Bytev [11] and O. M. Lavrentieva [50] considered problem (1.0.1)–(1.0.3) in a ring with two free boundaries, while the vector $\boldsymbol{v}_0$ was considered axially symmetric. In [11], it was shown that Ω_t can expand to infinity for $\sigma = 0$ in some cases. For $\sigma > 0$, it was established in [50] that Ω_t can turn into a circle for $t = t_0 < \infty$ and save this form for all $t > t_0$.

Problem (1.0.1)–(1.0.3) for a simply connected domain Ω_t with a closed boundary Γ_t was studied in the works of one of the authors. In the Sobolev–Slobodetskiĭ space, the existence of a global solution was proved for $\sigma > 0$

and initial data close to the equilibrium [78]. It was established in [79] that a solution to system (1.0.1)–(1.0.3) with $f = 0$ for $t \to \infty$ tends to a solution independent of t which describes the rotation of a fluid like a solid. The articles [82, 83] give a detailed proof of the solvability of problem (1.0.1)–(1.0.3) in the general case over a finite time interval depending on the data of the problem. A similar result in Hölder space for $\sigma > 0$ was obtained by this author together with I. Sh. Mogilevskiĭ [53, 54]. The chapter [93] provides a detailed overview of the results and some proofs.

As for the problem of the motion of two liquids, model nonstationary problems with given fixed interfaces were studied by V. Ya. Rivkind and N. B. Fridman [68]. In the complete statement, this problem was first studied in [12, 13, 15, 17], where local solvability was proved for it in the Sobolev–Slobodetskiĭ spaces with and without surface tension. The technique of the works for one fluid [82, 83] was used there. A little bit later, N. Tanaka, using the same method [78, 80], made an attempt to study global solvability for the problem with small data near an equilibrium position [96], but we believe that there are gaps in his proof. In Hölder classes, this problem was first considered in the works of the authors [16, 20, 28, 29], where its local solvability in the whole space $\mathbb{R}^3$ was obtained. In [21, 30], the existence of a global solution was proved without and with taking surface tension into account, respectively. A detailed presentation of this material is given in the dissertation [22].

We also mention the results of Yo. Giga and Sh. Takahashi [33, 95], as well as A. Nouri, F. Poupaud and I. Demay [55, 56], on the existence of global weak solutions for the Stokes and Navier–Stokes equations governing the motion of two incompressible fluids with different viscosities and densities, but without surface tension. We also note that in 1993, the book of D. D. Joseph and Yu. Renardy [39] was released dedicated to simultaneous dynamics of two immiscible liquids where problem statements, photographs of experiments, and their computer simulations are given.

In recent years, interest has grown in studying the problem of two-phase capillary fluid flow in various functional spaces. Researchers pose new questions in its analysis and point out its various aspects. In particular, H. Abels [1] estimated the Hausdorff measure of the interface, leaving open the question of the existence of a generalized solution to the problem. Next, Shibata and Shimizu studied the problem by operator methods in the anisotropic Sobolev spaces $W^{2,1}_{q,p}(\Omega^{\pm})$, $2 \leqslant n < q < \infty$, $2 < p < \infty$, and $\Omega^{\pm} \subset \mathbb{R}^n$. They proved the solvability of a model diffraction problem for the Stokes system [72]. A similar result for the nonlinear problem with an unknown interface was obtained in [73] under the assumption that the initial interface is given by the equation $x_n = \alpha(x')$, $x' \in \mathbb{R}^{n-1}$. Ja. Prüss and G. Simonett received unique local (in time) L_p—solvability and analyticity conditions for a solution of the nonlinear problem for $t > 0$ taking into account gravity and surface tension when $p > n + 3$ and the initial interface is close to the plane too [61].

However, the most attention is paid to the problem on the evolution of two fluids in a container, especially to the equilibrium stability problem (vector velocity field $\boldsymbol{v} = 0$, pressure is a step function, the interface is a sphere with an arbitrary center that does not touch the walls of the container). In [30] by us and independently in a series of works by Prüss et al. (in particular, in [40, 60], as well as in the monograph [62]) was shown that the rest state is exponentially stable in the following sense: for arbitrary initial data close enough to an equilibrium position, the problem has a unique solution defined for $\forall t > 0$ which tends exponentially to a rest position, maybe different from the original one. The proof was based on coercive estimates (i.e., estimates in a space of maximum regularity) of a solution to a linearized problem. In all of the above works, the problem with a free interface was reduced to a nonlinear system in two fixed regions by means of the Hanzawa transformation, but the details of the proof were essentially different. It should be noted, in particular, that in [30], the problem was studied in anisotropic Hölder spaces, whereas in [60] and [40], the main space was $W_p^{2,1}$, $p > n + 3$. In addition, the existence of a global solution to the problem was also obtained in the Sobolev spaces for $p > n = 3$ in [92].

In this book, we study the unsteady motion of two incompressible liquids separated by a closed unknown interface Γ_t. The fluids can occupy the entire space $\mathbb{R}^3$ or be bounded by a closed solid surface Σ where the adhesion condition is set. At the interface Γ_t, we take or do not into account the surface tension force. It is assumed that the surfaces Σ and Γ_t do not intersect at the initial moment $t = 0$.

1.1 Statement of the Problem, Definition of the Hölder Spaces

We give a mathematical statement of the problem governing the motion of two fluids filling the domains which are schematically depicted on Fig. 1.1.

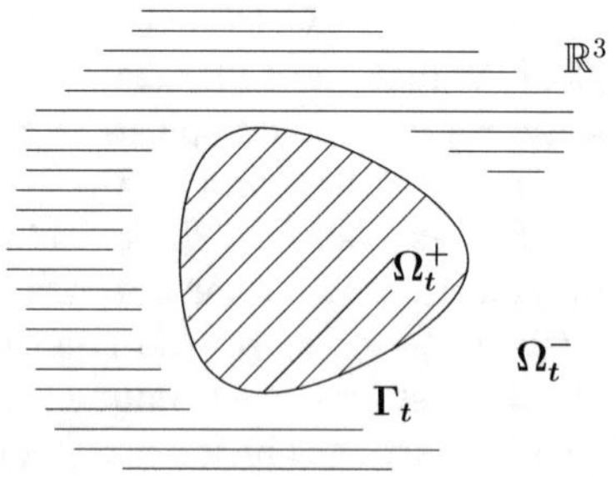

Fig. 1.1 Motion of a drop in the whole space

At the initial moment of time $t = 0$, let a fluid with viscosity $\nu^+ > 0$ and density $\rho^+ > 0$ occupy a bounded domain $\Omega_0^+ \subset \mathbb{R}^3$ and a fluid with viscosity $\nu^- > 0$ and density $\rho^- > 0$ fill the region Ω_0^- surrounding Ω_0^+. Denote $\partial\Omega_0^+$ by Γ_0.

For $t > 0$, it is necessary to find the interface Γ_t between the regions Ω_t^+ and Ω_t^-, as well as the velocity field $\boldsymbol{v}(x,t) = (v_1, v_2, v_3)$ and the pressure function $p(x,t)$ having no more than a power-law growth in x at infinity which satisfies the initial-boundary value problem as follows:

$$\begin{gathered}
\mathcal{D}_t\boldsymbol{v} + (\boldsymbol{v}\cdot\nabla)\boldsymbol{v} - \nu^{\pm}\nabla^2\boldsymbol{v} + \frac{1}{\rho^{\pm}}\nabla p = \boldsymbol{f}, \ \ \nabla\cdot\boldsymbol{v} = 0 \quad \text{in } \Omega_t^-\cup\Omega_t^+, \ \ t>0, \\
\boldsymbol{v}|_{t=0} = \boldsymbol{v}_0, \qquad \text{in} \quad \Omega_0^-\cup\Omega_0^+, \quad \boldsymbol{v}\xrightarrow[|x|\to\infty]{} 0, \qquad (1.1.1) \\
[\boldsymbol{v}]|_{\Gamma_t} \equiv \lim_{\substack{x\to x_0\in\Gamma_t,\\ x\in\Omega_t^+}} \boldsymbol{v}(x) - \lim_{\substack{x\to x_0\in\Gamma_t,\\ x\in\Omega_t^-}} \boldsymbol{v}(x) = 0, \quad [\mathbb{T}\boldsymbol{n}]|_{\Gamma_t} = \sigma H\boldsymbol{n}.
\end{gathered}$$

Here $\nu^{\pm}, \rho^{\pm}$ are the step functions of viscosity and density, respectively, $\boldsymbol{f}$ is a given field of mass forces, $\boldsymbol{v}_0$ is the initial velocity distribution, $\mathbb{T}$ is stress tensor with the elements

$$T_{ik} = -\delta_i^k p + \mu^{\pm}(\partial v_i/\partial x_k + \partial v_k/\partial x_i), \qquad i,k = 1,2,3;$$

$\mu^{\pm} = \nu^{\pm}\rho^{\pm}$, δ_i^k is the Kronecker delta, $\sigma \geqslant 0$ is surface tension coefficient, $\boldsymbol{n}$ is the outward normal to Ω_t^+, $H(x,t)$ is twice the mean curvature of Γ_t ($H < 0$ at the points of convexity of Γ_t toward Ω_t^-). We assume that in $\mathbb{R}^3$, the Cartesian coordinate system $\{x\}$ is introduced. The dot means the Cartesian scalar product.

Summation is implied over repeated indices from 1 to 3 if they are denoted by Latin letters and from 1 to 2 if they are done by Greek ones. We mark the vectors and the vector spaces by boldface letters.

In order to exclude mass transfer across the surface of Γ_t, we consider that Γ_t consists of the points $x(\xi,t)$ whose radius vector $\boldsymbol{x}(\xi,t)$ satisfies the Cauchy problem (1.0.3). Hence, $\Gamma_t = \{x(\xi,t)|\ \xi\in\Gamma_0\}$, $\Omega_t^{\pm} = \{x(\xi,t)|\ \xi\in\Omega_0^{\pm}\}$.

We pass from the Euler coordinates to the Lagrangian ones by the formula

$$\boldsymbol{x} = \boldsymbol{\xi} + \int_0^t \boldsymbol{u}(\xi,\tau)\,\mathrm{d}\tau \equiv X_{\boldsymbol{u}}(\xi,t) \qquad (1.1.2)$$

(here $\boldsymbol{u}(\xi,t)$ is the velocity field in the Lagrangian coordinates), and we use the well-known relation [98]

$$H\boldsymbol{n} = \Delta(t)\boldsymbol{x} \equiv \Delta(t)\boldsymbol{X}_{\boldsymbol{u}}(\xi,t), \qquad (1.1.3)$$

where $\Delta(t)$ denotes the Laplace–Beltrami operator on Γ_t. In the local coordinates s_1, s_2 introduced on the interface Γ_0, it has the form

$$\Delta(t) = \frac{1}{\sqrt{g}}\frac{\partial}{\partial s_\alpha} g^{\alpha\beta}\sqrt{g}\frac{\partial}{\partial s_\beta} \equiv g^{\alpha\beta}\frac{\partial^2}{\partial s_\alpha \partial s_\beta} + h^\beta \frac{\partial}{\partial s_\beta}, \tag{1.1.4}$$

where $\{g^{\alpha\beta}\}_{\alpha,\beta=1}^2$ is the inverse of metric tensor matrix $\{g_{\alpha\beta}\}_{\alpha,\beta=1}^2$,

$$g_{\alpha\beta} = \frac{\partial \boldsymbol{X_u}(\xi(s),t)}{\partial s_\alpha}\cdot\frac{\partial \boldsymbol{X_u}(\xi(s),t)}{\partial s_\beta}, \quad g = \det\{g_{\alpha\beta}\}_{\alpha,\beta=1}^2, \quad h^\beta = \frac{1}{\sqrt{g}}\frac{\partial}{\partial s_\alpha}(g^{\alpha\beta}\sqrt{g}).$$

As a result of these transformations and projection of the last boundary condition in (1.1.1) onto the tangent planes, first to Γ_t, and then to $\Gamma \equiv \Gamma_0$, we arrive at the problem for $\boldsymbol{u}$ and $q = p(X_{\boldsymbol{u}}, t)$ with the given interface Γ. If the angle between $\boldsymbol{n}$ and the outward normal $\boldsymbol{n}_0$ to Γ is acute, the resulting system is equivalent to the following one:

$$\begin{aligned}
&\mathcal{D}_t\boldsymbol{u} - \nu^\pm\nabla^2_{\boldsymbol{u}}\boldsymbol{u} + \frac{1}{\rho^\pm}\nabla_{\boldsymbol{u}} q = \boldsymbol{f}(X_{\boldsymbol{u}}, t),\\
&\nabla_{\boldsymbol{u}}\cdot\boldsymbol{u} = 0 \quad \text{in} \quad Q_T^\pm = \Omega_0^\pm \times (0,T),\\
&\boldsymbol{u}|_{t=0} = \boldsymbol{v}_0 \ \text{in}\ \Omega_0^+\cup\Omega_0^-, \quad \boldsymbol{u}\xrightarrow[|\xi|\to\infty]{} 0,\\
&[\boldsymbol{u}]|_{G_T} = 0, \ [\mu^\pm\Pi_0\Pi\mathbb{S}_{\boldsymbol{u}}(\boldsymbol{u})\boldsymbol{n}]|_{G_T} = 0 \quad (G_T \equiv \Gamma\times(0,T)),\\
&[\boldsymbol{n}_0\cdot\mathbb{T}_{\boldsymbol{u}}(\boldsymbol{u},q)\boldsymbol{n}]|_{G_T} - \sigma\boldsymbol{n}_0\cdot\Delta(t)X_{\boldsymbol{u}}|_{G_T} = 0.
\end{aligned} \tag{1.1.5}$$

The pressure $q(\xi,t)$ may have no more than a power growth as $|\xi| \to \infty$. In (1.1.5), we have used the following notation: $\nabla_{\boldsymbol{u}} = \mathbb{A}\nabla$, where $\mathbb{A}$ is the matrix of cofactors A_{ij} to the elements

$$a_{ij}(\xi,t) = \delta_i^j + \int_0^t \frac{\partial u_i}{\partial \xi_j}\mathrm{d}t'$$

of the Jacobian matrix of transformation (1.1.2); the vector $\boldsymbol{n}$ is related to $\boldsymbol{n}_0$ as follows:

$$\boldsymbol{n} = \frac{\mathbb{A}\boldsymbol{n}_0}{|\mathbb{A}\boldsymbol{n}_0|}; \tag{1.1.6}$$

$\Pi\boldsymbol{\omega} = \boldsymbol{\omega} - \boldsymbol{n}(\boldsymbol{n}\cdot\boldsymbol{\omega})$, $\Pi_0\boldsymbol{\omega} = \boldsymbol{\omega} - \boldsymbol{n}_0(\boldsymbol{n}_0\cdot\boldsymbol{\omega})$ are the projections of the vector $\boldsymbol{\omega}$ onto the tangent planes to Γ_t and to Γ, respectively; $\mathbb{T}_{\boldsymbol{u}}(\boldsymbol{w},q)$ is the tensor with the elements

$$(\mathbb{T}_{\boldsymbol{u}}(\boldsymbol{w},q))_{ij} = -\delta_j^i q + \mu^\pm(\mathbb{S}_{\boldsymbol{u}}(\boldsymbol{w}))_{ij}, \ (\mathbb{S}_{\boldsymbol{u}}(\boldsymbol{w}))_{ij} = A_{jk}\partial w_i/\partial\xi_k + A_{ik}\partial w_j/\partial\xi_k;$$

$H_0(\xi) = \boldsymbol{n}_0\cdot\Delta(0)\boldsymbol{\xi}$ is the doubled mean curvature of Γ.

The study of problem (1.1.5) is based on the analysis of the linearization of it. Therefore, we first consider the following linear problem for the Stokes system:

$$\mathcal{D}_t \boldsymbol{v} - \nu^{\pm}\Delta\boldsymbol{v} + \frac{1}{\rho^{\pm}}\nabla p = \boldsymbol{f},\ \nabla\cdot\boldsymbol{v} = r \quad \text{in } \Omega^- \cup \Omega^+,\ \ t > 0,$$

$$\boldsymbol{v}|_{t=0} = \boldsymbol{v}_0 \quad \text{in } \Omega^- \cup \Omega^+, \quad \boldsymbol{v} \xrightarrow[|x|\to\infty]{} 0,$$

$$[\boldsymbol{v}]|_\Gamma = 0, \qquad [\Pi_0 \mathbb{T}\boldsymbol{n}]|_\Gamma = \boldsymbol{b}, \qquad \boldsymbol{b}\cdot\boldsymbol{n} = 0, \tag{1.1.7}$$

$$[\boldsymbol{n}\cdot\mathbb{T}\boldsymbol{n}]\big|_\Gamma - \sigma\boldsymbol{n}\cdot\int_0^t \Delta_\Gamma \boldsymbol{v} dt' = b' + \sigma\int_0^t B dt',$$

where $\boldsymbol{f}, r, \boldsymbol{b}, b', B, \boldsymbol{v}_0$ are given functions. Here $\boldsymbol{n}$ denotes the outward unit normal to Γ.

The specifics of problem (1.1.7) is the presence of the integral term $\sigma\boldsymbol{n}\cdot\int_0^t \Delta_\Gamma \boldsymbol{v} dt'$ in the last boundary condition. For $\sigma > 0$, this problem is not coercive since the boundary condition contains two terms of different orders none of which can be considered as of higher order with respect to another. Nevertheless, the proof of solvability is carried out by constructing a regularizer on the basis of a priori estimates and the existence of an explicit solution to a problem with flat boundary, as in the case of linear parabolic equations [47].

Therefore, in Chapter 2, we analyze model problem (1.1.7), where the interface between the liquids is the plane $\{x_3 = 0\}$. The sign of the coefficient σ plays an essential role in this case: for $\sigma \leqslant 0$, the estimates obtained in Sect. 2.5 become impossible. The proof is based on theorems on Fourier multipliers in the Hölder spaces (Sect. 2.4; see also [28, 53]). In Chap. 3, we study a model problem with flat interface for $\sigma = 0$.

Problem (1.1.7) with a closed interface is considered in Chap. 4 where the solvability of the problem is proved in the Hölder classes of functions for any finite time interval. Further, in Sects. 5.2 and 5.3, estimates are obtained for linear and linearized problems in weighted Hölder spaces. In Sect. 5.4, we prove the local-in-time solvability of nonlinear problem (1.1.5). On this basis, we obtain the existence of a global solution for small initial data, first in the absence (Chap. 6) and then in the presence of surface tension (Chap. 7).

We also consider the motion of two fluids taking into account temperature factor. First we allow a dependence of the surface coefficient on temperature, and then we take into account temperature dependence of mass forces. Thus, the problem of thermocapillary convection is studied in Chap. 8, and the motion of two liquids in the Oberbeck–Boussinesq approximation is investigated in Chap. 9.

Finally, in the last chapters, the problem is analyzed in Sobolev–Slobodetskiĭ spaces. Local existence theorem is proved in Chap. 10. The global L_2 solvability of the problem is obtained without and with surface tension taken into account in Chaps. 11 and 12, respectively.

We recall the definition of the Hölder spaces.

Let Ω be a domain in $\mathbb{R}^n$, $n \in \mathbb{N}$; we set $Q_T = \Omega \times (0, T)$, $T > 0$. Assume that $\alpha \in (0, 1)$. The space $C^{\alpha,\alpha/2}(Q_T)$ means the class of the functions that are defined in Q_T and have finite norm

$$|f|_{Q_T}^{(\alpha,\alpha/2)} = |f|_{Q_T} + \langle f\rangle_{Q_T}^{(\alpha,\alpha/2)},$$

where

$$|f|_{Q_T} = \sup_{(x,t)\in Q_T} |f(x,t)|, \qquad \langle f\rangle_{Q_T}^{(\alpha,\alpha/2)} = \langle f\rangle_{x,Q_T}^{(\alpha)} + \langle f\rangle_{t,Q_T}^{(\alpha/2)},$$

$$\langle f\rangle_{x,Q_T}^{(\alpha)} = \sup_{x,y\in\Omega}\ \sup_{t\in(0,T)} |f(x,t) - f(y,t)||x-y|^{-\alpha},$$

$$\langle f\rangle_{t,Q_T}^{(\beta)} = \sup_{x\in\Omega}\ \sup_{t,\tau\in(0,T)} |f(x,t) - f(x,\tau)||t-\tau|^{-\beta},\ \beta \in (0,1).$$

Moreover, $|f|_{x,Q_T}^{(\alpha)} = |f|_{Q_T} + \langle f\rangle_{x,Q_T}^{(\alpha)}$, $|f|_{t,Q_T}^{(\beta)} = |f|_{Q_T} + \langle f\rangle_{t,Q_T}^{(\beta)}$.

We introduce the notation

$$\mathcal{D}_x^{\boldsymbol{r}} = \partial^{|\boldsymbol{r}|}/\partial x_1^{r_1}\ldots\partial x_n^{r_n}, \quad \boldsymbol{r} = (r_1, \ldots r_n), \quad r_i \geqslant 0,\ |\boldsymbol{r}| = r_1 + \cdots + r_n,$$
$$\mathcal{D}_t^s = \partial^s/\partial t^s, \quad s \in \mathbb{N}\cup\{0\}.$$

We equip the space $C^{k+\alpha,\frac{k+\alpha}{2}}(Q_T)$ with $k \in \mathbb{N}$ by the norm

$$|f|_{Q_T}^{(k+\alpha,\frac{k+\alpha}{2})} = \sum_{|\boldsymbol{r}|+2s\leqslant k} |\mathcal{D}_x^{\boldsymbol{r}}\mathcal{D}_t^s f|_{Q_T} + \langle f\rangle_{Q_T}^{(k+\alpha,\frac{k+\alpha}{2})},$$

where

$$\langle f\rangle_{Q_T}^{(k+\alpha,\frac{k+\alpha}{2})} = \sum_{|\boldsymbol{r}|+2s=k} \langle \mathcal{D}_x^{\boldsymbol{r}}\mathcal{D}_t^s f\rangle_{Q_T}^{(\alpha,\alpha/2)} + \sum_{|\boldsymbol{r}|+2s=k-1} \langle \mathcal{D}_x^{\boldsymbol{r}}\mathcal{D}_t^s f\rangle_{t,Q_T}^{(\frac{1+\alpha}{2})}.$$

The symbol $\mathring{C}^{k+\alpha,\frac{k+\alpha}{2}}(Q_T)$ denotes the subspace of $C^{k+\alpha,\frac{k+\alpha}{2}}(Q_T)$ which consists of functions f such that $\mathcal{D}_t^i f|_{t=0} = 0, i = 0, \ldots, \left[\frac{k+\alpha}{2}\right]$.

The space $C^{k+\alpha}(\Omega), k \in \mathbb{N}\cup\{0\}$, is a set of functions $f(x), x \in \Omega$, with the norm

$$|f|_{\Omega}^{(k+\alpha)} = \sum_{|\boldsymbol{r}|\leqslant k} |\mathcal{D}_x^{\boldsymbol{r}} f|_{\Omega} + \langle f\rangle_{\Omega}^{(k+\alpha)},$$

where

$$\langle f\rangle_{\Omega}^{(k+\alpha)} = \sum_{|\boldsymbol{r}|=k} \langle \mathcal{D}_x^{\boldsymbol{r}} f\rangle_{\Omega}^{(\alpha)} = \sup_{x,y\in\Omega} \sum_{|\boldsymbol{r}|=k} |\mathcal{D}_x^{\boldsymbol{r}} f(x) - \mathcal{D}_y^{\boldsymbol{r}} f(y)||x-y|^{-\alpha}.$$

The following Hölder semi-norms with $\alpha, \gamma \in (0,1)$ will be useful in the sequel:

$$|f|_{Q_T}^{(\gamma,1+\alpha)} = \langle f\rangle_{Q_T}^{(\gamma,1+\alpha)} + \langle f\rangle_{t,Q_T}^{(\frac{1+\alpha-\gamma}{2})},$$

where

$$\langle f\rangle_{Q_T}^{(\gamma,1+\alpha)} = \max_{t,\tau\in(0,T)} \max_{x,y\in\Omega} \frac{|f(x,t)-f(y,t)-f(x,\tau)+f(y,\tau)|}{|x-y|^{\gamma}|t-\tau|^{(1+\alpha-\gamma)/2}}.$$

It is shown (see [77]) that

$$\langle f\rangle_{Q_T}^{(\gamma,1+\alpha)} \leqslant c_1 \langle f\rangle_{Q_T}^{(1+\alpha,\frac{1+\alpha}{2})}.$$

It is said that $f \in C^{(\gamma,1+\alpha)}(Q_T)$ if

$$|f|_{Q_T} + |f|_{Q_T}^{(\gamma,1+\alpha)} < \infty.$$

Moreover, if f equals zero at $t=0$, then one writes $f \in \overset{\circ}{C}{}^{(\gamma,1+\alpha)}(Q_T)$.
Finally, the space $C^{\gamma,\mu}(Q_T)$ consists of functions with finite norm

$$|f|_{Q_T}^{(\gamma,\mu)} \equiv \langle f\rangle_{x,Q_T}^{(\gamma)} + |f|_{t,Q_T}^{(\mu)}, \qquad \gamma\in(0,1), \quad \mu\in[0,1),$$

where

$$|f|_{t,Q_T}^{(\mu)} = \begin{cases} |f|_{Q_T} + \langle f\rangle_{t,Q_T}^{(\mu)}, & \text{if } \mu>0, \\ |f|_{Q_T}, & \text{if } \mu=0. \end{cases}$$

The Hölder spaces of functions given on smooth surfaces, in particular, on Γ and on G_T, are introduced in a standard way, with the help of local maps and the partition of unity.

A vector valued function is said to be an element of a Hölder space if all of its components belong to this space, and its norm is defined as the maximal norm of the components.

If a vector $\boldsymbol{b}\in\boldsymbol{C}^{1+\alpha,\frac{1+\alpha}{2}}(G_T)$ and $\boldsymbol{b}\cdot\boldsymbol{n}=0$, we write $\boldsymbol{b}\in\boldsymbol{C}_{\boldsymbol{n}}^{1+\alpha,\frac{1+\alpha}{2}}(G_T)$. We set: $D_T=Q_T^-\cup Q_T^+$ and

$$|f|_{D_T}^{(k+\alpha,\frac{k+\alpha}{2})}=|f|_{Q_T^-}^{(k+\alpha,\frac{k+\alpha}{2})}+|f|_{Q_T^+}^{(k+\alpha,\frac{k+\alpha}{2})},$$

$$|f|_{\cup\Omega^\pm}^{(k+\alpha)}=|f|_{\Omega^-}^{(k+\alpha)}+|f|_{\Omega^+}^{(k+\alpha)}.$$

Chapter 2
A Model Problem with Plane Interface and with Positive Surface Tension Coefficient

Abstract In this chapter, we study problem (1.1.7) for the case of a flat interface. First, we construct an explicit solution to this problem with homogeneous equations in the space of Fourier images and prove its smoothness in the Holder classes of functions. Then we consider the inhomogeneous problem.

The results of this chapter were published by the authors in [28].

2.1 Auxiliary Propositions

We introduce the following notation: $D_T^{\pm} = \mathbb{R}^3_{\pm} \times (0,T)$, $D_T^3 \equiv \cup D_T^{\pm} = D_T^- \cup D_T^+$, $\mathbb{R}^2_T = \mathbb{R}^2 \times (0,T)$, $T \in (0,\infty]$.

We consider the Neumann and Dirichlet problems:

$$\Delta p = 0 \quad (x \in \mathbb{R}^3_+), \qquad \frac{\partial p}{\partial x_3}\Big|_{x_3=0} = p_0(x',t) = \sum_{\tau=1}^{2} \frac{\partial p_\tau(x',t)}{\partial x_\tau}, \tag{2.1.1}$$

$$\Delta q = 0 \quad (x \in \mathbb{R}^3_+), \qquad q\Big|_{x_3=0} = q_0(x',t), \tag{2.1.2}$$

assuming that p_0, p_τ decrease quite well at infinity and the function q_0 is at least bounded. Then solutions to these problems, as is known, are expressed in the form of potentials of single and double layers. The following proposition was proved in [76] (Lemma 7.2).

© The Author(s), under exclusive license to Springer Nature Switzerland AG 2021

I. V. Denisova, V. A. Solonnikov, *Motion of a Drop in an Incompressible Fluid*, Advances in Mathematical Fluid Mechanics,
https://doi.org/10.1007/978-3-030-70053-9_2

Lemma 2.1.1 *For a solution to problem* (2.1.1), *we have the estimate*

$$\langle \nabla p\rangle^{(\alpha,\frac{\alpha}{2})}_{D_T^+} \leqslant c\{\langle p_0\rangle^{(\alpha)}_{x,\mathbb{R}^2_T} + \sum_{\tau=1}^{2} |p_\tau|^{(\gamma,1+\alpha)}_{\mathbb{R}^2_T}\}, \qquad \forall \alpha,\gamma \in (0,1). \tag{2.1.3}$$

Corollary 2.1.1 *A solution to problem* (2.1.2) *is subjected to the inequalities*

$$\langle \nabla q\rangle^{(\alpha,\frac{\alpha}{2})}_{D_T^+} \leqslant c\{\langle \nabla' q_0\rangle^{(\alpha)}_{x,\mathbb{R}^2_T} + |q_0|^{(\gamma,1+\alpha)}_{\mathbb{R}^2_T}\}, \tag{2.1.4}$$

$$\langle q\rangle^{(\alpha,\frac{\alpha}{2})}_{D_T^+} \leqslant c\langle q_0\rangle^{(\alpha,\frac{\alpha}{2})}_{\mathbb{R}^2_T}. \tag{2.1.5}$$

Here $\nabla' = (\partial/\partial x_1, \partial/\partial x_2)$.

Proof of the Corollary Indeed, $q = \frac{\partial p}{\partial x_3}$, where p is a solution of the Neumann problem (2.1.1) with $p_0 = q_0$. By differentiating (2.1.1) with respect to x_τ, $\tau = 1, 2$, and applying the estimate (2.1.3), we obtain

$$\Big\langle \frac{\partial q}{\partial x_\tau}\Big\rangle^{(\alpha,\frac{\alpha}{2})}_{D_T^+} = \Big\langle \frac{\partial^2 p}{\partial x_3 \partial x_\tau}\Big\rangle^{(\alpha,\frac{\alpha}{2})}_{D_T^+} \leqslant c\Big\{\Big\langle \frac{\partial q_0}{\partial x_\tau}\Big\rangle^{(\alpha)}_{x,\mathbb{R}^2_T} + |q_0|^{(\gamma,1+\alpha)}_{\mathbb{R}^2_T}\Big\},$$

and

$$\Big\langle \frac{\partial q}{\partial x_3}\Big\rangle^{(\alpha,\frac{\alpha}{2})}_{D_T^+} \leqslant \sum_{\tau=1}^{2} \Big\langle \frac{\partial^2 p}{\partial x_\tau^2}\Big\rangle^{(\alpha,\frac{\alpha}{2})}_{D_T^+} \leqslant c\Big\{\sum_{\tau=1}^{2} \Big\langle \frac{\partial q_0}{\partial x_\tau}\Big\rangle^{(\alpha)}_{x,\mathbb{R}^2_T} + |q_0|^{(\gamma,1+\alpha)}_{\mathbb{R}^2_T}\Big\},$$

which implies (2.1.4).

Estimate (2.1.5) is obtained from the representation of $q(x,t)$ as a double-layer potential. The inequality

$$\langle q\rangle^{(\alpha)}_{\mathbb{R}^3_+} \leqslant c\langle q_0\rangle^{(\alpha)}_{\mathbb{R}^2}. \tag{2.1.6}$$

is well-known (see, e.g., [46]). In addition, it follows from the maximum principle that

$$\sup_{x\in\mathbb{R}^3_+} |q(x,t) - q(x,\tau)| \leqslant \sup_{x\in\mathbb{R}^2} |q_0(x',t) - q_0(x',\tau)|,$$

and hence, $\langle q\rangle^{(\alpha/2)}_{t,D_T^+} \leqslant c\langle q_0\rangle^{(\alpha/2)}_{t,\mathbb{R}^2_T}$. Inequality (2.1.5) follows from the last estimate and (2.1.6).

2.2 An Explicit Solution of a Homogeneous Model Problem

Below, a model problem with a plane interface is analyzed. First we consider diffraction problem for homogeneous Stokes equations:

$$\mathcal{D}_t \boldsymbol{v} - \nu^+ \Delta \boldsymbol{v} + \frac{1}{\rho^+}\nabla p = 0, \quad \nabla\cdot\boldsymbol{v} = 0 \quad \text{in} \quad D^+_\infty = \mathbb{R}^3_+ \times (0,\infty),$$

$$\mathcal{D}_t \boldsymbol{v} - \nu^- \Delta \boldsymbol{v} + \frac{1}{\rho^-}\nabla p = 0, \quad \nabla\cdot\boldsymbol{v} = 0 \quad \text{in} \quad D^-_\infty = \mathbb{R}^3_- \times (0,\infty),$$

$$\boldsymbol{v}\big|_{t=0} = \boldsymbol{v}_0 \quad \text{in} \quad \mathbb{R}^3_- \cup \mathbb{R}^3_+, \quad \boldsymbol{v} \xrightarrow[|x|\to\infty]{} 0, \tag{2.2.1}$$

$$[\boldsymbol{v}]\big|_{x_3=0} = 0, \quad -\Big[\mu^\pm\Big(\frac{\partial v_\beta}{\partial x_3} + \frac{\partial v_3}{\partial x_\beta}\Big)\Big]\Big|_{x_3=0} = b_\beta(x',t), \quad \beta = 1,2,$$

$$\Big[-p + 2\mu^\pm\frac{\partial v_3}{\partial x_3}\Big]\Big|_{x_3=0} + \sigma\Delta'\int\limits_0^t v_3\big|_{x_3=0}\mathrm{d}\tau =$$

$$= b_3 + \sigma\int\limits_0^t B\,\mathrm{d}\tau \equiv b_3' \quad \text{on} \quad \mathbb{R}^2_\infty.$$

Here $\Delta = \nabla\cdot\nabla$ is the Laplace operator, $\mathbb{R}^3_\pm = \{\pm x_3 > 0\}$, $\mathbb{R}^2_\infty = \mathbb{R}^2 \times (0,\infty)$, $x' = (x_1,x_2)$, $\Delta' = \partial^2/\partial x_1^2 + \partial^2/\partial x_2^2$, the positive constants $\nu^+,\nu^-,\rho^+,\rho^-$ are viscosity and density coefficients of the upper and lower fluids, respectively, $\mu^\pm$ is step function of the dynamic viscosity, $\mu^\pm = \nu^+\rho^+$ in $\mathbb{R}^3_+$, and $\mu^\pm = \nu^-\rho^-$ in $\mathbb{R}^3_-$. We assume that the given functions b_i, $i = 1,2,3$, and B vanish at $t = 0$ and decrease well enough as $|x| \to \infty$, and the pressure function p may grow in x no faster than as a power function.

We take the Fourier transform with respect to the tangent spatial variables $(x_1,x_2) = x'$ and the Laplace transform with respect to t according to the formula

$$\widetilde{f}(\xi,x_3,s) = \int\limits_0^\infty \mathrm{e}^{-st}\int\limits_{\mathbb{R}^2} f(x,t)\mathrm{e}^{-ix'\cdot\xi}\mathrm{d}x'\mathrm{d}t, \quad \operatorname{Re} s \geqslant 0, \quad \xi = (\xi_1,\xi_2). \tag{2.2.2}$$

Then problem (2.2.1) goes over into the system of ordinary differential equations

$$\begin{aligned}
&\frac{\mathrm{d}^2\widetilde{v}_\alpha}{\mathrm{d}x_3^2} - \Big(\frac{s}{\nu^\pm} + \xi^2\Big)\widetilde{v}_\alpha - \frac{\mathrm{i}\xi_\alpha}{\mu^\pm}\widetilde{p} = 0, \qquad \alpha = 1,2,\\
&\frac{\mathrm{d}^2\widetilde{v}_3}{\mathrm{d}x_3^2} - \Big(\frac{s}{\nu^\pm} + \xi^2\Big)\widetilde{v}_3 - \frac{1}{\mu^\pm}\frac{\mathrm{d}\widetilde{p}}{\mathrm{d}x_3} = 0, \qquad\qquad (2.2.3)\\
&\frac{\mathrm{d}\widetilde{v}_3}{\mathrm{d}x_3} + \mathrm{i}\xi_1\widetilde{v}_1 + \mathrm{i}\xi_2\widetilde{v}_2 = 0, \qquad\qquad \pm x_3 > 0,
\end{aligned}$$

with boundary conditions

$$\begin{aligned}
&\widetilde{\boldsymbol{v}} \xrightarrow[|x_3|\to\infty]{} 0, \quad [\widetilde{\boldsymbol{v}}]\big|_{x_3=0} = 0,\\
&-\Big[\mu^\pm\Big(\frac{\mathrm{d}\widetilde{v}_\alpha}{\mathrm{d}x_3} + \mathrm{i}\xi_\alpha\widetilde{v}_3\Big)\Big]\Big|_{x_3=0} = \widetilde{b}_\alpha, \quad \alpha = 1,2, \qquad\qquad (2.2.4)\\
&s\Big[-\widetilde{p}^\pm + 2\mu^\pm\frac{\mathrm{d}\widetilde{v}_3}{\mathrm{d}x_3}\Big]\Big|_{x_3=0} - \sigma\xi^2\widetilde{v}_3^+ = s\widetilde{b}_3 + \sigma\widetilde{B},
\end{aligned}$$

where $\xi^2 = \xi_1^2 + \xi_2^2$, $\widetilde{\boldsymbol{v}}^\pm = \lim\limits_{x_3\to 0\pm}\widetilde{\boldsymbol{v}}$, $\widetilde{p}^\pm = \lim\limits_{x_3\to 0\pm}\widetilde{p}$.

We construct an explicit solution to problem (2.2.3) and (2.2.4). We write the general solution of system (2.2.3) for two half-spaces:

$$\begin{aligned}
&\widetilde{\boldsymbol{v}} = C_1^\pm\begin{pmatrix}\pm r^\pm\\ 0\\ \mathrm{i}\xi_1\end{pmatrix}\mathrm{e}^{\mp r^\pm x_3} + C_2^\pm\begin{pmatrix}0\\ \pm r^\pm\\ \mathrm{i}\xi_2\end{pmatrix}\mathrm{e}^{\mp r^\pm x_3} + C_3^\pm\begin{pmatrix}\mathrm{i}\xi_1\\ \mathrm{i}\xi_2\\ \mp|\xi|\end{pmatrix}\mathrm{e}^{\mp|\xi|x_3},\\
&\widetilde{p} = -C_3^\pm\rho^\pm s\mathrm{e}^{\mp|\xi|x_3} \qquad\qquad \text{for } \pm x_3 > 0. \qquad (2.2.5)
\end{aligned}$$

Here $C_i^\pm$ are arbitrary constants, $r^\pm = \sqrt{\frac{s}{\nu^\pm} + \xi^2}$, $|\xi| = \sqrt{\xi_1^2 + \xi_2^2}$, and we assume that $|\arg\sqrt{z}| < \pi/2$ for $\forall z \in \mathbb{C}$.

We substitute solution (2.2.5) into the boundary conditions (2.2.4) and solve the resulting system. We write the solution in the form convenient for estimates

$$\begin{aligned}
&\widetilde{\boldsymbol{v}} = \boldsymbol{\omega}\mathrm{e}^{\mp r^\pm x_3} + \boldsymbol{V}^\pm e_1^\pm, \quad \pm x_3 > 0,\\
&\widetilde{p} = \mu^\pm(r^\pm + |\xi|)\Psi^\pm\mathrm{e}^{\mp|\xi|x_3}, \quad \pm x_3 > 0, \qquad\qquad (2.2.6)
\end{aligned}$$

where

$$\boldsymbol{\omega} = \begin{pmatrix}\omega_1\\ \omega_2\\ \omega_3\end{pmatrix}, \quad \boldsymbol{V}^\pm = \begin{pmatrix}\mathrm{i}\xi_1\\ \mathrm{i}\xi_2\\ \mp|\xi|\end{pmatrix}\Psi^\pm, \quad e_1^\pm = \frac{\mathrm{e}^{\mp r^\pm x_3} - \mathrm{e}^{\mp|\xi|x_3}}{r^\pm - |\xi|},$$

$$\Psi^{\pm} = \frac{1}{|\xi| P q}\Big\{|\xi| \widetilde{b}_3' s(\mp r^{\pm} q + |\xi| q') - \widetilde{d} s\big[(\rho^+ + \rho^-)s + |\xi| q \mp r^{\pm} q' + \frac{\sigma}{s}|\xi|^3\big]\Big\},$$

$$\omega_\tau = \frac{\widetilde{b}_\tau - \mathrm{i}\xi_\tau\{\mu^+\Psi^+ + \mu^-\Psi^- - (\mu^+ - \mu^-)\omega_3\}}{\mu^+ r^+ + \mu^- r^-}, \quad \tau = 1, 2, \tag{2.2.7}$$

$$\omega_3 = \frac{\widetilde{d} s q'}{P q} - \frac{|\xi| \widetilde{b}_3' s}{P}.$$

In the formulas (2.2.7), we have used the notation

$$\widetilde{d} = \mathrm{i}\xi_1 \widetilde{b}_1 + \mathrm{i}\xi_2 \widetilde{b}_2, \quad \widetilde{b}_3' = \widetilde{b}_3 + \frac{\sigma}{s}\widetilde{B},$$

$$P = (\rho^+ + \rho^-)s^2 + \frac{4|\xi| s}{q}\{|\xi|({\mu^+}^2 r^+ + {\mu^-}^2 r^-) + \mu^+\mu^-(r^+ r^- + |\xi|^2)\}$$
$$+\sigma|\xi|^3, \tag{2.2.8}$$
$$q = \mu^+(r^+ + |\xi|) + \mu^-(r^- + |\xi|), \quad q' = \mu^+(r^+ - |\xi|) - \mu^-(r^- - |\xi|).$$

Lemma 2.2.1 *If* $\operatorname{Re} s = a \geqslant a_0 > 0$, $\xi \in \mathbb{R}^2$, *then for the function* $P(\xi, s)$ (*see* (2.2.8)) *the estimates*

$$|P(\xi, s)| \geqslant c\big(|s|^2 + \xi^2|s| + \sigma|\xi|^3\big), \tag{2.2.9}$$

$$\Big|\frac{\partial P}{\partial \xi_\tau}\Big| \leqslant c\big(|s|^{3/2} + |s||\xi| + \sigma\xi^2\big), \tag{2.2.10}$$

$$\Big|\frac{\partial P}{\partial s}\Big| \leqslant c\big(|s| + \xi^2\big), \quad \Big|\frac{\partial^2 P}{\partial s \partial \xi_\tau}\Big| \leqslant c\big(|s| + \xi^2\big)^{1/2}, \quad \tau = 1, 2, \tag{2.2.11}$$

hold.

Besides

$$\frac{\partial P}{\partial \xi_1} = P_1 + \frac{\xi_1}{|\xi|} P_2,$$

and the functions

$$P_1(\xi, s) = \frac{\partial}{\partial \xi_1}\Big\{\frac{4|\xi| s}{q}\{|\xi|({\mu^+}^2 r^+ + {\mu^-}^2 r^-) + \mu^+\mu^-|\xi|^2\}\Big\} + 4\mu^+\mu^-|\xi| s \frac{\partial}{\partial \xi_1}\frac{r^+ r^-}{q}$$
$$+3\sigma|\xi|\xi_1,$$

$$P_2(\xi, s) = 4\mu^+\mu^- s \frac{r^+ r^-}{q}$$

satisfy the inequalities

$$
\begin{aligned}
&|P_1| \leqslant c\big(|s||\xi| + \sigma\xi^2\big), \qquad |P_2| \leqslant c|s|\big(|s| + \xi^2\big)^{1/2},\\
&\Big|\frac{\partial P_\beta}{\partial \xi_2}\Big| \leqslant c\big(|s| + \sigma|\xi|\big), \qquad\qquad (2.2.12)\\
&\Big|\frac{\partial P_\beta}{\partial s}\Big| \leqslant c\big(|s| + \xi^2\big)^{1/2}, \qquad \Big|\frac{\partial^2 P_\beta}{\partial s \partial \xi_2}\Big| \leqslant c, \quad \beta = 1,2.
\end{aligned}
$$

Corollary 2.2.1 *The following estimates*

$$
\Big|\frac{\partial^2 P}{\partial \xi_1 \partial \xi_2}\Big| \leqslant c\Big(|s|\Big[1 + \frac{|s|^{1/2}}{|\xi|}\Big] + \sigma|\xi|\Big), \qquad (2.2.13)
$$

$$
\Big|\frac{\partial^3 P}{\partial s \partial \xi_1 \partial \xi_2}\Big| \leqslant c\Big(1 + \frac{|s|^{1/2}}{|\xi|}\Big),
$$

$$
\begin{aligned}
&\sum_{\tau=1}^{2}\Big|\frac{\partial P}{\partial \xi_\tau}\Big||s|^{1/2} + \Big|\frac{\partial P}{\partial s}\Big||s| + \sum_{\tau=1}^{2}\Big|\frac{\partial^2 P}{\partial s \partial \xi_\tau}\Big||s|^{3/2} +\\
&+\Big(\Big|\frac{\partial^2 P}{\partial \xi_1 \partial \xi_2}\Big||s| + \Big|\frac{\partial^3 P}{\partial s \partial \xi_1 \partial \xi_2}\Big||s|^2\Big)\Big[1 + \frac{|s|^{1/2}}{|\xi|}\Big]^{-1} \leqslant c|P|, \qquad (2.2.14)
\end{aligned}
$$

$$
\sum_{\tau=1}^{2}\Big|\frac{\partial P}{\partial \xi_\tau}\Big||\xi| + \Big|\frac{\partial^2 P}{\partial \xi_1 \partial \xi_2}\Big||\xi|^2 \leqslant c|P|
$$

are true.

Proof of Lemma 2.1.1 Since for $\xi \in \mathbb{R}^2$, $\operatorname{Re} s = a > 0$,

$$
\operatorname{Re}\frac{{\mu^+}^2 r^+ + {\mu^-}^2 r^-}{q} > c_0(a_0) > 0, \quad \operatorname{Re}\frac{r^+ r^- + \xi^2}{q} > 0,
$$

c_0 does not depend on ξ and s, then

$$
\begin{aligned}
|P| \geqslant &|s| \operatorname{Re}\Big\{(\rho^+ + \rho^-)s + 4\xi^2\frac{{\mu^+}^2 r^+ + {\mu^-}^2 r^-}{q} + 4\mu^+\mu^-|\xi|\frac{r^+ r^- + \xi^2}{q} + \frac{\sigma|\xi|^3}{|s|^2}\bar{s}\Big\}\\
&\geqslant 4c_0\xi^2|s|.
\end{aligned}
$$

Next, for $|s| \geqslant \xi^2$, we have $\sigma|\xi|^3 \leqslant \sigma\xi^2\frac{|s|}{\sqrt{a}} \leqslant \frac{\sigma}{4c_0\sqrt{a}}|P|$,

$$
\begin{aligned}
(\rho^+ + \rho^-)|s|^2 &\leqslant |P| + 4\xi^2|s|\Big|\frac{{\mu^+}^2 r^+ + {\mu^-}^2 r^-}{q}\Big| + 4|\xi||s|\frac{\mu^+\mu^-|r^+ r^- + \xi^2|}{|q|} + \sigma|\xi|^3\\
&\leqslant |P| + c\big(|s|\xi^2 + |s|^{3/2}|\xi|\big) + \sigma|\xi|^3 \leqslant c\big(|P| + |s||P|^{1/2}\big),
\end{aligned}
$$

whence $|s|^2 \leqslant c|P|$.

If $|s| \leqslant \xi^2$, then

$$|s|^2 \leqslant |s|\xi^2 \leqslant \frac{1}{4c_0}|P|,$$

$$\sigma|\xi|^3 \leqslant (\rho^+ + \rho^-)|s|^2 + |P| + c_1\big(|s|\xi^2 + |s|^{3/2}|\xi|\big) \leqslant |P|\Big(\frac{\rho^+ + \rho^-}{4c_0} + 1 + \frac{c_1}{2c_0}\Big),$$

and inequality (2.2.9) is proved. Inequalities (2.2.10)–(2.2.12) are established by using elementary estimates of P, P_1, and P_2 and their derivatives. The lemma is proved. □

2.3 Theorems on the Fourier Multipliers in the Hölder Spaces

We evaluate the solution to problem (2.2.1) by the method applied by V. A. Solonnikov and I. Sh. Mogilevskiĭ [53] to a model problem arising in the study of the motion of an isolated liquid mass. It involves the application of the Fourier multiplier theorem in the Hölder spaces [36]. We note that O. A. Ladyzhenskaya [44] proved the theorem on the Fourier multipliers in the Hölder spaces in a simpler form. But in oder to apply it, one needs to adapt it to our case of the Stokes equations with the non-coercive boundary conditions. Here are some results from [53] and [36].

We consider functions $f(x', t)$ of the variables $x' = (x_1, x_2)$ and t given in $\mathbb{R}^3$ and vanishing for $t \leqslant 0$.

Let $\omega \in C_0^\infty(\mathbb{R})$, $\operatorname{supp}\omega \subset [0, 1]$, $\int_0^1 \omega(z)\,\mathrm{d}z = 1$, and let

$$\psi(z) = \sum_{k=1}^{n} \frac{(-1)^{k+1} n!}{k!(n-k)!k}\omega\Big(\frac{z}{k}\Big)$$

with some natural number n. Let us denote by $Y_\tau(h)$ $(h > 0)$ the averaging operator in the variable y_τ $(\tau = 1, 2)$ with the kernel $\frac{1}{h}\psi(\frac{z}{h})$:

$$Y_\tau(h) f(y', t) = \frac{1}{h}\int_{-\infty}^{\infty} f(y' - z\boldsymbol{e}_\tau, t)\psi\Big(\frac{z}{h}\Big)\,\mathrm{d}z,$$

(here $\boldsymbol{e}_1 = (1, 0)$, $\boldsymbol{e}_2 = (0, 1)$), and by $Y_0(h)$ the averaging operator in the variable t

$$Y_0(h) f(y', t) = \frac{1}{h}\int_{-\infty}^{\infty} f(y', t - z)\psi\Big(\frac{z}{h}\Big)\,\mathrm{d}z.$$

We set $Y(h) = Y_1(h)Y_2(h)Y_0(h^2)$ and $Y^*(h) = \prod_{k=0}^{\infty} Y(h2^{-k})$. As shown in [36] (see also the paper of K. K. Golovkin [35]), in the space $\mathring{C}^{l,l/2}(\mathbb{R}^2_\infty)$, the norm $\langle f\rangle^{l,l/2}_{\mathbb{R}^2_\infty}$ is equivalent to the norms

$$\|f\|^{(l,l/2)}_{\mathbb{R}^2_\infty,m_1,m_2,m_0} = \sum_{\tau=1}^{2} \sup_{h>0}\Big\{h^{-l+m_\tau}\Big|\frac{\partial^{m_\tau}}{\partial x_\tau^{m_\tau}}(Y^*(h)f)\Big|_{\mathbb{R}^2_\infty}\Big\} + \\ + \sup_{h>0}\Big\{h^{-l+2m_0}\Big|\frac{\partial^{m_0}}{\partial t^{m_0}}(Y^*(h)f)\Big|_{\mathbb{R}^2_\infty}\Big\}$$

with various m_τ satisfying the inequalities $m_1, m_2 > l$, $m_0 > l/2$.

We denote by Ff the Fourier transform of the function $f(x',t)$ in all variables, i.e., $\widetilde{f}(\xi,s) = Ff_a$, where

$$f_a = \mathrm{e}^{-at}f, \quad a = \operatorname{Re} s \geqslant 0 \text{ and } f_a = F^{-1}\widetilde{f}(\xi,s).$$

We consider the convolution $Kf = \int_0^\infty \int_{\mathbb{R}^2} K(x'-y',t-\tau)f(y',\tau)\,\mathrm{d}x'\,\mathrm{d}\tau$ the Fourier–Laplace image of which is given by

$$\widetilde{u}(\xi,s) = \widetilde{K}(\xi,s)\widetilde{f}(\xi,s). \tag{2.3.1}$$

As established in [53], the following theorem is a consequence of the equivalence of the norms $\langle f\rangle^{(l,l/2)}_{\mathbb{R}^2_\infty}$ and $\|f\|^{(l,l/2)}_{\mathbb{R}^2_\infty,m_1,m_2,m_0}$ [36].

Theorem 2.3.1 *Let $f \in \mathring{C}^{l,l/2}(\mathbb{R}^2_\infty)$ and let functions u and f be connected by the relation (2.3.1). If there exist nonnegative integers ν_0, ν_1, ν_2 and a positive constant c_0 independent of h such that*

$$\Gamma_h^{(\nu_j)}(\widetilde{K}) = \int_0^\infty \frac{\mathrm{d}y_0}{y_0^{3/2}} \int_0^\infty \frac{\mathrm{d}y_1}{y_1^{3/2}} \int_0^\infty \|\Delta_0(y_0)\Delta_1(y_1)\Delta_2(y_2)\{(\mathrm{i}\eta_j)^{(\nu_j)} \times \\ \times \widetilde{K}(\frac{\eta}{h}, s_h)\widehat{\psi}(\eta,\eta_0)\}\|_{L_2(\mathbb{R}^3)}\frac{\mathrm{d}y_2}{y_2^{3/2}} \leqslant c_0 h^\beta, \tag{2.3.2}$$

where $\beta > -l$, $s_h = a + \frac{\mathrm{i}\eta_0}{h^2}$, $\widehat{\psi} = F\psi$ is the Fourier image of the function $\psi(x',t) = \psi(x_1)\psi(x_2)\psi(t)$, $\Delta_j(y_j)g \equiv g(..,x_j+y_j,..) - g(..,x_j,..)$ is the finite difference of the function $g(x',t)$ in the variable x_j with a step y_j, $j = 0,1,2$ $(x_0 = t)$, then

$$\langle u_a\rangle^{(l+\beta,\frac{l+\beta}{2})}_{\mathbb{R}^2_\infty} \leqslant c_1\langle f_a\rangle^{(l,\frac{l}{2})}_{\mathbb{R}^2_\infty}.$$

The following theorem, which is a generalization of Theorem 2.2 from [53], gives some restrictions to the function $\widetilde{K}$ that guarantees that the inequalities (2.3.2) hold.

Theorem 2.3.2 *We assume that the function* $\widetilde{K}(\xi, s)$ *with* $\operatorname{Re} s \geqslant 0$ *satisfies the inequalities*

$$|\widetilde{K}(\xi, s)| \leqslant cR^{-m}, \quad \left|\frac{\partial \widetilde{K}}{\partial \xi_0}\right| \leqslant cR^{-m-2},$$
$$\left|\frac{\partial \widetilde{K}}{\partial \xi_\tau}\right| \leqslant cR^{-m-1}, \quad \left|\frac{\partial^2 \widetilde{K}}{\partial \xi_\tau \partial \xi_0}\right| \leqslant cR^{-m-3}, \tag{2.3.3}$$

where $\tau = 1, 2,\ R = \sqrt{|s| + \xi^2},\ m \geqslant 0$. *Let, in addition,*

$$\frac{\partial \widetilde{K}}{\partial \xi_1} = K_1(\xi, s) + \varphi(\xi) K_2(\xi, s),$$

where $\varphi(\xi)$ *is a* 0 *order homogeneous function of* ξ_1 *and* ξ_2, *which is smooth on the unit circle* $\{\xi_1^2 + \xi_2^2 = 1\}$, *and* K_1 *and* K_2 *are such that*

$$|K_q| \leqslant cR^{-m-1}, \quad \left|\frac{\partial K_q}{\partial \xi_2}\right| \leqslant cR^{-m-2},$$
$$\left|\frac{\partial K_q}{\partial \xi_0}\right| \leqslant cR^{-m-3}, \quad \left|\frac{\partial^2 K_q}{\partial \xi_0 \partial \xi_2}\right| \leqslant cR^{-m-4}, \qquad q = 1, 2. \tag{2.3.4}$$

Then, for $\nu_j > m + 2$ $(j = 1, 2)$ *and* $\nu_0 > \frac{m}{2} + \frac{5}{4}$, *one has*

$$\Gamma_h^{(\nu_j)}(\widetilde{K}) \leqslant ch^m \qquad (j = 0, 1, 2). \tag{2.3.5}$$

Proof The assumptions of the theorem imply that

$$\left|\frac{\partial^2 \widetilde{K}}{\partial \xi_1 \partial \xi_2}\right| \leqslant c_2(R^{-m-2} + |\xi|^{-1} R^{-m-1}) \leqslant 2c_2 |\xi|^{-1} R^{-m-1},$$
$$\left|\frac{\partial^3 \widetilde{K}}{\partial \xi_0 \partial \xi_1 \partial \xi_2}\right| \leqslant c_3(R^{-m-4} + |\xi|^{-1} R^{-m-3}) \leqslant 2c_3 |\xi|^{-1} R^{-m-3}. \tag{2.3.6}$$

The estimates (2.3.3) and (2.3.6) are enough to deduce (2.3.5) for $j = 1, 2$, repeating literally the proof of Theorem 2.2 from [53]. Therefore, we consider only the case $j = 0$. Following the proof of the above theorem, we represent $\Gamma_h^{(\nu_0)}(\widetilde{K})$ in the form

$$\Gamma_h^{(\nu_0)}(\widetilde{K}) = \sum_{i_0,i_1,i_2=1}^{2} \int_{d_{i_0}}^{d_{i_0+1}} \frac{dy_0}{y_0^{3/2}} \int_{d_{i_1}}^{d_{i_1+1}} \frac{dy_1}{y_1^{3/2}} \int_{d_{i_2}}^{d_{i_2+1}} \|\triangle_0(y_0)\triangle_1(y_1)\triangle_2(y_2)\{(i\eta_0)^{\nu_0} \times$$

$$\times \widetilde{K}(\frac{\eta}{h}, s_h)\widehat{\psi}(\eta,\eta_0)\}\|_{L_2(\mathbb{R}^3)} \frac{dy_2}{y_2^{3/2}} \equiv \sum_{i_0,i_1,i_2=1}^{2} I_{i_0 i_1 i_2},$$

where $s_h = a + \frac{i\eta_0}{h^2}$ and $d_1 = 0$, $d_2 = 1$, $d_3 = \infty$. We suppose that at least one of the indices $i_1 + 1$ or $i_2 + 1$ is equal to three. Then, as in Theorem 2.2 from [53], we can take advantage of the inequalities $\|\triangle_k(y_k)f\|_{L_2} \leqslant y_k \|\frac{\partial f}{\partial \eta_k}\|_{L_2}$ or of $\|\triangle_k(y_k)f\|_{L_2} \leqslant 2\|f\|_{L_2}$ (depending on whether y_k belongs to the interval $(0, 1)$ or $(1, \infty)$), and make use of estimates (2.3.3) (without resorting to the estimates of $\frac{\partial^2 \widetilde{K}}{\partial \xi_1 \partial \xi_2}$) to show that

$$|I_{i_0 i_1 i_2}| \leqslant c h^m.$$

In the case of $d_{i_1+1} = d_{i_2+1} = 1$, this argument is not enough since the estimates (2.3.6) do not guarantee the convergence of the integrals arising.

We consider separately the terms with $d_{i_1+1} = d_{i_2+1} = 1$. We have

$$\int_1^\infty \frac{dy_0}{y_0^{3/2}} \int_0^1 \frac{dy_1}{y_1^{3/2}} \int_0^1 \|\triangle_0(y_0)\triangle_1(y_1)\triangle_2(y_2)\{(i\eta_0)^{\nu_0}\widetilde{K}(\frac{\eta}{h}, s_h)\widehat{\psi}(\eta,\eta_0)\}\|_{L_2(\mathbb{R}^3)} \frac{dy_2}{y_2^{3/2}}$$

$$\leqslant 4 \int_0^1 \frac{dy_1}{y_1^{3/2}} \int_0^1 \|\triangle_1(y_1)\triangle_2(y_2)\{(i\eta_0)^{\nu_0}\widetilde{K}(\frac{\eta}{h}, s_h)\widehat{\psi}(\eta,\eta_0)\}\|_{L_2(\mathbb{R}^3)} \frac{dy_2}{y_2^{3/2}}, \quad (2.3.7)$$

$$\int_0^1 \frac{dy_0}{y_0^{3/2}} \int_0^1 \frac{dy_1}{y_1^{3/2}} \int_0^1 \|\triangle_0(y_0)\triangle_1(y_1)\triangle_2(y_2)\{(i\eta_0)^{\nu_0}\widetilde{K}(\frac{\eta}{h}, s_h)\widehat{\psi}(\eta,\eta_0)\}\|_{L_2(\mathbb{R}^3)} \frac{dy_2}{y_2^{3/2}}$$

$$\leqslant 2 \int_0^1 \frac{dy_1}{y_1^{3/2}} \int_0^1 \|\triangle_1(y_1)\triangle_2(y_2)\frac{\partial}{\partial \eta_0}\{(i\eta_0)^{\nu_0}\widetilde{K}(\frac{\eta}{h}, s_h)\widehat{\psi}(\eta,\eta_0)\}\|_{L_2} \frac{dy_2}{y_2^{3/2}}. \quad (2.3.8)$$

Let us evaluate the right-hand side in (2.3.7). By expressing the finite difference with respect to η_1 in terms of the derivative, we obtain

$$4 \int_0^1 \frac{dy_1}{y_1^{3/2}} \int_0^1 \|\triangle_1(y_1)\triangle_2(y_2)\{(i\eta_0)^{\nu_0}\widetilde{K}(\frac{\eta}{h}, s_h)\widehat{\psi}(\eta,\eta_0)\}\|_{L_2(\mathbb{R}^3)} \frac{dy_2}{y_2^{3/2}}$$

$$\leqslant 8 \int_0^1 \|\triangle_2(y_2)\frac{\partial}{\partial \eta_1}\{(i\eta_0)^{\nu_0}\widetilde{K}(\frac{\eta}{h}, s_h)\widehat{\psi}(\eta,\eta_0)\}\|_{L_2(\mathbb{R}^3)} \frac{dy_2}{y_2^{3/2}}$$

$$\leqslant 8 \int_0^1 \big\| \triangle_2(y_2)\big\{(\mathrm{i}\eta_0)^{\nu_0} \frac{\partial \widehat{\psi}}{\partial \eta_1} \widetilde{K}(\frac{\eta}{h}, s_h)\big\}\big\|_{L_2(\mathbb{R}^3)} \frac{\mathrm{d}y_2}{y_2^{3/2}} +$$

$$+8h^{-1} \int_0^1 \big\| \triangle_2(y_2)\big\{(\mathrm{i}\eta_0)^{\nu_0} \widehat{\psi}[K_1(\frac{\eta}{h}, s_h) + \varphi(\eta) K_2(\frac{\eta}{h}, s_h)]\big\}\big\|_{L_2} \frac{\mathrm{d}y_2}{y_2^{3/2}}.$$

In the first term of the right-hand side, we can again express the finite difference in terms of the derivative and, using (2.3.3), estimate its norm in terms of

$$16 \Big\| \frac{\partial}{\partial \eta_2} \{(\mathrm{i}\eta_0)^{\nu_0} \frac{\partial \widehat{\psi}}{\partial \eta_1} \widetilde{K}(\frac{\eta}{h}, s_h)\} \Big\|_{L_2(\mathbb{R}^3)}$$

$$\leqslant 16 c_4 \Big(\int_{\mathbb{R}^2} \int_{-\infty}^{\infty} |\eta_0|^{2\nu_0} \Big| \frac{\partial^2 \widehat{\psi}}{\partial \eta_1 \partial \eta_2} \Big|^2 \frac{\mathrm{d}\eta_0 \, \mathrm{d}\eta}{(|\eta_0| + \eta^2)^m} \Big)^{1/2} h^m +$$

$$+16 c_5 \Big(\int_{\mathbb{R}^2} \int_{-\infty}^{\infty} |\eta_0|^{2\nu_0} \Big| \frac{\partial \widehat{\psi}}{\partial \eta_1} \Big|^2 \frac{\mathrm{d}\eta_0 \, \mathrm{d}\eta}{(|\eta_0| + \eta^2)^{m+1}} \Big)^{1/2} h^m \leqslant c_6 h^m.$$

In the same way, by means of (2.3.4), we prove the estimate

$$8h^{-1} \int_0^1 \big\| \triangle_2(y_2)\big\{(\mathrm{i}\eta_0)^{\nu_0} \widehat{\psi} K_1(\frac{\eta}{h}, s_h)\big\}\big\|_{L_2} \frac{\mathrm{d}y_2}{y_2^{3/2}} \leqslant c h^m.$$

Finally, from the formula

$$\triangle_2(y_2) g(\eta)\varphi(\eta) = g(\eta)\triangle_2(y_2)\varphi(\eta) + \varphi(\eta + \boldsymbol{e}_2 y_2)\triangle_2(y_2) g(\eta),$$

where $\boldsymbol{e}_2 = (0, 1)$, it follows that

$$\big\| \triangle_2(y_2)\big\{(\mathrm{i}\eta_0)^{\nu_0} \widehat{\psi} \varphi(\eta) K_2(\frac{\eta}{h}, s_h)\big\}\big\|_{L_2} \leqslant$$

$$\leqslant \big\| (\mathrm{i}\eta_0)^{\nu_0} \widehat{\psi} K_2(\frac{\eta}{h}, s_h) \triangle_2(y_2)\varphi(\eta) \big\|_{L_2}$$

$$+ \sup_{\eta} |\varphi(\eta)| \big\| \triangle_2(y_2)\big\{(\mathrm{i}\eta_0)^{\nu_0} \widehat{\psi} K_2(\frac{\eta}{h}, s_h)\big\}\big\|_{L_2}.$$

The function $\varphi(\eta)$ is bounded; in addition, the formula

$$\triangle_2(y_2)\varphi(\eta) = \varphi(\eta + \boldsymbol{e}_2 y_2) - \varphi(\eta) = \int_l \nabla\varphi \cdot \mathrm{d}l,$$

where l is a contour connecting the points η and $\eta + \mathbf{e}_2 y_2$, having length $|l| \leqslant \pi y_2$ and a distance from the origin, equal to $\min(|\eta|, |\eta + \mathbf{e}_2 y_2|)$, implies the inequality

$$|\Delta_2(y_2)\varphi(\eta)| \leqslant |l| \max_{\eta \in l} |\nabla\varphi(\eta)| \leqslant c y_2 \Big(\frac{1}{|\eta|} + \frac{1}{|\eta + \mathbf{e}_2 y_2|}\Big).$$

Thus, for any $\alpha \in (1/2, 1)$

$$\begin{aligned} |\Delta_2(y_2)\varphi(\eta)| &\leqslant \big(2 \max_{\eta \in l} |\varphi(\eta)|\big)^{1-\alpha} \big(|l| \max_{\eta \in l} |\nabla\varphi(\eta)|\big)^{\alpha} \\ &\leqslant c y_2^{\alpha} \Big(\frac{1}{|\eta|^{\alpha}} + \frac{1}{|\eta + \mathbf{e}_2 y_2|^{\alpha}}\Big) \end{aligned} \tag{2.3.9}$$

and

$$8h^{-1} \int_0^1 \big\|\Delta_2(y_2)\{(i\eta_0)^{\nu_0} \widehat{\psi}\varphi K_2(\frac{\eta}{h}, s_h)\}\big\|_{L_2} \frac{dy_2}{y_2^{3/2}} \leqslant$$

$$\leqslant ch^m + ch^m \int_0^1 \frac{dy_2}{y_2^{3/2}} \Big(\int_{\mathbb{R}^2} \int_{-\infty}^{\infty} |\eta_0|^{2\nu_0} |\widehat{\psi}|^2 \Big(\frac{1}{|\eta|^{2\alpha}} + \frac{1}{|\eta + \mathbf{e}_2 y_2|^{2\alpha}}\Big) \times$$

$$\times \frac{d\eta_0\, d\eta}{(|\eta_0| + \eta^2)^{m+1}}\Big)^{1/2} \leqslant ch^m,$$

since

$$\int_{\mathbb{R}^2} \int_{-\infty}^{\infty} |\eta_0|^{2\nu_0} |\widehat{\psi}(\eta, \eta_0)|^2 \frac{d\eta_0\, d\eta}{|\eta + \mathbf{e}_2 y_2|^{2\alpha} (|\eta_0| + \eta^2)^{m+1}} \leqslant$$

$$\leqslant \int_{-\infty}^{\infty} \Big(\int_{|\eta| \leqslant |\eta + \mathbf{e}_2 y_2|} \frac{|\eta_0|^{2\nu_0} |\widehat{\psi}|^2\, d\eta}{|\eta|^{2\alpha} (|\eta_0| + \eta^2)^{m+1}} + \int_{|\eta| > |\eta + \mathbf{e}_2 y_2|} \frac{|\eta_0|^{2\nu_0} |\widehat{\psi}|^2\, d\eta}{|\eta + \mathbf{e}_2 y_2|^{2\alpha} (|\eta_0| + |\eta + \mathbf{e}_2 y_2|^2)^{m+1}} \Big) d\eta_0$$

$$\leqslant \int_{\mathbb{R}^2} \int_{-\infty}^{\infty} |\eta_0|^{2\nu_0} \Big(|\widehat{\psi}(\eta, \eta_0)|^2 + \sup_{|z| \leqslant 1} |\widehat{\psi}(\eta + z, \eta_0)|^2\Big) \frac{d\eta_0\, d\eta}{|\eta|^{2\alpha} (|\eta_0| + \eta^2)^{m+1}} \leqslant c.$$

The right-hand side of (2.3.8) is evaluated in the same way provided that ν_0 satisfies the hypothesis of the theorem (the presence of an extra derivative with respect to η_0 leads, in the worst case, only to an increase in the exponent $|\eta_0| + \eta^2$ in the denominator). The proof of inequalities (2.3.5) is complete. □

We now prove a statement similar to Theorem 2.3 in [53].

Theorem 2.3.3 *If $\widetilde{K}(\xi, s)$ satisfies the assumptions of Theorem 2.3.2 with $m \geqslant -3$, then for $\operatorname{Re} s \geqslant a_0 > 0$ and for large enough ν_j*

$$\left|\Gamma_h^{(\nu_j)}\Big(\frac{\widetilde{K}}{P}\Big)\right| \leqslant c_0 h^{m+3}, \qquad j = 0, 1, 2, \tag{2.3.10}$$

where the expression P is given by formula (2.2.8).

Proof If $j = 1, 2$, inequality (2.3.10) is established by using the estimates (2.3.3), (2.3.6), (2.2.13), and (2.2.14) just as in Theorem 2.3 in [53]. For $j = 0$, as we saw in the previous theorem, it suffices to estimate the sum

$$\int_0^1 \|\triangle_2(y_2)\frac{\partial^2}{\partial\eta_1\partial\eta_0}\{(\mathrm{i}\eta_0)^{\nu_0}\widehat{\psi}(\eta,\eta_0)\frac{\widetilde{K}(\frac{\eta}{h}, s_h)}{P(\frac{\eta}{h}, s_h)}\}\|_{L_2(\mathbb{R}^3)}\frac{dy_2}{y_2^{3/2}}$$
$$+4\int_0^1 \|\triangle_2(y_2)\frac{\partial}{\partial\eta_1}\{(\mathrm{i}\eta_0)^{\nu_0}\widehat{\psi}\frac{\widetilde{K}}{P}\}\|_{L_2(\mathbb{R}^3)}\frac{dy_2}{y_2^{3/2}}. \tag{2.3.11}$$

We consider the second term. We take advantage of the inequality

$$\left|\frac{P'_{\xi_\mu}(\frac{\eta}{h}, s_h)}{P(\frac{\eta}{h}, s_h)}\right| \leqslant c\frac{|s_h|^{3/2} + |s_h|\frac{|\eta|}{h} + \sigma\frac{|\eta|^2}{h^2}\frac{|s_h|^{1/2}}{\sqrt{a}}}{\left|P(\frac{\eta}{h}, s_h)\right|} \leqslant \frac{c}{|s_h|^{1/2}},$$

where $P'_{\xi_\mu} = \frac{\partial P(\xi,s)}{\partial\xi_\mu}$, $\mu = 0, 1, 2$, and also of the estimates

$$\frac{|\widetilde{K}(\frac{\eta}{h}, s_h)|}{|P(\frac{\eta}{h}, s_h)|} \leqslant c\frac{\left(|s_h| + \frac{\eta^2}{h^2}\right)^{-m/2}}{|s_h|^{3/2}a_0^{1/2} + \sigma\frac{|\eta|^3}{h^3}} \leqslant c\Big(|s_h| + \frac{\eta^2}{h^2}\Big)^{-\frac{m+3}{2}} \leqslant \frac{ch^{m+3}}{(|\eta_0| + \eta^2)^{\frac{m+3}{2}}},$$

$$\begin{aligned}\left|\frac{\partial}{\partial\eta_\mu}\frac{\widetilde{K}}{P}\right| &\leqslant h^{-1}\left(\frac{|\widetilde{K}'_{\xi_\mu}|}{|P|} + \frac{|\widetilde{K}|}{|P|}\frac{|P'_{\xi_\mu}|}{|P|}\right)\\ &\leqslant c(a_0)h^{m+3}\left(\frac{1}{(|\eta_0| + \eta^2)^{\frac{m+4}{2}}} + \frac{1}{(|\eta_0| + \eta^2)^{\frac{m+3}{2}}|\eta_0|^{1/2}}\right)\\ &\leqslant \frac{2c(a_0)h^{m+3}}{|\eta_0|^{1/2}(|\eta_0| + \eta^2)^{\frac{m+3}{2}}} \qquad (\mu = 1, 2),\end{aligned}$$

which is a consequence of (2.3.3) and (2.2.10). This implies the evaluation

$$\int_0^1 \|\triangle_2(y_2)(\mathrm{i}\eta_0)^{\nu_0}\widehat{\psi}'_{\eta_1}\frac{\widetilde{K}}{P}\|_{L_2(\mathbb{R}^3)}\frac{dy_2}{y_2^{3/2}} \leqslant 2\Big\|(\mathrm{i}\eta_0)^{\nu_0}\frac{\partial}{\partial\eta_2}\Big\{\widehat{\psi}'_{\eta_1}\frac{\widetilde{K}(\frac{\eta}{h}, s_h)}{P(\frac{\eta}{h}, s_h)}\Big\}\Big\|_{L_2(\mathbb{R}^3)}$$

$$\leqslant ch^{m+3}.$$

Next, we represent $\frac{\partial}{\partial\eta_1}\frac{\widetilde{K}}{P}$ in the form

$$\begin{aligned}\frac{\partial}{\partial\eta_1}\frac{\widetilde{K}(\frac{\eta}{h},s_h)}{P(\frac{\eta}{h},s_h)} &= h^{-1}\Big\{\frac{K_1(\frac{\eta}{h},s_h)}{P(\frac{\eta}{h},s_h)} - \frac{P_1(\frac{\eta}{h},s_h)}{P(\frac{\eta}{h},s_h)}\frac{\widetilde{K}}{P} + \\ &\quad +\varphi(\eta)\frac{K_2(\frac{\eta}{h},s_h)}{P(\frac{\eta}{h},s_h)} - \frac{\eta_1}{|\eta|}\frac{P_2(\frac{\eta}{h},s_h)}{P(\frac{\eta}{h},s_h)}\frac{\widetilde{K}}{P}\Big\} \\ &\equiv h^{-1}\Big\{M_1(\frac{\eta}{h},s_h) + \varphi(\eta)M_2(\frac{\eta}{h},s_h) + \frac{\eta_1}{|\eta|}M_3(\frac{\eta}{h},s_h)\Big\},\end{aligned}$$

and we note that functions M_i satisfy the inequalities

$$\begin{aligned}\big|M_i(\frac{\eta}{h},s_h)\big| &\leqslant \frac{ch^{m+4}}{|\eta_0|^{1/2}(|\eta_0|+\eta^2)^{\frac{m+3}{2}}} \\ \Big|\frac{\partial M_i(\frac{\eta}{h},s_h)}{\partial\eta_2}\Big| &\leqslant \frac{ch^{m+4}}{|\eta_0|(|\eta_0|+\eta^2)^{\frac{m+3}{2}}}, \qquad i=1,2,3, \qquad (2.3.12)\end{aligned}$$

which follow from (2.3.4), (2.2.9), and (2.2.12). By estimating $\Delta_2(y_2)\varphi$ and $\Delta_2(y_2)\frac{\eta_1}{|\eta|}$ by means of (2.3.9) for $\alpha\in(1/2,1)$, we obtain, due to (2.3.12),

$$\begin{aligned}&\int_0^1 \Big\|\Delta_2(y_2)(\mathrm{i}\eta_0)^{\nu_0}\widehat{\psi}\frac{\partial}{\partial\eta_1}\frac{\widetilde{K}}{P}\Big\|_{L_2(\mathbb{R}^3)}\frac{dy_2}{y_2^{3/2}} \leqslant \\ &\leqslant 2h^{-1}\Big\{\Big\|(\mathrm{i}\eta_0)^{\nu_0}\frac{\partial}{\partial\eta_2}(\widehat{\psi}M_1(\frac{\eta}{h},s_h))\Big\|_{L_2} + \max|\varphi|\Big\|(\mathrm{i}\eta_0)^{\nu_0}\frac{\partial}{\partial\eta_2}(\widehat{\psi}M_2)\Big\|_{L_2} + \\ &\quad +\Big\|(\mathrm{i}\eta_0)^{\nu_0}\frac{\partial}{\partial\eta_2}(\widehat{\psi}M_3(\frac{\eta}{h},s_h))\Big\|_{L_2}\Big\} + \\ &\quad +h^{-1}c\int_0^1\frac{dy_2}{y_2^{3/2-\alpha}}\Big\{\int_{\mathbb{R}^2}\int_{-\infty}^{\infty}|\eta_0|^{2\nu_0}|\widehat{\psi}|^2\Big(\frac{1}{|\eta|^{2\alpha}}+\frac{1}{|\eta+e_2y_2|^{2\alpha}}\Big)\times \\ &\quad \times\Big(\big|M_2(\frac{\eta}{h},s_h)\big|^2+\big|M_3(\frac{\eta}{h},s_h)\big|^2\Big)\,d\eta_0\,d\eta\Big\}^{1/2} \leqslant c(a_0)h^{m+3}.\end{aligned}$$

The first term of the right-hand side of (2.3.11) can be estimated similarly taking into account that

$$\Big|\frac{\partial}{\partial\eta_0}\frac{\widetilde{K}(\frac{\eta}{h},s_h)}{P(\frac{\eta}{h},s_h)}\Big| \leqslant h^{-2}\Big(\frac{|\widetilde{K}'_{\xi_\mu}|}{|P|}+\frac{|\widetilde{K}|}{|P|}\frac{|P'_{\xi_\mu}|}{|P|}\Big) \leqslant \frac{ch^{m+3}}{|\eta_0|(|\eta_0|+\eta^2)^{\frac{m+3}{2}}},$$

$$\Big|\frac{\partial M_i(\frac{\eta}{h}, s_h)}{\partial \eta_2}\Big| \leqslant \frac{ch^{m+4}}{|\eta_0|^{3/2}(|\eta_0|+\eta^2)^{\frac{m+3}{2}}},$$

$$\Big|\frac{\partial^2 M_i(\frac{\eta}{h}, s_h)}{\partial \eta_2 \partial \eta_0}\Big| \leqslant \frac{ch^{m+4}}{|\eta_0|^{2}(|\eta_0|+\eta^2)^{\frac{m+3}{2}}}.$$

□

2.4 An Estimate of the Solution of Problem (2.2.1)

Let us prove the following proposition.

Theorem 2.4.1 *Let the functions b_1, b_2, b_3, B in (2.2.1) decrease well enough as $|x| \to \infty$; moreover, $b_{\tau a} = b_\tau \mathrm{e}^{-at} \in \overset{\circ}{C}{}^{1+\alpha,\frac{1+\alpha}{2}}(\mathbb{R}^2_\infty)$, $\tau = 1,2$, $b_{3a} \in \overset{\circ}{C}{}^{(\gamma,1+\alpha)}(\mathbb{R}^2_\infty)$, $B_a \in \overset{\circ}{C}{}^{\alpha,\frac{\alpha}{2}}(\mathbb{R}^2_\infty)$ with $\alpha, \gamma \in (0,1)$, $a > 0$. Then for the functions $p_a = F^{-1}\widetilde{p}$ from (2.2.6) and $\boldsymbol{\omega}_a = F^{-1}\boldsymbol{\omega}$ from (2.2.7), the estimate*

$$\langle \boldsymbol{\omega}_a \rangle^{(2+\alpha,1+\frac{\alpha}{2})}_{\mathbb{R}^2_\infty} + \langle \nabla p_a \rangle^{(\alpha,\frac{\alpha}{2})}_{D^3_\infty} \leqslant c\Big\{\sum_{\tau=1}^{2} \langle b_{\tau a}\rangle^{(1+\alpha,\frac{1+\alpha}{2})}_{\mathbb{R}^2_\infty} + \langle \nabla' b_{3a}\rangle^{(\alpha)}_{x',\mathbb{R}^2_\infty}$$
$$+ \langle b_{3a}\rangle^{(\gamma,1+\alpha)}_{\mathbb{R}^2_\infty} + \langle B_a\rangle^{(\alpha,\frac{\alpha}{2})}_{\mathbb{R}^2_\infty}\Big\}. \quad (2.4.1)$$

holds. Here $x' = (x_1, x_2)$, $\nabla' = (\partial/\partial x_1, \partial/\partial x_2)$, $\boldsymbol{\omega} = (\omega_1, \omega_2, \omega_3)$.

Proof We suppose that $B = 0$ and write the functions $\frac{\partial p_a}{\partial x_\tau}$ $(\tau = 1,2)$ for $x_3 > 0$ in the form

$$\frac{\partial p_a}{\partial x_\tau} = F^{-1}[\mu^+(r^+ + |\xi|)\mathrm{i}\xi_\tau \Psi^+ \mathrm{e}^{-|\xi|x_3}] = u_\tau + \frac{\partial u}{\partial x_\tau},$$

where

$$u_\tau = F^{-1}\big[\widetilde{d}_\tau \frac{s}{P} L^+ \mathrm{e}^{-|\xi|x_3}\big],$$
$$u = -F^{-1}\big[\frac{\widetilde{d}}{|\xi|}\frac{s}{P} M^+ \mathrm{e}^{-|\xi|x_3}\big], \quad (2.4.2)$$
$$\widetilde{d}_\tau = \mathrm{i}\xi_\tau \widetilde{b}_3, \quad \tau = 1,2, \qquad \widetilde{d} = \mathrm{i}\xi_1\widetilde{b}_1 + \mathrm{i}\xi_2\widetilde{b}_2,$$
$$L^+ = \mu^+(r^+ + |\xi|)(-r^+ + |\xi|\frac{q'}{q}),$$
$$M^+ = \frac{\mu^+(r^+ + |\xi|)}{q}\{(\rho^+ + \rho^-)s + |\xi|q - r^+q' + \frac{\sigma}{s}|\xi|^3\}.$$

In addition

$$\frac{\partial p_a}{\partial x_3} = F^{-1}[-\mu^+(r^+ + |\xi|)|\xi|\Psi^+ e^{-|\xi|x_3}]$$
$$= -F^{-1}[\widetilde{d}_3 \frac{s}{P} L^+ e^{-|\xi|x_3} - \widetilde{d}\frac{s}{P} M^+ e^{-|\xi|x_3}]$$

where $\widetilde{d}_3 = |\xi|\widetilde{b}_3$.

We consider u_τ, u, $\frac{\partial p_a}{\partial x_3}$ as solutions to the Dirichlet and Neumann problems:

$$\Delta u_\tau = 0 \quad (x \in \mathbb{R}^3_+),\ u_\tau\big|_{x_3=0} = u^{(0)}_{\tau a}(x',t), \qquad \tau = 1,2,$$
$$\Delta u = 0 \quad (x \in \mathbb{R}^3_+),\ \frac{\partial u}{\partial x_3}\big|_{x_3=0} = u^{(0)}_a(x',t) = \sum_{\tau=1}^{2} \frac{\partial v^{(0)}_{\tau a}}{\partial x_\tau},$$
$$\Delta \frac{\partial p_a}{\partial x_3} = 0 \quad (x \in \mathbb{R}^3_+),\ \frac{\partial p_a}{\partial x_3}\big|_{x_3=0} = w^{(0)}_a(x',t),$$

in which

$$\begin{aligned} u^{(0)}_{\tau a} &= F^{-1}[\widetilde{d}_\tau \frac{s}{P} L^+], \\ v^{(0)}_{\tau a} &= -F^{-1}[\widetilde{b}_\tau \frac{s}{P} M^+], \\ w^{(0)}_a &= -F^{-1}[\widetilde{d}_3 \frac{s}{P} L^+ - \widetilde{d}\frac{s}{P} M^+]. \end{aligned} \tag{2.4.3}$$

By virtue of Lemma 2.1.1 and its corollary

$$\langle \nabla p_a \rangle^{(\alpha,\frac{\alpha}{2})}_{D^+_\infty} = \sum_{\tau=1}^{2} \Big(\langle u_{\tau a} \rangle^{(\alpha,\frac{\alpha}{2})}_{D^+_\infty} + \Big\langle \frac{\partial u}{\partial x_\tau} \Big\rangle^{(\alpha,\frac{\alpha}{2})}_{D^+_\infty} \Big) + \Big\langle \frac{\partial p_a}{\partial x_3} \Big\rangle^{(\alpha,\frac{\alpha}{2})}_{D^+_\infty}$$
$$\leqslant c\Big\{ \sum_{\tau=1}^{2} \Big(\langle u^{(0)}_{\tau a} \rangle^{(\alpha,\frac{\alpha}{2})}_{\mathbb{R}^2_\infty} + \langle v^{(0)}_{\tau a} \rangle^{(1+\alpha,\frac{1+\alpha}{2})}_{x',\mathbb{R}^2_\infty} \Big) + \langle w^{(0)}_a \rangle^{(\alpha,\frac{\alpha}{2})}_{\mathbb{R}^2_\infty} \Big\}. \tag{2.4.4}$$

To estimate the right-hand side (2.4.4), we use Theorems 2.3.1–2.3.3. We represent $\frac{s}{P}$ in the form

$$\frac{s}{P} = \frac{1}{Q} - \frac{\sigma|\xi|^3}{QP} \tag{2.4.5}$$

$$Q = (\rho^+ + \rho^-)s + \frac{4|\xi|}{q}\{|\xi|(\mu^{+^2} r^+ + \mu^{-^2} r^-) + \mu^+\mu^-(r^+ r^- + |\xi|^2)\}.$$

For $\operatorname{Re} s \geqslant 0$, the function Q satisfies the inequality

$$|Q| > c(|s| + \xi^2),$$

as well as the assumptions of Theorem 2.3.2 with $m = -2$, $\varphi = \frac{\xi_1}{|\xi|}$. In view of (2.4.5)

$$u_{\tau a}^{(0)} = F^{-1}\Big[\widetilde{d}_\tau \frac{L^+}{Q}\Big] - F^{-1}\Big[\widetilde{d}_\tau \frac{\sigma|\xi|^3 L^+}{QP}\Big].$$

Since functions $K' = \frac{L^+}{Q}$ and $K'' = \frac{\sigma|\xi|^3 L^+}{Q}$ satisfy the hypotheses of Theorem 2.3.2 with $m_1 = 0$ and $m_2 = -3$, respectively, due to Theorems 2.3.1–2.3.3, we have

$$\langle u_{\tau a}^{(0)} \rangle_{\mathbb{R}^2_\infty}^{(\alpha, \frac{\alpha}{2})} \leqslant c \langle d_{\tau a} \rangle_{\mathbb{R}^2_\infty}^{(\alpha, \frac{\alpha}{2})}, \quad \tau = 1, 2, \tag{2.4.6}$$

where $d_{\tau a} = F^{-1}\widetilde{d}_\tau$. Similarly, from the representations

$$v_{\tau a}^{(0)} = -F^{-1}\Big[\widetilde{b}_\tau \frac{M_1^+}{Q}\Big] + F^{-1}\Big[\widetilde{b}_\tau \frac{\sigma|\xi|^3 M_1^+}{QP}\Big]$$
$$- F^{-1}\Big[\widetilde{b}_\tau \frac{\sigma|\xi|^3 \mu^+ (r^+ + |\xi|)}{qP}\Big],$$

$$w_a^{(0)} = -F^{-1}\Big[\widetilde{d}_3 \frac{L^+}{Q}\Big] + F^{-1}\Big[\widetilde{d} \frac{M_1^+}{Q}\Big] + F^{-1}\Big[\widetilde{d}_3 \frac{\sigma|\xi|^3 L^+}{QP} - \widetilde{d} \frac{\sigma|\xi|^3 M_1^+}{QP}\Big]$$
$$+ F^{-1}\Big[\widetilde{d} \frac{\sigma|\xi|^3 \mu^+ (r^+ + |\xi|)}{qP}\Big],$$

where $M_1^+ = \mu^+(r^+ + |\xi|)\Big\{(\rho^+ + \rho^-)\frac{s}{q} + |\xi| - \frac{r^+ q'}{q}\Big\}$, one can conclude that

$$\sum_{\tau=1}^{2} \langle v_{\tau a}^{(0)} \rangle_{\mathbb{R}^2_\infty}^{(1+\alpha, \frac{1+\alpha}{2})} \leqslant c \sum_{\tau=1}^{2} \langle b_{\tau a} \rangle_{\mathbb{R}^2_\infty}^{(1+\alpha, \frac{1+\alpha}{2})},$$

$$\langle w_a^{(0)} \rangle_{\mathbb{R}^2_\infty}^{(\alpha, \frac{\alpha}{2})} \leqslant c\big(\langle d_{3a} \rangle_{\mathbb{R}^2_\infty}^{(\alpha, \frac{\alpha}{2})} + \langle d_a \rangle_{\mathbb{R}^2_\infty}^{(\alpha, \frac{\alpha}{2})}\big) \tag{2.4.7}$$
$$\leqslant c\Big(\langle d_{3a} \rangle_{\mathbb{R}^2_\infty}^{(\alpha, \frac{\alpha}{2})} + \sum_{\tau=1}^{2} \langle b_{\tau a} \rangle_{\mathbb{R}^2_\infty}^{(1+\alpha, \frac{1+\alpha}{2})}\Big),$$

here $d_a = F^{-1}\widetilde{d} = \sum_{\tau=1}^{2} \frac{\partial b_{\tau a}}{\partial x_\tau}$. Finally, we note that $(d_{1a}, d_{2a}, -d_{3a}) \equiv F^{-1}\big[(\mathrm{i}\xi_1, \mathrm{i}\xi_2, -|\xi|)\widetilde{b}_3\big] = \nabla U_a\big|_{x_3=0}$, where $U_a = F^{-1}\big[\mathrm{e}^{-|\xi| x_3}\widetilde{b}_3\big]$ is a harmonic function equal to b_{3a} at $x_3 = 0$. Therefore, due to estimate (2.1.4)

$$\sum_{j=1}^{3}\langle d_{ja}\rangle_{\mathbb{R}^2_\infty}^{(\alpha,\frac{\alpha}{2})} \leqslant c\{\langle\nabla' b_{3a}\rangle_{x',\mathbb{R}^2_\infty}^{(\alpha)} + \langle b_{3a}\rangle_{\mathbb{R}^2_\infty}^{(\gamma,1+\alpha)}\}. \tag{2.4.8}$$

By combining the proven estimates, we obtain

$$\langle\nabla p_a\rangle_{D^+_\infty}^{(\alpha,\frac{\alpha}{2})} \leqslant c\Big\{\sum_{\tau=1}^{2}\langle b_{\tau a}\rangle_{\mathbb{R}^2_\infty}^{(1+\alpha,\frac{1+\alpha}{2})} + \langle\nabla' b_{3a}\rangle_{x',\mathbb{R}^2_\infty}^{(\alpha)} + \langle b_{3a}\rangle_{\mathbb{R}^2_\infty}^{(\gamma,1+\alpha)}\Big\}.$$

In the same way, it can be shown that

$$\langle\nabla p_a\rangle_{D^-_\infty}^{(\alpha,\frac{\alpha}{2})} \leqslant c\Big\{\sum_{\tau=1}^{2}\langle b_{\tau a}\rangle_{\mathbb{R}^2_\infty}^{(1+\alpha,\frac{1+\alpha}{2})} + \langle\nabla' b_{3a}\rangle_{x',\mathbb{R}^2_\infty}^{(\alpha)} + \langle b_{3a}\rangle_{\mathbb{R}^2_\infty}^{(\gamma,1+\alpha)}\Big\}.$$

We consider now the function

$$\widetilde{\omega}_3 = \sum_{\tau=1}^{2} \mathrm{i}\xi_\tau \widetilde{b}_\tau\Big(\frac{1}{Q} - \frac{\sigma|\xi|^3}{QP}\Big)\frac{q'}{q} - \widetilde{d}_3\Big(\frac{1}{Q} - \frac{\sigma|\xi|^3}{QP}\Big).$$

Due to Theorems 2.3.1–2.3.3,

$$\begin{aligned}\langle\omega_{3a}\rangle_{\mathbb{R}^2_\infty}^{(2+\alpha,1+\frac{\alpha}{2})} &\leqslant c\Big\{\sum_{\tau=1}^{2}\langle b_{\tau a}\rangle_{\mathbb{R}^2_\infty}^{(1+\alpha,\frac{1+\alpha}{2})} + \langle d_{3a}\rangle_{\mathbb{R}^2_\infty}^{(\alpha,\frac{\alpha}{2})}\Big\} \\ &\leqslant c\Big\{\sum_{\tau=1}^{2}\langle b_{\tau a}\rangle_{\mathbb{R}^2_\infty}^{(1+\alpha,\frac{1+\alpha}{2})} + \langle\nabla' b_{3a}\rangle_{x',\mathbb{R}^2_\infty}^{(\alpha)} + \langle b_{3a}\rangle_{\mathbb{R}^2_\infty}^{(\gamma,1+\alpha)}\Big\}.\end{aligned} \tag{2.4.9}$$

Finally, (2.2.7) and Theorem 2.3.2 imply that

$$\begin{aligned}\sum_{\tau=1}^{2}\langle\omega_{\tau a}\rangle_{\mathbb{R}^2_\infty}^{(2+\alpha,1+\frac{\alpha}{2})} \leqslant c\Big\{\sum_{\tau=1}^{2}\Big(\langle b_{\tau a}\rangle_{\mathbb{R}^2_\infty}^{(1+\alpha,\frac{1+\alpha}{2})} + \langle F^{-1}[\mathrm{i}\xi_\tau(\mu^+\Psi^+ + \mu^-\Psi^-)]\rangle_{\mathbb{R}^2_\infty}^{(1+\alpha,\frac{1+\alpha}{2})}\Big) \\ + \langle\omega_{3a}\rangle_{\mathbb{R}^2_\infty}^{(2+\alpha,1+\frac{\alpha}{2})}\Big\}.\end{aligned} \tag{2.4.10}$$

We evaluate the second term in parentheses using the results obtained above. Indeed, by virtue of Theorem 2.3.2 and formulas (2.4.2) and (2.4.3)

$$\begin{aligned}\langle\mu^+F^{-1}[\mathrm{i}\xi_\tau\Psi^+]\rangle_{\mathbb{R}^2_\infty}^{(1+\alpha,\frac{1+\alpha}{2})} &\leqslant \Big\langle F^{-1}\Big[\widetilde{d}_\tau\frac{s}{P}\frac{L^+}{r^+ + |\xi|}\Big]\Big\rangle_{\mathbb{R}^2_\infty}^{(1+\alpha,\frac{1+\alpha}{2})} \\ &+ \Big\langle F^{-1}\Big[\frac{\mathrm{i}\xi_\tau}{|\xi|}\widetilde{d}\frac{s}{P}\frac{M^+}{r^+ + |\xi|}\Big]\Big\rangle_{\mathbb{R}^2_\infty}^{(1+\alpha,\frac{1+\alpha}{2})}\end{aligned}$$

$$\leqslant c\Big\{\Big\langle F^{-1}\Big[\widetilde{d}_\tau \frac{s}{P} L^+\Big]\Big\rangle^{(\alpha,\frac{\alpha}{2})}_{\mathbb{R}^2_\infty} + \Big\langle F^{-1}\Big[\frac{\mathrm{i}\xi_\tau}{|\xi|}\widetilde{d}\frac{s}{P}M^+\Big]\Big\rangle^{(\alpha,\frac{\alpha}{2})}_{\mathbb{R}^2_\infty}\Big\} =$$

$$= c\Big\{\langle u^{(0)}_{\tau a}\rangle^{(\alpha,\frac{\alpha}{2})}_{\mathbb{R}^2_\infty} + \Big\langle \frac{\partial u}{\partial x_\tau}\Big\rangle^{(\alpha,\frac{\alpha}{2})}_{\mathbb{R}^2_\infty}\Big\},$$

whence, by (2.4.4) and (2.4.6)–(2.4.8), we conclude that

$$\sum_{\tau=1}^{2}\langle \mu^+ F^{-1}[\mathrm{i}\xi_\tau \Psi^+]\rangle^{(1+\alpha,\frac{1+\alpha}{2})}_{\mathbb{R}^2_\infty} \leqslant c\Big\{\sum_{\tau=1}^{2}\langle b_{\tau a}\rangle^{(1+\alpha,\frac{1+\alpha}{2})}_{\mathbb{R}^2_\infty} + \\ + \langle \nabla' b_{3a}\rangle^{(\alpha)}_{x',\mathbb{R}^2_\infty} + \langle b_{3a}\rangle^{(\gamma,1+\alpha)}_{\mathbb{R}^2_\infty}\Big\}. \tag{2.4.11}$$

Semi-norm $\langle \mu^- F^{-1}[\mathrm{i}\xi_\tau \Psi^-]\rangle^{(1+\alpha,\frac{1+\alpha}{2})}_{\mathbb{R}^2_\infty}$ can be evaluated in a similar way. From inequalities (2.4.9)–(2.4.11), it follows that $\omega_{\tau a}$ and $\tau = 1,2$ also satisfy inequality (2.4.9).

Thereby estimate (2.4.1) is proved in the case when $b_3' = b_3,\ B = 0$.

Let now $b_3 = 0,\ b_1 = b_2 = 0$. Then for $x_3 > 0$

$$\frac{\partial p_a}{\partial x_\tau} = F^{-1}\Big[\frac{\mathrm{i}\xi_\tau L^+}{P}\sigma \widetilde{B}\mathrm{e}^{-|\xi|x_3}\Big], \quad \tau = 1,2,$$

$$\frac{\partial p_a}{\partial x_3} = -F^{-1}\Big[\frac{|\xi| L^+}{P}\sigma \widetilde{B}\mathrm{e}^{-|\xi|x_3}\Big],$$

$$\widetilde{\omega}_3 = \frac{\sigma|\xi|\widetilde{B}}{P}$$

$$\widetilde{\omega}_\tau = -\frac{\mathrm{i}\xi_\tau}{\mu^+ r^+ + \mu^- r^-}\Big(\mu^+(-r^+ + |\xi|\frac{q'}{q}) + \mu^-(r^- + |\xi|\frac{q'}{q})\Big)\frac{\sigma\widetilde{B}}{P}, \quad \tau = 1,2.$$

Therefore, the corollary of Lemma 2.1.1 and Theorem 2.3.3 imply that

$$\langle \nabla p_a\rangle^{(\alpha,\frac{\alpha}{2})}_{D^3_\infty} \leqslant \sigma c\Big\{\sum_{\tau=1}^{2}\Big\langle F^{-1}\Big[\frac{\mathrm{i}\xi_\tau L^+}{P}\widetilde{B}\Big]\Big\rangle^{(\alpha,\frac{\alpha}{2})}_{\mathbb{R}^2_\infty} + \Big\langle F^{-1}\Big[\frac{|\xi| L^+}{P}\widetilde{B}\Big]\Big\rangle^{(\alpha,\frac{\alpha}{2})}_{\mathbb{R}^2_\infty}\Big\}$$

$$\leqslant \sigma c\langle B_a\rangle^{(\alpha,\frac{\alpha}{2})}_{\mathbb{R}^2_\infty},$$

$$\sum_{j=1}^{3}\langle \omega_{ja}\rangle^{(2+\alpha,1+\frac{\alpha}{2})}_{\mathbb{R}^2_\infty} \leqslant \sigma c\langle B_a\rangle^{(\alpha,\frac{\alpha}{2})}_{\mathbb{R}^2_\infty}.$$

□

Now we can derive an estimate of the solution to problem (2.2.1).

Theorem 2.4.2 *Let $b_1,\, b_2 \in \overset{\circ}{C}{}^{1+\alpha,\frac{1+\alpha}{2}}(\mathbb{R}^2_T)$, $T < \infty$, $b_3 \in \overset{\circ}{C}{}^{(\gamma,1+\alpha)}(\mathbb{R}^2_T)$, $B \in \overset{\circ}{C}{}^{\alpha,\frac{\alpha}{2}}(\mathbb{R}^2_T)$ with $\alpha, \gamma \in (0,1)$, $\gamma \leqslant \alpha$. Then problem (2.2.1) has a solution $(\boldsymbol{v}, p)$ such that $\boldsymbol{v} \in \overset{\circ}{C}{}^{2+\alpha,1+\frac{\alpha}{2}}(D^3_T)$, $p \in \overset{\circ}{C}{}^{(\gamma,1+\alpha)}(D^3_T)$, $\nabla p \in \overset{\circ}{C}{}^{\alpha,\frac{\alpha}{2}}(D^3_T)$, and the growth of function p at $|x| \to \infty$ is bounded by the degree of $|x|$ less than the unit. In addition, the inequality*

$$
\begin{aligned}
&\langle \boldsymbol{v}\rangle^{(2+\alpha,1+\frac{\alpha}{2})}_{D^3_T} + \langle \nabla p\rangle^{(\alpha,\frac{\alpha}{2})}_{D^3_T} + |p|^{(\gamma,1+\alpha)}_{D^3_T} \qquad (2.4.12)\\
&\leqslant c(T)\Big\{\sum_{\tau=1}^{2} \langle b_\tau\rangle^{(1+\alpha,\frac{1+\alpha}{2})}_{\mathbb{R}^2_T} + \langle \nabla' b_3\rangle^{(\alpha)}_{x',\mathbb{R}^2_T} + \langle b_3\rangle^{(\gamma,1+\alpha)}_{\mathbb{R}^2_T} + \sigma\langle B\rangle^{(\alpha,\frac{\alpha}{2})}_{\mathbb{R}^2_T}\Big\}\\
&\equiv c(T)M(T),
\end{aligned}
$$

holds, where $c(T)$ depends on T exponentially.

Proof We extend b_1, b_2, b_3, B first by zero to the region $t < 0$ and then to the region $t > T$ with the conservation of class (i.e., so that their norms are estimated in $\mathbb{R}^2_\infty$ in terms of the norms in $\mathbb{R}^2_T$, the same ones as in the right-hand side of (2.4.12)).

Next, we multiply these functions by the cutoff function $\zeta(\frac{x'}{R})$ ($\zeta \in C^\infty_0(\mathbb{R}^2)$, $\zeta = 1$ for $|x| \leqslant 1$, $R > 0$). Let $b^R_j = b_j\zeta(\frac{\cdot}{R})$, $j = 1,2,3$; $B^R = B\zeta(\frac{\cdot}{R})$. We note that the functions b^R_j and B^R satisfy the assumptions of Theorem 2.4.1, so that problem (2.2.1) has a solution $\boldsymbol{v}^R = FL^{-1}\widetilde{\boldsymbol{v}}^R$, $p^R = FL^{-1}\widetilde{p}^R$, where $\widetilde{\boldsymbol{v}}^R$ and $\widetilde{p}^R$ are defined by formulas (2.2.6) and (2.2.7). The functions $\boldsymbol{\omega}^R_a = F^{-1}\widetilde{\boldsymbol{v}}^R|_{x_3=0}$ and $p^R_a = F^{-1}\widetilde{p}^R$ are subjected to inequality (2.4.1).

Let us consider $\widetilde{\boldsymbol{v}}^R$ as a solution to the first initial-boundary value problem

$$
\begin{aligned}
&\mathcal{D}_t\boldsymbol{v}^R - \nu^\pm\nabla^2\boldsymbol{v}^R = -\frac{1}{\rho^\pm}\mathrm{e}^{at}\nabla p^R_a,\\
&\boldsymbol{v}^R|_{t=0} = 0, \qquad \boldsymbol{v}^R|_{x_3=0} = \boldsymbol{\omega}^R_a\mathrm{e}^{at}.
\end{aligned}
$$

Due to the well-known Schauder estimate for its solution and because of the inequality, one has (2.4.1):

$$
\begin{aligned}
\langle \boldsymbol{v}^R\rangle^{(2+\alpha,1+\frac{\alpha}{2})}_{D^3_T} &\leqslant c\Big\{\langle \mathrm{e}^{at}\nabla p^R_a\rangle^{(\alpha,\frac{\alpha}{2})}_{D^3_T} + \langle \mathrm{e}^{at}\boldsymbol{\omega}^R_a\rangle^{(2+\alpha,1+\frac{\alpha}{2})}_{\mathbb{R}^2_T}\Big\}\\
&\leqslant c(T)\Big\{\langle \nabla p^R_a\rangle^{(\alpha,\frac{\alpha}{2})}_{D^3_T} + \langle \boldsymbol{\omega}^R_a\rangle^{(2+\alpha,1+\frac{\alpha}{2})}_{\mathbb{R}^2_T}\Big\}\\
&\leqslant c(T)\big(M(T) + N(R,T)\big), \qquad (2.4.13)
\end{aligned}
$$

where $N(R,T)$ contains the Hölder norms of the cutoff function $\zeta(\frac{x'}{R})$ and tends to zero as $R \to \infty$.

In order to estimate $\langle p^R\rangle^{(\gamma,1+\alpha)}_{D^3_T}$, we use the fact that, for $\forall t < T$, p^R is a solution to diffraction problem

$$\frac{1}{\rho^{\pm}}\Delta p^R = 0 \qquad \text{in} \quad \mathbb{R}^3_{\pm}, \tag{2.4.14}$$

$$[p^R]\big|_{x_3=0} = -b_3^R + 2\Big[\mu^{\pm}\frac{\partial v_3^R}{\partial x_3}\Big]\Big|_{x_3=0} + \sigma\int\limits_0^t (\Delta' v_3^R - B^R)\mathrm{d}\tau = d_1^R \quad \text{on} \quad \mathbb{R}^2,$$

$$\Big[\frac{1}{\rho^{\pm}}\frac{\partial p^R}{\partial x_3}\Big]\Big|_{x_3=0} = [\nu^{\pm}\Delta v_3^R]\big|_{x_3=0}.$$

We represent p^R as the sum $p^R = p_1^R + p_2^R$, where $p_1^R = 0$ for $x_3 < 0$, and for $x_3 \geqslant 0$, it is a solution to the Dirichlet problem

$$\frac{1}{\rho^+}\Delta p_1^R = 0, \qquad p_1^R\big|_{x_3=0} = d_1^R,$$

p_2^R is a harmonic function in $\mathbb{R}^3_- \cup \mathbb{R}^3_+$, continuous for $x_3 = 0$, and satisfies the condition

$$\Big[\frac{1}{\rho^{\pm}}\frac{\partial p_2^R}{\partial x_3}\Big]\Big|_{x_3=0} = -\frac{1}{\rho^+}\frac{\partial p_1^R}{\partial x_3}\Big|_{x_3=0} + [\nu^{\pm}\Delta v_3^R]\big|_{x_3=0} = d_2^R.$$

The functions p_1^R and p_2^R are expressed as double- and single-layer potentials, respectively:

$$p_1^R(x) = 2\int\limits_{\mathbb{R}^2}\frac{\partial\mathcal{E}(x,y')}{\partial y_3}d_1^R(y')\mathrm{d}y', \qquad x_3 \geqslant 0,$$

$$p_2^R(x) = \frac{2\rho^+\rho^-}{\rho^+ + \rho^-}\int\limits_{\mathbb{R}^2}\mathcal{E}(x,y')d_2^R(y')\mathrm{d}y', \qquad x \in \mathbb{R}^3_- \cup \mathbb{R}^3_+. \tag{2.4.15}$$

Here $\mathcal{E}(x,y) = -1/(4\pi|x-y|)$ is the fundamental solution of the Laplace equation in $\mathbb{R}^3$.

Note that d_2^R can be written as the sum $d_2^R(x') = \sum_{\tau=1}^2 \frac{\partial D_\tau^R(x')}{\partial x_\tau}$. This becomes evident if we take into account the equalities

$$\frac{\partial p_1^R(x)}{\partial x_3} = -2\int\limits_{\mathbb{R}^2}\frac{\partial^2\mathcal{E}(x',y')}{\partial y_3^2}d_1^R(y')\mathrm{d}y' = 2\sum_{\tau=1}^2\int\limits_{\mathbb{R}^2}\frac{\partial^2\mathcal{E}(x',y')}{\partial y_\tau^2}d_1^R(y')\mathrm{d}y',$$

and

$$\frac{\partial v_3^R}{\partial x_3} = -\sum_{\tau=1}^{2} \frac{\partial v_\tau^R(x')}{\partial x_\tau}.$$

We set

$$D_\tau^R = -\frac{2}{\rho^+} \int\limits_{\mathbb{R}^2} \frac{\partial \mathcal{E}(x', y')}{\partial y_\tau} d_1^R(y')\mathrm{d}y' + \left[\nu^\pm \left(\frac{\partial v_3^R}{\partial x_\tau} - \frac{\partial v_\tau^R}{\partial x_3}\right)\right]\Big|_{x_3=0}.$$

After integration by parts, we obtain

$$p_2^R(x) = \frac{2\rho^+\rho^-}{2\pi(\rho^+ + \rho^-)} \int\limits_{\mathbb{R}^2} \sum_{\tau=1}^{2} \frac{\partial}{\partial y_\tau} \frac{1}{|x-y|} D_\tau^R(y')\mathrm{d}y'.$$

By Lemma 7.1 from [76] for any $t < T$, we have

$$\langle p_1^R \rangle_{\mathbb{R}^3_+}^{(\gamma)} \leqslant c \langle d_1^R \rangle_{\mathbb{R}^2}^{(\gamma)},$$

$$\langle p_2^R \rangle_{\mathbb{R}^3_- \cup \mathbb{R}^3_+}^{(\gamma)} \leqslant c \sum_{\tau=1}^{2} \langle D_\tau^R \rangle_{\mathbb{R}^2}^{(\gamma)} \leqslant c\big(\langle d_1^R \rangle_{\mathbb{R}^2}^{(\gamma)} + \langle \boldsymbol{v}^R \rangle_{\mathbb{R}^3_- \cup \mathbb{R}^3_+}^{(1+\gamma)} \big). \tag{2.4.16}$$

The above arguments are true also for $\Delta_t(h)p(x,t) = p(x,t+h) - p(x,t)$; therefore

$$\langle \Delta_t(h) p^R \rangle_{\mathbb{R}^3_- \cup \mathbb{R}^3_+}^{(\gamma)} \leqslant c\big(\langle \Delta_t(h) d_1^R \rangle_{\mathbb{R}^2}^{(\gamma)} + \langle \Delta_t(h) \boldsymbol{v}^R \rangle_{\mathbb{R}^3_- \cup \mathbb{R}^3_+}^{(1+\gamma)} \big)$$

$$\leqslant c\Big(\langle \Delta_t(h) b_3^R \rangle_{\mathbb{R}^2}^{(\gamma)} + \langle \Delta_t(h) \boldsymbol{v}^R \rangle_{\mathbb{R}^3_- \cup \mathbb{R}^3_+}^{(1+\gamma)} + \sigma \langle \int\limits_t^{t+h} (\Delta' v_3^R\big|_{x_3=0} - B^R)\mathrm{d}t' \rangle_{\mathbb{R}^2}^{(\gamma)} \Big)$$

for any $\gamma \leqslant \alpha$, $h \in (0,T)$, and $t \in (0, T-h)$. By dividing this inequality by $h^{\frac{1+\alpha-\gamma}{2}}$ and taking the supremum over $t \in (0, T-h)$ and $h \in (0,T)$, we conclude that for $\langle p^R \rangle_{D_T^3}^{(\gamma,1+\alpha)}$, estimate (2.4.13) is true with some other constant $c(T)$ and $\gamma \leqslant \alpha$.

Further, by taking into account the zero initial values of $\boldsymbol{v}^R$ and ∇p^R and using the equation for them, we conclude that inequalities of the type (2.4.13) are also valid for the complete norms $|\boldsymbol{v}|_{D_T^3}^{(2+\alpha, 1+\frac{\alpha}{2})}$ and $|\nabla p|_{D_T^3}^{(\alpha, \frac{\alpha}{2})}$.

We consider now the difference $q^R(x,t) = p^R(x,t) - p(x_0,t)$, where x_0 is some point in $\mathbb{R}^3_-$. Since q^R satisfies problem (2.4.14), from (2.4.16), it follows that

$$\langle q^R\rangle^{(\gamma)}_{x,D^3_T} \leqslant c(T)\big(M(T)+N(R,T)+|\boldsymbol{v}^R|^{(2+\alpha,1+\frac{\alpha}{2})}_{D^3_T}\big). \tag{2.4.17}$$

In view of the inequalities

$$|q^R(x,t)| \leqslant |x-x_0|^\gamma \langle p^R(\cdot,t)\rangle^{(\gamma)}_{\cup\mathbb{R}^3_\pm} \equiv |x-x_0|^\gamma \langle q^R(\cdot,t)\rangle^{(\gamma)}_{\cup\mathbb{R}^3_\pm}, \tag{2.4.18}$$

we have

$$\langle q^R\rangle^{(\frac{1+\alpha-\gamma}{2})}_{(0,T)} \leqslant |x-x_0|^\gamma \langle q^R\rangle^{(\gamma,1+\alpha)}_{D^3_T}. \tag{2.4.19}$$

The estimates (2.4.13) and (2.4.16)–(2.4.19) imply the compactness of the set $\{\boldsymbol{v}^R, q^R, \mathcal{D}^{\boldsymbol{r}}_x\boldsymbol{v}^R, \mathcal{D}^k_t\boldsymbol{v}^R, \mathcal{D}^{\boldsymbol{k}}_x q^R\}_{R>0}$, $k=1$, $|\boldsymbol{k}|=1$, and $|\boldsymbol{r}|\leqslant 2$, on any bounded range of argument values. Therefore, choosing a convergent $\{\boldsymbol{v}^{R'}, q^{R'}\}$ subsequence and passing to the limit with respect to $R'\to\infty$ in (2.2.1) and in inequalities (2.4.13) and (2.4.17)–(2.4.19), we make sure that the limit functions $\boldsymbol{v}=\lim \boldsymbol{v}^{R'}$ and $p=\lim q^{R'}$ satisfy system (2.2.1) and belong to the corresponding Hölder spaces and inequality (2.4.12) is true for them; in addition, the estimates are valid for p:

$$\max_{t\in(0,T)} |p(x,t)| \leqslant c(T)|x-x_0|^\gamma,$$

$$\langle p(x)\rangle^{(\frac{1+\alpha-\gamma}{2})}_{(0,T)} \leqslant |x-x_0|^\gamma \langle p\rangle^{(\gamma,1+\alpha)}_{D^3_T}, \quad \gamma\leqslant\alpha. \tag{2.4.20}$$

The theorem is proved. □

Remark 2.4.1 *In all the above arguments, one can replace the parameter α by $k+\alpha$ with any natural k.*

2.5 The Problem for the Inhomogeneous Stokes System

We consider now the inhomogeneous problem

$$\mathcal{D}_t\boldsymbol{v} - \nu^+\Delta\boldsymbol{v} + \frac{1}{\rho^+}\nabla p = \boldsymbol{f}, \quad \nabla\cdot\boldsymbol{v} = g \quad \text{in} \quad D^+_T,$$

$$\mathcal{D}_t\boldsymbol{v} - \nu^-\Delta\boldsymbol{v} + \frac{1}{\rho^-}\nabla p = \boldsymbol{f}, \quad \nabla\cdot\boldsymbol{v} = g \quad \text{in} \quad D^-_T,$$

$$\boldsymbol{v}\big|_{t=0} = \boldsymbol{v}_0 \quad \text{in} \quad \mathbb{R}^3_-\cup\mathbb{R}^3_+, \quad \boldsymbol{v}\xrightarrow[|x|\to\infty]{} 0, \tag{2.5.1}$$

$$[\boldsymbol{v}]\big|_{x_3=0} = 0, \quad -\left[\mu^\pm\left(\frac{\partial v_\beta}{\partial x_3} + \frac{\partial v_3}{\partial x_\beta}\right)\right]\Big|_{x_3=0} = b_\beta(x', t), \ \ \beta = 1, 2;$$

$$\left[-p + 2\mu^\pm \frac{\partial v_3}{\partial x_3}\right]\Big|_{x_3=0} + \sigma\Delta' \int_0^t v_3\big|_{x_3=0} \mathrm{d}\tau =$$

$$= b_3 + \sigma \int_0^t B \,\mathrm{d}\tau \equiv b_3' \quad \text{on } \mathbb{R}^2_T.$$

Concerning the pressure p, we assume weak power growth for $|x| \to \infty$. The functions $\boldsymbol{f}$, g, b_i, $i = 1, 2, 3$, B are given. We suppose compatibility conditions to be satisfied:

$$\nabla \cdot \boldsymbol{v}_0 = g(x, 0), \quad x \in \mathbb{R}^3_- \cup \mathbb{R}^3_+,$$

$$[\boldsymbol{v}_0]\big|_{x_3=0} = 0, \quad -\left[\mu^\pm\left(\frac{\partial v_{0\beta}}{\partial x_3} + \frac{\partial v_{03}}{\partial x_\beta}\right)\right]\Big|_{x_3=0} = b_\beta(x', 0), \tag{2.5.2}$$

$$\left[\nu^\pm \Delta v_{0\beta} - \frac{1}{\rho^\pm}\frac{\partial p(x, 0)}{\partial x_\beta} + f_\beta(x, 0)\right]\Big|_{x_3=0} = 0, \quad \beta = 1, 2;$$

The last of these conditions arises from the need to make compatible the jump of the derivative $\mathcal{D}_t \boldsymbol{v}$ with zero at $t = 0$: $[\mathcal{D}_t \boldsymbol{v}]|_{x_3=0, t=0} = 0$. The normal part of this equality $[\mathcal{D}_t v_3]|_{x_3=0, t=0} = 0$ holds if $p_0 \equiv p(x, 0)$ is a solution to the problem

$$\frac{1}{\rho^\pm}\Delta p_0 = \left(\nabla \cdot \boldsymbol{f} + \nu^\pm \Delta g - \mathcal{D}_t g\right)\big|_{t=0} \quad \text{in} \quad \mathbb{R}^3_- \cup \mathbb{R}^3_+,$$

$$[p_0]\big|_{x_3=0} = 2\left[\mu^\pm \frac{\partial v_{03}}{\partial x_3}\right]\Big|_{x_3=0} - b_3|_{t=0}, \tag{2.5.3}$$

$$\left[\frac{1}{\rho^\pm}\frac{\partial p_0}{\partial x_3}\right]\Big|_{x_3=0} = \left[f_3(x, 0) + \nu^\pm \Delta v_{03}\right]\big|_{x_3=0}.$$

In what follows, we need the following statement proved in [77] (Lemma 4.3).

Lemma 2.5.1 *For any $w \in C^{2+\alpha, 1+\frac{\alpha}{2}}(D^+_T)$, $T < \infty$, the inequalities*

$$c_1 \langle w \rangle^{(2+\alpha, 1+\frac{\alpha}{2})}_{D^+_T} \leqslant \sum_{|\boldsymbol{s}|=2} \langle \mathcal{D}^{\boldsymbol{s}}_x w \rangle^{(\alpha)}_{x, D^+_T} + \langle \mathcal{D}_t w \rangle^{(\frac{\alpha}{2})}_{t, D^+_T} +$$

$$+\langle w|_{t=0}\rangle^{(2+\alpha)}_{\mathbb{R}^3_+} + \langle \mathcal{D}_t w|_{t=0}\rangle^{(\alpha)}_{\mathbb{R}^3_+} \leqslant c_2\langle w\rangle^{(2+\alpha,1+\frac{\alpha}{2})}_{D^+_T}$$

hold, and c_1, c_2 *do not depend on* w.

We state now the main result of this chapter.

Theorem 2.5.1 *For some* $T < \infty$, *let* $\boldsymbol{f} \in \boldsymbol{C}^{\alpha,\frac{\alpha}{2}}(D^3_T)$, $g \in C^{1+\alpha,\frac{1+\alpha}{2}}(D^3_T)$, $g = \nabla\boldsymbol{h}$, $\boldsymbol{h} \in \boldsymbol{C}^{2+\alpha,1+\alpha/2}(D^3_T)$, $[h_3]\big|_{x_3=0} = 0$, $\boldsymbol{v}_0 \in \boldsymbol{C}^{2+\alpha}(\mathbb{R}^3_- \cup \mathbb{R}^3_+)$, $b_\tau \in C^{1+\alpha,\frac{1+\alpha}{2}}(\mathbb{R}^2_T)$, $\tau = 1,2$, $B \in C^{\alpha,\frac{\alpha}{2}}(\mathbb{R}^2_T)$, *and* $b_3 \in C^{(\gamma,1+\alpha)}(\mathbb{R}^2_T)$, *i.e., the bounded function* b_3 *is such that the semi-norms* $\langle\nabla' b_3\rangle^{(\alpha)}_{x',\mathbb{R}^2_T}$ *and* $\langle b_3\rangle^{(\gamma,1+\alpha)}_{\mathbb{R}^2_T}$ *are finite with* $\alpha,\gamma \in (0,1)$, $\gamma \leqslant \alpha$. *We assume also that compatibility conditions* (2.5.2) *are satisfied and the following representation is true:*

$$\boldsymbol{f} - \mathcal{D}_t\boldsymbol{h} = \nabla\cdot\mathbb{H}, \tag{2.5.4}$$

where $\mathbb{H}$ *is a tensor with components* H_{ik}, $i,k = 1,2,3$, *having finite semi-norm* $\langle H_{ik}\rangle^{(\varkappa,1+\alpha)}_{D^3_T}$ *with some* $\varkappa \in (0,1)$, $\varkappa \leqslant \gamma$, $\nabla\cdot\mathbb{H} \equiv \left\{\sum_{k=1}^3 \partial H_{ik}/\partial x_k\right\}_{i=1}^3$. *In addition, suppose that all the given functions are sufficiently well decreasing as* $|x| \to \infty$.

Then problem (2.5.1) *has a solution* $(\boldsymbol{v},p)$ *such that* $\boldsymbol{v} \in C^{2+\alpha,1+\frac{\alpha}{2}}(D^3_T)$, $\nabla p \in C^{\alpha,\frac{\alpha}{2}}(D^3_T)$, *and*

$$\begin{aligned}
&\langle\boldsymbol{v}\rangle^{(2+\alpha,1+\frac{\alpha}{2})}_{D^3_T} + \langle\nabla p\rangle^{(\alpha,\frac{\alpha}{2})}_{D^3_T} + \langle p\rangle^{(\gamma,1+\alpha)}_{D^3_T} \\
&\leqslant c_1(T)\Big\{\langle\boldsymbol{f}\rangle^{(\alpha,\alpha/2)}_{D^3_T} + \langle g\rangle^{(1+\alpha,\frac{1+\alpha}{2})}_{D^3_T} + \langle\mathcal{D}_t\boldsymbol{h}\rangle^{(\alpha/2)}_{t,D^3_T} + |\boldsymbol{v}_0|^{(2+\alpha)}_{\cup\mathbb{R}^3_\pm} + \langle\mathbb{H}\rangle^{(\varkappa,1+\alpha)}_{D^3_T} + \\
&\quad + \sum_{\tau=1}^2 \langle b_\tau\rangle^{(1+\alpha,\frac{1+\alpha}{2})}_{\mathbb{R}^2_T} + \langle\nabla' b_3\rangle^{(\alpha)}_{x',\mathbb{R}^2_T} + \langle b_3\rangle^{(\gamma,1+\alpha)}_{\mathbb{R}^2_T} + \sigma\langle B\rangle^{(\alpha,\frac{\alpha}{2})}_{\mathbb{R}^2_T}\Big\};
\end{aligned} \tag{2.5.5}$$

moreover, the velocity vector $\boldsymbol{v}$ *is uniquely determined, and the pressure* p *is done in the class of functions of weak power growth, up to a bounded time function.*

Remark 2.5.1 *By using interpolation inequalities for the norms of functions in Hölder spaces and equations* (2.5.1) *at* $t = 0$, *it is not difficult to make sure that under the assumptions of Theorem* 2.5.1, *there are estimates of the full Hölder norms* $|\boldsymbol{v}|^{(2+\alpha,1+\frac{\alpha}{2})}_{D^3_T}$ *and* $|\nabla p|^{(\alpha,\frac{\alpha}{2})}_{D^3_T}$ *in terms of the full norms of the right-hand sides in* (2.5.5).

Proof We prove Theorem 2.5.1 in several stages.

I. The Reduction to a Problem Whose Data Is Consistent with Zero at $t = 0$

We consider the solution p_0 of problem (2.5.3). From the assumptions of the theorem it follows that

$$\frac{1}{\rho^\pm}\Delta p_0(x) = \nabla\cdot\big(\boldsymbol{f}(x,0) + \nu^\pm\nabla g(x,0) - \mathcal{D}_t\boldsymbol{h}(x,0)\big) \equiv \nabla\cdot\boldsymbol{z}(x).$$

We represent p_0 in the form $p_0 = p_1 + p_2$, where p_1 is a solution to the Dirichlet problems:

$$\frac{1}{\rho^+}\Delta p_1 = \nabla\cdot\boldsymbol{z} \qquad \text{in} \quad \mathbb{R}^3_+,$$

$$p_1|_{x_3\to+0} = 2\left[\mu^\pm\frac{\partial v_{03}}{\partial x_3}\right]\Big|_{x_3=0} - b_3(x',0) \equiv q_1$$

and

$$\frac{1}{\rho^-}\Delta p_1 = \nabla\cdot\boldsymbol{z} \qquad \text{in} \quad \mathbb{R}^3_-,$$

$$p_1|_{x_3\to-0} = 0;$$

and p_2 is given by formula (2.4.15), where the role of d_2^R is played by the function $-\left[\frac{1}{\rho^\pm}\frac{\partial p_1}{\partial x_3}\right]\Big|_{x_3=0} + [f_3 + \nu^\pm\Delta v_{03}]\big|_{x_3=0} \equiv q_2$. It is easy to see that

$$\langle\nabla p_1\rangle^{(\alpha)}_{\mathbb{R}^3_+} \leqslant c\big(\langle\boldsymbol{z}\rangle^{(\alpha)}_{\mathbb{R}^3_+} + \langle q_1\rangle^{(1+\alpha)}_{\mathbb{R}^2}\big),$$

$$\langle\nabla p_1\rangle^{(\alpha)}_{\mathbb{R}^3_-} \leqslant c\langle\boldsymbol{z}\rangle^{(\alpha)}_{\mathbb{R}^3_-},$$

$$\langle\nabla p_2\rangle^{(\alpha)}_{\mathbb{R}^3_+\cup\mathbb{R}^3_-} \leqslant c\langle q_2\rangle^{(\alpha)}_{\mathbb{R}^2},$$

whence we obtain

$$\begin{aligned}\langle\nabla p_0\rangle^{(\alpha)}_{\mathbb{R}^3_+\cup\mathbb{R}^3_-} &\leqslant c\big(\langle\boldsymbol{z}\rangle^{(\alpha)}_{\mathbb{R}^3_+\cup\mathbb{R}^3_-} + \langle q_1\rangle^{(1+\alpha)}_{\mathbb{R}^2} + \langle q_2\rangle^{(\alpha)}_{\mathbb{R}^2}\big)\\ &\leqslant c\big(\langle g(\cdot,0)\rangle^{(1+\alpha)}_{\cup\mathbb{R}^3_\pm} + \langle\boldsymbol{f}(\cdot,0)\rangle^{(\alpha)}_{\cup\mathbb{R}^3_\pm} + \langle\mathcal{D}_t\boldsymbol{h}(\cdot,0)\rangle^{(\alpha)}_{\cup\mathbb{R}^3_\pm} + \langle b_3(x,0)\rangle^{(1+\alpha)}_{\mathbb{R}^2}\big).\end{aligned} \tag{2.5.6}$$

We extend the function p_0 in D^3_T so that the continuation $P_0 \in C^{1+\alpha,\frac{1+\alpha}{2}}(D^3_T)$ satisfies the initial condition

$$P_0|_{t=0} = p_0 \quad \text{in} \quad \mathbb{R}^3_+ \cup \mathbb{R}^3_-$$

and inequality

$$\langle P_0 \rangle^{(1+\alpha, \frac{1+\alpha}{2})}_{D^3_T} \leqslant c \langle \nabla p_0 \rangle^{(\alpha)}_{\cup \mathbb{R}^3_\pm}. \tag{2.5.7}$$

Now we construct a vector field $\boldsymbol{U} \in \boldsymbol{C}^{2+\alpha, 1+\alpha/2}(D^3_T)$ such that

$$\boldsymbol{U}\big|_{t=0} = \boldsymbol{v}_0, \quad [\boldsymbol{U}]\big|_{x_3=0} = 0, \quad \mathcal{D}_t \boldsymbol{U}\big|_{t=0} = \nu^\pm \Delta \boldsymbol{v}_0 - \frac{1}{\rho^\pm} \nabla p_0 + \boldsymbol{f}\big|_{t=0}.$$

One can take as $\boldsymbol{U}|_{D^-_T}$ the restriction $\boldsymbol{U}^-|_{D^-_T}$ of a solution to the Cauchy problem

$$\mathcal{D}_t \boldsymbol{U}^- - \nu^- \Delta \boldsymbol{U}^- = \boldsymbol{u}^{(1)}, \qquad \boldsymbol{U}^-\big|_{t=0} = \boldsymbol{u}^{(0)} \qquad \text{in} \quad \mathbb{R}^3,$$

in which $\boldsymbol{u}^{(0)} \in \boldsymbol{C}^{2+\alpha}(\mathbb{R}^3)$ is the continuation of $\boldsymbol{v}^0|_{\mathbb{R}^3_-}$ in $\mathbb{R}^3$ with preservation of class; $\boldsymbol{u}^{(1)}$ is a solution of the Cauchy problem for the homogeneous heat equation:

$$\mathcal{D}_t \boldsymbol{u}^{(1)} - \Delta \boldsymbol{u}^{(1)} = 0, \qquad \boldsymbol{u}^{(1)}\big|_{t=0} = \boldsymbol{u}^{(1)}_0 \qquad \text{in} \quad \mathbb{R}^3,$$

and $\boldsymbol{u}^{(1)}_0$ is an extension of the vector-functions $-\frac{1}{\rho^-}\nabla p_0 + \boldsymbol{f}\big|_{t=0}$ from $\mathbb{R}^3_-$ into $\mathbb{R}^3$ with preservation of class.

The restriction $\boldsymbol{U}|_{D^+_T} \equiv \boldsymbol{U}^+$ will be found as a solution to the initial-boundary value problem

$$\mathcal{D}_t \boldsymbol{U}^+ - \nu^+ \Delta \boldsymbol{U}^+ = \boldsymbol{u}^{(2)} \qquad \text{in} \quad D^+_T,$$

$$\boldsymbol{U}^+\big|_{t=0} = \boldsymbol{v}_0 \qquad \text{in} \quad \mathbb{R}^3_+, \qquad \boldsymbol{U}^+\big|_{x_3=0} = \boldsymbol{U}^-\big|_{x_3=0}, \tag{2.5.8}$$

where $\boldsymbol{u}^{(2)}$ satisfies the Cauchy problem

$$\mathcal{D}_t \boldsymbol{u}^{(2)} - \Delta \boldsymbol{u}^{(2)} = 0 \qquad \text{in} \quad \mathbb{R}^3_T = \mathbb{R}^3 \times (0, T),$$
$$\boldsymbol{u}^{(2)}\big|_{t=0} = \boldsymbol{u}^{(2)}_0 \qquad \text{in} \quad \mathbb{R}^3;$$

$\boldsymbol{u}^{(2)}_0 \in \boldsymbol{C}^\alpha(\mathbb{R}^3)$; moreover , $\boldsymbol{u}^{(2)}|_{\mathbb{R}^3_+} = \big(\boldsymbol{f}\big|_{t=0} - \frac{1}{\rho^+}\nabla p_0\big)\big|_{x_3>0}$. We note that in system (2.5.8), the compatibility conditions

$$\boldsymbol{U}^-\big|_{t=0, x_3=0} = \boldsymbol{v}_0\big|_{x_3=0},$$

$$\mathcal{D}_t \boldsymbol{U}^+\big|_{t=0,\, x_3=0} \equiv (\nu^+ \Delta \boldsymbol{v}_0 - \frac{1}{\rho^+} \nabla p_0 + \boldsymbol{f})\big|_{t=0,\, x_3 \to +0} =$$

$$= \mathcal{D}_t \boldsymbol{U}^- \big|_{t=0, x_3=0} \equiv (\nu^- \Delta \boldsymbol{v}_0 - \frac{1}{\rho^-} \nabla p_0 + \boldsymbol{f})\big|_{t=0,\, x_3 \to -0}$$

hold. (The last equality follows from conditions (2.5.2) and (2.5.3).) Therefore, the results of § 4, 5 of Chapter IV of [47], where similar constructions were carried out, imply the estimate

$$\langle \boldsymbol{U} \rangle_{D_T^3}^{(2+\alpha, 1+\frac{\alpha}{2})} \leqslant c\Big(\langle \boldsymbol{v}_0 \rangle_{\cup \mathbb{R}^3_\pm}^{(2+\alpha)} + \langle \nabla p_0 \rangle_{\cup \mathbb{R}^3_\pm}^{(\alpha)} + \langle \boldsymbol{f}|_{t=0} \rangle_{\cup \mathbb{R}^3_\pm}^{(\alpha)} \Big). \tag{2.5.9}$$

We introduce the new unknown functions $\boldsymbol{V} = \boldsymbol{v} - \boldsymbol{U}$ and $s = p - P_0$, for which, due to system (2.5.1), we obtain the problem

$$\mathcal{D}_t \boldsymbol{V} - \nu^\pm \Delta \boldsymbol{V} + \frac{1}{\rho^\pm} \nabla s = \boldsymbol{f} - \mathcal{D}_t \boldsymbol{U} + \nu^\pm \Delta \boldsymbol{U} - \frac{1}{\rho^\pm} \nabla P_0 \equiv \boldsymbol{F},$$

$$\nabla \cdot \boldsymbol{V} = g - \nabla \cdot \boldsymbol{U} \equiv r, \qquad r = \nabla \cdot \boldsymbol{R}, \qquad \boldsymbol{R} = \boldsymbol{h} - \boldsymbol{U} \qquad \text{in} \quad D_T^3,$$

$$\boldsymbol{V}\big|_{t=0} = 0 \quad \text{in} \quad \mathbb{R}^3_- \cup \mathbb{R}^3_+, \quad \boldsymbol{V} \xrightarrow[|x| \to \infty]{} 0, \tag{2.5.10}$$

$$[\boldsymbol{V}]\Big|_{x_3=0} = 0, \; -\Big[\mu^\pm \Big(\frac{\partial V_\beta}{\partial x_3} + \frac{\partial V_3}{\partial x_\beta} \Big)\Big]\Big|_{x_3=0} = b_\beta + \Big[\mu^\pm \Big(\frac{\partial U_\beta}{\partial x_3} + \frac{\partial U_3}{\partial x_\beta} \Big)\Big]\Big|_{x_3=0} \equiv a_\beta, \; \beta = 1, 2;$$

$$\Big[- s + 2\mu^\pm \frac{\partial V_3}{\partial x_3} \Big]\Big|_{x_3=0} + \sigma \Delta' \int_0^t V_3\big|_{x_3=0} \mathrm{d}\tau = a_3 + \sigma \int_0^t A \, \mathrm{d}\tau \quad \text{on} \quad \mathbb{R}^2_T.$$

Here

$$a_3 \equiv b_3 + \Big[p_0 - 2\mu^\pm \frac{\partial U_3}{\partial x_3} \Big]\Big|_{x_3=0}, \qquad A = B - \Delta' U_3\big|_{x_3=0},$$

in addition, $\boldsymbol{F} \in \mathring{\boldsymbol{C}}^{\alpha, \frac{\alpha}{2}}(D_T^3)$, $r \in \mathring{C}^{1+\alpha, \frac{1+\alpha}{2}}(D_T^3)$, $\boldsymbol{R} \in \boldsymbol{C}^{2+\alpha, 1+\frac{\alpha}{2}}(D_T^3)$, $[R_3]|_{x_3=0} = 0$, $a_\beta \in \mathring{C}^{1+\alpha, \frac{1+\alpha}{2}}(\mathbb{R}^2_T)$, $a_3 \in \mathring{C}^{(\gamma, 1+\alpha)}(\mathbb{R}^2_T)$, and $A \in C^{\alpha, \frac{\alpha}{2}}(\mathbb{R}^2_T)$.

Since function A, in the general case, does not vanish at $t = 0$, we consider instead of it $A' = A + \mathcal{D}_t D$, where $D \in C^{2+\alpha, 1+\frac{\alpha}{2}}(\mathbb{R}^2_T)$, $D|_{t=0} = 0$, $\mathcal{D}_t D|_{t=0} = -A|_{t=0} \equiv \big(- B + \Delta' U_3\big|_{x_3=0}\big)|_{t=0}$, and

$$|D|_{\mathbb{R}^2_T}^{(2+\alpha, 1+\frac{\alpha}{2})} \leqslant c\big(|B|_{\mathbb{R}^2_T}^{(\alpha, \frac{\alpha}{2})} + |\boldsymbol{v}_0|_{\cup \mathbb{R}^3_\pm}^{(2+\alpha)} \big). \tag{2.5.11}$$

In this case, a_3 goes over into $a_3' = a_3 - \sigma D \in \mathring{C}^{(\gamma, 1+\alpha)}(\mathbb{R}^2_T)$.

We also note that in view of relation (2.5.4)

$$\boldsymbol{F} - \mathcal{D}_t \boldsymbol{R} = \nabla \cdot \mathbb{M}, \tag{2.5.12}$$

where $\mathbb{M}$ is the tensor with the components $M_{kj} = H_{kj} + \nu^{\pm} \frac{\partial U_k}{\partial x_j} - \frac{1}{\rho^{\pm}} \delta_j^k P_0$.

II. The Reduction of Problem (2.5.10) to a Homogeneous System Like (2.2.1)

We represent $\boldsymbol{V}$ as a sum $\boldsymbol{w} + \boldsymbol{w}' + \boldsymbol{u}$, where $\boldsymbol{w}$ is a solution to the problem

$$\begin{aligned} &\mathcal{D}_t \boldsymbol{w} - \nu^{\pm} \Delta \boldsymbol{w} = \boldsymbol{F} \qquad \text{in} \quad D_T^{\pm}, \\ &[\boldsymbol{w}]\big|_{x_3=0} = 0, \qquad \boldsymbol{w}\big|_{t=0} = 0; \end{aligned} \tag{2.5.13}$$

$\boldsymbol{w}' = \nabla \Phi$; Φ is a solution to the Poisson equation

$$\Delta \Phi = r - \nabla \boldsymbol{w} \equiv r' \qquad \text{in} \quad \mathbb{R}^3, \tag{2.5.14}$$

and $\boldsymbol{u}$ together with $q = s - \mu^{\pm} r' + \rho^{\pm} \mathcal{D}_t \Phi$ solves problem (2.2.1) with

$$\begin{aligned} b_\beta &= a_\beta + \left[\mu^{\pm} \left(\frac{\partial (w_\beta + w'_\beta)}{\partial x_3} + \frac{\partial (w_3 + w'_3)}{\partial x_\beta} \right)\right]\bigg|_{x_3=0} \equiv a'_\beta, \quad \beta = 1, 2; \\ b_3 &= a'_3 - \left[2\mu^{\pm} \frac{\partial (w_3 + w'_3)}{\partial x_3} - \rho^{\pm} \mathcal{D}_t \Phi + \mu^{\pm} r'\right]\bigg|_{x_3=0} \\ &\quad + \sigma \int_0^t \left(A' - \Delta'(w_3 + w'_3)\big|_{x_3=0}\right) \mathrm{d}\tau = a''_3 + \sigma \int_0^t A'' \, \mathrm{d}\tau. \end{aligned} \tag{2.5.15}$$

Since for system (2.5.13) compatibility conditions

$$[\boldsymbol{w}]\big|_{x_3=0,\, t=0} = 0, \qquad [\mathcal{D}_t \boldsymbol{w}]\big|_{x_3=0,\, t=0} = 0,$$

are satisfied, the inequality

$$\langle \boldsymbol{w} \rangle_{D_T^3}^{(2+\alpha, 1+\frac{\alpha}{2})} \leqslant c \langle \boldsymbol{F} \rangle_{D_T^3}^{(\alpha, \frac{\alpha}{2})} \tag{2.5.16}$$

holds [47].

The solution Φ of equation (2.5.14) is given by the formula

$$\Phi(x, t) = \int_{\mathbb{R}^3} \mathcal{E}(x, y) r'(y, t) \mathrm{d}y = - \int_{\mathbb{R}^3} \nabla_y \mathcal{E}(x, y) \cdot (\boldsymbol{R} - \boldsymbol{w}) \mathrm{d}y \tag{2.5.17}$$

because $[R_3]|_{x_3=0} = 0$.

The well-known estimates of the Hölder norms of the second derivatives of the Newtonian potential in the half-space [76] imply that for any $t \in (0, T)$, $\frac{\partial^2 \boldsymbol{w}'}{\partial x_j \partial x_\tau} \in C^\alpha(\mathbb{R}^3_- \cup \mathbb{R}^3_+)$, and

$$\Big\langle \frac{\partial^2 \boldsymbol{w}'}{\partial x_j \partial x_\tau} \Big\rangle^{(\alpha)}_{\cup \mathbb{R}^3_\pm} \leqslant c \Big\langle \frac{\partial r'}{\partial x_\tau} \Big\rangle^{(\alpha)}_{\cup \mathbb{R}^3_\pm}, \qquad \tau = 1,2, \quad j = 1,2,3.$$

The derivative of $\frac{\partial^2 \boldsymbol{w}'}{\partial^2 x_3}$ can be estimated taking into account the Poisson equation; therefore

$$\sum_{|\boldsymbol{\beta}|=2} \langle \mathcal{D}_x^{\boldsymbol{\beta}} \boldsymbol{w}' \rangle^{(\alpha)}_{x, D^3_T} \leqslant c \langle \nabla r' \rangle^{(\alpha)}_{x, D^3_T}. \tag{2.5.18}$$

Next

$$\begin{aligned} \mathcal{D}_t \boldsymbol{w}'(x,t) &= \nabla_x \int\limits_{\mathbb{R}^3} \nabla_y \mathcal{E}(x,y) \cdot (\mathcal{D}_t \boldsymbol{w} - \mathcal{D}_t \boldsymbol{R}) \mathrm{d}y \\ &= \nabla_x \int\limits_{\mathbb{R}^3} \nabla_y \mathcal{E}(x,y) \cdot \boldsymbol{d}(y,t) \mathrm{d}y, \end{aligned} \tag{2.5.19}$$

where $\boldsymbol{d} = \boldsymbol{F} + \nu^\pm \Delta \boldsymbol{w} - \mathcal{D}_t \boldsymbol{R} = \Big\{ \sum_{j=1}^3 \frac{\partial D_{kj}}{\partial x_j} \Big\}_{k=1}^3$, $D_{kj} = M_{kj} + \nu^\pm \frac{\partial w_k}{\partial x_j}$. This follows from equality (2.5.12).

We introduce the following notation: $B^\pm_\rho(x) = B_\rho(x) \cap \mathbb{R}^3_\pm$, $B_\rho(x) = \{y \in \mathbb{R}^3 \big| |x-y| < \rho\}$.

We fix an arbitrary point $x \in \mathbb{R}^3_- \cup \mathbb{R}^3_+$ and choose the parameter $\rho > 0$. Let, for definiteness, $x \in \mathbb{R}^3_+$. Then taking into account that $[d_3]|_{x_3=0} = 0$, one writes

$$\begin{aligned} \mathcal{D}_t \boldsymbol{w}'(x,t) = &- \int\limits_{B^+_\rho(x)} \nabla_x \mathcal{E}(x,y) \nabla_y \cdot \big(\boldsymbol{d}(y,t) - \boldsymbol{d}(x,t)\big) \mathrm{d}y \\ &- \int\limits_{\mathbb{R}^3 \setminus B_\rho(x)} \nabla_x \mathcal{E}(x,y) \nabla_y \cdot \boldsymbol{d}(y,t) \mathrm{d}y - \int\limits_{B^-_\rho(x)} \nabla_x \mathcal{E}(x,y) \nabla_y \cdot (\boldsymbol{d}(y,t) - \boldsymbol{d}(x',t)) \mathrm{d}y, \end{aligned}$$

where $x' = (x_1, x_2, 0)$. Integrating by parts and cancelling surface integrals, we have

$$\begin{aligned} \mathcal{D}_t \boldsymbol{w}'(x,t) = &\int\limits_{B^+_\rho(x)} \nabla_x \nabla_y \mathcal{E}(x,y) \cdot (\boldsymbol{d}(y,t) - \boldsymbol{d}(x,t)) \mathrm{d}y \\ &+ \int\limits_{\partial B^+_\rho(x)} \nabla_x \mathcal{E} \boldsymbol{n}^+(y) \cdot \boldsymbol{d}(x,t) \mathrm{d}S + \int\limits_{B^-_\rho(x)} \nabla_x \nabla_y \mathcal{E} \cdot \big(\boldsymbol{d}(y,t) - \boldsymbol{d}(x',t)\big) \mathrm{d}y \end{aligned}$$

$$+\int\limits_{\partial B_\rho^-(x)} \nabla_x \mathcal{E}(x,y)\boldsymbol{n}^-(y)\cdot \boldsymbol{d}(x',t)\mathrm{d}S + \int\limits_{\cup\mathbb{R}^3_\pm\setminus B_\rho(x)} \nabla_x\nabla_y\mathcal{E}(x,y)\cdot \boldsymbol{d}(y,t)\mathrm{d}y$$

($\boldsymbol{n}^\pm(y)$ are the outward normals to $B_\rho^\pm(x)$ at the point y).

According to similar calculations for the half-space [77], the integral over the exterior domain $\cup\mathbb{R}^3_\pm \setminus B_\rho(x)$ is equal to

$$\int\limits_{\cup\mathbb{R}^3_\pm\setminus B_\rho(x)} \nabla_x\nabla_y\mathcal{E}(x,y)\cdot \boldsymbol{d}(y,t)\mathrm{d}y = -\int\limits_{\mathbb{R}^3_+\setminus B_\rho(x)} \nabla_x \sum_{l,k=1}^3 \frac{\partial^2\mathcal{E}(x,y)}{\partial y_l\partial y_k}\big(D_{lk}(y,t)-D_{lk}(x,t)\big)\mathrm{d}y$$

$$-\int\limits_{\mathbb{R}^3_-\setminus B_\rho(x)} \nabla_x \frac{\partial^2\mathcal{E}(x,y)}{\partial y_l\partial y_k}\big(D_{lk}(y,t)-D_{lk}(x',t)\big)\mathrm{d}y + \int\limits_{\partial\left(\mathbb{R}^3_+\setminus B_\rho(x)\right)} \nabla_x \frac{\partial\mathcal{E}(x,y)}{\partial y_l}\big(D_{lk}(y,t)$$

$$-D_{lk}(x,t)\big)n^+_{0k}(y)\mathrm{d}S + \int\limits_{\partial\left(\mathbb{R}^3_-\setminus B_\rho(x)\right)} \nabla_x \frac{\partial\mathcal{E}(x,y)}{\partial y_l}\big(D_{lk}(y,t)-D_{lk}(x',t)\big)n^-_{0k}(y)\mathrm{d}S$$

where $\{n^\pm_{0k}\}_{k=1}^3$ are the components of the outward normals to $\mathbb{R}^3_\pm \setminus B_\rho(x)$, respectively.

Taking into account the last relation, the increment in t of the vector $\mathcal{D}_t\boldsymbol{w}'$ can be estimated as follows:

$$|\triangle_t(h)\mathcal{D}_t\boldsymbol{w}'(x,t)| \leqslant 2\langle\boldsymbol{d}\rangle^{(\alpha)}_{x,D^3_T} \int\limits_{B_\rho(x)} |\nabla_x\nabla_y\mathcal{E}||x-y|^\alpha\mathrm{d}y +$$

$$+|\triangle_t(h)\boldsymbol{d}(x,t)|\Big|\int\limits_{\partial B_\rho^+(x)} \boldsymbol{n}^+\nabla_x\mathcal{E}\mathrm{d}S\Big| + |\triangle_t(h)\boldsymbol{d}(x',t)|\Big|\int\limits_{\partial B_\rho^-(x)} \boldsymbol{n}^-\nabla_x\mathcal{E}\mathrm{d}S\Big|+$$

$$+\sup_{y\in\mathbb{R}^3_+}\frac{|\triangle_t(h)D_{lk}(y,t)-\triangle_t(h)D_{lk}(x,t)|}{|x-y|^\varkappa}\Big\{\int\limits_{\mathbb{R}^3_+\setminus B_\rho(x)}\Big|\nabla_x\frac{\partial^2\mathcal{E}(x,y)}{\partial y_l\partial y_k}\Big||x-y|^\varkappa\mathrm{d}y +$$

$$+\int\limits_{\partial\left(\mathbb{R}^3_+\setminus B_\rho(x)\right)}\Big|\nabla_x\frac{\partial\mathcal{E}(x,y)}{\partial y_l}\Big||x-y|^\varkappa\mathrm{d}S\Big\}+$$

$$+\sup_{y\in\mathbb{R}^3_-}\frac{|\triangle_t(h)D_{lk}(y,t)-\triangle_t(h)D_{lk}(x',t)|}{|x'-y|^\varkappa}\Big\{\int\limits_{\mathbb{R}^3_-\setminus B_\rho(x)}\Big|\nabla_x\frac{\partial^2\mathcal{E}(x,y)}{\partial y_l\partial y_k}\Big||x-y|^\varkappa\mathrm{d}y +$$

$$+\int\limits_{\partial\left(\mathbb{R}^3_-\setminus B_\rho(x)\right)}\Big|\nabla_x\frac{\partial\mathcal{E}(x,y)}{\partial y_l}\Big||x-y|^\varkappa\mathrm{d}S\Big\}.$$

Surface integrals $\Big|\int\limits_{\partial B_\rho^\pm(x)} \boldsymbol{n}^\pm \nabla_x \mathcal{E} \mathrm{d}S\Big|$ are bounded; therefore, setting $\rho = \sqrt{h}$, we deduce that

$$|\triangle_t(h)\mathcal{D}_t \boldsymbol{w}'(x,t)| \leqslant ch^{\alpha/2}\{\langle \boldsymbol{d}\rangle_{D_T^3}^{(\alpha,\alpha/2)} + \max_{l,k=1,2,3}\langle D_{lk}\rangle_{D_T^3}^{(\varkappa,1+\alpha)}\}. \tag{2.5.20}$$

Hence, due to the arbitrariness of h, the necessary inequality for $\langle \mathcal{D}_t \boldsymbol{w}'\rangle_{t,\mathbb{R}_T^3}^{(\frac{\alpha}{2})}$ holds.

Since $\boldsymbol{w}'|_{t=0} = 0$, to obtain the estimate of the higher-order semi-norm of $\boldsymbol{w}'$ by virtue of Lemma 2.5.1, it remains to evaluate only $\langle \mathcal{D}_t \boldsymbol{w}'|_{t=0}\rangle_{\mathbb{R}^3}^{(\alpha)}$. From (2.5.19) it follows that

$$\langle \mathcal{D}_t \boldsymbol{w}'|_{t=0}\rangle_{\mathbb{R}^3}^{(\alpha)} \leqslant c_1 \langle \boldsymbol{d}|_{t=0}\rangle_{\cup\mathbb{R}_\pm^3}^{(\alpha)} = c_1 \langle \mathcal{D}_t \boldsymbol{R}|_{t=0}\rangle_{\cup\mathbb{R}_\pm^3}^{(\alpha)}.$$

Therefore, taking into account the previous inequalities, in particular (2.5.18), we have

$$\begin{aligned}\langle \boldsymbol{w}'\rangle_{D_T^3}^{(2+\alpha,1+\alpha/2)} \leqslant c\Big\{&\langle \nabla g'\rangle_{x,D_T^3}^{(\alpha)} + \langle \boldsymbol{F}\rangle_{D_T^3}^{(\alpha,\alpha/2)} + \\ &+ \langle \mathcal{D}_t \boldsymbol{R}\rangle_{D_T^3}^{(\alpha,\alpha/2)} + \langle \boldsymbol{w}\rangle_{D_T^3}^{(2+\alpha,1+\alpha/2)} + \max_{l,k=1,2,3}\langle M_{lk}\rangle_{D_T^3}^{(\varkappa,1+\alpha)}\Big\}.\end{aligned} \tag{2.5.21}$$

III. The Estimate of the Solution $(\mathbf{v},\mathbf{p})$ of Problem (2.5.1).

We estimate $(\boldsymbol{u},q)$ by applying Theorem 2.4.2 to system (2.2.1) and (2.5.15). The functions in (2.5.15) vanish at $t = 0$. This follows from (2.5.10), (2.5.15), and (2.5.3) and also from the conditions (2.5.2). So

$$\begin{aligned}\mathcal{D}_t\Phi(x,0) &= \int\limits_{\mathbb{R}^3}\mathcal{E}(x,y)\mathcal{D}_t(r - \nabla \boldsymbol{w})\mathrm{d}y\Big|_{t=0} = \int\limits_{\mathbb{R}^3}\mathcal{E}\mathcal{D}_t(g - \nabla \boldsymbol{U})\Big|_{t=0}\mathrm{d}y \\ &= \int\limits_{\mathbb{R}^3}\mathcal{E}(x,y)\big(\nu^\pm \Delta g(y,0) - \nabla\cdot \nu^\pm \Delta \boldsymbol{v}_0(y)\big)\mathrm{d}y = 0.\end{aligned}$$

Let us verify the boundedness of the norms of given functions (2.5.15). By means of the interpolation inequalities, we conclude

$$\begin{aligned}&\sum_{\tau=1}^{2}\langle a_\tau'\rangle_{\mathbb{R}_T^2}^{(1+\alpha,\frac{1+\alpha}{2})} + \langle \nabla' a_3''\rangle_{x',\mathbb{R}_T^2}^{(\alpha)} + \langle a_3''\rangle_{\mathbb{R}_T^2}^{(\gamma,1+\alpha)} + \sigma\langle A''\rangle_{\mathbb{R}_T^2}^{(\alpha,\frac{\alpha}{2})} \\ &\quad \leqslant c\Big\{\sum_{\tau=1}^{2}\langle a_\tau\rangle_{\mathbb{R}_T^2}^{(1+\alpha,\frac{1+\alpha}{2})} + \langle \nabla' a_3'\rangle_{x',\mathbb{R}_T^2}^{(\alpha)} + \langle a_3'\rangle_{\mathbb{R}_T^2}^{(\gamma,1+\alpha)} + \sigma\langle A'\rangle_{\mathbb{R}_T^2}^{(\alpha,\frac{\alpha}{2})} + \langle \boldsymbol{w}'\rangle_{D_T^3}^{(2+\alpha,1+\frac{\alpha}{2})}\end{aligned}$$

$$+\langle \boldsymbol{w}\rangle^{(2+\alpha,1+\frac{\alpha}{2})}_{D^3_T} + \langle r'\rangle^{(1+\alpha,\frac{1+\alpha}{2})}_{D^3_T} + \langle \nabla'\mathcal{D}_t\Phi\rangle^{(\alpha)}_{x',\mathbb{R}^2_T} + \langle \mathcal{D}_t\Phi\rangle^{(\gamma,1+\alpha)}_{\mathbb{R}^2_T}\Big\}. \quad (2.5.22)$$

It is evident that

$$\langle \nabla'\mathcal{D}_t\Phi\rangle^{(\alpha)}_{x',\mathbb{R}^2_T} \leqslant c\langle \boldsymbol{w}'\rangle^{(2+\alpha,1+\frac{\alpha}{2})}_{D^3_T}.$$

The semi-norm $\langle \mathcal{D}_t\Phi\rangle^{(\gamma,1+\alpha)}_{\mathbb{R}^2_T}$ is estimated on the basis of the equality

$$\mathcal{D}_t\Phi(x,t) = \int\limits_{\mathbb{R}^3} \nabla_y\mathcal{E}(x,y)\cdot \boldsymbol{d}(y,t)\mathrm{d}y =$$

$$= \frac{\partial}{\partial x_l}\int\limits_{\mathbb{R}^3_-\cup\mathbb{R}^3_+} \frac{\partial\mathcal{E}(x,y)}{\partial y_k} D_{lk}(y,t)\mathrm{d}y - \frac{\partial}{\partial x_l}\int\limits_{\mathbb{R}^2} \mathcal{E}(x,y')[D_{l3}]\big|_{y_3=0}\mathrm{d}y',$$

where the vector $\boldsymbol{d}$ and the tensor $\{D_{lk}\}$ are the same as in (2.5.19). The estimate of the singular integrals yields

$$\langle \mathcal{D}_t\Phi\rangle^{(\varkappa)}_{\mathbb{R}^3} \leqslant c\max_{l,k=1,2,3}\langle D_{lk}\rangle^{(\varkappa)}_{\mathbb{R}^3_-\cup\mathbb{R}^3_+}, \qquad \forall t\in(0,T).$$

If one applies this inequality to $\triangle_t(h)\mathcal{D}_t\Phi$ and divides both parts by $h^{\frac{1+\alpha-\varkappa}{2}}$, then, after taking the supremum in $t\in(0,T-h)$ and in $h\in(0,T)$, one arrives at the estimate

$$\langle \mathcal{D}_t\Phi\rangle^{(\varkappa,1+\alpha)}_{\mathbb{R}^3_T} \leqslant c\max_{l,k=1,2,3}\langle D_{lk}\rangle^{(\varkappa,1+\alpha)}_{D^3_T}. \quad (2.5.23)$$

Since $\varkappa\leqslant\gamma$, this implies the necessary inequality for $\langle \mathcal{D}_t\Phi\rangle^{(\gamma,1+\alpha)}_{\mathbb{R}^2_T}$. Therefore, taking into account (2.5.16), (2.5.21), and (2.5.22), we obtain, by Theorem 2.4.2, the existence of a solution $(\boldsymbol{u},q)$ to problem (2.2.1) and (2.5.15) such that $\boldsymbol{u}\in\mathring{C}^{2+\alpha,1+\frac{\alpha}{2}}(D^3_T)$, $q\in\mathring{C}^{(\gamma,1+\alpha)}(D^3_T)$, $\nabla q\in\mathring{C}^{\alpha,\frac{\alpha}{2}}(D^3_T)$, and

$$\langle \boldsymbol{u}\rangle^{(2+\alpha,1+\frac{\alpha}{2})}_{D^3_T} + \langle \nabla q\rangle^{(\alpha,\frac{\alpha}{2})}_{D^3_T} + \langle q\rangle^{(\gamma,1+\alpha)}_{D^3_T} \leqslant c(T)\Big\{\langle \boldsymbol{F}\rangle^{(\alpha,\alpha/2)}_{D^3_T} + \langle g\rangle^{(1+\alpha,\frac{1+\alpha}{2})}_{D^3_T}$$

$$+\langle \mathcal{D}_t\boldsymbol{R}\rangle^{(\alpha,\alpha/2)}_{D^3_T} + \sum_{\tau=1}^{2}\langle a_\tau\rangle^{(1+\alpha,\frac{1+\alpha}{2})}_{\mathbb{R}^2_T} + \langle \nabla' a_3\rangle^{(\alpha)}_{x',\mathbb{R}^2_T} + \langle a_3\rangle^{(\gamma,1+\alpha)}_{\mathbb{R}^2_T}$$

$$+\sigma\langle A\rangle^{(\alpha,\frac{\alpha}{2})}_{\mathbb{R}^2_T} + |D|^{(2+\alpha,1+\frac{\alpha}{2})}_{\mathbb{R}^2_T} + \max_{l,k=1,.,3}\langle H_{lk}\rangle^{(\varkappa,1+\alpha)}_{D^3_T}\Big\} \equiv c(T)G.$$

Consequently, problem (2.5.10) is also solvable in the Hölder classes, and its solution $(\boldsymbol{V},s)$ has the properties $\boldsymbol{V}\in\mathring{C}^{2+\alpha,1+\frac{\alpha}{2}}(D^3_T)$, $s\in\mathring{C}^{(\gamma,1+\alpha)}(D^3_T)$, and $\nabla s \in \mathring{C}^{\alpha,\frac{\alpha}{2}}(D_T^3)$, and, by virtue of (2.5.16), (2.5.21), and (2.5.23), the inequality

$$\langle \boldsymbol{V}\rangle_{D_T^3}^{(2+\alpha,1+\frac{\alpha}{2})} + \langle \nabla s\rangle_{D_T^3}^{(\alpha,\frac{\alpha}{2})} + \langle s\rangle_{D_T^3}^{(\gamma,1+\alpha)} \leqslant cG$$

holds, whence the existence of $\boldsymbol{U}$ and P_0 and estimates (2.5.7), (2.5.9), (2.5.6), and (2.5.11) imply the solvability of problem (2.5.1) and inequality (2.5.5) for its solution $(\boldsymbol{v},p)$. We note that the growth of functions q, s, and p at ∞ is limited by a small degree of $|x|$ due to Theorem 2.4.2.

The uniqueness of the solution $(\boldsymbol{v},p)$ is established in the same way as it was done in [53] for the half-space problem. We repeat this argument.

We suppose that $(\boldsymbol{v},p)$ is a solution to problem (2.5.1) with zero functions in all the right-hand sides such that $\boldsymbol{v} \in \mathring{C}^{2+\alpha,1+\frac{\alpha}{2}}(D_T^3)$, $p \in \mathring{C}^{(\gamma,1+\alpha)}(D_T^3)$, $\nabla p \in \mathring{C}^{\alpha,\frac{\alpha}{2}}(D_T^3)$, and the growth of p for big values $|x|$ is bounded by $|x|^\beta$, $\beta < 1$. We set $\boldsymbol{v}^N = \boldsymbol{v}\zeta^N(x)$, $p^N = p\zeta^N(x)$, and $N > 0$, where $\zeta^N(x) = \zeta(\frac{x}{N})$, $\zeta \in C_0^\infty(\mathbb{R}^3)$, $\zeta(x) = 1$ if $|x| < 1$, and $\zeta(x) = 0$ if $|x| > 2$. Then

$$\mathcal{D}_t\boldsymbol{v}^N - \nu^\pm\Delta\boldsymbol{v}^N + \frac{1}{\rho^\pm}\nabla p^N = \boldsymbol{f}^N, \quad \nabla\cdot\boldsymbol{v}^N = g^N,$$

$$\boldsymbol{v}^N\big|_{t=0} = 0, \quad \boldsymbol{v}^N = 0, \quad p^N = 0 \quad \text{for } |x| > 2N, \tag{2.5.24}$$

$$[\boldsymbol{v}^N]\big|_{x_3=0} = 0, \quad -\Big[\mu^\pm\Big(\frac{\partial v_\beta^N}{\partial x_3} + \frac{\partial v_3^N}{\partial x_\beta}\Big)\Big]\Big|_{x_3=0} = b_\beta^N(x',t), \quad \beta = 1,2;$$

$$\Big[-p^N + 2\mu^\pm\frac{\partial v_3^N}{\partial x_3}\Big]\Big|_{x_3=0} + \sigma\Delta'\int_0^t v_3^N\big|_{x_3=0}\mathrm{d}\tau = b_3^N + \sigma\int_0^t B^N\,\mathrm{d}\tau.$$

Here $\boldsymbol{f}^N = \frac{1}{\rho^\pm}p\nabla\zeta^N - 2\nu^\pm(\nabla\zeta^N\cdot\nabla)\boldsymbol{v} - \nu^\pm\boldsymbol{v}\Delta\zeta^N$, $g^N = \boldsymbol{v}\cdot\nabla\zeta^N$, $b_\beta^N = -\Big[\mu^\pm\Big(v_\beta\frac{\partial\zeta^N}{\partial x_3} + v_3\frac{\partial\zeta^N}{\partial x_\beta}\Big)\Big]\Big|_{x_3=0}$, $b_3^N = 2\Big[\mu^\pm\frac{\partial\zeta^N}{\partial x_3}v_3\Big]\Big|_{x_3=0}$, $B^N = 2\nabla'\zeta^N\cdot\nabla' v_3 + v_3\Delta'\zeta^N\big|_{x_3=0}$, $g^N = \nabla\cdot\boldsymbol{h}^N$, $\boldsymbol{h}^N = -\nabla\int_{\mathbb{R}^3}\frac{\boldsymbol{v}\cdot\nabla\zeta^N}{4\pi|x-y|}dy$. The relation $\boldsymbol{f}^N - \mathcal{D}_t\boldsymbol{h}^N = \nabla\cdot\mathbb{H}^N$ will be satisfied if we take as $\mathbb{H}^N$ the tensor with the following components:

$$H_{kj}^N(x,t) = -\frac{1}{4\pi}\Big\{\frac{\partial}{\partial x_j}\int_{\mathbb{R}^3}\frac{f_k^N(y,t)}{|x-y|}\mathrm{d}y - \delta_j^k\int_{\mathbb{R}^3}\frac{\nabla\zeta^N(y,t)}{|x-y|}\cdot(\nu^\pm\Delta\boldsymbol{v} - \frac{1}{\rho^\pm}\nabla p)\,\mathrm{d}y\Big\}$$

$$= -\frac{1}{4\pi}\Big\{\frac{\partial}{\partial x_j}\int_{\mathbb{R}^3}\frac{f_k^N(y,t)}{|x-y|}\mathrm{d}y + \delta_j^k\int_{\mathbb{R}^3}\Big(\frac{\partial}{\partial y_i}\frac{\nu^\pm}{|x-y|}\frac{\partial\boldsymbol{v}}{\partial y_i}\cdot\nabla\zeta^N$$

$$+\frac{\nu^{\pm}}{|x-y|}\frac{\partial \boldsymbol{v}}{\partial y_i}\cdot\frac{\partial \nabla\zeta^N}{\partial y_i}-\frac{1}{\rho^{\pm}}p\Big(\nabla\frac{1}{|x-y|}\cdot\nabla\zeta^N-\frac{\Delta\zeta^N}{|x-y|}\Big)\Big)\,\mathrm{d}y$$
$$-\delta_j^k\int\limits_{\mathbb{R}^2}\frac{1}{|x-y|}\Big[\nu^{\pm}\frac{\partial \boldsymbol{v}}{\partial y_3}\cdot\nabla\zeta^N-\frac{1}{\rho^{\pm}}p\frac{\partial\zeta^N}{\partial y_3}\Big]\Big|_{x_3=0}\mathrm{d}y\Big\}.$$

We estimate the norms of the right-hand sides of systems (2.5.24). Under the above assumption on the growth of p at ∞, the pressure p is a unique solution to problem (2.4.14), since in our case $b_3 = 0$, the semi-norm $\langle p\rangle^{(\gamma_1,1+\alpha)}_{D^3_T}$ is finite for p, and inequalities (2.4.20) hold with any γ_1 : $0<\gamma_1\leqslant\alpha$. Therefore, choosing $\gamma_1<\varkappa$, we have

$$|\triangle_t(h)\big(H^N_{kj}(x,t)-H^N_{kj}(x_0,t)\big)|\leqslant$$
$$\leqslant c\Big(\langle p\rangle^{(\gamma_1,1+\alpha)}_{D^3_T}N^{\gamma_1}h^{\frac{1+\alpha-\gamma_1}{2}}+h^{\frac{1+\alpha-\varkappa}{2}}\langle\nabla\boldsymbol{v}\rangle^{(\frac{1+\alpha-\varkappa}{2})}_{t,D^3_T}\Big)\frac{\rho^{\varkappa}}{N^{\varkappa}}$$
$$\leqslant ch^{\frac{1+\alpha-\varkappa}{2}}\rho^{\varkappa}\Big(\frac{T^{\varkappa-\gamma_1}}{N^{\varkappa-\gamma_1}}\langle p\rangle^{(\gamma_1,1+\alpha)}_{D^3_T}+\frac{1}{N^{\varkappa}}\langle\boldsymbol{v}\rangle^{(2+\alpha,1+\frac{\alpha}{2})}_{D^3_T}\Big).$$

Next

$$\langle\boldsymbol{f}^N\rangle^{(\alpha,\frac{\alpha}{2})}_{D^3_T}\leqslant c(T)\Big(\frac{1}{N}|\boldsymbol{v}|^{(1+\alpha,\frac{1+\alpha}{2})}_{D^3_T}+\frac{1}{N^{1+\alpha}}|p|_{\operatorname{supp}\zeta^N\times(0,T)}+\frac{1}{N}\langle p\rangle^{(\alpha,\frac{\alpha}{2})}_{\operatorname{supp}\zeta^N\times(0,T)}\Big)$$
$$\leqslant c(T)\Big(\frac{1}{N}|\boldsymbol{v}|^{(1+\alpha,\frac{1+\alpha}{2})}_{D^3_T}+N^{\gamma_1-1-\alpha}+\frac{1}{N^{\alpha}}|\nabla p|_{D^3_T}$$
$$+N^{\gamma_1-1}T^{\frac{1-\gamma_1}{2}}\langle p\rangle^{(\gamma_1,1+\alpha)}_{D^3_T}\Big).$$

Taking into account Remark 2.5.1, we can show in a similar way that the norms of g^N, b_i^N, B^N, and $\boldsymbol{h}^N$ in the right-hand sides of (2.5.5) are bounded and tend to zero as $N\to\infty$. The results of Sect. 10.4 imply the uniqueness of a compactly supported solution $(\boldsymbol{v}^N,p^N)$ and the coincidence of it with the one constructed above. Applying inequality (2.5.5) to this solution and letting N go to ∞, we obtain that $\boldsymbol{v}=0$, and $p=c(t)$ is a function depending only on time. □

Remark 2.5.2 *The restriction on the behavior of the given functions at infinity can be removed if one carries out arguments similar to those which are given at the end of Sect.* 2.4 *for the system with homogeneous equations.*

Chapter 3
The Model Problem Without Surface Tension Forces

Abstract We study the case of the whole space where the interface is a plane and $\sigma = 0$. The estimates of the solution to problem (2.2.1) carried out in the previous chapter cannot be obtained in the absence of surface tension by the Fourier multipliers. But this complicated method can be replaced in this case by a simpler technique. Namely, we consider the solution of problem (2.2.1) as a solution to the vector-valued heat equation. The necessary estimates are obtained by evaluating convolution integrals that convert the right-hand sides of the boundary conditions in the Stokes problem into that in the initial-boundary value problem for the heat equation.

The results of this chapter are first published in [20]. We will use them in Chap. 6 to prove the local and global solvability of the nonlinear problem for $\sigma = 0$.

3.1 Statement of the Problem and Formulation of Existence Theorem

So, let velocity vector field $\boldsymbol{v} = (v_1, v_2, v_3)$ and pressure function p satisfy the diffraction problem for the Stokes system:

$$\mathcal{D}_t\boldsymbol{v} - \nu^+\Delta\boldsymbol{v} + \frac{1}{\rho^+}\nabla p = \boldsymbol{f}, \quad \nabla\cdot\boldsymbol{v} = g \quad \text{in } D^+_\infty = \mathbb{R}^3_+ \times (0,\infty),$$

$$\mathcal{D}_t\boldsymbol{v} - \nu^-\Delta\boldsymbol{v} + \frac{1}{\rho^-}\nabla p = \boldsymbol{f}, \quad \nabla\cdot\boldsymbol{v} = g \quad \text{in } D^-_\infty = \mathbb{R}^3_- \times (0,\infty),$$

$$\boldsymbol{v}\big|_{t=0} = \boldsymbol{v}_0 \quad \text{in } \mathbb{R}^3_- \cup \mathbb{R}^3_+, \quad \boldsymbol{v} \xrightarrow[|x|\to\infty]{} 0, \tag{3.1.1}$$

© The Author(s), under exclusive license to Springer Nature Switzerland AG 2021

I. V. Denisova, V. A. Solonnikov, *Motion of a Drop in an Incompressible Fluid*, Advances in Mathematical Fluid Mechanics,
https://doi.org/10.1007/978-3-030-70053-9_3

$$[\boldsymbol{v}]\big|_{x_3=0}=0,\quad -\left[\mu^{\pm}\left(\frac{\partial v_\beta}{\partial x_3}+\frac{\partial v_3}{\partial x_\beta}\right)\right]\bigg|_{x_3=0}=b_\beta(x',t),\quad \beta=1,2;$$

$$\left[-p+2\mu^{\pm}\frac{\partial v_3}{\partial x_3}\right]\bigg|_{x_3=0}=b_3(x',t)\quad \text{on } \mathbb{R}^2_\infty.$$

As for pressure function, we assume no more than a power-law growth of it at infinity in the spatial coordinates. The functions $\boldsymbol{f}$, g, b_i, $i=1,2,3$ are given.

We state the main result of this chapter.

Theorem 3.1.1 *Let* $\alpha,\gamma\in(0,1)$ *and* $T\in(0,\infty)$. *We assume that* $\boldsymbol{f}\in \boldsymbol{C}^{\alpha,\alpha/2}(D^3_T)$, $g\in C^{1+\alpha,\frac{1+\alpha}{2}}(D^3_T)$, $b_\beta\in C^{1+\alpha,\frac{1+\alpha}{2}}(\mathbb{R}^2_T)$, $\beta=1,2$, $b_3\in C^{1+\alpha,0}(\mathbb{R}^2_T)\cap C^{(\gamma,1+\alpha)}(\mathbb{R}^2_T)$, *and* $\boldsymbol{v}_0\in \boldsymbol{C}^{2+\alpha}(\mathbb{R}^3_-\cup\mathbb{R}^3_+)$. *In addition, we suppose that* g *is representable in the form*

$$g=\nabla\cdot\boldsymbol{h}\quad \text{with } \boldsymbol{h}\in \boldsymbol{C}^{2+\alpha,0}(D^3_T),\ \mathcal{D}_t\boldsymbol{h}\in \boldsymbol{C}^{\alpha,0}(D^3_T),\ [h_3]|_{x_3=0}=0,$$

and

$$\mathcal{D}_t\boldsymbol{h}-\boldsymbol{f}=\nabla\cdot\mathbb{H}\quad \text{with}\,\mathbb{H}=\{H_{ik}\}^3_{i,k=1},\quad H_{ik}\in C^{(\gamma,1+\alpha)}(D^3_T),$$

where $\nabla\cdot\mathbb{H}\equiv\sum_{k=1}^3\partial H_{ik}/\partial\xi_k$, $i=1,2,3$, *and that compatibility conditions* (2.5.2) *hold.*

We assume also $p_0(x)=p(x,0)$ *to be a solution to problem* (2.5.3).

Then problem (3.1.1) *has a solution* $(\boldsymbol{v},p)$ *with the following properties:* $\boldsymbol{v}\in \boldsymbol{C}^{2+\alpha,1+\alpha/2}(D^3_T)$, $p\in C^{(\gamma,1+\alpha)}(D^3_T)$, $\nabla p\in C^{\alpha,\alpha/2}(D^3_T)$, *and*

$$\begin{aligned}
&|\boldsymbol{v}|^{(2+\alpha,1+\alpha/2)}_{D^3_T}+|\nabla p|^{(\alpha,\alpha/2)}_{D^3_T}+\langle p\rangle^{(\gamma,1+\alpha)}_{D^3_T}\\
&\quad\leqslant c\Big\{|\boldsymbol{f}|^{(\alpha,\alpha/2)}_{D^3_T}+|\boldsymbol{h}|^{(2+\alpha)}_{x,D^3_T}+|\mathcal{D}_t\boldsymbol{h}|^{(\alpha)}_{x,D^3_T}+\langle\mathbb{H}\rangle^{(\gamma,1+\alpha)}_{D^3_T}+|\boldsymbol{v}_0|^{(2+\alpha)}_{\cup\mathbb{R}^3_\pm}\\
&\qquad+\sum_{\beta=1}^2|b_\beta|^{(1+\alpha,\frac{1+\alpha}{2})}_{\mathbb{R}^2_T}+|b_3|^{(1+\alpha)}_{x',\mathbb{R}^2_T}+\langle b_3\rangle^{(\gamma,1+\alpha)}_{\mathbb{R}^2_T}\Big\}\equiv cF(T).
\end{aligned}\tag{3.1.2}$$

Velocity vector field $\boldsymbol{v}$ *is uniquely determined, and pressure function* p *is defined in the class of functions of weak power growth up to a bounded function of time.*

3.2 Preliminary Considerations

In this section, we study diffraction problem for the vector heat equation

$$\begin{aligned} &\mathcal{D}_t \boldsymbol{v} - \nu^{\pm}\Delta \boldsymbol{v} = \boldsymbol{f} \quad \text{in} \quad D^{\pm}_{\infty}, \\ &\boldsymbol{v}\big|_{t=0} = 0 \quad \text{in} \quad \mathbb{R}^3_- \cup \mathbb{R}^3_+, \qquad \boldsymbol{v} \xrightarrow[|x|\to\infty]{} 0, \\ &[\boldsymbol{v}]\big|_{x_3=0} = 0, \quad -\left[\mu^{\pm}\frac{\partial \boldsymbol{v}}{\partial x_3}\right]\bigg|_{x_3=0} = \boldsymbol{d}(x',t) \quad \text{on} \quad \mathbb{R}^2_{\infty}. \end{aligned} \tag{3.2.1}$$

First, we consider diffraction problem for the homogeneous scalar heat equation

$$\begin{aligned} &\mathcal{D}_t \theta - \nu^{\pm}\Delta \theta = 0 \quad \text{in} \quad D^{\pm}_{\infty}, \\ &\theta\big|_{t=0} = 0 \quad \text{in} \quad \mathbb{R}^3_- \cup \mathbb{R}^3_+, \qquad \theta \xrightarrow[|x|\to\infty]{} 0, \\ &[\theta]\big|_{x_3=0} = 0, \quad -\left[\mu^{\pm}\frac{\partial \theta}{\partial x_3}\right]\bigg|_{x_3=0} = b(x',t) \quad \text{on} \quad \mathbb{R}^2_{\infty}. \end{aligned} \tag{3.2.2}$$

We take the Fourier transform in tangent variables $(x_1, x_2) = x'$ and Laplace in t by formula (2.2.2).

For estimating kernels of convolution integrals and their derivatives, we use Proposition 6.1 from [88] (see also [86], Proposition 2.1):

Lemma 3.2.1 *We suppose that the kernel $\widetilde{K}(\zeta, s) = \widetilde{K}(\zeta_1, \dots, \zeta_n, s)$ is defined for all complex $\zeta_j = \xi_j + \mathrm{i}\eta_j$, $j = 1, \dots, n$, and s such that*

$$\operatorname{Re} s + \varkappa|\operatorname{Im} s| \geqslant -\delta\xi^2, \quad |\eta| \leqslant \delta_1|\xi|, \qquad \xi \neq 0, \tag{3.2.3}$$

(where $\varkappa$, δ, δ_1 are positive constants) is analytic in domain (3.2.3), homogeneous, i. e., $\widetilde{K}(\lambda\zeta, \lambda^2 s) = \lambda^{m-l}\widetilde{K}(\zeta, s)$, and is subjected to the inequality

$$|\widetilde{K}(\zeta, s)| \leqslant \frac{c|\xi|^m}{(|s| + |\xi|^2)^{l/2}},$$

where l and m are integers, $l > 0$, $m > -n$.

Then for the Fourier–Laplace original of $\widetilde{K}$ (3.2.5), the estimate

$$\left|\mathcal{D}_t^k \mathcal{D}_x^{\boldsymbol{i}} K(x,t)\right| \leqslant \frac{ct^{l/2-1-k}}{(|x|^2 + t)^{\frac{m+n+|\boldsymbol{i}|}{2}}}, \quad k \geqslant 0, \quad |\boldsymbol{i}| \geqslant 0, \tag{3.2.4}$$

holds, moreover, $K(x,t) = 0$ for $t < 0$.

We will also need Theorem A.1 from [10].

Theorem 3.2.1 *We suppose that function* $\widetilde{V}(\zeta, s, x_n)$, $x_n > 0$, *has the following properties*:

(1) It is holomorphic with respect to s *and* $\zeta_j = \xi_j + \mathrm{i}\eta_j$ *in domain* (3.2.3).
(2) It is homogeneous, i.e., $\widetilde{V}(\lambda\zeta, \lambda^2 s, x_n) = \lambda^d \widetilde{V}(\zeta, s, \lambda x_n)$, $\lambda > 0$, $x_n > 0$.
(3) It satisfies the inequality

$$|\widetilde{V}(\zeta, s, x_n)| \leqslant c\mathrm{e}^{-ax_n}, \quad a > 0,$$

for any ζ, s *in domain* (3.2.3) *provided that* $|s| + \xi^2 = 1$.

Then for the kernel $V(x,t) = (FL)^{-1}\big(\widetilde{V}(\xi, s, x_n)\big)$, *we have the estimate*

$$|V(x,t)| \leqslant ct^{-\frac{n+1+d}{2}} \mathrm{e}^{-ax^2/t}, \qquad x \in \mathbb{R}^n_+, \quad t > 0,$$

and if $t < 0$, *then* $V(x,t) = 0$.

In this theorem, we have used the notation of the inverse Fourier–Laplace transform

$$K(x',t) = (FL)^{-1}\big(\widetilde{K}(\xi, s)\big) = \frac{1}{(2\pi)^3 \mathrm{i}} \int_{\mathrm{Re}s=a>0} \mathrm{e}^{st} \int_{\mathbb{R}^2} \mathrm{e}^{\mathrm{i}\boldsymbol{x}'\cdot\boldsymbol{\xi}} \widetilde{K}(\xi, s)\mathrm{d}\xi \mathrm{d}s, \tag{3.2.5}$$

$x' = (x_1, x_2)$.

We also consider the convolution integral

$$u(x,t) = \int_{-\infty}^{\infty} \int_{\mathbb{R}^n} G(x-y, t-t')\varphi(y,t')\mathrm{d}y\mathrm{d}t',$$

and we formulate a particular case of Theorem 5 in [34] which gives us Hölder norm estimates for it.

Theorem 3.2.2 *Let for some integer* $l > \mu > 0$

$$\int_{-\infty}^{\infty} \int_{\mathbb{R}^n} | \triangle_i^l (h) G(x,t)| \mathrm{d}x \mathrm{d}t < ch^\mu,$$

and let for some $r > \mu$, $r \notin \mathbb{N}$,

$$\sup_{h\in(0,\infty)} \left| \frac{\varepsilon(h)}{h^r} \int_{-\infty}^{\infty} \int_{\mathbb{R}^n} | \triangle_i^l (h) G(x,t)| \mathrm{d}x \mathrm{d}t \right| < \infty.$$

Then, if $|\varphi|_{\mathbb{R}^n_\infty} < \infty$, *we have*

$$\left\langle \frac{\partial^{r'} u}{\partial x_i^{r'}} \right\rangle^{(r-r')}_{x_i,\mathbb{R}^n_\infty} \leqslant c\langle\varphi\rangle^{(r-\mu)}_{x_i,\mathbb{R}^n_\infty},$$

where $r' \equiv [r]$ *is the integral part of* r*;*

$$\varepsilon(y) = \begin{cases} 0, & y \leqslant 1 \\ 1, & y > 1; \end{cases}$$

$\triangle_i^l(h)$ *means the* l*th order increment operator with respect to* x_i, $i = 0, 1, \ldots, n$, $x_0 \equiv t$.

Moreover, we apply two estimates proved in [85] (the corollary to Lemma 3.1); the first one was established also by A. Lunardi in [52].

Lemma 3.2.2 *Each function* $w \in \boldsymbol{C}^{2+\alpha,0}(G_T)$ *such that* $\mathcal{D}_t w \in \boldsymbol{C}^{\alpha,0}(G_T)$ *satisfies the inequalities*

$$\langle\nabla w\rangle^{(\frac{1+\alpha}{2})}_{t,G_T} \leqslant c\big\{\langle w\rangle^{(2+\alpha)}_{x,G_T} + \langle\mathcal{D}_t w\rangle^{(\alpha)}_{x,G_T}\big\}, \tag{3.2.6}$$

$$\langle\nabla\nabla w\rangle^{(\frac{\alpha}{2})}_{t,G_T} \leqslant c\big\{\langle w\rangle^{(2+\alpha)}_{x,G_T} + \langle\mathcal{D}_t w\rangle^{(\alpha)}_{x,G_T}\big\}, \tag{3.2.7}$$

where $G_T = D^3_T$ *or* $G_T = \mathbb{R}^2_T$.

Remark 3.2.1 *In the case of homogeneous heat equation, we can express* $\mathcal{D}_t\theta$ *from the equation in* (3.2.2)*, therefore*

$$\langle\mathcal{D}_t\theta\rangle^{(\alpha/2)}_{t,D^3_T} \leqslant \nu^{\pm}\langle\Delta\theta\rangle^{(\alpha/2)}_{t,D^3_T}.$$

Then from (3.2.6), (3.2.7) *it follows that the semi-norm* $\langle\theta\rangle^{(2+\alpha,1+\frac{\alpha}{2})}_{D^3_T}$ *of a solution is equivalent to the sum* $\langle\theta\rangle^{(2+\alpha)}_{x,D^3_T} + \langle\mathcal{D}_t\theta\rangle^{(\alpha)}_{x,D^3_T}$.

Theorem 3.2.3 *If* $b \in \mathring{C}^{1+\alpha,\frac{1+\alpha}{2}}(\mathbb{R}^2_T)$ *with* $\alpha \in (0,1)$, $T \in (0,\infty)$, *and* b *decreases at infinity as a power function, then there exists a solution* $\theta \in \mathring{C}^{2+\alpha,1+\frac{\alpha}{2}}(D^3_T)$ *of problem* (3.2.2) *and*

$$\langle\theta\rangle^{(2+\alpha,1+\frac{\alpha}{2})}_{D^3_T} \leqslant c\langle b\rangle^{(1+\alpha,\frac{1+\alpha}{2})}_{\mathbb{R}^2_T}. \tag{3.2.8}$$

Proof We extend $b(x',t)$ into $\mathbb{R}$ in t by setting first $b(x',t) = 0$ for $t < 0$ and then $b(x',T+t) = 3b(x',T-t) - 2b(x',T-2t)$ for $t > T$. The general solution of the heat equation transformed by (2.2.2) is

$$\widetilde{\theta}(\xi, x_3, s) = C^{\pm}\mathrm{e}^{\mp r^{\pm}x_3}, \quad \pm x_3 > 0,$$

where $r^{\pm} = \sqrt{\frac{s}{\nu^{\pm}} + \xi^2}$, $\xi^2 = \xi_1^2 + \xi_2^2$. We substitute this solution into the transformed boundary conditions:

$$C^+ - C^- = 0, \qquad \mu^+ r^+ C^+ + \mu^- r^- C^- = \widetilde{b},$$

whence we find

$$\widetilde{\theta}(\xi, x_3, s) = \frac{\widetilde{b}}{\mu^+ r^+ + \mu^- r^-} \mathrm{e}^{\mp r^{\pm} x_3}, \quad \pm x_3 > 0.$$

This function satisfies the assumptions of Theorem 3.2.1 with the multiplier $\widetilde{K} = \frac{1}{\mu^+ r^+ + \mu^- r^-}$ and $n = 3$, $d = -1$. Consequently

$$\begin{aligned} \big|\mathcal{D}_{x'}^{\boldsymbol{i}} (FL)^{-1}\big(\widetilde{K}(\xi, s)\mathrm{e}^{-r^+ x_3}\big)\big| &\leqslant ct^{-\frac{3+|\boldsymbol{i}|}{2}} \mathrm{e}^{-\frac{ax^2}{t}} \\ &\leqslant \frac{c}{(|x|^2 + t)^{\frac{3+|\boldsymbol{i}|}{2}}}, \quad |\boldsymbol{i}| = 0, 1, 2, \quad (3.2.9) \\ \big|\mathcal{D}_t (FL)^{-1}\big(\widetilde{K}(\xi, s)\mathrm{e}^{-r^+ x_3}\big)\big| &\leqslant ct^{-\frac{3+2}{2}} \mathrm{e}^{-\frac{ax^2}{t}} \leqslant \frac{c}{t(|x|^2 + t)^{\frac{3}{2}}} \end{aligned}$$

for $a > 0$, $x \in \mathbb{R}^3_+$, $t > 0$. When $t < 0$, one has $K(x', t) = 0$.

We observe that (3.2.9) at $x_3 = 0$ implies

$$\int_0^{\infty} \int_{\mathbb{R}^2} |\Delta_{\beta}^2(h) K(x', t)| \mathrm{d}x' \mathrm{d}t \leqslant ch, \qquad \beta = 1, 2, \qquad (3.2.10)$$

$$\int_{-\infty}^{\infty} \int_{\mathbb{R}^2} |\Delta_t(h) K(x', t)| \mathrm{d}x' \mathrm{d}t \leqslant ch^{1/2}, \qquad (3.2.11)$$

where $\Delta_t(h)K(x', t) = K(x', t + h) - K(x', t)$ is the increment of K with respect to t and $\Delta_{\beta}^2(h)K$ is the second-order increment of K with respect to x_β. Indeed

$$\begin{aligned} \int_{-\infty}^{\infty} \int_{\mathbb{R}^2} |\Delta_t(h) K(x', t)| \mathrm{d}x' \mathrm{d}t &\leqslant \int_{-h}^{h} \int_{\mathbb{R}^2} |K(x', t + h)| \mathrm{d}x' \mathrm{d}t \\ &+ \int_0^h \int_{\mathbb{R}^2} |K(x', t)| \mathrm{d}x' \mathrm{d}t + \int_0^h \int_h^{\infty} \int_{\mathbb{R}^2} |\mathcal{D}_t K(x', t + \eta)| \mathrm{d}x' \mathrm{d}t \mathrm{d}\eta \\ &\leqslant \int_0^{2h} \frac{\mathrm{d}t}{\sqrt{t}} + \int_0^h \frac{\mathrm{d}\eta}{\sqrt{h + \eta}} \leqslant c\sqrt{h}. \end{aligned}$$

Inequality (3.2.10) is proved in a similar way.

Next, we apply Theorem 3.2.2 by K. K. Golovkin to θ. By this theorem with $r = 2 + \alpha$, $\mu = 1$, inequality (3.2.10) leads to

$$\Big\langle \frac{\partial^2\theta}{\partial x_\beta^2}\Big|_{x_3=0}\Big\rangle^{(\alpha)}_{x_\beta,\mathbb{R}^2_T} \leqslant c\Big\langle \frac{\partial b}{\partial x_\beta}\Big\rangle^{(\alpha)}_{x_\beta,\mathbb{R}^2_T}, \quad \beta=1,2. \tag{3.2.12}$$

Moreover, if $r=1+\alpha/2$, $\mu=1/2$, estimate (3.2.11) yields

$$\big\langle \mathcal{D}_t\theta|_{x_3=0}\big\rangle^{(\alpha/2)}_{t,\mathbb{R}^2_T} \leqslant c\big\langle b\big\rangle^{(\frac{1+\alpha}{2})}_{t,\mathbb{R}^2_T}. \tag{3.2.13}$$

In accordance with the theorem on equivalent norms in the Hölder spaces [35] applied to $\overset{\circ}{C}{}^{2+\alpha,1+\frac{\alpha}{2}}(\mathbb{R}^2_T)$ (see Lemma 2.5.1 with respect to $\mathbb{R}^2_T$), inequalities (3.2.12), (3.2.13) imply the estimate

$$\big\langle \theta|_{x_3=0}\big\rangle^{(2+\alpha,1+\frac{\alpha}{2})}_{\mathbb{R}^2_T} \leqslant c\big\langle b\big\rangle^{(1+\alpha,\frac{1+\alpha}{2})}_{\mathbb{R}^2_T}. \tag{3.2.14}$$

Thus, $\theta(x',0,t)$ is a smooth function on the interface. We can consider it as a boundary value of solutions of the Dirichlet problems in two half-spaces $\{\pm x_3>0\}$. Hence, we can express $\theta(x,t)$ in each half-space as a double-layer heat potential:

$$\theta(x,t)=\begin{cases} -2\nu^+\int_0^t\int_{\mathbb{R}^3}\frac{\partial\Gamma_{\nu^+}(x-y',t-\tau)}{\partial x_3}\theta(y',0,\tau)\mathrm{d}y'\mathrm{d}\tau, & x_3>0,\\ 2\nu^-\int_0^t\int_{\mathbb{R}^3}\frac{\partial\Gamma_{\nu^-}(x-y',t-\tau)}{\partial x_3}\theta(y',0,\tau)\mathrm{d}y'\mathrm{d}\tau, & x_3<0,\end{cases}$$

where $y'=(y_1,y_2)$, $\Gamma_\nu(x,t)=\mathrm{e}^{\frac{-|x|^2}{4\nu t}}/(4\pi\nu t)^{\frac{3}{2}}$, is the fundamental solution of the heat equation. The well-known estimate [47]

$$\big\langle \theta\big\rangle^{(2+\alpha,1+\frac{\alpha}{2})}_{D^3_T} \leqslant c\big\langle \theta|_{x_3=0}\big\rangle^{(2+\alpha,1+\frac{\alpha}{2})}_{\mathbb{R}^2_T}, \tag{3.2.15}$$

together with (3.2.14), yields (3.2.8).

The equality $\theta\big|_{t=0}=0$ follows from the continuity of the inverse Laplace transform of $\widetilde{\theta}$ at $t=0$; and $\mathcal{D}_t\theta\big|_{t=0}=0$ is a corollary of $\theta\big|_{t=0}=0$ and the equation in (3.2.2). Consequently, $\theta\in\overset{\circ}{C}{}^{2+\alpha,1+\frac{\alpha}{2}}(D^3_T)$. □

Finally, we analyze system (3.2.1).

Theorem 3.2.4 *Let $\alpha\in(0,1)$, $T\in(0,\infty)$. We also assume that $\boldsymbol{f}\in\boldsymbol{C}^{\alpha,0}(D^3_T)$ and $[\boldsymbol{f}]\big|_{x_3=0}\in\overset{\circ}{\boldsymbol{C}}{}^{\alpha,\alpha/2}(\mathbb{R}^2_T)$, $\boldsymbol{d}\in\overset{\circ}{\boldsymbol{C}}{}^{1+\alpha,\frac{1+\alpha}{2}}(\mathbb{R}^2_T)$. Then problem (3.2.1) has a unique solution $\boldsymbol{v}\in\boldsymbol{C}^{2+\alpha,0}(D^3_T)$ with $\mathcal{D}_t\boldsymbol{v}\in\boldsymbol{C}^{\alpha,0}(D^3_T)$, and the inequality*

$$\langle\boldsymbol{v}\rangle^{(2+\alpha)}_{x,D^3_T}+\langle\mathcal{D}_t\boldsymbol{v}\rangle^{(\alpha)}_{x,D^3_T}\leqslant c\big\{\langle\boldsymbol{f}\rangle^{(\alpha)}_{x,D^3_T}+\langle[\boldsymbol{f}]\big|_{x_3=0}\rangle^{(\alpha/2)}_{t,\mathbb{R}^2_T}+\langle\boldsymbol{d}\rangle^{(1+\alpha,\frac{1+\alpha}{2})}_{\mathbb{R}^2_T}\big\} \tag{3.2.16}$$

holds.

Proof We construct a solution of (3.2.1) in the form

$$\boldsymbol{v} = \begin{cases} \boldsymbol{u}^+ + \boldsymbol{\theta} & \text{in } \mathbb{R}^3_+, \\ \boldsymbol{u}^- + \boldsymbol{\theta} & \text{in } \mathbb{R}^3_-. \end{cases}$$

Here $\boldsymbol{u}^+$ is a solution of the Cauchy problem

$$\mathcal{D}_t \boldsymbol{u}^+ - \nu^+ \Delta \boldsymbol{u}^+ = \boldsymbol{f}^+ \quad \text{in} \quad \mathbb{R}^3_T,$$
$$\boldsymbol{u}^+\big|_{t=0} = 0,$$

where $\boldsymbol{f}^+$ is an extension of the restriction $\boldsymbol{f}|_{\mathbb{R}^3_+\times(0,T)}$ into $\mathbb{R}^3_T$ with preservation of class. The vector-valued function $\boldsymbol{u}^-$ is a solution of the Dirichlet problem

$$\mathcal{D}_t \boldsymbol{u}^- - \nu^- \Delta \boldsymbol{u}^- = \boldsymbol{f}^- \quad \text{in} \quad \mathbb{R}^3_- \times (0,T),$$
$$\boldsymbol{u}^-\big|_{t=0} = 0, \quad \boldsymbol{u}^-\big|_{x_3=0} = \boldsymbol{u}^+\big|_{x_3=0} \tag{3.2.17}$$

with $\boldsymbol{f}^- = \boldsymbol{f}|_{\mathbb{R}^3_-\times(0,T)}$. Finally, $\boldsymbol{\theta}$ is a solution of diffraction problem (3.2.1) for the homogeneous heat equation with $\boldsymbol{H} \equiv \mu^+\partial\boldsymbol{u}^+/\partial x_3 - \mu^-\partial\boldsymbol{u}^-/\partial x_3 + \boldsymbol{d}$ in place of $\boldsymbol{d}$.

The vector $\boldsymbol{u}^+$ is volume heat potential:

$$\boldsymbol{u}^+(x,t) = \int_0^t \int_{\mathbb{R}^3} \Gamma_{\nu^+}(x-y, t-\tau)\boldsymbol{f}^+(y,\tau)\mathrm{d}y\mathrm{d}\tau.$$

By virtue of estimates obtained in [47] (Sec. 2 of Chap. IV), we arrive at the inequality

$$\langle \boldsymbol{u}^+\rangle^{(2+\alpha)}_{x,\mathbb{R}^3_T} + \langle \mathcal{D}_t\boldsymbol{u}^+\rangle^{(\alpha)}_{x,\mathbb{R}^3_T} \leqslant c\langle \boldsymbol{f}^+\rangle^{(\alpha)}_{x,\mathbb{R}^3_T} \leqslant c\langle \boldsymbol{f}\rangle^{(\alpha)}_{x,D^3_T} \tag{3.2.18}$$

(see also [85]).

The solution of (3.2.17) may be expressed in the form $\boldsymbol{u}^- = \boldsymbol{u}^{-(1)} + \boldsymbol{u}^{-(2)}$, where

$$\boldsymbol{u}^{-(1)}(x,t) = \int_0^t \int_{\mathbb{R}^3} \Gamma_{\nu^-}(x-y, t-\tau)\boldsymbol{f}^-(y,\tau)\mathrm{d}y\mathrm{d}\tau, \quad x_3 < 0,$$

with smoothly extended $\boldsymbol{f}^-$ into the whole space, while $\boldsymbol{u}^{-(2)}$ is the double-layer heat potential

$$\boldsymbol{u}^{-(2)}(x,t) = 2\nu^- \int_0^t \int_{\mathbb{R}^2} \frac{\partial\Gamma_{\nu^-}(x-y,t-\tau)}{\partial x_3}(\boldsymbol{u}^+ - \boldsymbol{u}^{-(1)})(y,\tau)\Big|_{y_3=0}\mathrm{d}y'\mathrm{d}\tau.$$

For $\boldsymbol{u}^{-(1)}$ we have estimate similar to (3.2.18), while $\boldsymbol{u}^{-(2)}$ can be estimated as follows:

$$\left\langle \boldsymbol{u}^{-(2)}\right\rangle^{(2+\alpha)}_{x,\mathbb{R}^3_-\times(0,T)} + \left\langle \mathcal{D}_t\boldsymbol{u}^{-(2)}\right\rangle^{(\alpha)}_{x,\mathbb{R}^3_-\times(0,T)} \leqslant c\left\langle \boldsymbol{u}^+ - \boldsymbol{u}^{-(1)}\right\rangle^{(2+\alpha,1+\alpha/2)}_{\mathbb{R}^2_T}. \tag{3.2.19}$$

Indeed, for estimating $\boldsymbol{u}^{-(2)}$, it is impossible to reduce norm exponent for the boundary function with respect to t. It follows from the fact that $\boldsymbol{u}^{-(2)}$ satisfies homogeneous heat equation, the complete norm of whose solution is equivalent to the sum of semi-norms in the left-hand side of (3.2.19) (see Remark 3.2.1). Thus, (3.2.15) for $\boldsymbol{u}^{-(2)}$ implies that (3.2.19) is a sharp estimate.

We note that compatibility condition for (3.2.17) is fulfilled:

$$[\mathcal{D}_t\boldsymbol{u}^{\pm}(x,0)]\big|_{x_3=0} = [\boldsymbol{f}^{\pm}(x,0)]\big|_{x_3=0} = 0.$$

Taking the equations on the interface into account, we can write

$$\left\langle \mathcal{D}_t\boldsymbol{u}^+ - \mathcal{D}_t\boldsymbol{u}^{-(1)}\right\rangle^{(\alpha/2)}_{t,\mathbb{R}^2_T} \leqslant \left\langle [\boldsymbol{f}]\big|_{x_3=0}\right\rangle^{(\alpha/2)}_{t,\mathbb{R}^2_T} + \left\langle \nu^+\Delta\boldsymbol{u}^+\right\rangle^{(\alpha/2)}_{t,\mathbb{R}^2_T} + \left\langle \nu^-\Delta\boldsymbol{u}^{-(1)}\right\rangle^{(\alpha/2)}_{t,\mathbb{R}^2_T}.$$

Consequently, by (3.2.6), (3.2.7), and (3.2.18) for $\boldsymbol{u}^+$ and $\boldsymbol{u}^{-(1)}$, we conclude

$$\begin{aligned} \left\langle \boldsymbol{u}^+ - \boldsymbol{u}^{-(1)}\right\rangle^{(2+\alpha,1+\alpha/2)}_{\mathbb{R}^2_T} &\equiv \left\langle \nabla\nabla\big(\boldsymbol{u}^+ - \boldsymbol{u}^{-(1)}\big)\right\rangle^{(\alpha,\alpha/2)}_{\mathbb{R}^2_T} + \left\langle \nabla\big(\boldsymbol{u}^+ - \boldsymbol{u}^{-(1)}\big)\right\rangle^{(\frac{1+\alpha}{2})}_{t,\mathbb{R}^2_T} \\ &\quad + \left\langle \mathcal{D}_t\big(\boldsymbol{u}^+ - \boldsymbol{u}^{-(1)}\big)\right\rangle^{(\alpha,\alpha/2)}_{\mathbb{R}^2_T} \\ &\leqslant c\Big\{\left\langle \boldsymbol{f}\right\rangle^{(\alpha)}_{x,D^3_T} + \left\langle [\boldsymbol{f}]\big|_{x_3=0}\right\rangle^{(\alpha/2)}_{t,\mathbb{R}^2_T}\Big\}. \end{aligned} \tag{3.2.20}$$

Since $\boldsymbol{H}$ satisfies the zero initial condition, the vector $\boldsymbol{\theta}$ can be evaluated by Theorem 3.2.3. By virtue of (3.2.6) and (3.2.18)–(3.2.20), we deduce that

$$\langle\boldsymbol{\theta}\rangle^{(2+\alpha,1+\frac{\alpha}{2})}_{D^3_T} \leqslant c\langle\boldsymbol{H}\rangle^{(1+\alpha,\frac{1+\alpha}{2})}_{\mathbb{R}^2_T} \leqslant c\Big\{\left\langle \boldsymbol{f}\right\rangle^{(\alpha)}_{x,D^3_T} + \left\langle [\boldsymbol{f}]\big|_{x_3=0}\right\rangle^{(\alpha/2)}_{t,\mathbb{R}^2_T} + \langle\boldsymbol{d}\rangle^{(1+\alpha,\frac{1+\alpha}{2})}_{\mathbb{R}^2_T}\Big\}.$$

By summing all the above estimates, we arrive at inequality (3.2.16).

We prove now the uniqueness of the solution. Let $\boldsymbol{V} \in \boldsymbol{C}^{2+\alpha,1+\alpha/2}(D^3_T)$ be a solution of (3.2.1) with $\boldsymbol{f} = 0$ and $\boldsymbol{d} = 0$.

We introduce a cutoff function $\zeta_R(x) \in C^{2+\alpha}(\mathbb{R}^3)$ such that

$$\zeta_R(x) = \begin{cases} 1, & |x| < R, \\ 0, & |x| \geqslant 2R, \end{cases} \tag{3.2.21}$$

and $|\nabla\zeta|_{\mathbb{R}^3} \leqslant cR^{-1}$, $|\nabla\nabla\zeta|_{\mathbb{R}^3} \leqslant cR^{-2}$, and $\langle\nabla\nabla\zeta\rangle^{(\alpha)}_{\mathbb{R}^3} \leqslant cR^{-(2+\alpha)}$. The vector-field $\boldsymbol{V}_R \equiv \boldsymbol{V}\zeta_R$ satisfies the relations

$$\mathcal{D}_t\boldsymbol{V}_R - \nu^{\pm}\Delta\boldsymbol{V}_R = -2\nu^{\pm}(\nabla\zeta_R\cdot\nabla)\boldsymbol{V} - \nu^{\pm}\boldsymbol{V}\Delta\zeta_R \equiv \boldsymbol{f}_R \quad \text{in} \quad D_T^{\pm},$$

$$\boldsymbol{V}_R\big|_{t=0} = 0 \quad \text{in} \quad \mathbb{R}^3_- \cup \mathbb{R}^3_+, \qquad \boldsymbol{V}_R \xrightarrow[|x|\to\infty]{} 0,$$

$$[\boldsymbol{V}_R]\big|_{x_3=0} = 0, \quad -\Big[\mu^{\pm}\frac{\partial\boldsymbol{V}_R}{\partial x_3}\Big]\Big|_{x_3=0} = -\Big[\mu^{\pm}\boldsymbol{V}\frac{\partial\zeta_R}{\partial x_3}\Big]\Big|_{x_3=0} \equiv \boldsymbol{d}_R.$$

Since $\boldsymbol{V}_R$, $\boldsymbol{f}_R$, and $\boldsymbol{d}_R$ are compactly supported, we can express $\boldsymbol{V}_R$ in terms of $\boldsymbol{f}_R$ and $\boldsymbol{d}_R$ as made above and apply inequality (3.2.16). In view of the estimate

$$\langle\boldsymbol{f}_R\rangle^{(\alpha,\alpha/2)}_{D_T^3} + \langle\boldsymbol{d}_R\rangle^{(1+\alpha,\frac{1+\alpha}{2})}_{\mathbb{R}^2_T} \leqslant c\{|\boldsymbol{V}|^{(1+\alpha,\frac{1+\alpha}{2})}_{D_T^3} + |\nabla\boldsymbol{V}|^{(\alpha,\alpha/2)}_{D_T^3}\}R^{-1}$$

and of the boundedness of the norms of $\boldsymbol{V}$ and $\nabla\boldsymbol{V}$, we conclude that

$$\langle\boldsymbol{V}_R\rangle^{(2+\alpha,1+\alpha/2)}_{D_T^3} \leqslant cR^{-1}$$

Letting R go to ∞, we see that $\boldsymbol{V} = 0$. □

3.3 The Homogeneous Problem: An Explicit Solution

We consider now problem (3.1.1) with $\boldsymbol{f} = 0$, $g = 0$ and $\boldsymbol{v}_0 = 0$:

$$\mathcal{D}_t\boldsymbol{v} - \nu^{+}\Delta\boldsymbol{v} + \frac{1}{\rho^{+}}\nabla p = 0, \quad \nabla\cdot\boldsymbol{v} = 0 \quad \text{in } D^{+}_{\infty} = \mathbb{R}^3_+ \times (0,\infty),$$

$$\mathcal{D}_t\boldsymbol{v} - \nu^{-}\Delta\boldsymbol{v} + \frac{1}{\rho^{-}}\nabla p = 0, \quad \nabla\cdot\boldsymbol{v} = 0 \quad \text{in } D^{-}_{\infty} = \mathbb{R}^3_- \times (0,\infty),$$

$$\boldsymbol{v}\big|_{t=0} = 0 \quad \text{in} \quad \mathbb{R}^3_- \cup \mathbb{R}^3_+, \quad \boldsymbol{v} \xrightarrow[|x|\to\infty]{} 0, \tag{3.3.1}$$

$$[\boldsymbol{v}]\big|_{x_3=0} = 0, \quad -\Big[\mu^{\pm}\Big(\frac{\partial v_\beta}{\partial x_3} + \frac{\partial v_3}{\partial x_\beta}\Big)\Big]\Big|_{x_3=0} = a_\beta(t,x'), \quad \beta = 1,2;$$

$$\Big[-p + 2\mu^{\pm}\frac{\partial v_3}{\partial x_3}\Big]\Big|_{x_3=0} = a_3(t,x') \quad \text{on } \mathbb{R}^2_{\infty}.$$

We admit no more than a power-law growth of the pressure at infinity in the spatial coordinates.

Theorem 3.3.1 *Let* $a_\beta \in \overset{\circ}{C}{}^{1+\alpha,\frac{1+\alpha}{2}}(\mathbb{R}^2_T)$, $\beta = 1,2$, $a_3 \in \overset{\circ}{C}{}^{1+\alpha,0}(\mathbb{R}^2_T) \cap C^{(\gamma,1+\alpha)}(\mathbb{R}^2_T)$ *for some* $\alpha,\gamma \in (0,1)$ *and* $T > 0$.

Then problem (3.3.1) *has a solution* $(\boldsymbol{v}, p)$ *with the properties as follows:* $\boldsymbol{v} \in \overset{\circ}{\boldsymbol{C}}{}^{2+\alpha,1+\frac{\alpha}{2}}(D^3_T)$, $p \in \overset{\circ}{C}{}^{(\gamma,1+\alpha)}(D^3_T)$, $\nabla p \in \overset{\circ}{C}{}^{\alpha,\alpha/2}(D^3_T)$, *and*

$$\langle \boldsymbol{v}\rangle^{(2+\alpha,1+\frac{\alpha}{2})}_{D^3_T} + \langle \nabla p\rangle^{(\alpha,\alpha/2)}_{D^3_T} + \langle p\rangle^{(\gamma,1+\alpha)}_{D^3_T} \leqslant c\Big\{\sum_{\beta=1}^{2}\langle a_\beta\rangle^{(1+\alpha,\frac{1+\alpha}{2})}_{\mathbb{R}^2_T} + \langle a_3\rangle^{(1+\alpha)}_{x',\mathbb{R}^2_T} + \langle a_3\rangle^{(\gamma,1+\alpha)}_{\mathbb{R}^2_T}\Big\}. \quad (3.3.2)$$

The growth of the function p *as* $|x| \to \infty$ *is bounded by* $|x|^\gamma$.

We assume that the given functions a_i decrease as power functions with respect to x at infinity. We extend a_i first in the domain $t < 0$ taking them equal to zero and then in the domain $t > T$ with preservation of class. Now we can take the Fourier–Laplace transform (2.2.2) in problem (3.3.1). Then it goes over into the system of ordinary differential equations with the unknown functions $\widetilde{\boldsymbol{v}}$, $\widetilde{p}$

$$\begin{aligned}
&\frac{\mathrm{d}^2\widetilde{v}_\beta}{\mathrm{d}x_3^2} - \Big(\frac{s}{\nu^\pm} + \xi^2\Big)\widetilde{v}_\beta - \frac{\mathrm{i}\xi_\beta}{\mu^\pm}\widetilde{p} = 0, \qquad \beta = 1,2,\\
&\frac{\mathrm{d}^2\widetilde{v}_3}{\mathrm{d}x_3^2} - \Big(\frac{s}{\nu^\pm} + \xi^2\Big)\widetilde{v}_3 - \frac{1}{\mu^\pm}\frac{\mathrm{d}\widetilde{p}}{\mathrm{d}x_3} = 0,\\
&\frac{\mathrm{d}\widetilde{v}_3}{\mathrm{d}x_3} + \mathrm{i}\xi_\beta\widetilde{v}_\beta = 0, \qquad \pm x_3 > 0,
\end{aligned} \quad (3.3.3)$$

with the boundary conditions

$$\begin{aligned}
&\widetilde{\boldsymbol{v}} \xrightarrow[|x_3|\to\infty]{} 0, \qquad [\widetilde{\boldsymbol{v}}]\big|_{x_3=0} = 0,\\
&-\Big[\mu^\pm\Big(\frac{\mathrm{d}\widetilde{v}_\beta}{\mathrm{d}x_3} + \mathrm{i}\xi_\beta\widetilde{v}_3\Big)\Big]\Big|_{x_3=0} = \widetilde{a}_\beta, \quad \beta = 1,2,\\
&\Big[-\widetilde{p} + 2\mu^\pm\frac{\mathrm{d}\widetilde{v}_3}{\mathrm{d}x_3}\Big]\Big|_{x_3=0} = \widetilde{a}_3.
\end{aligned} \quad (3.3.4)$$

The solution of system (3.3.3) and (3.3.4) can be derived from (2.2.6) and (2.2.7), if one sets $\sigma = 0$:

$$\widetilde{\boldsymbol{v}} = \boldsymbol{\omega} \mathrm{e}^{\mp r^{\pm} x_3} + \boldsymbol{V}^{\pm} \mathrm{e}_1^{\pm}, \quad \pm x_3 > 0,$$
$$\widetilde{p} = -\mu^{\pm}(r^{\pm} + |\xi|)\Psi^{\pm} \mathrm{e}^{\mp|\xi| x_3}, \quad \pm x_3 > 0, \tag{3.3.5}$$

where as above $r^{\pm} = \sqrt{\frac{s}{\nu^{\pm}} + \xi^2}$, $|\xi| = \sqrt{\xi_1^2 + \xi_2^2}$, $|\arg \sqrt{z}| < \pi/2$ for any $z \in \mathbb{C}$,

$$\boldsymbol{\omega} = \begin{pmatrix} \omega_1 \\ \omega_2 \\ \omega_3 \end{pmatrix}, \quad \boldsymbol{V}^{\pm} = \begin{pmatrix} \mathrm{i}\xi_1 \\ \mathrm{i}\xi_2 \\ \mp|\xi| \end{pmatrix} \Psi^{\pm}, \quad e_1^{\pm} = \frac{\mathrm{e}^{\mp r^{\pm} x_3} - \mathrm{e}^{\mp|\xi| x_3}}{r^{\pm} - |\xi|},$$

$$\Psi^{\pm} = \frac{1}{|\xi| P q} \Big\{ |\xi| \widetilde{a}_3 (\mp r^{\pm} q + |\xi| q') - \sum_{\tau=1}^{2} \mathrm{i}\xi_\tau \widetilde{a}_\tau [(\rho^+ + \rho^-)s + |\xi| q \mp r^{\pm} q'] \Big\},$$

$$\omega_\tau = \frac{\widetilde{a}_\tau - \mathrm{i}\xi_\tau \{\mu^+ \Psi^+ + \mu^- \Psi^- - (\mu^+ - \mu^-)\omega_3\}}{\mu^+ r^+ + \mu^- r^-}, \quad \tau = 1, 2,$$

$$\omega_3 = \frac{\sum_{\tau=1}^{2} \mathrm{i}\xi_\tau \widetilde{a}_\tau q'}{Pq} - \frac{|\xi| \widetilde{a}_3}{P},$$

but now P has the form

$$P = (\rho^+ + \rho^-)s + \frac{4|\xi|}{q} \{ |\xi| ({\mu^+}^2 r^+ + {\mu^-}^2 r^-) + \mu^+ \mu^- (r^+ r^- + \xi^2) \},$$

where

$$q = \mu^+(r^+ + |\xi|) + \mu^-(r^- + |\xi|), \quad q' = \mu^+(r^+ - |\xi|) - \mu^-(r^- - |\xi|).$$

Lemma 3.3.1 *For any $s \in \mathbb{C}$ such that*

$$\operatorname{Re} s + \varkappa |\operatorname{Im} s| \geqslant -\delta \xi^2, \qquad \xi \neq 0,$$

with small positive $\varkappa, \delta$, we have:

$$c_1(|s| + \xi^2) \leqslant |P(\xi, s)| \leqslant c_2(|s| + \xi^2). \tag{3.3.6}$$

This inequality remains valid (maybe with other constants) if ξ is changed by $\zeta = \xi_\beta + \mathrm{i}\eta_\beta$, $\beta = 1, 2$, and $\xi, \eta \in \mathbb{R}^2$, such that

$$|\eta| \leqslant \delta_1 |\xi|,$$

where $\delta_1 > 0$ is small enough.

Proof We introduce the notation

$$Q = \frac{ab}{a+b}, \qquad a = \mu^+ r^+ + \mu^- |\xi|, \quad b = \mu^- r^- + \mu^+ |\xi|.$$

We note that $q = a + b$ and $P = (\rho^+ + \rho^-)s + 4|\xi|Q$.

In order to prove (3.3.6), we estimate the expression $\operatorname{Re} P + \varkappa_1 |\operatorname{Im} P|$ from below.

It is obvious that

$$c_3(|s|^{1/2} + |\xi|) \leqslant |r^\pm| \leqslant c_4(|s|^{1/2} + |\xi|).$$

We consider two cases. First, let $\operatorname{Re} s > 0$. The arguments of the complex values a, b and q have a single sign and belong to the interval $(\frac{-\pi}{4}, \frac{\pi}{4})$. Therefore, $\arg Q \in (\frac{-\pi}{4}, \frac{\pi}{4})$ too, hence, $\operatorname{Re} Q > 0$, $|\operatorname{Im} Q| \leqslant \operatorname{Re} Q$. One deduces

$$\begin{aligned}\operatorname{Re} P + \varkappa_1 |\operatorname{Im} P| &\geqslant (\rho^+ + \rho^-) \min(1, \varkappa_1)|s| + 4|\xi|\{\operatorname{Re} Q - \varkappa_1 |\operatorname{Im} Q|\} \\ &\geqslant c|s| + 4|\xi|(1 - \varkappa_1)\operatorname{Re} Q. \end{aligned} \tag{3.3.7}$$

When $\operatorname{Re} s \leqslant 0$, we have

$$|\operatorname{Im} s| \geqslant \frac{|s| - \delta\xi^2}{1 + \varkappa}.$$

On the other hand

$$\operatorname{Re}\Big(\frac{s}{\nu^\pm} + \xi^2\Big) + \varkappa \Big|\operatorname{Im}\Big(\frac{s}{\nu^\pm} + \xi^2\Big)\Big| \geqslant \Big(\xi^2 - \frac{\delta}{\nu^\pm}\xi^2\Big) > 0$$

for $\delta < \min\{\nu^+, \nu^-\}$ and $\xi \neq 0$.

We suppose, without restriction of generality, that the arguments of r^+, r^- are nonnegative. (In the case of non-positive arguments of r^+, r^-, the picture will be mirror-symmetric.) Then $\arg r^\pm$, $\arg a, \arg b, \arg Q \in (0, \pi/4 + \beta/2)$, $\beta = \arctan \varkappa$, and

$$\begin{aligned}\frac{|\operatorname{Im} Q|}{\operatorname{Re} Q} &\leqslant \tan\Big(\frac{1}{2}\max\{\arg {r^+}^2, \arg {r^-}^2\}\Big) \leqslant \tan\Big(\frac{1}{2}\big(\frac{\pi}{2} + \beta\big)\Big) \\ &= \Big(1 + \frac{\tan\beta}{\sqrt{1 + \tan^2\beta}}\Big)\sqrt{1 + \tan^2\beta} = \sqrt{1 + \varkappa^2} + \varkappa \leqslant 2,\end{aligned}$$

provided that $\varkappa$ is sufficiently small. This implies that

$$|\operatorname{Im} Q| \leqslant 2\operatorname{Re} Q.$$

Therefore, we obtain

$$
\begin{aligned}
&\operatorname{Re} P + \varkappa_1 |\operatorname{Im} P| \geqslant \\
&\quad \geqslant (\rho^+ + \rho^-)\Big\{ - \delta|\xi|^2 + \frac{\varkappa_1 - \varkappa}{\varkappa + 1}(|s| - \delta\xi^2)\Big\} + 4|\xi|(\operatorname{Re} Q - \varkappa_1 |\operatorname{Im} Q|) \\
&\quad \geqslant (\rho^+ + \rho^-)\Big\{ \frac{\varkappa_1 - \varkappa}{\varkappa + 1}|s| - \frac{\varkappa_1 + 1}{\varkappa + 1}\delta\xi^2 \Big\} + 4|\xi|(1 - 2\varkappa_1)\operatorname{Re} Q. \quad (3.3.8)
\end{aligned}
$$

If $|b| \leqslant |a|$, then

$$
\operatorname{Re} Q \geqslant |Q| \cos\left(\frac{\pi}{4} + \frac{\beta}{2}\right) \geqslant \frac{\sqrt{2}|a||b|}{2|a+b|} \geqslant \frac{\sqrt{2}|b|}{4} \geqslant c_5\{|s|^{1/2} + |\xi|\}. \quad (3.3.9)
$$

Otherwise, a and b change places.

We choose $\varkappa < \varkappa_1 < \frac{1}{2}$ and δ also small enough, then the estimate

$$
\operatorname{Re} P + \varkappa_1 |\operatorname{Im} P| \geqslant c_6(|s| + |\xi|^2)
$$

follows from inequalities (3.3.7)–(3.3.9), which gives us the left-hand side of (3.3.6). The fact that $|P|$ is bounded above by $c(|s| + |\xi|^2)$ is evident.

The same inequalities for function $P(\zeta, s)$ with $\zeta = \xi + i\eta$, $|\eta| \leqslant \delta_1|\xi|$, follow from the estimate

$$
|P(\zeta, s) - P(\xi, s)| \leqslant c_7|\eta||\xi| \leqslant c_7\delta_1\xi^2
$$

with small δ_1. □

3.4 The Proof of Theorem 3.3.1

Proof We have seen that if a_i decrease well enough, problem (3.3.1) has a solution $(\boldsymbol{v}, p)$ whose Fourier–Laplace image is given by (3.3.5). Let us show that for the smooth given functions, the original of solution (3.3.5) possesses the properties that theorem statement prescribes. To this end, we consider $\boldsymbol{v}$ as a solution of diffraction problem (3.2.1) with $\boldsymbol{f}$ and $\boldsymbol{d}$ as follows:

$$
\boldsymbol{f} = -\frac{1}{\rho^\pm}\nabla p, \quad (3.4.1)
$$

$$
d_\beta = a_\beta + \left[\mu^\pm \frac{\partial v_3}{\partial x_\beta}\right]\Big|_{x_3=0}, \quad \beta = 1, 2, \qquad d_3 = \left[\mu^\pm \sum_{\tau=1}^{2} \frac{\partial v_\tau}{\partial x_\tau}\right]\Big|_{x_3=0}.
$$

(In the last equality, we have used the solenoidality of $\boldsymbol{v}$.) Pressure function $p(x, t)$ is a solution (in a weak sense) of diffraction problem for the Laplace equation

$$\frac{1}{\rho^\pm}\Delta p = 0 \quad \text{in} \quad \mathbb{R}^3_- \cup \mathbb{R}^3_+,$$

$$[p]|_{x_3=0} = -a_3 - 2d_3 \equiv \pi_0, \qquad \left[\frac{1}{\rho^\pm}\frac{\partial p}{\partial x_3}\right]\bigg|_{x_3=0} = \left[\nu^\pm \Delta v_3\right]\big|_{x_3=0} \equiv \pi_3. \quad (3.4.2)$$

In order to apply Theorem 3.2.4 to (3.2.1), we have to verify that the functions (3.4.1) are sufficiently regular. □

3.4.1 The Analysis of $\boldsymbol{d}$

We express $\widetilde{\boldsymbol{d}}$ in terms of the known functions $\widetilde{a}_j$, $j = 1, 2, 3$, and the boundary values of solution (3.3.5):

$$\widetilde{d}_\beta = \widetilde{a}_\beta + \left[\mu^\pm \mathrm{i}\xi_\beta \widetilde{v}_3\right]\big|_{x_3=0} = \widetilde{a}_\beta + \mathrm{i}\xi_\beta(\mu^+ - \mu^-)\omega_3, \qquad \beta = 1, 2,$$

$$\widetilde{d}_3 = \left[\mu^\pm \sum_{\tau=1}^{2} \mathrm{i}\xi_\tau \widetilde{v}_\tau\right]\big|_{x_3=0} = (\mu^+ - \mu^-)\sum_{\tau=1}^{2} \mathrm{i}\xi_\tau \omega_\tau. \qquad (3.4.3)$$

We take the inverse Fourier–Laplace transform by (3.2.5)

$$d_\beta = a_\beta + (\mu^+ - \mu^-)\frac{\partial}{\partial x_\beta}\left\{\sum_{\tau=1}^{2} \frac{\partial}{\partial x_\tau} K_1 * a_\tau + K_1' * a_3\right\}, \quad \beta = 1, 2,$$

$$d_3 = (\mu^+ - \mu^-)\sum_{\tau=1}^{2} \frac{\partial}{\partial x_\tau}\left\{(K_2 + K_3) * a_\tau + \frac{\partial}{\partial x_\tau} K_2' * a_3\right\},$$

where $K * a$ means the convolution $\int_0^t \int_{\mathbb{R}^2} K(x' - y', t - t')a(y', t')\mathrm{d}y'\mathrm{d}t'$, $y' = (y_1, y_2)$, and the kernels $K_1, K_1', K_2, K_2', K_3$ are the originals of

$$\widetilde{K_1} = -\frac{q'}{Pq}, \quad \widetilde{K_1'} = -\frac{|\xi|}{P}, \quad \widetilde{K_2} = \frac{1}{\mu^+ r^+ + \mu^- r^-}, \quad \widetilde{K_2'} = \frac{q'}{Pq},$$

$$\widetilde{K_3} = \frac{-|\xi|\{(\mu^+\rho^- + \mu^-\rho^+)s + (\mu^+ + \mu^-)|\xi|q + 2\mu^+\mu^-(r^+r^- - \xi^2)\}}{(\mu^+ r^+ + \mu^- r^-)Pq},$$

respectively; P, q, and q' are given by formulas in (3.3.3).

In view of (3.3.6), it is easily seen that $\widetilde{K_1}$, $\widetilde{K_2'}$ satisfy the hypotheses of Lemma 3.2.1 with $m = 0$, $l = 2$, and $\widetilde{K_1'}$ and $\widetilde{K_3}$ do it with $m = 1$, $l = 2$. Therefore, we have

$$\left|\mathcal{D}_t^k \mathcal{D}_{x'}^{\boldsymbol{i}} K_1(x',t)\right| + \left|\mathcal{D}_t^k \mathcal{D}_{x'}^{\boldsymbol{i}} K_2'(x',t)\right| \leqslant \frac{c}{t^k(|x'|^2+t)^{\frac{2+|\boldsymbol{i}|}{2}}}, \quad k=0,1, \ |\boldsymbol{i}| \geqslant 0, \tag{3.4.4}$$

$$\begin{aligned} &\left|\mathcal{D}_t^k \mathcal{D}_{x'}^{\boldsymbol{i}} K_1'(x',t)\right| + \left|\mathcal{D}_t^k \mathcal{D}_{x'}^{\boldsymbol{i}} K_3(x',t)\right| + \left|\mathcal{D}_t^k \mathcal{D}_{x'}^{\boldsymbol{i}} K_2(x',t)\right| \\ &\qquad \leqslant \frac{c}{t^k(|x'|^2+t)^{\frac{3+|\boldsymbol{i}|}{2}}}, \quad k=0,1, \quad |\boldsymbol{i}| \geqslant 0, \end{aligned} \tag{3.4.5}$$

Inequality (3.4.5) for K_2 follows from (3.2.9). All these kernels vanish when $t<0$. It is evident that convolution integrals with the kernels $\mathcal{D}_{x'}K_1$, $\mathcal{D}_{x'}K_2'$, K_1', K_3, K_2 are convergent because of power-law decays of a_j and $\nabla' a_j$ at infinity. Inequalities (3.4.4) and (3.4.5) for these kernels with $k=0$ yield integral estimate (3.2.10) for them which, by Theorem 3.2.3, gives us

$$\langle \nabla' d_i \rangle_{x',\mathbb{R}_t^2}^{(\alpha)} \leqslant c \sum_{j=1}^{3} \langle \nabla' a_j \rangle_{x',\mathbb{R}_t^2}^{(\alpha)}, \qquad t<T, \quad i=1,2,3, \tag{3.4.6}$$

where $\nabla' = (\partial/\partial x_1, \partial/\partial x_2)$.

Next, we need to estimate $\langle d_j \rangle_{t,\mathbb{R}_T^2}^{(\frac{1+\alpha}{2})}$, $j=1,2,3$. The first equality in (3.4.3) can be written in the form

$$\widetilde{d}_\beta = \widetilde{a}_\beta + (\mu^+ - \mu^-)\mathrm{i}\xi_\beta \left[\widetilde{K_1}\widetilde{\varphi}_2 + \frac{\widetilde{K_1'}}{|\xi|}\widetilde{\varphi}_3\right] \equiv \sum_{i=1}^{3} \widetilde{A}_{i\beta}, \qquad \beta=1,2,$$

where $\widetilde{\varphi}_2 = \sum_{\tau=1}^{2} \mathrm{i}\xi_\tau \widetilde{a}_\tau$, $\widetilde{\varphi}_3 = |\xi|\widetilde{a}_3$.

The kernels $K_{2\beta}'' = (FL)^{-1}(\mathrm{i}\xi_\beta \widetilde{K_1})$ and $K_{3\beta}'' = (FL)^{-1}(\mathrm{i}\xi_\beta \widetilde{K_1'}/|\xi|)$ satisfy Lemma 3.2.1 with $m=1$, $l=2$, which yields inequality (3.4.5) for them. These kernels are derivatives with respect to x_β; therefore, $\int_{\mathbb{R}^2} K_{i\beta}''(x'-y',t-t')\mathrm{d}y' = 0$, $i=2,3$, and $\beta=1,2$, and we can write

$$\begin{aligned} &A_{i\beta}(x',t+h) - A_{i\beta}(x',t) \\ &= \int_{t-h}^{t+h} \int_{\mathbb{R}^2} K_{i\beta}''(x'-y',t+h-t')(\varphi_i(y',t') - \varphi_i(x',t'))\mathrm{d}y'\mathrm{d}t' \\ &\quad - \int_{t-h}^{t} \int_{\mathbb{R}^2} K_{i\beta}''(x'-y',t-t')(\varphi_i(y',t') - \varphi_i(x',t'))\mathrm{d}y'\mathrm{d}t' \\ &\quad + \int_0^h \int_0^{t-h} \int_{\mathbb{R}^2} \mathcal{D}_t K_{i\beta}''(x'-y',t+\lambda-t')(\varphi_i(y',t') - \varphi_i(x',t'))\mathrm{d}y'\mathrm{d}t'\mathrm{d}\lambda. \end{aligned}$$

If $h < t, t + h < T$, then by (3.4.5) for $K''_{i\beta}$ with $|\boldsymbol{i}| = 0$, $k = 0, 1$, we deduce

$$\begin{aligned}
&|A_{i\beta}(x', t+h) - A_{i\beta}(x', t)| \\
&\leqslant c \sup_{t' < t+h} \langle \varphi_i(\cdot, t')\rangle^{(\alpha)}_{\mathbb{R}^2} \bigg\{ \int_{t-h}^{t+h} \int_{\mathbb{R}^2} \frac{|x'-y'|^\alpha \mathrm{d}y'\mathrm{d}t'}{(|x'-y'|^2 + t + h - t')^{3/2}} \\
&\qquad + \int_{t-h}^{t} \int_{\mathbb{R}^2} \frac{|x'-y'|^\alpha \mathrm{d}y'\mathrm{d}t'}{(|x'-y'|^2 + t - t')^{3/2}} \\
&\qquad + \int_0^h \int_0^{t-h} \int_{\mathbb{R}^2} \frac{|x'-y'|^\alpha \mathrm{d}y'\mathrm{d}t'\mathrm{d}\lambda}{(t+\lambda-t')(|x'-y'|^2 + t + \lambda - t')^{3/2}} \bigg\} \\
&\leqslant c \sup_{t' < t+h} \langle \varphi_i(\cdot, t')\rangle^{(\alpha)}_{\mathbb{R}^2} h^{\frac{1+\alpha}{2}}, \quad i = 2, 3, \quad \beta = 1, 2.
\end{aligned}$$

When $h > t$, we obtain

$$\begin{aligned}
|A_{i\beta}(x', t+h) - A_{i\beta}(x', t)| &\leqslant |A_{i\beta}(x', t+h)| + |A_{i\beta}(x', t)| \\
&\leqslant c\langle \varphi_i\rangle^{(\alpha)}_{x', \mathbb{R}^2_T} (t+h)^{\frac{1+\alpha}{2}} \leqslant c_1 \langle \varphi_i\rangle^{(\alpha)}_{x', \mathbb{R}^2_T} h^{\frac{1+\alpha}{2}},
\end{aligned}$$

$i = 2, 3$, $\beta = 1, 2$.

Obviously

$$\langle \varphi_2\rangle^{(\alpha)}_{x', \mathbb{R}^2_T} \leqslant \sum_{\tau=1}^{2} \langle a_\tau\rangle^{(1+\alpha)}_{x', \mathbb{R}^2_T}.$$

As to $\varphi_3(x', t) = (FL)^{-1}(|\xi|\widetilde{a}_3)$, we can interpret it as the boundary value of the function

$$\varphi_3(x, t) = 2\frac{\partial^2}{\partial x_3^2} \int_{\mathbb{R}^2} \mathcal{E}(x - y') a_3(y', t)\mathrm{d}y',$$

where $\mathcal{E}(x) = \frac{-1}{4\pi|x|}$. The following estimate is well-known:

$$\langle \varphi_3(\cdot, 0, t)\rangle^{(\alpha)}_{\mathbb{R}^2} \leqslant c\langle a_3(\cdot, t)\rangle^{(1+\alpha)}_{\mathbb{R}^2}. \tag{3.4.7}$$

Thus

$$\langle d_\beta\rangle^{(\frac{1+\alpha}{2})}_{t, \mathbb{R}^2_T} \leqslant \sum_{i=1}^{3} \langle A_{i\beta}\rangle^{(\frac{1+\alpha}{2})}_{t, \mathbb{R}^2_T} \leqslant \langle a_\beta\rangle^{(\frac{1+\alpha}{2})}_{t, \mathbb{R}^2_T} + c\sum_{j=1}^{3} \langle a_j\rangle^{(1+\alpha)}_{x', \mathbb{R}^2_T}, \quad \beta = 1, 2. \tag{3.4.8}$$

We estimate now function d_3 rewriting it as follows:

$$d_3 = (\mu^+ - \mu^-) \sum_{\tau=1}^{2} \Big\{ \frac{\partial K_2 * a_\tau}{\partial x_\tau} + (FL)^{-1}\Big(\frac{\mathrm{i}\xi_\tau \widetilde{K_3}}{|\xi|}\Big) * (FL)^{-1}(|\xi|\widetilde{a}_\tau) + \frac{\partial K_2'}{\partial x_\tau} * \frac{\partial a_3}{\partial x_\tau} \Big\}.$$

The semi-norms $\langle\cdot\rangle_{t,\mathbb{R}^2_T}^{(\frac{1+\alpha}{2})}$ of the second and third terms in the brackets are estimated in the same way as above. As to the term with K_2, we can interpret it as a tangential derivative of the boundary value of Θ_τ, a solution to diffraction problem (3.2.2) with the boundary function a_τ instead of b. Since $[\Theta_\tau]|_{x_3=0} = 0$, $\tau = 1, 2$, the function $\sum_{\tau=1}^{2} \partial\Theta_\tau/\partial x_\tau$ is also continuous when passing across the plane $\{x_3 = 0\}$. Moreover, compatibility conditions $a_\tau|_{t=0} = 0$, $\tau = 1, 2$, are satisfied; therefore, by Theorem 3.2.3, we have

$$\Big\langle \sum_{\tau=1}^{2} \frac{\partial\Theta_\tau}{\partial x_\tau} \Big\rangle_{t,\mathbb{R}^2_T}^{(\frac{1+\alpha}{2})} \leqslant \sum_{\tau=1}^{2} \langle\Theta_\tau\rangle_{D^3_T}^{(2+\alpha,1+\frac{\alpha}{2})} \leqslant c \sum_{\tau=1}^{2} \langle a_\tau\rangle_{\mathbb{R}^2_T}^{(1+\alpha,\frac{1+\alpha}{2})}.$$

Hence, we obtain

$$\langle d_3\rangle_{t,\mathbb{R}^2_T}^{(\frac{1+\alpha}{2})} \leqslant c\Big\{ \sum_{\tau=1}^{2} \langle a_\tau\rangle_{\mathbb{R}^2_T}^{(1+\alpha,\frac{1+\alpha}{2})} + \langle a_3\rangle_{x',\mathbb{R}^2_T}^{(1+\alpha)} \Big\}. \tag{3.4.9}$$

Finally, a consequence of estimates (3.4.6)–(3.4.9) is

$$\langle d_j\rangle_{\mathbb{R}^2_T}^{(1+\alpha,\frac{1+\alpha}{2})} \leqslant c\Big\{ \sum_{\tau=1}^{2} \langle a_\tau\rangle_{\mathbb{R}^2_T}^{(1+\alpha,\frac{1+\alpha}{2})} + \langle a_3\rangle_{x',\mathbb{R}^2_T}^{(1+\alpha)} \Big\}, \quad j = 1, 2, 3. \tag{3.4.10}$$

Next, we verify that $\boldsymbol{f} = -\frac{1}{\rho^\pm}\nabla p \in \boldsymbol{C}^{\alpha,0}(D^3_T)$ and $[\boldsymbol{f}]\big|_{x_3=0} \in \mathring{\boldsymbol{C}}^{\alpha,\alpha/2}(\mathbb{R}^2_T)$. To this end, we prove that $\boldsymbol{\pi} \equiv \big[\frac{1}{\rho^\pm}\nabla p\big]\big|_{x_3=0} \in \mathring{\boldsymbol{C}}^{\alpha,\alpha/2}(\mathbb{R}^2_T)$.

3.4.2 The Estimate of $\boldsymbol{\pi}$

Let us first consider the tangential part of $\boldsymbol{\pi}$. By (3.3.5), its Fourier–Laplace image is expressed in the form

$$\begin{aligned} \widetilde{\pi}_\beta &= \big[\mathrm{i}\xi_\beta\widetilde{p}/\rho^\pm\big]\big|_{x_3=0} = -\big[\mathrm{i}\xi_\beta\nu^\pm(r^\pm + |\xi|)\Psi^\pm \mathrm{e}^{\mp|\xi|x_3}\big]\big|_{x_3=0} \\ &= -\mathrm{i}\xi_\beta\big(\nu^+(r^+ + |\xi|)\Psi^+ - \nu^-(r^- + |\xi|)\Psi^-\big), \quad \beta = 1, 2. \end{aligned}$$

We select only the terms containing $sr^\pm$ and $(r^\pm)^2$. They are included in the summand $-\mathrm{i}\xi_\beta\Big(\nu^+r^+\Psi^+ - \nu^-r^-\Psi^-\Big)$:

$$\sum_{\tau=1}^{2} \frac{\mathrm{i}\xi_\beta \mathrm{i}\xi_\tau \widetilde{a}_\tau}{|\xi| qP} \Big\{ (\rho^+ + \rho^-) s \big(\nu^+ r^+ - \nu^- r^-\big) - \big(\nu^+ (r^+)^2 + \nu^- (r^-)^2\big) q' \Big\}$$
$$+ \frac{\mathrm{i}\xi_\beta \widetilde{a}_3}{P} \Big\{ \nu^+ (r^+)^2 + \nu^- (r^-)^2 \Big\}$$
$$= \sum_{\tau=1}^{2} \frac{-\mathrm{i}\xi_\beta \mathrm{i}\xi_\tau \widetilde{a}_\tau}{qP} \Big\{ \frac{(\rho^+ - \rho^-) s \big(\nu^+ r^+ + \nu^- r^-\big)}{|\xi|} - 2(\mu^+ - \mu^-) s + (\nu^+ + \nu^-)|\xi| q' \Big\}$$
$$+ \frac{2\mathrm{i}\xi_\beta \widetilde{a}_3}{\rho^+ + \rho^-} \Big\{ 1 - \frac{4|\xi| Q + (\rho^+ + \rho^-)(\nu^+ + \nu^-)|\xi|^2}{P} \Big\}.$$

We set $u_\tau^\pm = (FL)^{-1}\big(r^\pm \widetilde{a}_\tau\big)$, $\tau = 1, 2$. These functions are the boundary values of the second derivatives of the single-layer heat potentials:

$$u_\tau^\pm (x', 0, t) = 2\nu^\pm \frac{\partial^2}{\partial x_3^2} \int_0^t \int_{\mathbb{R}^2} \Gamma_{\nu^\pm}(x - y', t - t') \widetilde{a}_\tau (y', t') \mathrm{d}y' \mathrm{d}t' \Big|_{x_3 = 0}, \quad \tau = 1, 2.$$

One well knows the classical estimate for $u_\tau^\pm$ on the interface:

$$\langle u_\tau^\pm \rangle_{x', \mathbb{R}_T^2}^{(\alpha)} \leqslant c^\pm \langle a_\tau \rangle_{\mathbb{R}_T^2}^{(1+\alpha, \frac{1+\alpha}{2})}, \quad \tau = 1, 2, \tag{3.4.11}$$

(see [47], chap. IV).

The multiplier $\widetilde{K}' = \sum_{\tau=1}^{2} \frac{\mathrm{i}\xi_\tau (\rho^+ + \rho^-) s}{|\xi| qP}$ satisfies Lemma 3.2.1 with $m = 0$, $l = 1$. We observe that, in this case, inequality (3.2.4) for $k = 0$ becomes

$$\big|\mathcal{D}_x^{\boldsymbol{i}} K'(x, t)\big| \leqslant \frac{c}{t^{1/2} (|x|^2 + t)^{\frac{2+|\boldsymbol{i}|}{2}}}, \quad |\boldsymbol{i}| \geqslant 0.$$

It is not very difficult to verify that this estimate implies (3.2.10) (see Remark 2.2 from [86]). Hence, for $\partial B^\pm / \partial x_\beta = (FL)^{-1}\big(\sum_{\tau=1}^{2} \mathrm{i}\xi_\beta \widetilde{K}' \widetilde{u}_\tau^\pm\big)$, $\beta = 1, 2$, we deduce

$$\Big\langle \frac{\partial B^\pm}{\partial x_\beta} \Big\rangle_{x', \mathbb{R}_T^2}^{(\alpha)} \leqslant c \sum_{\tau=1}^{2} \langle u_\tau^\pm \rangle_{x', \mathbb{R}_T^2}^{(\alpha)}, \quad \beta = 1, 2. \tag{3.4.12}$$

The semi-norm $\langle \partial B^\pm / \partial x_\beta \rangle_{t, \mathbb{R}_T^2}^{(\frac{\alpha}{2})}$ is estimated in a way similar to evaluating the semi-norm $\langle A_{i\beta} \rangle_{t, \mathbb{R}_T^2}^{(\frac{1+\alpha}{2})}$ above. Namely, if $h < t, t + h < T$, estimates (3.2.4) for $\mathrm{i}\xi_\beta \widetilde{K}'$, with $|\boldsymbol{i}| = 0$ and $k = 0, 1$, give

$$\Big| \frac{\partial B^\pm}{\partial x_\beta}(x', t + h) - \frac{\partial B^\pm}{\partial x_\beta}(x', t) \Big|$$

$$\leqslant c \sup_{t'<t+h} \Big\langle \sum_{\tau=1}^{2} u_\tau^\pm(\cdot,t') \Big\rangle_{\mathbb{R}^2}^{(\alpha)} \bigg\{ \int_{t-h}^{t+h} \int_{\mathbb{R}^2} \frac{|x'-y'|^\alpha \mathrm{d}y'\mathrm{d}t'}{(t+h-t')^{1/2}(|x'-y'|^2+t+h-t')^{3/2}}$$

$$+ \int_{t-h}^{t} \int_{\mathbb{R}^2} \frac{|x'-y'|^\alpha \mathrm{d}y'\mathrm{d}t'}{(t-t')^{1/2}(|x'-y'|^2+t-t')^{3/2}}$$

$$+ \int_0^h \int_0^{t-h} \int_{\mathbb{R}^2} \frac{|x'-y'|^\alpha \mathrm{d}y'\mathrm{d}t'\mathrm{d}\lambda}{(t+\lambda-t')^{3/2}(|x'-y'|^2+t+\lambda-t')^{3/2}} \bigg\}$$

$$\leqslant c\langle u_\tau^\pm\rangle_{x',\mathbb{R}_T^2}^{(\alpha)} \bigg\{ \int_{t-h}^{t+h} \frac{\mathrm{d}t'}{(t+h-t')^{1/2+(1-\alpha)/2}} + \int_{t-h}^{t} \frac{\mathrm{d}t'}{(t-t')^{1-\alpha/2}}$$

$$+ \int_0^h \int_0^{t-h} \frac{\mathrm{d}t'\mathrm{d}\lambda}{(t+\lambda-t')^{2-\alpha/2}} \bigg\}$$

$$\leqslant c \sum_{\tau=1}^{2} \Big\langle u_\tau^\pm \Big\rangle_{x',\mathbb{R}_T^2}^{(\alpha)} h^{\frac{\alpha}{2}}, \quad \beta = 1,2.$$

If $h > t$, we obtain

$$\Big|\frac{\partial B^\pm}{\partial x_\beta}(x',t+h) - \frac{\partial B^\pm}{\partial x_\beta}(x',t)\Big| \leqslant c\langle u_\tau^\pm\rangle_{x',\mathbb{R}_T^2}^{(\alpha)} (t+h)^{\frac{\alpha}{2}} \leqslant c_1\langle u_\tau^\pm\rangle_{x',\mathbb{R}_T^2}^{(\alpha)} h^{\frac{\alpha}{2}},$$

$\beta = 1,2$. Consequently

$$\Big\langle \frac{\partial B^\pm}{\partial x_\beta} \Big\rangle_{t,\mathbb{R}_T^2}^{(\alpha/2)} \leqslant c \sum_{\tau=1}^{2} \langle u_\tau^\pm \rangle_{x',\mathbb{R}_T^2}^{(\alpha)}, \quad \beta = 1,2,$$

which together with (3.4.12) and (3.4.11) leads us to the estimate

$$\langle \nabla' B^\pm \rangle_{\mathbb{R}_T^2}^{(\alpha,\alpha/2)} \leqslant c \sum_{\tau=1}^{2} \langle a_\tau \rangle_{\mathbb{R}_T^2}^{(1+\alpha,\frac{1+\alpha}{2})}.$$

It is easily seen that the semi-norm $\langle\cdot\rangle_{\mathbb{R}_T^2}^{(\alpha,\alpha/2)}$ of the original of the term $\sum_{\tau=1}^{2} \frac{\mathrm{i}\xi_\beta \mathrm{i}\xi_\tau \widetilde{a}_\tau}{qP}\{2(\mu^+-\mu^-)s - (\nu^++\nu^-)|\xi|q'\}$ is estimated from above by $c\sum_{\tau=1}^{2}\Big\langle\frac{\partial a_\tau}{\partial x_\beta}\Big\rangle_{x',\mathbb{R}_T^2}^{(\alpha)}$.

As to a_3, it is clear that the norm $\Big\langle (FL)^{-1}\Big(\frac{2\mathrm{i}\xi_\beta\widetilde{a}_3}{\rho^++\rho^-}\Big)\Big\rangle_{\mathbb{R}_T^2}^{(\alpha,\alpha/2)}$ is equivalent to $\langle\nabla' a_3\rangle_{\mathbb{R}_T^2}^{(\alpha,\alpha/2)}$. Next, since the original of $|\xi|\widetilde{a}_3 \equiv \widetilde{\varphi_3}$ is subjected to inequality (3.4.7) and the multiplier $\frac{8Q}{(\rho^+ + \rho^-)P}$ satisfies Lemma 3.2.1 with $m = 0$, $l = 1$, the original of the term $\frac{8\mathrm{i}\xi_\beta Q|\xi|\widetilde{a}_3}{(\rho^+ + \rho^-)P}$ is evaluated in the same way as $\frac{\partial B^\pm}{\partial x_\beta}$, i.e.,

$$\Big\langle \frac{\partial}{\partial x_\beta}(FL)^{-1}\Big(\frac{8|\xi|Q\widetilde{a}_3}{(\rho^+ + \rho^-)P}\Big)\Big\rangle^{(\alpha,\alpha/2)}_{\mathbb{R}^2_T} \leqslant c \sup_{t\in(0,T)} \langle \varphi_3(\cdot,0,t)\rangle^{(\alpha)}_{\mathbb{R}^2} \leqslant c\langle a_3\rangle^{(1+\alpha)}_{x',\mathbb{R}^2_T},$$

$$\beta = 1,2.$$

The other terms in $\widetilde{\pi}_\beta$ contain ξ^2 as a multiplier. If we give $\mathrm{i}\xi_\tau$ to $\widetilde{a}_j$, $j = 1,2,3$, the kernels have the form $\mathrm{i}\xi_\tau \widetilde{K}_j$, $\tau = 1,2$, where $\widetilde{K}_j$ satisfy Lemma 3.2.1 again with $m = 0$, $l = 1$. We evaluate π_β as above. Their semi-norms are bounded above by $c\sum_{j=1}^3 \langle \nabla' a_j\rangle^{(\alpha)}_{x',\mathbb{R}^2_T}$.

Summing all these estimates, we conclude that

$$\langle \pi_\beta\rangle^{(\alpha,\alpha/2)}_{\mathbb{R}^2_T} \leqslant c\Big\{\sum_{\tau=1}^2 \langle a_\tau\rangle^{(1+\alpha,\frac{1+\alpha}{2})}_{\mathbb{R}^2_T} + \langle \nabla' a_3\rangle^{(\alpha,\alpha/2)}_{\mathbb{R}^2_T}\Big\}, \quad \beta = 1,2. \tag{3.4.13}$$

By virtue of (3.3.1), we can write

$$\pi_3 = \Big[\nu^\pm\Big(\nabla'^2 v_3 - \frac{\partial}{\partial x_3}\sum_{\tau=1}^2 \frac{\partial v_\tau}{\partial x_\tau}\Big)\Big]\Big|_{x_3=0} \tag{3.4.14}$$

($\nabla'^2 = \nabla' \cdot \nabla'$). The Fourier–Laplace image of π_3 is

$$\begin{aligned}
\widetilde{\pi}_3 &= -(\nu^+ - \nu^-)\xi^2\omega_3 - \Big[\nu^\pm \frac{\partial}{\partial x_3}\Big(\sum_{\tau=1}^2 \mathrm{i}\xi_\tau \widetilde{v}_\tau\Big)\Big]\Big|_{x_3=0} \\
&= -(\nu^+ - \nu^-)\xi^2\omega_3 + \sum_{\tau=1}^2 \mathrm{i}\xi_\tau\big(\nu^+ r^+\omega_\tau + \nu^+\mathrm{i}\xi_\tau\Psi^+ + \nu^- r^-\omega_\tau + \nu^-\mathrm{i}\xi_\tau\Psi^-\big) \\
&= \sum_{\tau=1}^2 \mathrm{i}\xi_\tau \frac{\nu^+ r^+ + \nu^- r^-}{\mu^+ r^+ + \mu^- r^-}\widetilde{a}_\tau - \xi^2\Big\{(\nu^+ - \nu^-)\omega_3 + \nu^+\Psi^+ + \nu^-\Psi^- \\
&\qquad - \frac{\nu^+ r^+ + \nu^- r^-}{\mu^+ r^+ + \mu^- r^-}\big(\mu^+\Psi^+ + \mu^-\Psi^- - (\mu^+ - \mu^-)\omega_3\big)\Big\}.
\end{aligned} \tag{3.4.15}$$

We interpret the original of $\mathrm{i}\xi_\tau \frac{\nu^+ r^+ + \nu^- r^-}{\mu^+ r^+ + \mu^- r^-}\widetilde{a}_\tau$ as a tangential derivative of the function $-\big[\nu^\pm \frac{\partial \Theta_\tau}{\partial x_3}\big]\big|_{x_3=0}$, where Θ_τ is as above. Then by Theorem 3.2.3

$$\left\langle \frac{\partial}{\partial x_\tau}\left[\nu^\pm \frac{\partial \Theta_\tau}{\partial x_3}\right]\Big|_{x_3=0}\right\rangle^{(\alpha,\frac{\alpha}{2})}_{\mathbb{R}^2_T} \leqslant \langle \Theta_\tau\rangle^{(2+\alpha,1+\frac{\alpha}{2})}_{D^3_T} \leqslant c\langle a_\tau\rangle^{(1+\alpha,\frac{1+\alpha}{2})}_{\mathbb{R}^2_T}, \quad \tau=1,2. \tag{3.4.16}$$

All the other terms in (3.4.15) contain ξ^2 as a multiplier, and their semi-norms are bounded above by $c\langle \nabla' a_j\rangle^{(\alpha)}_{x',\mathbb{R}^2_T}$, $j=1,2,3$. This together with (3.4.16) leads us to the inequality

$$\langle \pi_3\rangle^{(\alpha,\frac{\alpha}{2})}_{\mathbb{R}^2_T} \leqslant c\Big\{\sum_{\tau=1}^{2}\langle a_\tau\rangle^{(1+\alpha,\frac{1+\alpha}{2})}_{\mathbb{R}^2_T} + \langle a_3\rangle^{(1+\alpha)}_{x',\mathbb{R}^2_T}\Big\}. \tag{3.4.17}$$

We note that that the zero initial condition for π follows from the fact that $\Delta \boldsymbol{v}|_{t=0}=0$.

3.4.3 The Estimate of the Vector $f = -\frac{1}{\rho^\pm}\nabla p$ in D^3_T

As a solution of problem (3.4.2), the pressure can be represented similarly to Sect. 2.5 in the form $p=p_1+p_2$, where $p_1=0$ for $x_3<0$, and for $x_3>0$, it is a solution of the Dirichlet problem

$$\frac{1}{\rho^+}\Delta p_1 = 0 \quad \text{in } \mathbb{R}^3_+, \qquad p_1|_{x_3=0}=\pi_0;$$

p_2 is a harmonic function in $\mathbb{R}^3_-\cup\mathbb{R}^3_+$, continuous when passing across $\{x_3 = 0\}$, and satisfies the condition

$$\left[\frac{1}{\rho^\pm}\frac{\partial p_2}{\partial x_3}\right]\Big|_{x_3=0} = \pi_3 - \frac{1}{\rho^+}\frac{\partial p_1}{\partial x_3}\Big|_{x_3=0} \equiv h_0.$$

The functions p_1 and p_2 are expressed by the double- and single-layer potentials, respectively:

$$p_1(x,t) = 2\int_{\mathbb{R}^2}\frac{\partial \mathcal{E}(x-y')}{\partial x_3}\pi_0(y',t)\mathrm{d}y', \qquad x_3\geqslant 0,$$

$$p_2(x,t) = \frac{2\rho^+\rho^-}{\rho^++\rho^-}\int_{\mathbb{R}^2}\mathcal{E}(x-y')h_0(y',t)\mathrm{d}y', \qquad x\in\mathbb{R}^3_-\cup\mathbb{R}^3_+,$$

for which the estimates

$$\langle \nabla p_1\rangle^{(\alpha)}_{x,D^3_T} \leqslant c\langle \pi_0\rangle^{(1+\alpha)}_{x',\mathbb{R}^2_T},$$

$$\langle \nabla p_2 \rangle^{(\alpha)}_{x,D^3_T} \leqslant c \langle h_0 \rangle^{(\alpha)}_{x',\mathbb{R}^2_T}$$

are known. And, by taking (3.4.6) and (3.4.17) into account, we obtain

$$\langle \nabla p \rangle^{(\alpha)}_{x,D^3_T} \leqslant c \Big\{ \sum_{\tau=1}^{2} \langle a_\tau \rangle^{(1+\alpha,\frac{1+\alpha}{2})}_{\mathbb{R}^2_T} + \langle \nabla' a_3 \rangle^{(\alpha)}_{x',\mathbb{R}^2_T} \Big\}. \tag{3.4.18}$$

Now, we can apply Theorem 3.2.4 to (3.2.1) with right-hand side (3.4.1). We note that the compatibility conditions $d_i(x',0) = 0$, $i = 1,2,3$, follow from the zero initial conditions $\boldsymbol{v}|_{t=0} = 0$, $a_i(x',0) = 0$, $i = 1,2,3$. Finally, in view of (3.4.13), (3.4.17), and (3.4.18), we deduce

$$\langle \boldsymbol{v} \rangle^{(2+\alpha)}_{x,D^3_T} + \langle \mathcal{D}_t \boldsymbol{v} \rangle^{(\alpha)}_{x,D^3_T} \leqslant c \Big\{ \sum_{\tau=1}^{2} \langle a_\tau \rangle^{(1+\alpha,\frac{1+\alpha}{2})}_{\mathbb{R}^2_T} + \langle \nabla' a_3 \rangle^{(\alpha,\alpha/2)}_{\mathbb{R}^2_T} \Big\}. \tag{3.4.19}$$

Moreover, in view of (3.4.14) and

$$\frac{\partial p_1(x,t)}{\partial x_3}\Big|_{x_3=0} = 2 \int_{\mathbb{R}^2} \frac{\partial^2 \mathcal{E}(x-y')}{\partial x_3^2} \pi_0(y',t)\mathrm{d}y' \Big|_{x_3=0}$$
$$= -2 \int_{\mathbb{R}^2} \sum_{\beta=1}^{2} \frac{\partial^2 \mathcal{E}(x'-y')}{\partial x_\beta^2} \pi_0(y',t)\mathrm{d}y',$$

h_0 can be written in the form $h_0 = \sum_{\beta=1}^{2} \frac{\partial G_\beta}{\partial x_\beta}$, where

$$G_\beta(x',t) = \frac{2}{\rho^+} \int_{\mathbb{R}^2} \frac{\partial \mathcal{E}(x'-y')}{\partial x_\beta} \pi_0(y',t)\mathrm{d}y' + \Big[\nu^\pm \Big(\frac{\partial v_3}{\partial x_\beta} - \frac{\partial v_\beta}{\partial x_3} \Big) \Big] \Big|_{x_3=0}, \quad \beta = 1,2.$$

Therefore, by the results of Sec. 7 in [76], one has

$$\langle p_1 \rangle^{(\gamma)}_{\mathbb{R}^3_+} \leqslant c \langle \pi_0 \rangle^{(\gamma)}_{\mathbb{R}^2}, \qquad \gamma \in (0,1),$$

$$\langle p_2 \rangle^{(\gamma)}_{\mathbb{R}^3_- \cup \mathbb{R}^3_+} \leqslant c \sum_{\beta=1}^{2} \langle G_\beta \rangle^{(\gamma)}_{\mathbb{R}^2} \leqslant c \Big\{ \langle \pi_0 \rangle^{(\gamma)}_{\mathbb{R}^2} + \langle \nabla \boldsymbol{v} \rangle^{(\gamma)}_{\mathbb{R}^3_- \cup \mathbb{R}^3_+} \Big\} \tag{3.4.20}$$

for any $t < T$. By applying these inequalities to the increments of p_1 and p_2 with respect of t, we deduce that

$$\langle \Delta_t(h) p \rangle^{(\gamma)}_{\mathbb{R}^3_- \cup \mathbb{R}^3_+} \leqslant c \Big\{ \langle \Delta_t(h) \pi_0 \rangle^{(\gamma)}_{\mathbb{R}^2} + \langle \Delta_t(h) \nabla \boldsymbol{v} \rangle^{(\gamma)}_{\mathbb{R}^3_- \cup \mathbb{R}^3_+} \Big\},$$

any $h \in (0, T)$ and $t \in (0, T-h)$. Dividing this inequality by $h^{\frac{1+\alpha-\gamma}{2}}$ and taking supremum over $t \in (0, T-h)$ and $h \in (0, T)$, we conclude that

$$\langle p\rangle_{D_T^3}^{(\gamma,1+\alpha)} \leqslant c\Big\{\langle \pi_0\rangle_{\mathbb{R}_T^2}^{(\gamma,1+\alpha)} + \langle \nabla \boldsymbol{v}\rangle_{D_T^3}^{(\gamma,1+\alpha)}\Big\}$$
$$\leqslant c\Big\{\langle a_3\rangle_{\mathbb{R}_T^2}^{(\gamma,1+\alpha)} + \langle d_3\rangle_{\mathbb{R}_T^2}^{(1+\alpha,\frac{1+\alpha}{2})} + \langle \nabla \boldsymbol{v}\rangle_{D_T^3}^{(1+\alpha,\frac{1+\alpha}{2})}\Big\}.$$

Due to (3.4.10), (3.2.6), and (3.4.19), we arrive at the final estimate

$$\langle p\rangle_{D_T^3}^{(\gamma,1+\alpha)} \leqslant c\Big\{\sum_{\beta=1}^{2}\langle a_\beta\rangle_{\mathbb{R}_T^2}^{(1+\alpha,\frac{1+\alpha}{2})} + \langle \nabla' a_3\rangle_{\mathbb{R}_T^2}^{(\alpha,\frac{\alpha}{2})} + \langle a_3\rangle_{\mathbb{R}_T^2}^{(\gamma,1+\alpha)}\Big\}. \tag{3.4.21}$$

Next, we note that

$$\langle \nabla p\rangle_{t,D_T^3}^{(\alpha/2)} \leqslant c\Big\{\langle \nabla p\rangle_{x,D_T^3}^{(\alpha)} + \langle p\rangle_{D_T^3}^{(\gamma,1+\alpha)}\Big\}. \tag{3.4.22}$$

Indeed, by Golovkin's theorem on the equivalent norms in the Hölder spaces (Theorem 1 from [34], see also [35]), we have

$$\Big\langle \frac{\partial p}{\partial x_i}\Big\rangle_{t,D_T^3}^{(\alpha/2)} \leqslant \sup_{(x,t)\in D_T^3}\ \sup_{h_1,h_2\in(0,\infty)} \left|\frac{\triangle_t(h_1)\,\triangle_{x_i}^2(h_2)p(x,t)}{h_1^{\alpha/2}h_2}\right|$$
$$\leqslant 2 \sup_{(x,t)\in D_T^3}\ \sup_{h_1,h_2;\,h_1<h_2^2} \left|\frac{\triangle_t(h_1)\,\triangle_{x_i}(h_2)p(x,t)}{h_1^{\frac{\alpha}{2}}h_1^{\frac{1-\gamma}{2}}h_2^{\gamma}}\right|$$
$$+2 \sup_{(x,t)\in D_T^3}\ \sup_{h_2<\sqrt{h_1}} \left|\frac{\triangle_{x_i}^2(h_2)p(x,t)}{h_2^{1+\alpha}}\right|$$
$$\leqslant 2\langle p\rangle_{D_T^3}^{(\gamma,1+\alpha)} + 2\Big\langle \frac{\partial p}{\partial x_i}\Big\rangle_{x_i,D_T^3}^{(\alpha)}, \qquad i=1,2,3; \quad \gamma\in(0,1).$$

Hence, by using the first equation in (3.3.1) and inequality (3.2.7), we obtain

$$\langle \mathcal{D}_t\boldsymbol{v}\rangle_{t,D_T^3}^{(\alpha/2)} \leqslant c\Big\{\langle \Delta\boldsymbol{v}\rangle_{t,D_T^3}^{(\alpha/2)} + \langle \nabla p\rangle_{t,D_T^3}^{(\alpha/2)}\Big\}$$
$$\leqslant c\Big\{\langle \boldsymbol{v}\rangle_{x,D_T^3}^{(2+\alpha)} + \langle \mathcal{D}_t\boldsymbol{v}\rangle_{x,D_T^3}^{(\alpha)} + \langle \nabla p\rangle_{x,D_T^3}^{(\alpha)} + \langle p\rangle_{D_T^3}^{(\gamma,1+\alpha)}\Big\},$$

which together with (3.2.6), (3.2.7), (3.4.19) and (3.4.21) implies inequality (3.3.2).

If the data do not decrease at infinity but have necessary norms, we can approximate them by functions with power-law decay, obtain uniform estimates, and then pass to the limit as has been done in Chap. 2 ([28]). In addition, following Sect. 2.4, we can prove that the pressure p is bounded by a power function:

$$\max_{t\in(0,T)} |p(x,t)| \leqslant c(T)|x-x_0|^{\gamma},$$

$$\langle p(x)\rangle_{(0,T)}^{(\frac{1+\alpha-\gamma}{2})} \leqslant |x-x_0|^{\gamma}\langle p\rangle_{D_T^3}^{(\gamma,1+\alpha)}, \quad \gamma\in(0,1),$$

where x_0 is a point in $\mathbb{R}^3_-$.

3.5 The Proof of Theorem 3.1.1

Proving Theorem 3.1.1, we basically repeat the arguments of the proof of Theorem 2.4.1 from Chap. 2. First, we reduce problem (3.1.1) to a problem with zero initial data. In Chap. 2, we have constructed a vector $\boldsymbol{U}\in C^{2+\alpha,1+\alpha/2}(D_T)$ such that

$$\boldsymbol{U}\big|_{t=0}=\boldsymbol{v}_0, \quad [\boldsymbol{U}]\big|_{x_3=0}=0, \quad \mathcal{D}_t\boldsymbol{U}\big|_{t=0}=\nu^{\pm}\Delta\boldsymbol{v}_0-\frac{1}{\rho^{\pm}}\nabla p_0+\boldsymbol{f}\big|_{t=0}, \tag{3.5.1}$$

$$\langle\boldsymbol{U}\rangle_{D_T^3}^{(2+\alpha,1+\alpha/2)} \leqslant c\Big\{\langle\boldsymbol{v}_0\rangle_{\cup\mathbb{R}^3_\pm}^{(2+\alpha)}+\langle p_0\rangle_{\cup\mathbb{R}^3_\pm}^{(1+\alpha)}+\langle\boldsymbol{f}|_{t=0}\rangle_{\cup\mathbb{R}^3_\pm}^{(\alpha)}\Big\}, \tag{3.5.2}$$

and a function $P_0\in C^{1+\alpha,\frac{1+\alpha}{2}}(D_T^3)$ such that $P_0|_{t=0}=p_0$ in $\mathbb{R}^3_+\cup\mathbb{R}^3_-$,

$$\langle P_0\rangle_{D_T^3}^{(1+\alpha,\frac{1+\alpha}{2})} \leqslant c\langle p_0\rangle_{\cup\mathbb{R}^3_\pm}^{(1+\alpha)}. \tag{3.5.3}$$

Being a solution to problem (2.5.3), p_0 has been estimated in Chap. 2 (inequality (2.5.6)) as follows:

$$\langle p_0\rangle_{\cup\mathbb{R}^3_\pm}^{(1+\alpha)} \leqslant c\Big\{\langle\boldsymbol{f}(\cdot,0)\rangle_{\cup\mathbb{R}^3_\pm}^{(\alpha)}+\langle g(\cdot,0)\rangle_{\cup\mathbb{R}^3_\pm}^{(1+\alpha)}+\langle\mathcal{D}_t\boldsymbol{h}(\cdot,0)\rangle_{\cup\mathbb{R}^3_\pm}^{(\alpha)} + \langle\boldsymbol{v}_0\rangle_{\cup\mathbb{R}^3_\pm}^{(2+\alpha)}+\langle b_3(\cdot,0)\rangle_{\mathbb{R}^2}^{(1+\alpha)}\Big\}. \tag{3.5.4}$$

We introduce the new functions $\boldsymbol{V}=\boldsymbol{v}-\boldsymbol{U}$, $s=p-P_0$ which satisfy the problem

$$\mathcal{D}_t\boldsymbol{V}-\nu^{\pm}\Delta\boldsymbol{V}+\frac{1}{\rho^{\pm}}\nabla s=\boldsymbol{f}-\mathcal{D}_t\boldsymbol{U}+\nu^{\pm}\Delta\boldsymbol{U}-\frac{1}{\rho^{\pm}}\nabla P_0\equiv\boldsymbol{F},$$

$$\nabla\cdot\boldsymbol{V}=g-\nabla\cdot\boldsymbol{U}\equiv g' \qquad \text{in} \quad D_T^{\pm}=\mathbb{R}^3_\pm\times(0,T),$$

$$\boldsymbol{V}\big|_{t=0}=0 \quad \text{in} \quad \mathbb{R}^3_-\cup\mathbb{R}^3_+, \quad \boldsymbol{V}\xrightarrow[|x|\to\infty]{}0, \quad [\boldsymbol{V}]\big|_{x_3=0}=0, \tag{3.5.5}$$

$$-\left[\mu^{\pm}\left(\frac{\partial V_\beta}{\partial x_3}+\frac{\partial V_3}{\partial x_\beta}\right)\right]\Big|_{x_3=0}=b_\beta+\left[\mu^{\pm}\left(\frac{\partial U_\beta}{\partial x_3}+\frac{\partial U_3}{\partial x_\beta}\right)\right]\Big|_{x_3=0}\equiv d_\beta,$$

$$\beta=1,2,$$

$$\left[-s+2\mu^{\pm}\frac{\partial V_3}{\partial x_3}\right]\Big|_{x_3=0}=b_3+\left[P_0-2\mu^{\pm}\frac{\partial U_3}{\partial x_3}\right]\Big|_{x_3=0}\equiv d_3 \text{ on } \mathbb{R}^2_T.$$

We note that $g'=\nabla\cdot\boldsymbol{h}'$ and $\boldsymbol{h}'=\boldsymbol{h}-\boldsymbol{U}$ and that $\boldsymbol{F}\in\mathring{\boldsymbol{C}}^{\alpha,\alpha/2}(D^3_T)$, $g'\in\mathring{\boldsymbol{C}}^{1+\alpha,\frac{1+\alpha}{2}}(D^3_T)$, $d_\beta\in\mathring{C}^{1+\alpha,\frac{1+\alpha}{2}}(\mathbb{R}^2_T)$, $\beta=1,2$, and $d_3\in\mathring{C}^{1+\alpha,0}(\mathbb{R}^2_T)\cap\mathring{\boldsymbol{C}}^{(\gamma,1+\alpha)}(\mathbb{R}^2_T)$ because of compatibility conditions (2.5.2), (2.5.3), and (3.5.1)–(3.5.3).

We find a solution of (3.5.5) in the form $\boldsymbol{V}=\boldsymbol{w}+\boldsymbol{w}'+\boldsymbol{u}$, $s=q+\mu^{\pm}g'-\rho^{\pm}\mathcal{D}_t\Phi$, where $\boldsymbol{w}$ is a solution of the problem

$$\begin{aligned}&\mathcal{D}_t\boldsymbol{w}-\nu^{\pm}\Delta\boldsymbol{w}=\boldsymbol{F}\quad\text{in}\quad D^{\pm}_T,\\&\boldsymbol{w}\big|_{t=0}=0,\qquad[\boldsymbol{w}]\big|_{x_3=0}=0;\end{aligned}\tag{3.5.6}$$

the vector field $\boldsymbol{w}'=\nabla\Phi$, where Φ solves diffraction problem for the Poisson equation

$$\begin{aligned}&\Delta\Phi=g'-\nabla\cdot w\equiv g''\quad\text{in}\quad\mathbb{R}^3_-\cup\mathbb{R}^3_+,\\&[\boldsymbol{\Phi}]\big|_{x_3=0}=0,\qquad\left[\frac{\partial\Phi}{\partial x_3}\right]\Big|_{x_3=0}=0;\end{aligned}\tag{3.5.7}$$

and the pair $(\boldsymbol{u},q)$ is a solution of homogeneous problem (3.3.1) with

$$\begin{aligned}&a_\tau=d_\tau+\left[\mu^{\pm}\left(\frac{\partial(w_\tau+w'_\tau)}{\partial x_3}+\frac{\partial(w_3+w'_3)}{\partial x_\tau}\right)\right]\Big|_{x_3=0},\quad\tau=1,2,\\&a_3=d_3-\left[2\mu^{\pm}\frac{\partial(w_3+w'_3)}{\partial x_3}-\mu^{\pm}g''+\rho^{\pm}\mathcal{D}_t\Phi\right]\Big|_{x_3=0}.\end{aligned}\tag{3.5.8}$$

For problem (3.5.6), compatibility conditions

$$[\boldsymbol{w}]\big|_{x_3=0,t=0}=0,\quad[\mathcal{D}_t\boldsymbol{w}]\big|_{x_3=0,t=0}=0$$

hold; therefore, its solution is represented, as in the proof of Theorem 3.2.4, in the form

$$\boldsymbol{w} = \begin{cases} \boldsymbol{u}^+ & \text{in } \mathbb{R}^3_+, \\ \boldsymbol{u}^- & \text{in } \mathbb{R}^3_-. \end{cases}$$

From estimates (3.2.18)–(3.2.20) for $\boldsymbol{u}^\pm$, it follows the inequality

$$\langle \boldsymbol{w} \rangle^{(2+\alpha)}_{x,D^3_T} + \langle \mathcal{D}_t \boldsymbol{w} \rangle^{(\alpha)}_{x,D^3_T} \leqslant c\big\{ \langle \boldsymbol{F} \rangle^{(\alpha)}_{x,D^3_T} + \langle [\boldsymbol{F}]\big|_{x_3=0} \rangle^{(\alpha/2)}_{t,\mathbb{R}^2_T} \big\}. \tag{3.5.9}$$

Diffraction problem (3.5.7) can be reduced to the Dirichlet and Neumann problems in the half-space $\{x_3 > 0\}$ by introducing two new unknown functions

$$\psi(x', x_3) = \Phi(x', x_3) - \Phi(x', -x_3),$$
$$\chi(x', x_3) = \Phi(x', x_3) + \Phi(x', -x_3).$$

For these functions we have

$$\Delta \psi = g''_+ - g''_- \quad \text{in} \quad \mathbb{R}^3_+, \qquad \psi\big|_{x_3=0} = 0 \tag{3.5.10}$$

and

$$\Delta \chi = g''_+ + g''_- \quad \text{in} \quad \mathbb{R}^3_+, \qquad \frac{\partial \chi}{\partial x_3}\Big|_{x_3=0} = 0, \tag{3.5.11}$$

where $g''_+ = g''\big|_{\mathbb{R}^3_+}$ and $g''_-(x', x_3) = g''(x', -x_3)$, $x_3 > 0$. We express the functions ψ and χ in the terms of the Green functions for problems (3.5.10) and (3.5.11):

$$\psi(x,t) = \int_{\mathbb{R}^3_+} G_1(x,y)(g''_+ - g''_-)(y,t)\mathrm{d}y,$$

$$\chi(x,t) = \int_{\mathbb{R}^3_+} G_2(x,y)(g''_+ + g''_-)(y,t)\mathrm{d}y,$$

where $G_1(x,y) = \mathcal{E}(x-y) - \mathcal{E}(x-y^*)$, $G_2(x,y) = \mathcal{E}(x-y) + \mathcal{E}(x-y^*)$, and $y^* = (y', -y_3)$. By Lemma 7.1 in [76], we obtain the inequalities

$$\langle \nabla \psi \rangle^{(2+\alpha)}_{\mathbb{R}^3_+} \leqslant c\big\langle \nabla(g''_+ - g''_-) \big\rangle^{(\alpha)}_{\mathbb{R}^3_+} \leqslant c\langle \nabla g'' \rangle^{(\alpha)}_{\mathbb{R}^3_- \cup \mathbb{R}^3_+}, \tag{3.5.12}$$
$$\langle \nabla \chi \rangle^{(2+\alpha)}_{\mathbb{R}^3_+} \leqslant c\big\langle \nabla(g''_+ + g''_-) \big\rangle^{(\alpha)}_{\mathbb{R}^3_+} \leqslant c\langle \nabla g'' \rangle^{(\alpha)}_{\mathbb{R}^3_- \cup \mathbb{R}^3_+}.$$

By taking the supremum in (3.5.12) over $t \in (0,T)$, we arrive at the required estimates for the semi-norms of $\nabla\psi$, $\nabla\chi$ with respect to the spacial variables. We set

$$\Phi(x',x_3) = \begin{cases} \frac{1}{2}\big(\psi(x)+\chi(x)\big) & \text{in } x_3 \geqslant 0, \\ \frac{1}{2}\big(\chi(x^*)-\psi(x^*)\big) & \text{in } x_3 \leqslant 0, \end{cases}$$

where $x^* = (x', -x_3)$.

Thus, for $\boldsymbol{w}' = \nabla\Phi$, we deduce

$$\langle \boldsymbol{w}' \rangle^{(2+\alpha)}_{x,D^3_T} \leqslant c\langle \nabla(g'')\rangle^{(\alpha)}_{x,D^3_T} \leqslant c\langle \boldsymbol{h}''\rangle^{(2+\alpha)}_{x,D^3_T}, \tag{3.5.13}$$

where $g'' = \nabla\cdot\boldsymbol{h}''$ and $\boldsymbol{h}'' = \boldsymbol{h} - \boldsymbol{U} - \boldsymbol{w}$.

Moreover, we need to evaluate the semi-norm $\langle \mathcal{D}_t \boldsymbol{w}'\rangle^{(\alpha)}_{x,D_T}$. We can do it if we differentiate solution gradients to problems (3.5.10) and (3.5.11) with respect to time and then integrate by parts

$$\frac{\partial \nabla\psi(x,t)}{\partial t} = \nabla \int_{\mathbb{R}^3_+} \Big\{ \sum_{\beta=1}^{2} \frac{\partial G_1(x,y)}{\partial x_\beta} \frac{\partial (h''^{+}_\beta - h''^{-}_\beta)(y,t)}{\partial t} + \frac{\partial G_2(x,y)}{\partial x_3} \frac{\partial (h''^{+}_3 + h''^{-}_3)(y,t)}{\partial t} \Big\} dy,$$

$$\frac{\partial \nabla\chi(x,t)}{\partial t} = \nabla \int_{\mathbb{R}^3_+} \Big\{ \sum_{\beta=1}^{2} \frac{\partial G_2(x,y)}{\partial x_\beta} \frac{\partial (h''^{+}_\beta + h''^{-}_\beta)(y,t)}{\partial t} + \frac{\partial G_1(x,y)}{\partial x_3} \frac{\partial (h''^{+}_3 - h''^{-}_3)(y,t)}{\partial t} \Big\} dy,$$

the surface integrals vanishing because of zero values of $G_1(x,y)\big|_{x_3=0}$ and $(h''^{+}_3 - h''^{-}_3)\big|_{x_3=0} = [h_3]\big|_{x_3=0}$. The vectors $\boldsymbol{h}''^{\pm}$ are defined in the same way as $g''_{\pm}$.

As above, we have

$$\begin{aligned} \langle \mathcal{D}_t \boldsymbol{w}'\rangle^{(\alpha)}_{x,D^3_T} &\leqslant \langle \mathcal{D}_t\nabla\psi\rangle^{(\alpha)}_{x,\mathbb{R}^3_+\times(0,T)} + \langle \mathcal{D}_t\nabla\chi\rangle^{(\alpha)}_{x,\mathbb{R}^3_+\times(0,T)} \leqslant c\langle \mathcal{D}_t\boldsymbol{h}''\rangle^{(\alpha)}_{x,D^3_T} \\ &\leqslant c\Big\{\langle \mathcal{D}_t\boldsymbol{h}\rangle^{(\alpha)}_{x,D^3_T} + \langle \mathcal{D}_t\boldsymbol{U}\rangle^{(\alpha)}_{x,D^3_T} + \langle \mathcal{D}_t\boldsymbol{w}\rangle^{(\alpha)}_{x,D^3_T}\Big\}. \end{aligned} \tag{3.5.14}$$

In order to estimate the solution $(\boldsymbol{u}, q)$ of problem (3.3.1) with boundary functions (3.5.8), we apply Theorem 3.3.1. We note that $a_i\big|_{t=0} = 0$, $i = 1,2,3$. Indeed, d_i and $i = 1,2,3$ vanish when $t = 0$; in addition, $\boldsymbol{w}|_{t=0} = \boldsymbol{w}'|_{t=0} = 0$,

$g''|_{t=0} = 0$, and $\mathcal{D}_t\Phi(x,0) = 0$, in view of the relation

$$\mathcal{D}_t g''\big|_{t=0} = \mathcal{D}_t g\big|_{t=0} - \nabla\cdot\big(\nu^{\pm}\Delta \boldsymbol{v}_0 - \frac{1}{\rho\pm}\nabla p_0 + \boldsymbol{f}\big|_{t=0}\big) = 0$$

which follows from (3.5.1) and (2.5.3).

Now we estimate the norms of the functions a_i:

$$\sum_{\beta=1}^{2}\langle a_\beta\rangle_{\mathbb{R}^2_T}^{(1+\alpha,\frac{1+\alpha}{2})} + \big\langle\nabla' a_3\big\rangle_{x',\mathbb{R}^2_T}^{(\alpha)} + \big\langle a_3\big\rangle_{\mathbb{R}^2_T}^{(\gamma,1+\alpha)}$$

$$\leqslant \sum_{\beta=1}^{2}\langle b_\beta\rangle_{\mathbb{R}^2_T}^{(1+\alpha,\frac{1+\alpha}{2})} + \big\langle\nabla' b_3\big\rangle_{x',\mathbb{R}^2_T}^{(\alpha)} + \big\langle b_3\big\rangle_{\mathbb{R}^2_T}^{(\gamma,1+\alpha)} + c\Big\{\langle \boldsymbol{U}\rangle_{D^3_T}^{(2+\alpha,1+\alpha/2)}$$

$$+ \langle P_0\rangle_{D^3_T}^{(1+\alpha,\frac{1+\alpha}{2})} + \langle\nabla(\boldsymbol{w}+\boldsymbol{w}')\rangle_{D^3_T}^{(1+\alpha,\frac{1+\alpha}{2})} + \langle g'\rangle_{x,D^3_T}^{(1+\alpha)} + \langle g'\rangle_{\mathbb{R}^2_T}^{(\gamma,1+\alpha)}$$

$$+ \langle \mathcal{D}_t\Phi\rangle_{x,D^3_T}^{(1+\alpha)} + \langle \mathcal{D}_t\Phi\rangle_{D^3_T}^{(\gamma,1+\alpha)}\Big\}. \tag{3.5.15}$$

It is clear that

$$\langle g'\rangle_{\mathbb{R}^2_T}^{(\gamma,1+\alpha)} \leqslant \langle g'\rangle_{D^3_T}^{(1+\alpha,\frac{1+\alpha}{2})}, \quad \langle \mathcal{D}_t\Phi\rangle_{x,D^3_T}^{(1+\alpha)} \leqslant \langle \mathcal{D}_t\boldsymbol{w}'\rangle_{x,D^3_T}^{(\alpha)}. \tag{3.5.16}$$

It remains to evaluate the semi-norm $\langle \mathcal{D}_t\Phi\rangle_{D^3_T}^{(\gamma,1+\alpha)}$. We do it by estimating the semi-norms of ψ and χ. The derivative $\mathcal{D}_t\boldsymbol{h}''$ has divergence form $\nabla\cdot\mathbb{H}''$:

$$\mathcal{D}_t\boldsymbol{h}'' = \mathcal{D}_t\boldsymbol{h} - \mathcal{D}_t\boldsymbol{U} - \mathcal{D}_t\boldsymbol{w} = \mathcal{D}_t\boldsymbol{h} - \nu^{\pm}\Delta\boldsymbol{w} - \boldsymbol{f} - \nu^{\pm}\Delta\boldsymbol{U} + \frac{1}{\rho^{\pm}}\nabla P_0,$$

i.e., the tensor $\mathbb{H}''$ has the elements $H''_{ij} = H_{ij} - \nu^{\pm}\partial(w_i+U_i)/\partial x_j + \frac{1}{\rho^{\pm}}\delta^i_j P_0$, where $\{\delta^i_j\}^3_{i,j=1}$ are the Kronecker deltas. Hence

$$\frac{\partial\psi(x,t)}{\partial t} = \int_{\mathbb{R}^3_+}\Big\{\sum_{\beta=1}^{2}\frac{\partial G_1(x,y)}{\partial x_\beta}\frac{\partial(h''^{+}_\beta - h''^{-}_\beta)(y,t)}{\partial t}$$

$$+ \frac{\partial G_2(x,y)}{\partial x_3}\frac{\partial(h''^{+}_3 + h''^{-}_3)(y,t)}{\partial t}\Big\}\mathrm{d}y$$

$$= \sum_{\beta,\mu=1}^{2}\frac{\partial}{\partial x_\beta}\int_{\mathbb{R}^3_+}\Big\{\frac{\partial G_1(x,y)}{\partial x_\mu}(H''^{+}_{\beta\mu} - H''^{-}_{\beta\mu})(y,t)$$

$$+\frac{\partial G_2(x,y)}{\partial x_3}(H''^{+}_{\beta 3}+H''^{-}_{\beta 3})(y,t)\Big\}\mathrm{d}y$$

$$+\frac{\partial}{\partial x_3}\int_{\mathbb{R}^3_+}\Big\{\sum_{\mu=1}^{2}\frac{\partial G_2(x,y)}{\partial x_\mu}(H''^{+}_{3\mu}+H''^{-}_{3\mu})(y,t)$$

$$+\frac{\partial G_1(x,y)}{\partial x_3}(H''^{+}_{33}-H''^{-}_{33})(y,t)\Big\}\mathrm{d}y$$

$$-\frac{\partial}{\partial x_3}\int_{\mathbb{R}^2}G_2(x,y')[H''_{33}]\big|_{y_3=0}\mathrm{d}y',$$

where $H''^{+}_{ij}\equiv H''_{ij}\big|_{\mathbb{R}^3_+}$, $H''^{-}_{ij}(x',x_3,t)\equiv H''_{ij}(x',-x_3,t)$, and $x_3>0$. We can express the function $\mathcal{D}_t\chi$ in a similar way. By Lemma 7.1 in [76], we estimate first the semi-norms

$$\langle\mathcal{D}_t\psi(\cdot,t)\rangle^{(\gamma)}_{\mathbb{R}^3_+}+\langle\mathcal{D}_t\chi(\cdot,t)\rangle^{(\gamma)}_{\mathbb{R}^3_+}\leqslant c\sum_{l,k=1}^{3}\langle H''_{lk}(\cdot,t)\rangle^{(\gamma)}_{\mathbb{R}^3_-\cup\mathbb{R}^3_+},\quad\forall t\in(0,T).$$

By applying then this inequality to the increments $\Delta_t(h)\mathcal{D}_t\psi$ and $\Delta_t(h)\mathcal{D}_t\chi$, dividing them by $h^{\frac{1+\alpha-\gamma}{2}}$, and taking supremum over $h\in(0,T)$ and $t\in(0,T-h)$, we obtain

$$\langle\mathcal{D}_t\Phi\rangle^{(\gamma,1+\alpha)}_{D^3_T}\leqslant\langle\mathcal{D}_t\psi\rangle^{(\gamma,1+\alpha)}_{D^3_T}+\langle\mathcal{D}_t\chi\rangle^{(\gamma,1+\alpha)}_{D^3_T}\leqslant c\,\langle\mathbb{H}''\rangle^{(\gamma,1+\alpha)}_{D^3_T}.\tag{3.5.17}$$

Now by Theorem 3.3.1, we conclude that there exists a solution $(\boldsymbol{u},q)$ of problem (3.3.1) and (3.5.8) such that $\boldsymbol{u}\in\mathring{\boldsymbol{C}}^{2+\alpha,1+\alpha/2}(D^3_T)$, $q\in\mathring{C}^{(\gamma,1+\alpha)}(D^3_T)$, and $\nabla q\in\mathring{\boldsymbol{C}}^{\alpha,\alpha/2}(D^3_T)$, and for which the inequalities

$$\langle\boldsymbol{u}\rangle^{(2+\alpha,1+\alpha/2)}_{D^3_T}+\langle\nabla q\rangle^{(\alpha)}_{x,D^3_T}+\langle q\rangle^{(\gamma,1+\alpha)}_{D^3_T}$$

$$\leqslant c\Big\{\sum_{\beta=1}^{2}\langle a_\beta\rangle^{(1+\alpha,\frac{1+\alpha}{2})}_{\mathbb{R}^2_T}+\langle\nabla' a_3\rangle^{(\alpha)}_{x',\mathbb{R}^2_T}+\langle a_3\rangle^{(\gamma,1+\alpha)}_{\mathbb{R}^2_T}\Big\}\leqslant cF(T)\tag{3.5.18}$$

hold. Here $cF(T)$ is the right-hand side of (3.1.2). In the last inequality of (3.5.18), we have taken estimates (3.5.2)–(3.5.4) and (3.5.13)–(3.5.17) into account.

Hence, there also exists a pair $(\boldsymbol{V},s)$ which solves problem (3.5.5). Inequalities (3.5.9), (3.5.13), (3.5.14), and (3.5.18) imply that

$$\langle \boldsymbol{V}\rangle^{(2+\alpha)}_{x,D^3_T} + \langle \mathcal{D}_t \boldsymbol{V}\rangle^{(\alpha)}_{x,D^3_T} \leqslant cF(T). \tag{3.5.19}$$

As to the pressure s, one has for it

$$\begin{aligned}\langle \nabla s\rangle^{(\alpha)}_{x,D^3_T} + \langle s\rangle^{(\gamma,1+\alpha)}_{D^3_T} &\leqslant \langle \nabla q\rangle^{(\alpha)}_{x,D^3_T} + \langle q\rangle^{(\gamma,1+\alpha)}_{D^3_T}\\ &\quad + c\Big\{\langle g'\rangle^{(1+\alpha,\frac{1+\alpha}{2})}_{D^3_T} + \big\langle \nabla \mathcal{D}_t \Phi\big\rangle^{(\alpha)}_{x,D^3_T} + \langle \mathcal{D}_t \Phi\rangle^{(\gamma,1+\alpha)}_{D^3_T}\Big\}\\ &\leqslant cF(T) \end{aligned} \tag{3.5.20}$$

due to (3.5.2), (3.5.3), (3.5.14), and (3.5.17). Making use of inequality (3.4.22) for s and of the first equation in (3.5.5), as well as of (3.5.19) and (3.5.20), we deduce that

$$\big\langle \mathcal{D}_t \boldsymbol{V}\big\rangle^{(\alpha/2)}_{t,D^3_T} \leqslant cF(T).$$

Consequently, the vector field $\boldsymbol{V} \in \mathring{\boldsymbol{C}}^{2+\alpha,1+\alpha/2}(D^3_T)$, and the function $s \in \mathring{C}^{(\gamma,1+\alpha)}(D^3_T)$ and $\nabla s \in \mathring{\boldsymbol{C}}^{\alpha,\alpha/2}(D^3_T)$. Thus, problem (3.1.1) is solvable. By virtue of (3.5.2) and (3.5.3) and of the equivalence of the semi-norm $\langle\cdot\rangle^{(2+\alpha,1+\frac{\alpha}{2})}_{D^3_T}$ and the complete norm $|\cdot|^{(2+\alpha,1+\alpha/2)}_{D^3_T}$ in the space $\mathring{\boldsymbol{C}}^{2+\alpha,1+\alpha/2}$, we conclude that estimate (3.1.2) holds for $(\boldsymbol{v},p)$. We note that, as a consequence of Theorem 3.3.1, the growth of the functions q, s, and p at ∞ is bounded by$|x|^\gamma$, $\gamma < 1$.

The uniqueness of the solution $(\boldsymbol{v},p)$ is proved by multiplying a solution to problem (3.1.1) with zero data by cutoff function (3.2.21), by analogy with the proof of solution uniqueness for problem (3.2.1). Unique solvability of the problem follows from the existence of a unique solution with compact support to it (see Sect. 10.2) which coincides with the solution constructed above. Obviously, system (3.1.1) and inequality (3.1.2) admit the fact that the pressure function is not defined uniquely but up to a bounded function depending only on t. This procedure has been realized in Chap. 2 for a system like (3.1.1) but containing an additional term with the coefficient of surface tension in the last boundary condition.

Chapter 4
A Linear Problem with Closed Interface Under Nonnegative Surface Tension

Abstract In this chapter, the solvability of linear problem (1.1.7) is investigated. A single existence theorem in the Hölder spaces is formulated for all values $\sigma \geqslant 0$.

4.1 Auxiliary Propositions: The Statement of Results

Here are three Lemmas used in the following sections. They concern the Newtonian potential

$$u(x,t) = \int_\Omega f(y,t)\mathcal{E}(x,y)\mathrm{d}y, \tag{4.1.1}$$

where $\mathcal{E}(x,y) = -1/(4\pi|x-y|)$ is the fundamental solution of the Laplace equation in $\mathbb{R}^3$. When Ω is unbounded, we assume that the function $f(x,t)$ decreases as $x \to \infty$ so quickly that the integral (4.1.1) is convergent. The following statement is true (Lemma 2.4 in [54]).

Lemma 4.1.1 *We assume that the support of f lies in the ball $B_\lambda(x_0) = \{|x-x_0| \leqslant \lambda\}$, where $x_0 \in \Omega$, $\lambda \leqslant \lambda_0 < \infty$, $Q_T \equiv \Omega \times (0,T)$, $T > 0$. Let $\alpha, \gamma \in (0,1)$, $\gamma < \alpha$. Then*

$$|u|_{Q_T}^{(\alpha,\frac{\alpha}{2})} \leqslant c_2(\lambda_0)\lambda^{2-\alpha}\big(|f|_{Q_T} + \langle f\rangle_{t,Q_T}^{(\alpha/2)}\big),$$

$$|\nabla u|_{Q_T}^{(\alpha,\frac{\alpha}{2})} \leqslant c_3\lambda^{1-\alpha}\big(|f|_{Q_T} + \langle f\rangle_{t,Q_T}^{(\alpha/2)}\big),$$

$$|u|_{Q_T}^{(\gamma,1+\alpha)} + \lambda|\nabla u|_{Q_T}^{(\gamma,1+\alpha)} \leqslant c_4\lambda^{2-\gamma}\langle f\rangle_{t,Q_T}^{(\frac{1+\alpha-\gamma}{2})}.$$

These inequalities follow from the evident estimate

$$|u(\cdot,t)|_\Omega + \lambda|\nabla u(\cdot,t)|_\Omega + \lambda^\alpha|u(\cdot,t)|_\Omega^{(\alpha)} + \lambda^{1+\alpha}|\nabla u(\cdot,t)|_\Omega^{(\alpha)} \leqslant c\lambda^2|f(\cdot,t)|_\Omega,$$

© The Author(s), under exclusive license to Springer Nature Switzerland AG 2021

I. V. Denisova, V. A. Solonnikov, *Motion of a Drop in an Incompressible Fluid*, Advances in Mathematical Fluid Mechanics,
https://doi.org/10.1007/978-3-030-70053-9_4

applied to $u(x,t)$ or to the difference $u(x,t) - u(x,t')$.

Now let Ω be a domain with a compact boundary S. We consider single-layer potential

$$w(x,t) = \int_S \psi(y,t)\mathcal{E}(x,y)\mathrm{d}S(y).$$

Lemma 4.1.2 *Let $f \in C^\gamma(\Omega)$, $\psi \in C^\gamma(\Omega)$, $\gamma < 1$, then*

$$\langle u\rangle_\Omega^{(2+\gamma)} \leqslant c|f|_\Omega^{(\gamma)}, \qquad |w|_\Omega^{(1+\gamma)} \leqslant c|\psi|_S^{(\gamma)}. \tag{4.1.2}$$

Moreover, if $f(x) = \nabla \cdot \boldsymbol{R}(x) + f_0(x)$ with supp $f_0 \subset B_\lambda(x_0)$, $x_0 \in \Omega$, *and* $\langle \boldsymbol{R}\rangle_\Omega^{(\beta)} < \infty$, $\beta < 1$, *then*

$$|\mathcal{D}_x^{\boldsymbol{\mu}} u(x)|_\Omega \leqslant c\big\{\langle f\rangle_\Omega^{(\gamma)} + (1+\lambda^\varepsilon)|f_0|_\Omega + \langle \boldsymbol{R}\rangle_\Omega^{(\beta)}\big\}, \tag{4.1.3}$$

where $|\boldsymbol{\mu}| = 2$, $\varepsilon \in (0,3)$.

Inequalities (4.1.2) are well-known, and (4.1.3) was proved in [76].

Remark 4.1.1 *It is easy to see that for two domains Ω^- and Ω^+, separated by a closed interface Γ, inequalities* (4.1.2) *imply the estimates*

$$\langle u\rangle_{x,D_T}^{(2+\gamma)} \leqslant c|f|_{x,D_T}^{(\gamma)}, \qquad |w|_{x,D_T}^{(1+\gamma)} \leqslant c|\psi|_{x,G_T}^{(\gamma)}. \tag{4.1.4}$$

Here and in the following $D_T \equiv (\Omega^+ \cup \Omega^-) \times (0,T)$, $G_T \equiv \Gamma \times (0,T)$; $u(x,t) = \int_{\Omega^+\cup\Omega^-} f(y,t)\mathcal{E}(x,y)\mathrm{d}y$.

Inequalities (4.1.2) and (4.1.3) applied to the difference $u(x,t) - u(x,t')$ yield

$$\langle \mathcal{D}_x^{\boldsymbol{\mu}} u\rangle_{D_T}^{(\gamma,1+\alpha)} \leqslant c_9|f|_{D_T}^{(\gamma,1+\alpha)}, \qquad |\nabla w|_{D_T}^{(\gamma,1+\alpha)} \leqslant c_{10}|\psi|_{G_T}^{(\gamma,1+\alpha)},$$

$$\langle \mathcal{D}_x^{\boldsymbol{\mu}} u\rangle_{t,D_T}^{(\frac{1+\alpha-\gamma}{2})} \leqslant c_{11}\big(|f|_{D_T}^{(\gamma,1+\alpha)} + \langle \boldsymbol{R}\rangle_{D_T}^{(\gamma,1+\alpha)}\big) \tag{4.1.5}$$

In addition

$$\langle \nabla w\rangle_{t,D_T}^{(\alpha/2)} \leqslant c_{10}\big|\langle \psi\rangle_{(0,T)}^{(\alpha/2)}\big|_\Gamma^{(\beta)}, \qquad \langle \mathcal{D}_x^{\boldsymbol{\mu}} u\rangle_{t,D_T}^{(\alpha/2)} \leqslant c_{11}\big|\langle f_0\rangle_{(0,T)}^{(\alpha/2)}\big|_{\cup\Omega^\pm}^{(\beta)}, \ \forall \beta < 1, \tag{4.1.6}$$

Lemma 4.1.3 *Let vector $\boldsymbol{h} \in C^{\alpha,\alpha/2}(D_T)$ be representable in the form $\boldsymbol{h} = \nabla \cdot \mathbb{H}$, the components of the tensor $\mathbb{H}$ being such that $H_{ik} \in C^{(\gamma,1+\alpha)}(D_T)$, $i,k = 1,2,3$, $\alpha, \gamma \in (0,1)$. Then for the vector-field*

$$\boldsymbol{g} = \nabla_x \int_{\Omega^-\cup\Omega^+} \nabla_y \mathcal{E} \cdot \boldsymbol{h}\mathrm{d}y$$

the estimate

$$\langle \boldsymbol{g}\rangle_{t,D_T}^{(\alpha/2)} \leqslant c_{12}\big(\langle \boldsymbol{h}\rangle_{D_T}^{(\alpha,\alpha/2)} + \max_{i,k} \langle H_{ik}\rangle_{D_T}^{(\gamma,1+\alpha)}\big)$$

holds.

This Lemma is deduced in the same way as a similar result obtained in Chap. 2 for functions (2.5.19) in the case of a flat interface between Ω^- and Ω^+ (inequality (2.5.20)).

We consider now to the system (1.1.7). First, we assume that there exist two representations

$$r = \nabla \cdot \boldsymbol{R} \text{ and } \mathcal{D}_t \boldsymbol{R} - \boldsymbol{f} \equiv \boldsymbol{g} = \nabla \cdot \mathbb{M} = \sum_{k=1}^{3} \partial M_{ik}/\partial x_k. \tag{4.1.7}$$

Second, we suppose that compatibility conditions

$$[\boldsymbol{v}_0]|_\Gamma = 0,\ \nabla \cdot \boldsymbol{v}_0(x) = r(x,0),\ x \in \Omega^- \cup \Omega^+,$$

$$[\mu^\pm \Pi_0 \mathbb{S}(\boldsymbol{v}_0(x))\boldsymbol{n}]|_{x\in\Gamma} = \boldsymbol{b}(x,0),\ x \in \Gamma, \tag{4.1.8}$$

$$\big[\Pi_0\big(\boldsymbol{f}(x,0) - \frac{1}{\rho^\pm}\nabla p(x,0) + \nu^\pm \Delta \boldsymbol{v}_0(x)\big)\big]\big|_{x\in\Gamma} = 0$$

hold. The last of these conditions appears due to the need of making the jump of the derivative $\mathcal{D}_t \boldsymbol{v}$ compatible with zero at $t=0:\ [\mathcal{D}_t \boldsymbol{v}]|_\Gamma = 0$. The normal part of this equality $[\mathcal{D}_t \boldsymbol{v}\cdot \boldsymbol{n}]|_{\Gamma,t=0} = 0$ is valid if we assume that $p(x,0) \equiv p_0$ is a solution to the problem

$$\frac{1}{\rho^\pm}\Delta p_0(x) = \nabla \cdot \big(\boldsymbol{f}(x,0) + \nu^\pm \nabla r(x,0) - \mathcal{D}_t \boldsymbol{R}(x,0)\big) \equiv \nabla \cdot \boldsymbol{h},$$

$$[p_0]|_\Gamma = \Big[2\mu^\pm \frac{\partial \boldsymbol{v}_0}{\partial \boldsymbol{n}} \cdot \boldsymbol{n}\Big]\Big|_\Gamma - b'|_{t=0} \equiv p_{00}, \tag{4.1.9}$$

$$\Big[\frac{1}{\rho^\pm}\frac{\partial p_0}{\partial \boldsymbol{n}}\Big]\Big|_\Gamma = \big[\boldsymbol{n}\cdot\big(\boldsymbol{f}(x,0) + \nu^\pm \Delta \boldsymbol{v}_0\big)\big]\big|_\Gamma \equiv p_{01} \qquad \Big(\frac{\partial \psi}{\partial \boldsymbol{n}} \equiv \nabla\psi \cdot \boldsymbol{n}\Big).$$

Theorem 4.1.1 *Let the three above assumptions be fulfilled, and, in addition, for* $\alpha, \gamma \in (0,1)$ *and* $T < \infty$, *let* $\Gamma \in C^{2+\alpha}$, $\boldsymbol{f} \in \boldsymbol{C}^{\alpha,\alpha/2}(D_T)$, $r \in C^{1+\alpha,\frac{1+\alpha}{2}}(D_T)$, $\boldsymbol{R} \in \boldsymbol{C}^{2+\alpha,1+\alpha/2}(D_T)$, $[\boldsymbol{R}\cdot\boldsymbol{n}]\big|_{G_T} = 0$, $\boldsymbol{v}_0 \in \boldsymbol{C}^{2+\alpha}(\Omega^- \cup \Omega^+)$, $\boldsymbol{b} \in \boldsymbol{C}_{\boldsymbol{n}}^{1+\alpha,\frac{1+a}{2}}(G_T)$, $b' \in C^{(\gamma,1+\alpha)}(G_T)$, $B \in C^{\alpha,\alpha/2}(G_T)$, *and tensor* $\mathbb{M}$ *elements have finite semi-norms* $|M_{ik}|_{D_T}^{(\gamma,1+\alpha)}$, $\langle M_{ik}\rangle_{x,D_T}^{(\gamma)}$, $i,k = 1,2,3$. *We suppose also that all the known functions decrease well enough for* $|x| \to \infty$

(for example, in power-law way). Then for a solution $(\boldsymbol{v}, p)$ *to problem* (1.1.7) *such that* $\boldsymbol{v} \in \boldsymbol{C}^{2+\alpha,1+\alpha/2}(D_T)$, $\nabla p \in \boldsymbol{C}^{\alpha,\alpha/2}(D_T)$, $p \in C^{(\gamma,1+\alpha)}(B_T)$, *the inequality*

$$\begin{aligned}
&|\boldsymbol{v}|_{D_T}^{(2+\alpha,1+\alpha/2)} + |\nabla p|_{D_T}^{(\alpha,\alpha/2)} + \langle p\rangle_{D_T}^{(\gamma,1+\alpha)} + |p|_{t,B_T}^{(\frac{1+\alpha-\gamma}{2})} \\
&\leqslant c_{13}(T)\big(|\boldsymbol{f}|_{D_T}^{(\alpha,\frac{\alpha}{2})} + |r|_{D_T}^{(1+\alpha,\frac{1+\alpha}{2})} + |\boldsymbol{R}|_{D_T}^{(2+\alpha,1+\frac{\alpha}{2})} + |\boldsymbol{b}|_{G_T}^{(1+\alpha,\frac{1+\alpha}{2})} + |b'|_{G_T}^{(\gamma,1+\alpha)} \\
&+|b'|_{G_T} + |\nabla_\tau b'|_{G_T}^{(\alpha,\frac{\alpha}{2})} + \sigma|B|_{G_T}^{(\alpha,\frac{\alpha}{2})} + |\mathbb{M}|_{D_T}^{(\gamma,1+\alpha)} + \langle\mathbb{M}\rangle_{x,D_T}^{(\gamma)} + |\boldsymbol{v}_0|_{\cup\Omega^\pm}^{(2+\alpha)}\big) \\
&\equiv c_{13}(T)F(T) \qquad (4.1.10)
\end{aligned}$$

holds, where $\nabla_\tau = \Pi\nabla, c_{13}(T)$ *is a nondecreasing function of* T, $B_T = B_1 \times (0,T)$, *and* B_1 *is a ball containing* $\overline{\Omega^+}$.

Theorem 4.1.2 *We assume that* $\sigma \geqslant 0$ *and the hypotheses of Theorem* 4.1.1 *are fulfilled. Let, in addition,* $\gamma < \alpha$ *when* $\sigma > 0$. *Then system* (1.1.7) *has a solution* $\boldsymbol{v} \in \boldsymbol{C}^{2+\alpha,1+\alpha/2}(D_T)$, $p \in C^{(\gamma,1+\alpha)}(B_T)$ *with the bounded seminorm* $\langle p\rangle_{D_T}^{(\gamma,1+\alpha)}$ *and* $\nabla p \in \boldsymbol{C}^{\alpha,\alpha/2}(D_T)$; *moreover, the velocity vector field* $\boldsymbol{v}$ *is determined uniquely, and the pressure* p *is defined in the class of functions of weak power-law growth up to a smooth bounded time-dependent function.*

4.2 A Priori Estimates of the Solution of Problem (1.1.7)

In this subsection, we prove Theorem 4.1.1. In order to derive estimate (4.1.10), we use Schauder's localization method. Local estimates of solution near the boundary Γ are obtained in the same way as in the case of a single fluid [54]. They are based on the fact that for the model problem (1.1.7) with $\Gamma = \{x_3 = 0\}$, the unique solvability and the estimate of a solution have already been established (Theorems 2.5.1 and 3.1.1).

For estimating a solution in domains which do not intersect the boundary Γ, we need a similar theorem for the Cauchy problem [77]

$$\begin{aligned}
&\mathcal{D}_t\boldsymbol{u} - \nu\Delta\boldsymbol{u} + \frac{1}{\rho}\nabla q = \boldsymbol{f}, \\
&\nabla\cdot\boldsymbol{u} = r \quad \text{in } \mathbb{R}^3_T = \mathbb{R}^3 \times (0,T), \qquad (4.2.1)\\
&\boldsymbol{u}|_{t=0} = \boldsymbol{u}_0.
\end{aligned}$$

Theorem 4.2.1 *Let* $\nu, \rho > 0$, $\boldsymbol{f} \in \boldsymbol{C}^{\alpha,\alpha/2}(\mathbb{R}^3_T)$, $r \in C^{1+\alpha,\frac{1+\alpha}{2}}(\mathbb{R}^3_T)$, $\boldsymbol{u}_0 \in \boldsymbol{C}^{2+\alpha}(\mathbb{R}^3)$ *with* $\alpha \in (0,1)$, $T < \infty$, $\nabla\cdot\boldsymbol{u}_0 = r(x,0)$. *We assume that the representations*

$$\mathcal{D}_t r - \nabla \cdot \boldsymbol{f} = \nabla \cdot \boldsymbol{g},\ \boldsymbol{g} = \nabla \cdot \mathbb{G} \equiv \{\sum_{k=1}^{3} \partial G_{ik}/\partial x_k\}_{i=1}^{3}, \tag{4.2.2}$$

Exist; moreover, $\boldsymbol{g} \in \boldsymbol{C}^{\alpha,\alpha/2}(\mathbb{R}^3_T)$, *and for* $\mathbb{G}$, *the semi-norm* $\langle \mathbb{G} \rangle^{(\gamma,1+\alpha)}_{\mathbb{R}^3_T}$ *is finite with* $\gamma \in (0,1)$. *Then the Cauchy problem* (4.2.1) *has a solution* $(\boldsymbol{u}, q)$ *such that* $\boldsymbol{u} \in \boldsymbol{C}^{2+\alpha,1+\alpha/2}(\mathbb{R}^3_T)$, $q \in C^{(\gamma,1+\alpha)}(\mathbb{R}^3_T)$, *and* $\nabla q \in \boldsymbol{C}^{\alpha,\alpha/2}(\mathbb{R}^3_T)$, *and the inequality*

$$\langle \boldsymbol{u} \rangle^{(2+\alpha,1+\alpha/2)}_{\mathbb{R}^3_T} + \langle \nabla q \rangle^{(\alpha,\alpha/2)}_{\mathbb{R}^3_T} \leqslant c_2 \{ \langle \boldsymbol{f} \rangle^{(\alpha,\alpha/2)}_{\mathbb{R}^3_T} + \langle \boldsymbol{u}_0 \rangle^{(2+\alpha)}_{\mathbb{R}^3} + \langle r \rangle^{(1+\alpha,\frac{1+\alpha}{2})}_{\mathbb{R}^3_T} + \langle \boldsymbol{g} \rangle^{(\alpha,\alpha/2)}_{\mathbb{R}^3_T} + \langle \mathbb{G} \rangle^{(\gamma,1+\alpha)}_{\mathbb{R}^3_T} \} \tag{4.2.3}$$

holds with a constant c_2 *independent of* T. *The solution of* (4.2.1) *is unique in the above class up to* $(\boldsymbol{w} = \boldsymbol{\eta} t,\ s = -\rho \boldsymbol{\eta} \cdot \boldsymbol{x})$ *with an arbitrary constant vector* $\boldsymbol{\eta}$.

Proof of Theorem 4.1.1

1°. First, we estimate the higher-order semi-norms of $\boldsymbol{v}$ and p.

We suppose the Cartesian coordinates are introduced so that the origin is inside Ω^+. We consider the ball $B_d = \{|x| < d\}$ with a radius d such that $\overline{\Omega+} \subset B_d$. We multiply system (1.1.7) by the cutoff function η: $\eta = 0$ inside B_d and $\eta = 1$ outside $B_{d+\lambda}$, where $\lambda \in (0,1)$, such that $|\nabla^k \eta| \leqslant c_3 \lambda^{-k}$. For the new unknown functions $\boldsymbol{u} = \boldsymbol{v}\eta$ and $q = p\eta$, we obtain the problem

$$\begin{aligned} \mathcal{D}_t \boldsymbol{u} - \nu^- \Delta \boldsymbol{u} + \frac{1}{\rho^-} \nabla q &= \boldsymbol{f}\eta - \nu^-(\boldsymbol{v}\Delta\eta + 2(\nabla\eta \cdot \nabla)\boldsymbol{v}) + \frac{1}{\rho^-} p \nabla \eta \\ &\equiv \boldsymbol{f}\eta + \boldsymbol{\psi} \equiv \boldsymbol{f}^{(1)}, \\ \nabla \cdot \boldsymbol{u} &= r\eta + \boldsymbol{v} \cdot \nabla \eta = r^{(1)} \quad \text{in } \mathbb{R}^3_T, \\ \boldsymbol{u}|_{t=0} &= \boldsymbol{v}_0 \eta. \end{aligned} \tag{4.2.4}$$

Let us verify that for system (4.2.4) all the assumptions of Theorem 4.2.1 are satisfied. Indeed, there are representations of type (4.2.2):

$$\mathcal{D}_t r^{(1)} - \nabla \cdot \boldsymbol{f}^{(1)} = \nabla \cdot \boldsymbol{g}^{(1)}, \qquad \boldsymbol{g}^{(1)} = \nabla \cdot \mathbb{G}^{(1)} \equiv \Big\{ \sum_{k=1}^{3} \partial G^{(1)}_{ik} / \partial x_k \Big\}_{i=1}^{3},$$

where

$$\nabla \cdot \boldsymbol{g}^{(1)} = \nabla \cdot \big(\mathcal{D}_t \boldsymbol{R}\eta - \boldsymbol{f}\eta - \boldsymbol{\psi} \big) + \nabla\eta \cdot \big(\nu^- \Delta \boldsymbol{v} - \frac{1}{\rho^-} \nabla p + \boldsymbol{f} - \mathcal{D}_t \boldsymbol{R} \big)$$

$$= \nabla\cdot\Big\{\mathcal{D}_t\boldsymbol{R}\eta - \boldsymbol{f}\eta - \boldsymbol{\psi} - \frac{1}{\rho^-}p\nabla\eta + \nu^-\big(\nabla\cdot\boldsymbol{v}\nabla\eta - \nabla\eta\times\nabla\times\boldsymbol{v}\big)\Big\}$$
$$+\nabla\eta\cdot(\boldsymbol{f} - \mathcal{D}_t\boldsymbol{R}) + \Delta\eta(\frac{1}{\rho^-}p - \nu^- r).$$

In the last relation, we have used the chain of equalities

$$\nu^-\Delta\boldsymbol{v}\cdot\nabla\eta = \nu^-\nabla\eta\cdot\big(\nabla(\nabla\cdot\boldsymbol{v}) - \nabla\times\nabla\times\boldsymbol{v}\big)$$
$$= \nu^-\nabla\cdot(\nabla\cdot\boldsymbol{v}\nabla\eta) - \nu^-\nabla\cdot\boldsymbol{v}\Delta\eta - \nu^-\nabla\cdot(\nabla\eta\times\nabla\times\boldsymbol{v}),$$

where the sign "×" means the vector product. Hence

$$\boldsymbol{g}^{(1)} = \boldsymbol{g}\eta + \nu^-\big(\boldsymbol{v}\Delta\eta + 2(\nabla\eta\cdot\nabla)\boldsymbol{v} + r\nabla\eta - \nabla\eta\times\nabla\times\boldsymbol{v}\big) - \frac{2}{\rho^-}p\nabla\eta$$
$$+\nabla\int_{\mathbb{R}^3}\mathcal{E}(x,y)\{(-\boldsymbol{g})\cdot\nabla_y\eta + (\frac{1}{\rho^-}p - \nu^- r)\Delta_y\eta\}\mathrm{d}y. \tag{4.2.5}$$

Here and below $\mathcal{E}(x,y)$ is the fundamental solution of the Laplace equation in $\mathbb{R}^3$.

The tensor $\mathbb{G}^{(1)}$ has the components as follows:

$$G_{ij}^{(1)} = M_{ij}\eta - \delta_j^i\int_{\mathbb{R}^3}\mathcal{E}(x,y)\{(\nabla\cdot\mathbb{M})\cdot\nabla\eta - (\frac{1}{\rho^-}p - \nu^- r)\Delta\eta\}\mathrm{d}y$$
$$+\frac{\partial}{\partial x_j}\int_{\mathbb{R}^3}\mathcal{E}(x,y)\Big\{-\sum_k M_{ik}\frac{\partial\eta}{\partial y_k} + \nu^-\big[v_i\Delta\eta + 2(\nabla\eta\cdot\nabla)v_i$$
$$+r\frac{\partial\eta}{\partial y_i} - (\nabla\eta\times\nabla\times\boldsymbol{v})_i\big] - \frac{2}{\rho^-}p\frac{\partial\eta}{\partial y_i}\Big\}\mathrm{d}y. \tag{4.2.6}$$

We note that the support of the derivatives of the cutoff function η is contained in ring cylinder $K_T = (B_{d+\lambda}\backslash B_d)\times(0,T)$ of thickness λ. By taking into account the dependence of the derivatives η on λ and the Hölder norm estimates of the product of two functions, in view of formulas (4.2.5) and (4.2.6), we obtain the inequalities

$$\langle\boldsymbol{g}^{(1)}\rangle_{D_T}^{(\alpha,\alpha/2)} \leqslant c_4\big\{\lambda^{-\alpha}\big(|\mathcal{D}_t\boldsymbol{R}|_{D_T}^{(\alpha,\alpha/2)} + |\boldsymbol{f}|_{D_T}^{(\alpha,\alpha/2)}\big) + \lambda^{-2-\alpha}|\boldsymbol{v}|_{K_T}^{(\alpha,\alpha/2)}$$
$$+\lambda^{-2}\big(|\nabla\boldsymbol{v}|_{K_T}^{(\alpha,\alpha/2)} + |r|_{K_T}^{(\alpha,\alpha/2)} + |p|_{K_T}^{(\alpha,\alpha/2)}\big)\big\}, \tag{4.2.7}$$
$$\langle G_{ij}^{(1)}\rangle_{D_T}^{(\gamma,1+\alpha)} \leqslant c_5\big\{\lambda^{-2}\big(|M_{ij}|_{D_T}^{(\gamma,1+\alpha)} + \langle r\rangle_{t,K_T}^{(\frac{1+\alpha-\gamma}{2})} + |\boldsymbol{v}|_{K_T}^{(1+\alpha,\frac{1+\alpha}{2})}\big)$$
$$+\lambda^{-1}\big(\langle\nabla\boldsymbol{v}\rangle_{t,K_T}^{(\frac{1+\alpha-\gamma}{2})} + \langle p\rangle_{t,K_T}^{(\frac{1+\alpha-\gamma}{2})}\big)\big\}.$$

For estimating the potentials in (4.2.5) and (4.2.6), we have used the compactness of the supports of the integrands and Lemma 4.1.1.

The other norms of the functions in the right-hand sides in (4.2.4) are estimated in a similar way, for example,

$$\langle \boldsymbol{f}^{(1)}\rangle_{D_T}^{(\alpha,\alpha/2)} \leqslant c_6\{\lambda^{-\alpha}|\boldsymbol{f}|_{D_T}^{(\alpha,\alpha/2)}+\lambda^{-1-\alpha}\big(|\boldsymbol{v}|_{K_T}^{(1+\alpha,\frac{1+\alpha}{2})}+|p|_{K_T}^{(\alpha,\alpha/2)}\big)+\lambda^{-2-\alpha}|\boldsymbol{v}|_{K_T}^{(\alpha,\alpha/2)}\}. \tag{4.2.8}$$

Now it is evident that one may apply Theorem 4.2.1 to system (4.2.4). Inequality (4.2.3) together with (4.2.7) and (4.2.8) gives us

$$\begin{aligned}
&\langle \boldsymbol{v}\rangle_{D_T\setminus B_T}^{(2+\alpha,1+\alpha/2)} + \langle \nabla p\rangle_{D_T\setminus B_T}^{(\alpha,\alpha/2)}\\
&\quad\leqslant c_7\big(\langle \boldsymbol{u}\rangle_{D_T}^{(2+\alpha,1+\alpha/2)} + \langle \nabla q\rangle_{D_T}^{(\alpha,\alpha/2)}\big)\\
&\quad\leqslant c_8(T)\Big\{|\boldsymbol{f}|_{D_T}^{(\alpha,\alpha/2)} + |\boldsymbol{v}_0|_{\cup\Omega^\pm}^{(2+\alpha)} + |r|_{D_T}^{(1+\alpha,\frac{1+\alpha}{2})} + |\mathcal{D}_t\boldsymbol{R}|_{D_T}^{(\alpha,\alpha/2)}\\
&\qquad+|\mathbb{M}|_{D_T}^{(\gamma,1+\alpha)} + \lambda^{-2-\alpha}\Big(|\boldsymbol{v}|_{K_T}^{(1+\alpha,\frac{1+\alpha}{2})} + |p|_{K_T}^{(\alpha,\alpha/2)} + \langle p\rangle_{t,K_T}^{(\frac{1+\alpha-\gamma}{2})}\Big)\Big\},
\end{aligned}$$

where $B_T \equiv B_{d+\lambda}\times(0,T)$. If one takes in system (4.2.4) as η a cutoff function with the support in the ball $B_{2\lambda}(x_0) = \{|x-x_0| < 2\lambda\}$ which does not intersect the interface Γ, one can similarly obtain the estimates in the balls $B_\lambda(x_0)$. The evaluations near Γ are deduced in the same way as in the case of a bounded volume of a single fluid (see inequalities (3.11) in [54]). Next, by considering a finite covering of $B_{d+2\lambda}$ by the balls $B_\lambda(x_0)$, on the basis of the above arguments, we conclude that

$$\begin{aligned}
&\langle \boldsymbol{v}\rangle_{D_T}^{(2+\alpha,1+\alpha/2)} + \langle \nabla p\rangle_{D_T}^{(\alpha,\alpha/2)}\\
&\quad\leqslant c_9(T)\Big\{F(T) + \lambda|p|_{B_{1T}}^{(\gamma,1+\alpha)} + \lambda^{-2-\alpha}\Big(\sum_{|\boldsymbol{\beta}|+2k\leqslant 2}|\mathcal{D}_t^k D_x^{\boldsymbol{\beta}}\boldsymbol{v}|_{D_T} + |\boldsymbol{v}|_{x,D_T}^{(1+\alpha)}\\
&\qquad+\langle\nabla\boldsymbol{v}\rangle_{t,D_T}^{(\frac{1+\alpha-\gamma}{2})} + |\nabla p|_{B_{1T}} + |p|_{B_{1T}} + \langle p\rangle_{t,B_{1T}}^{(\frac{1+\alpha-\gamma}{2})}\Big)\Big\}
\end{aligned} \tag{4.2.9}$$

with a nondecreasing function $c_9(T)$. (Here it has been taken into account that

$$\langle p\rangle_{t,B_{1T}}^{(\frac{\alpha}{2})} \leqslant T^{\frac{1-\gamma}{2}}\langle p\rangle_{t,B_{1T}}^{(\frac{1+\alpha-\gamma}{2})};$$

$B_{1T} \equiv B_{d+2\lambda}\times(0,T)$.)

2°. Now we evaluate pressure semi-norms $\langle p\rangle_{D_T}^{(\gamma,1+\alpha)}$ and $\langle p\rangle_{t,B_{1T}}^{(\frac{1+\alpha-\gamma}{2})}$.

We introduce the function $P_0(x,t) = p_0(x)$ for any $t > 0$, where $p_0(x)$ is a solution of problem (4.1.9), i.e., it is the initial pressure at $t = 0$.

Next, we analyze the initial-boundary value problem for the heat equation

$$\mathcal{D}_t \boldsymbol{w} - \nu^{\pm} \Delta \boldsymbol{w} = \boldsymbol{f} - \frac{1}{\rho^{\pm}} \nabla P_0 \quad \text{in } D_T, \tag{4.2.10}$$

$$[\boldsymbol{w}]|_{\Gamma} = 0, \ \boldsymbol{w}|_{t=0} = \boldsymbol{v}_0.$$

From (4.1.8) and (4.1.9), it follows that the compatibility conditions are satisfied for this problem:

$$[\boldsymbol{w}]|_{\Gamma,t=0} = 0, \ [\mathcal{D}_t \boldsymbol{w}]|_{\Gamma,t=0} = 0.$$

By easy considerations, using the results for the first initial-boundary value problem [47], we verify that a solution to (4.2.10) exists and is subjected to the estimate

$$|\boldsymbol{w}|_{D_T}^{(2+\alpha,1+\alpha/2)} \leqslant c_{10}\{|\boldsymbol{f}|_{D_T}^{(\alpha,\alpha/2)} + |\nabla p_0|_{\cup\Omega^{\pm}}^{(\alpha)} + |\boldsymbol{v}_0|_{\cup\Omega^{\pm}}^{(2+\alpha)}\}. \tag{4.2.11}$$

We set $\boldsymbol{w}' = \nabla\Phi$, where Φ satisfies the Poisson equation

$$\Delta\Phi = r - \nabla \cdot \boldsymbol{w} = r' \quad \text{in } \mathbb{R}^3_T. \tag{4.2.12}$$

We fix two arbitrary points $x^- \in \Omega^-$ and $x^+ \in \Omega^+$. By (1.1.7), (4.1.9), (4.2.10), and (4.2.12), it is easily seen that $p = P_0 - P_0(x^{\pm}) + q + \mu^{\pm} r' - \rho^{\pm}\big(\mathcal{D}_t\Phi - \mathcal{D}_t\Phi(x^{\pm},t)\big)$, where q is a solution to diffraction problem

$$\frac{1}{\rho^{\pm}} \Delta q = 0 \quad \text{in } D_T,$$

$$[q]|_{\Gamma} = -\big[P_0 - P_0(x^{\pm})\big]\big|_{\Gamma} + 2\big[\mu^{\pm}(\partial \boldsymbol{v}/\partial \boldsymbol{n})_{\boldsymbol{n}}\big]\big|_{\Gamma} - \sigma \int_0^t (\Delta_{\Gamma} \boldsymbol{v})_{\boldsymbol{n}} dt' \tag{4.2.13}$$

$$+\big[\rho^{\pm}\big(\mathcal{D}_t\Phi - \mathcal{D}_t\Phi(x^{\pm},t)\big) - \mu^{\pm} r'\big]\big|_{\Gamma} - b^l - \sigma \int_0^t B dt' \equiv q_0,$$

$$\Big[\frac{1}{\rho^{\pm}} \frac{\partial q}{\partial \boldsymbol{n}}\Big]\Big|_{\Gamma} = \big[\nu^{\pm}(\Delta \boldsymbol{u})_{\boldsymbol{n}}\big]\big|_{\Gamma} \equiv q_1.$$

Here $\psi_{\boldsymbol{n}} \equiv \boldsymbol{\psi} \cdot \boldsymbol{n}$, and $\boldsymbol{u} \equiv \boldsymbol{v} - \boldsymbol{w} - \boldsymbol{w}'$; moreover, $\nabla \cdot \boldsymbol{u} = 0$.

We estimate solution semi-norm $|q|_{D_T}^{(\gamma,1+\alpha)}$ of problem (4.2.13). To this end, we represent q in the form of the sum $q = \rho^{\pm} s_1 + s_2$, where the functions s_1 and s_2 are solutions to the problems

$$\Delta s_1 = 0, \ [\rho^{\pm} s_1]|_{\Gamma} = q_0, \quad \Big[\frac{\partial s_1}{\partial \boldsymbol{n}}\Big]\Big|_{\Gamma} = 0, \quad s_1 \xrightarrow[|x|\to\infty]{} 0, \tag{4.2.14}$$

$$\frac{1}{\rho^\pm}\Delta s_2 = 0, \quad [s_2]|_\Gamma = 0, \quad \Big[\frac{1}{\rho^\pm}\frac{\partial s_2}{\partial \boldsymbol{n}}\Big]\Big|_\Gamma = q_1, \quad s_2 \xrightarrow[|x|\to\infty]{} 0. \tag{4.2.15}$$

A solution to (4.2.14) can be represented in the form of the double-layer potential

$$s_1(x) = -\int_\Gamma \frac{\partial \mathcal{E}(x,y)}{\partial \boldsymbol{n}_y}\mu(y)\mathrm{d}\Gamma(y), \ x \in \mathbb{R}^3.$$

Here $\frac{\partial}{\partial \boldsymbol{n}_y}$ is the derivative with respect to the outward normal to Γ at the point y, and $\mu(y)$ is the unknown density which, in view of the first boundary condition in (4.2.14), has to satisfy the equation

$$\mu(x) = \frac{2}{\rho^+ + \rho^-}q_0(x) + \frac{2(\rho^+ - \rho^-)}{\rho^+ + \rho^-}\int_\Gamma \frac{\partial \mathcal{E}(x,y)}{\partial \boldsymbol{n}_y}\mu(y)\mathrm{d}\Gamma(y). \tag{4.2.16}$$

By analogy, a solution to problem (4.2.15) is given by the single-layer potential

$$s_2(x) = \int_\Gamma \mathcal{E}(x,y)\omega(y)\mathrm{d}\Gamma(y), \qquad x \in \mathbb{R}^3,$$

for whose density, the second boundary condition in (4.2.15) implies the equation

$$\omega(x) = \frac{2p^+\rho^-}{\rho^+ + \rho^-}q_1(x) + \frac{2(\rho^+ - \rho^-)}{\rho^+ + \rho^-}\int_\Gamma \frac{\partial \mathcal{E}}{\partial \boldsymbol{n}_x}\omega(y)\mathrm{d}\Gamma(y). \tag{4.2.17}$$

The kernel $\mathcal{K}(x,y) = -\dfrac{\partial \mathcal{E}(x,y)}{\partial \boldsymbol{n}_y}$ of integral equation (4.2.16) has a weak singularity. By considering homogeneous equation (4.2.16) and problem (4.2.14) corresponding to it, one can verify that $\varkappa = -\dfrac{2(\rho^+ - \rho^-)}{\rho^+ + \rho^-}$ is not an eigenvalue of $\mathcal{K}$. Therefore, as a result of several iterations, we obtain an integral equation with the Hilbert–Schmidt kernel $\mathcal{K}_n$, a solution of which is expressed in terms of the resolvent

$$\mu(x) = P(x) + \varkappa^n \int_\Gamma \mathcal{R}_t^n(x,y)P(y)\mathrm{d}\Gamma(y), \ x \in \Gamma, \tag{4.2.18}$$

where

$$P(x) = \frac{2q_0(x)}{\rho^+ + \rho^-} + \sum_{j=1}^{n-1}\varkappa^i \int_\Gamma \mathcal{K}_i(x,y)\frac{2q_0(y)}{\rho^+ + \rho^-}\mathrm{d}\Gamma(y),$$

$\mathcal{K}_i(x,y)$ are iterated kernels of the order i.The resolvent $\mathcal{R}^n_\varkappa$ corresponding to the kernel $\mathcal{K}_n$ is, in turn, a weakly polar kernel, and a sufficiently large number of iterations provide the estimate

$$|\mathcal{R}^n_\varkappa(x,y)| \leqslant c_{11}|x-y|^{\alpha-2}$$

(see [57]), whence one can conclude that

$$|\mu|_\Gamma \leqslant c_{12}|q_0|_\Gamma, \quad \langle\mu\rangle^{(\gamma)}_\Gamma \leqslant c_{13}|q_0|^{(\gamma)}_\Gamma, \tag{4.2.19}$$

and consequently

$$|s_1|_{\cup\Omega^\pm} \leqslant c_{14}|q_0|_\Gamma, \quad \langle s_1\rangle^{(\gamma)}_{\cup\Omega^\pm} \leqslant c_{15}|q_0|^{(\gamma)}_\Gamma. \tag{4.2.20}$$

The same arguments are true also for Eq. (4.2.17) with the conjugate kernel $\mathcal{K}^*(x,y) = \mathcal{K}(y,x)$ and $\varkappa_1 = -\varkappa$. For functions $\omega(x)$, there is a representation (4.2.18), where $\mathcal{K}^*$ plays the role of $\mathcal{K}$, and the number $\varkappa_1$ does the role of $\varkappa$; one should substitute there also $\rho^+\rho^- q_1$ instead of q_0.

For estimating $|s_2|^{(\beta)}_{\cup\Omega^\pm}$, we repeat the arguments from § 6 in [76] which were developed for the evaluation of a solution to the Neumann problem. By multiplying Eq. (4.2.17) by $\partial\mathcal{E}(z,x)/\partial\boldsymbol{n}_z$ and integrating it over $\Gamma(x)$, we verify that the function $\theta(x) = \displaystyle\int_\Gamma \frac{\partial\mathcal{E}(x,y)}{\partial\boldsymbol{n}_x}\omega(y)\mathrm{d}\Gamma(y)$ satisfies a similar equation:

$$\theta(x) = \frac{2\rho^+\rho^-}{\rho^+ + \rho^-}Q_1(x) + x_1\int_\Gamma \frac{\partial\mathcal{E}}{\partial\boldsymbol{n}_x}\theta(y)\mathrm{d}\Gamma(y)$$

with $Q_1(x) = \displaystyle\int_\Gamma \frac{\partial\mathcal{E}}{\partial\boldsymbol{n}_x} q_1(y)\mathrm{d}\Gamma(y)$. Therefore for it, estimates (4.2.19) hold. In particular,

$$|\theta|_\Gamma \leqslant c_{12}|Q_1|_\Gamma. \tag{4.2.21}$$

Now we use the fact that $\boldsymbol{n}\cdot(\Delta\boldsymbol{u}) = -\boldsymbol{n}\cdot(\nabla\times\nabla\times\boldsymbol{u})$ by virtue of the solenoidality of the vector $\boldsymbol{u}$. In view of the Stokes formula

$$\int_\Gamma \psi(\nabla\times\boldsymbol{a})_{\boldsymbol{n}}\mathrm{d}\Gamma = -\int_\Gamma (\nabla\psi\times\boldsymbol{a})_{\boldsymbol{n}}\mathrm{d}\Gamma$$

we conclude that

$$Q_1(x) = \int_\Gamma \Big(\frac{\partial}{\partial\boldsymbol{n}_x}\nabla_y\mathcal{E}(x,y)\Big)\times\big(\boldsymbol{a}(y)-\boldsymbol{a}(x)\big)\cdot\boldsymbol{n}(y)\mathrm{d}\Gamma(y)$$

and

$$s_2(x) = \frac{2\rho^+\rho^-}{\rho^+ + \rho^-}\int_\Gamma \nabla_y \mathcal{E} \times \big(\boldsymbol{a}(y) - \boldsymbol{a}(x)\big)_{\boldsymbol{n}} \mathrm{d}\Gamma(y) + \varkappa_1 \int_\Gamma \mathcal{E}(x,y)\theta(y)\mathrm{d}\Gamma(y), \tag{4.2.22}$$

where $\boldsymbol{a} = [\nu^\pm \nabla \times \boldsymbol{u}]\big|_\Gamma$. These representation formulas imply the estimates

$$|Q_1|_\Gamma \leqslant c_{16}\langle \nabla \boldsymbol{u}\rangle^{(\beta)}_{\cup\Omega^\pm},$$
$$|s_2|^{(\beta)}_{\cup\Omega^\pm} \leqslant c_{12}\big(\langle \nabla \boldsymbol{u}\rangle^{(\beta)}_{\cup\Omega^\pm} + |\theta|_\Gamma\big) \;\text{ with } \forall \beta \in (0,1).$$

This together with (4.2.21) yields the inequality

$$|s_2|^{(\beta)}_{\cup\Omega^\pm} \leqslant c_{18}\langle \nabla \boldsymbol{u}\rangle^{(\beta)}_{\cup\Omega^\pm} \;\text{ with }\; \forall \beta \in (0,1),$$

whence, taking (4.2.20) into account, we deduce that

$$|q|_{D_T} \leqslant c_{19}\big(|q_0|_{G_T} + \langle \nabla \boldsymbol{u}\rangle^{(\beta)}_{x,D_T}\big), \tag{4.2.23}$$
$$\langle q\rangle^{(\gamma)}_{x,D_T} \leqslant c_{20}\big(|q_0|^{(\gamma)}_{x,G_T} + \langle \nabla \boldsymbol{u}\rangle^{(\gamma)}_{x,D_T}\big).$$

By repeating the same arguments for functions $\frac{s_{1,2}(x,t) - s_{1,2}(x,t')}{|t-t'|^{\frac{1+\alpha-\gamma}{2}}}$, we obtain

$$\langle q\rangle^{(\frac{1+\alpha-\gamma}{2})}_{t,D_T} \leqslant c_{19}\Big(\langle q_0\rangle^{(\frac{1+\alpha-\gamma}{2})}_{t,G_T} + |\langle \nabla \boldsymbol{u}\rangle^{(\frac{1+\alpha-\gamma}{2})}_{(0,T)}|^{(\beta)}_{\cup\Omega^\pm}\Big), \;\forall \beta < 1, \tag{4.2.24}$$

$$\langle q\rangle^{(\gamma,1+\alpha)}_{D_T} \leqslant c_{20}\Big(|q_0|^{(\gamma,1+\alpha)}_{G_T} + \langle \nabla \boldsymbol{u}\rangle^{(\gamma,1+\alpha)}_{D_T}\Big). \tag{4.2.25}$$

Now we consider the integral $J = -\sigma \displaystyle\int_0^t \big(\boldsymbol{n}\cdot \Delta_\Gamma \boldsymbol{v} + B\big)\mathrm{d}t'$ included in q_0 which has been defined in (4.2.13). In [54], it was shown that inequality

$$|J|^{(\gamma,1+\alpha)}_{G_T} \leqslant c_{21}\Big(\langle \nabla_\tau J\rangle^{(\alpha/2)}_{t,G_T} + |\mathcal{D}_t J|_{G_T}\Big)$$

holds. Due to the boundary condition in (1.1.7), we conclude that

$$\langle \nabla_\tau J\rangle^{(\alpha/2)}_{t,G_T} \leqslant c_{22}\big(\langle \nabla_\tau b'\rangle^{(\alpha/2)}_{t,G_T} + \langle \nabla p\rangle^{(\alpha/2)}_{t,D_T} + \langle \boldsymbol{v}\rangle^{(2+\alpha,1+\alpha/2)}_{D_T}\big).$$

Therefore inequality (4.2.25) can be continued as follows:

$$\langle q\rangle^{(\gamma,1+\alpha)}_{D_T} \leqslant c_{23}\Big\{\langle \nabla p\rangle^{(\alpha,\alpha/2)}_{D_T} + |\boldsymbol{v}|^{(2+\alpha,1+\alpha/2)}_{D_T} + \langle \nabla_\tau b'\rangle^{(\alpha/2)}_{t,G_T} + |B|_{G_T}$$
$$+|b'|^{(\gamma,1+\alpha)}_{G_T} + |r'|^{(1+\alpha,\frac{1+\alpha}{2})}_{D_T} + \big|\big[\rho^\pm\big(\mathcal{D}_t\Phi - \mathcal{D}_t\Phi(x^\pm,t)\big)\big]\big|_\Gamma\big|^{(\gamma,1+\alpha)}_{G_T}$$

$$+|\nabla \boldsymbol{w}'|_{D_T}^{(\gamma,1+\alpha)} + |\boldsymbol{w}|_{D_T}^{(2+\alpha,1+\alpha/2)}\Big\}. \qquad (4.2.26)$$

Taking Eq. (4.2.12) into account, we represent the gradient of the vector $\boldsymbol{w}'$ as a volume potential:

$$\nabla \boldsymbol{w}'(x) = \nabla_x \int_{\mathbb{R}^3} \nabla_x \mathcal{E}(x,y) r'(y) \mathrm{d}y = \nabla_x \nabla_x \int_{\mathbb{R}^3} \mathcal{E} \nabla_y \cdot (\boldsymbol{R} - \boldsymbol{w}) \mathrm{d}y.$$

Estimates (4.1.5) of the singular integral imply the inequality

$$|\nabla \boldsymbol{w}'|_{D_T}^{(\gamma,1+\alpha)} \leqslant c_{24}\Big(|r|_{D_T}^{(1+\alpha,\frac{1+a}{2})} + \langle \boldsymbol{R}\rangle_{x,D_T}^{(\gamma)} + \langle \boldsymbol{R}\rangle_{D_T}^{(\gamma,1+\alpha)} + |\boldsymbol{w}|_{D_T}^{(2+\alpha,1+\alpha/2)}\Big). \qquad (4.2.27)$$

Next,

$$\mathcal{D}_t \Phi(x,t) = \int_{\mathbb{R}^3} \mathcal{E}(x,y) \mathcal{D}_t r'(y,t) \mathrm{d}y = \int_{\mathbb{R}^3} \nabla_y \mathcal{E} \cdot (\mathcal{D}_t \boldsymbol{R} - \mathcal{D}_t \boldsymbol{w}) \mathrm{d}y.$$

We express $\mathcal{D}_t w$ from the equation in (4.2.10) and make use of (4.1.7). After integration by parts, we have

$$\mathcal{D}_t \Phi = -\nabla_x \int_{\mathbb{R}^3} \nabla_y \mathcal{E} \cdot \mathbb{D} \mathrm{d}y + \nabla_x \int_\Gamma \mathcal{E}[\mathbb{D}\boldsymbol{n}]|_\Gamma \mathrm{d}\Gamma,$$

where $\mathbb{D} = \mathbb{M} - \nu^\pm \nabla \boldsymbol{w} - \mathbb{I} P_0/\rho^\pm$, $\mathbb{I}$ is the identity matrix. It should be noted that $\int_{\mathbb{R}^3} \nabla_y \nabla_y \mathcal{E}(x,y) \cdot \mathbb{I} P_0(y) \mathrm{d}y = P_0(x)$. The other terms in the volume integral are also summable; the surface integral contains a singular kernel and a kernel with weak singularity. By virtue of inequality (4.1.4), we obtain

$$\langle \mathcal{D}_t \Phi\rangle_{x,D_T}^{(\gamma)} \leqslant c_{25}\big(\langle \mathbb{M}\rangle_{x,D_T}^{(\gamma)} + \langle \nabla \boldsymbol{w}\rangle_{x,D_T}^{(\gamma)} + \langle p_0\rangle_{\cup\Omega^\pm}^{(\gamma)}\big), \qquad (4.2.28)$$

$$\langle \mathcal{D}_t \Phi\rangle_{D_T}^{(\gamma,1+\alpha)} \leqslant c_{25}\big(\langle \mathbb{M}\rangle_{D_T}^{(\gamma,1+\alpha)} + |\boldsymbol{w}|_{D_T}^{(2+\alpha,1+\alpha/2)}\big). \qquad (4.2.29)$$

The estimate of the vector $\boldsymbol{w}'$ is proved in a similar way:

$$\langle \nabla\nabla \boldsymbol{w}'\rangle_{x,D_T}^{(\alpha)} \leqslant c_{26}|\nabla r'|_{x,D_T}^{(\alpha)} \leqslant c_{27}\big(|r|_{D_T}^{(1+\alpha,\frac{1+\alpha}{2})} + |\boldsymbol{w}|_{D_T}^{(2+\alpha,1+\alpha/2)}\big). \qquad (4.2.30)$$

In view of the evident inequality

$$|\mathcal{D}_t \Phi(x,t) - \mathcal{D}_t \Phi(x^\pm,t)| \leqslant |x - x^\pm|^\beta \langle D_t \Phi(\cdot,t)\rangle_{\cup\Omega^\pm}^{(\beta)}, \quad \forall \beta < 1, \qquad (4.2.31)$$

and estimate (4.2.28), we conclude that the difference $\mathcal{D}_t \Phi(x,t) - \mathcal{D}_t \Phi(x^\pm,t)$ has a weak power-law growth and is bounded on any bounded set, in particular, on the surface Γ. Consequently, for $\beta = \gamma$, one deduces that

$$\langle[\rho^{\pm}(\mathcal{D}_t\Phi(x,t)-\mathcal{D}_t\Phi(x^{\pm},t))]|_{\Gamma}\rangle_{t,G_T}^{(\frac{1+\alpha-\gamma}{2})} \leqslant c_{28}\langle\mathcal{D}_t\Phi\rangle_{D_T}^{(\gamma,1+\alpha)}.$$

Therefore, inequalities (4.2.26)–(4.2.29) and (4.2.11) imply the estimate

$$\langle q\rangle_{D_T}^{(\gamma,1+\alpha)} \leqslant c_{29}\big(\langle\nabla p\rangle_{D_T}^{(\alpha,\alpha/2)} + |\boldsymbol{v}|_{D_T}^{(2+\alpha,1+\alpha/2)} + |\nabla p_0|_{\cup\Omega^{\pm}}^{(\alpha)} + F(T)\big).$$

For the pressure $p = P_0 - P_0(x^{\pm}) + q + \mu^{\pm}r' - \rho^{\pm}\big(\mathcal{D}_t\Phi - \mathcal{D}_t\Phi(x^{\pm},t)\big)$, we have the same inequality:

$$\langle p\rangle_{D_T}^{(\gamma,1+\alpha)} \leqslant c_{30}\big(\langle\nabla p\rangle_{D_T}^{(\alpha,\alpha/2)} + |\boldsymbol{v}|_{D_T}^{(2+\alpha,1+\alpha/2)} + |\nabla p_0|_{\cup\Omega^{\pm}}^{(\alpha)} + F(T)\big). \qquad (4.2.32)$$

In addition, by (4.2.23) and (4.2.24) with $\beta<\gamma$, by (4.2.31) with $\beta=\gamma$, and a similar inequality for P_0, we obtain that

$$\langle p\rangle_{t,B_{1T}}^{(\frac{1+\alpha-\gamma}{2})} + |p|_{B_{1T}} \leqslant c_{31}\Big\{|\boldsymbol{v}|_{D_T}^{(2+\nu,1+\nu/2)} + \langle p_0\rangle_{\cup\Omega^{\pm}}^{(\gamma)} + |\nabla p_0|_{\cup\Omega^{\pm}}^{(\alpha)} + F(T)\Big\} \qquad (4.2.33)$$

with $\nu = \alpha-\gamma+\beta<\alpha$.

The function $p_0(x)$ satisfying problem (4.1.9) can be estimated in the same way as it has been done for q if we represent it as the sum $p_0 = s_0 + \rho^{\pm}s_1 + s_2$. Here s_0 is a solution of the Poisson equation

$$\frac{1}{\rho^{\pm}}\Delta s_0 = \nabla\cdot\boldsymbol{h},$$

which is expressed in terms of the Newtonian potential

$$s_0(x) = \int_{\cup\Omega^{\pm}} \mathcal{E}(x,y)\rho^{\pm}\nabla_y\cdot\boldsymbol{h}\,dy = -\int_{\cup\Omega^{\pm}} \rho^{\pm}\nabla_y\mathcal{E}\cdot\boldsymbol{h}\,dy + \int_{\Gamma}\mathcal{E}[\rho^{\pm}\boldsymbol{h}\cdot\boldsymbol{n}]\big|_{\Gamma}d\Gamma(y);$$

and s_1, s_2 are solutions to problems (4.2.14) and (4.2.15) with $q_0 = p_{00}$ and $q_1 = p_{01} - [\partial s_0/\partial\boldsymbol{n}]\big|_{\Gamma}$, respectively.

Taking into account the divergence form of $\boldsymbol{h} = \nabla\cdot\mathbb{H}$, where $\mathbb{H}(x) = -\mathbb{M}(x,0) + \nu^{\pm}\mathbb{I}r(x,0)$, as well as the decay of $\boldsymbol{h}$ and $\mathbb{H}$ at infinity, we conclude that

$$\langle s_0\rangle_{\cup\Omega^{\pm}}^{(\gamma)} + |\nabla s_0|_{\cup\Omega^{\pm}} \leqslant c_{32}\big(\langle\mathbb{M}(\cdot,0)\rangle_{\cup\Omega^{\pm}}^{(\gamma)} + \langle r(\cdot,0)\rangle_{\cup\Omega^{\pm}}^{(\gamma)} + |\boldsymbol{h}|_{\cup\Omega^{\pm}}^{(\gamma)}\big), \qquad (4.2.34)$$

$$\langle\nabla s_0\rangle_{\cup\Omega^{\pm}}^{(\alpha)} \leqslant c_{33}|\boldsymbol{h}|_{\cup\Omega^{\pm}}^{(\alpha)} \leqslant c_{34}F(T). \qquad (4.2.35)$$

By (4.2.20), for the function s_1, we deduce

$$\langle s_1\rangle_{\cup\Omega^{\pm}}^{(\gamma)} \leqslant c_{15}|p_{00}|_{\Gamma}^{(\gamma)} \leqslant c_{35}F(T), \qquad (4.2.36)$$

$$|\nabla s_1|^{(\alpha)}_{\Omega^\pm} \leqslant c_{36}|p_{00}|^{(1+\alpha)}_{\Gamma} \leqslant c_{37}F(T). \tag{4.2.37}$$

It is obvious that the function s_2 also satisfies the inequality

$$\langle s_2\rangle^{(\gamma)}_{\cup\Omega^\pm} + |\nabla s_2|^{(\alpha)}_{\cup\Omega^\pm} \leqslant c_{38}|q_1|^{(\alpha)}_{\cup\Omega^\pm} \leqslant c_{39}F(T). \tag{4.2.38}$$

Summing up inequalities (4.2.34)–(4.2.38), we obtain

$$\langle p_0\rangle^{(\gamma)}_{\cup\Omega^\pm} + |\nabla p_0|^{(\alpha)}_{\cup\Omega^\pm} \leqslant c_{40}F(T), \tag{4.2.39}$$

which makes it possible to continue inequality (4.2.33) as follows:

$$\langle p\rangle^{(\frac{1+\alpha-\gamma}{2})}_{t,B_{1T}} + |p|_{B_{1T}} \leqslant c_{41}\big(|\boldsymbol{v}|^{(2+\nu,1+\nu/2)}_{D_T} + F(T)\big), \quad \nu<\alpha. \tag{4.2.40}$$

3°. We complete the estimates similarly to paper [54].

We return to estimate (4.2.9). By using the well-known interpolation inequality

$$|\nabla p|_{B_{1T}} \leqslant \varepsilon^\alpha\langle\nabla p\rangle^{(\alpha)}_{x,B_{1T}} + c_{42}(\varepsilon)|p|_{B_{1T}}, \tag{4.2.41}$$

as well as (4.2.32), (4.2.39), and (4.2.40), we conclude that

$$\begin{aligned}
&\langle\boldsymbol{v}\rangle^{(2+\alpha,1+\alpha/2)}_{D_T} + \langle\nabla p\rangle^{(\alpha,\alpha/2)}_{D_T}\\
&\quad\leqslant c_{43}(T)\Big\{F(T) + \big(\lambda+\frac{\varepsilon^\alpha}{\lambda^{2+\alpha}}\big)\big(\langle\nabla p\rangle^{(\alpha,\alpha/2)}_{D_T}\\
&\qquad+\langle\boldsymbol{v}\rangle^{(2+\alpha,1+\alpha/2)}_{D_T}\big) + \frac{1}{\lambda^{2+\alpha}}|\boldsymbol{v}|^{(2+\nu,1+\nu/2)}_{D_T}\Big\}, \quad \nu<\alpha,
\end{aligned}$$

c_{43} being a nondecreasing function of T.

The derivatives of the velocities $\boldsymbol{v}$ with respect to spatial variables are estimated by inequalities of the type of (4.2.41), and $\mathcal{D}_t\boldsymbol{v}$ is done on the basis of Lemma 2.1 in [54]:

$$|\mathcal{D}_t\boldsymbol{v}|_{D_T} \leqslant \varepsilon_1^\alpha\langle\mathcal{D}_t\boldsymbol{v}\rangle^{(\alpha/2)}_{t,D_T} + c_{44}(\varepsilon_1,T)\big(|\boldsymbol{v}|_{D_T} + |\mathcal{D}_t\boldsymbol{v}(\cdot,0)|_{\cup\Omega^\pm}\big), \tag{4.2.42}$$

the boundedness of $\mathcal{D}_t\boldsymbol{v}\big|_{t=0}$ follows from the equation in (1.1.7) at $t=0$. The Hölder constants of higher-order derivatives of $\boldsymbol{v}$ can be estimated by means of multiplicative inequalities and the Young inequality. For example

$$\langle\nabla\boldsymbol{v}\rangle^{(\frac{1+\nu}{2})}_{t,D_T} \leqslant \big(\langle\nabla\boldsymbol{v}\rangle^{(\frac{1+\alpha}{2})}_{t,D_T}\big)^{\frac{1+\nu}{1+\alpha}}\big(2|\nabla\boldsymbol{v}|_{D_T}\big)^{\frac{\alpha-\nu}{1+\alpha}}$$

$$\leqslant \varepsilon_2^{\frac{1+\alpha}{1+\nu}} \langle \nabla \boldsymbol{v} \rangle_{t,D_T}^{(\frac{1+\alpha}{2})} + 2\varepsilon_2^{-\frac{1+\alpha}{\alpha-\nu}} |\nabla \boldsymbol{v}|_{D_T}.$$

We choose first small enough λ and then ε, ε_1, and ε_2. By the above, we get

$$\langle \boldsymbol{v} \rangle_{D_T}^{(2+\alpha,1+\alpha/2)} + \langle \nabla p \rangle_{D_T}^{(\alpha,\alpha/2)} \leqslant c_{45}(T)\big(F(T) + |\boldsymbol{v}|_{D_T}\big). \tag{4.2.43}$$

Since in this inequality $c_{45}(T)$ is a nondecreasing function of T, it is valid for D_t for any $t < T$ with the same constant $c_{45}(T)$. The same is also true for estimate (4.2.42). Hence

$$|\boldsymbol{v}|_{D_t} \leqslant |\boldsymbol{v}_0|_{\cup\Omega^\pm} + \int_0^t |\mathcal{D}_t \boldsymbol{v}(\cdot,t')|_{\cup\Omega^\pm} \mathrm{d}t' \leqslant c_{46}(t)\Big(F(t) + \int_0^t |\boldsymbol{v}|_{D_{t'}} \mathrm{d}t'\Big).$$

By applying the Gronwall Lemma, we arrive at

$$|\boldsymbol{v}|_{D_t} \leqslant c_{47}(t)F(t),$$

which together with inequalities (4.2.43), (4.2.32), (4.2.39), and (4.2.40) completes the proof of estimate (4.1.10).

4.3 The Solvability of Problem (1.1.7): Constructing a Regularizer

Proof of Theorem 4.1.2

1°. First we prove the solvability of problem (1.1.7) when $\boldsymbol{f} = \boldsymbol{0}$, $r = 0$, $\boldsymbol{v}_0 = \boldsymbol{0}$ and the given functions in the boundary conditions vanish at $t = 0$, i.e., $\boldsymbol{b} \in \mathring{\boldsymbol{C}}_{\boldsymbol{n}}^{1+\alpha,\frac{1+a}{2}}(G_T)$, $b' \in \mathring{C}^{(\gamma,1+\alpha)}(G_T)$, $B \in \mathring{C}^{\alpha,\alpha/2}(G_T)$. The proof will include the construction of a regularizer [47] of this problem and follow the outline of Chap. 4 in paper [54] for the case of one fluid.

Let $\widehat{C}^{\gamma,\alpha}(D_T)$ be the set of function $q(x,t)$, $(x,t) \in D_T$ having finite semi-norm

$$|q|_{\widehat{C}^{1+\alpha}(D_T)} = |\nabla q|_{D_T}^{(\alpha,\alpha/2)} + |q|_{B_T}^{(\gamma,1+\alpha)}.$$

We write system (1.1.7) in operator form

$$\mathcal{A}(\boldsymbol{v}, p) = \boldsymbol{D},$$

where the given term $\boldsymbol{D} = (\boldsymbol{b}, b', B)$; moreover $\boldsymbol{b} \cdot \boldsymbol{n} = 0$. First of all, we construct a linear continuous operator

$$\mathcal{R} : \mathring{\boldsymbol{C}}_{\boldsymbol{n}}^{1+\alpha, \frac{1+\alpha}{2}}(G_T) \times \mathring{C}^{(\gamma, 1+\alpha)}(G_T) \times \mathring{C}^{\alpha, \alpha/2}(G_T) \to \mathring{\boldsymbol{C}}^{2+\alpha, 1+\alpha/2}(D_T) \times \widehat{C}^{\gamma, \alpha}(D_T),$$

inverting operator equation

$$\mathcal{A}(\boldsymbol{v}, p) = (\mathcal{I} + \mathcal{M})\boldsymbol{D}$$

with some linear operator $\mathcal{M}$ (here $\mathcal{I}$ is the identity operator), i.e., $(\boldsymbol{v}, p) = \mathcal{R}\boldsymbol{D}$. Then we will show that that $\mathcal{M}$ satisfies the inequality

$$\|\mathcal{M}\boldsymbol{D}\| \leqslant \varepsilon \|\boldsymbol{D}\| + c_1(\varepsilon) \int_0^t \|\boldsymbol{D}\| \mathrm{d}t', \ \varepsilon < 1,$$

whence it follows that the operator $\mathcal{I} + \mathcal{M}$ is invertible on the entire time interval $[0, T]$. The latter can be proved by successive approximations. Indeed, let us consider the equation

$$\boldsymbol{D} + \mathcal{M}\boldsymbol{D} = \boldsymbol{D}_0.$$

We set $\boldsymbol{D}^{(0)} = \boldsymbol{D}_0, \boldsymbol{D}^{(m+1)} = \boldsymbol{D}_0 - \mathcal{M}\boldsymbol{D}^{(m)}$, $m \geqslant 0$.

$$\|\boldsymbol{D}^{(m+1)} - \boldsymbol{D}^{(m)}\| \leqslant \varepsilon \|\boldsymbol{D}^{(m)} - \boldsymbol{D}^{(m-1)}\| + c_1(\varepsilon) \int_0^t \|\boldsymbol{D}^{(m)} - \boldsymbol{D}^{(m-1)}\| \mathrm{d}t'.$$

After summing up these inequalities over m from 1 to ∞, with the help of the Gronwall lemma, we establish that $\sum_{m=1}^{\infty} \|\boldsymbol{D}^{(m+1)} - \boldsymbol{D}^{(m)}\| < \infty$ which means the convergence of the successive approximations $\boldsymbol{D}^{(m)}$.

The uniqueness of the right inverse operator $a^{-1} = \mathcal{R}(\mathcal{I} + \mathcal{M})^{-1}$ follows from a priori estimate (4.1.10), so it is the inverse of the operator $\mathcal{A}$.

(a) **Construction of the regularizer $\mathcal{R}$.** Let a family of infinitely smooth cutoff functions $\{\zeta_j\}_{j \leqslant N_\lambda}$ be a partition of unity subordinated to the covering of Γ by the balls $K_{j,\lambda} = \{|x - \xi_j| < \lambda\}$ of a sufficiently small radius λ with centers in some points $\xi_j \in \Gamma$. Let also η_j be smooth cutoff functions with supports in $K_{j,2\lambda}$ such that $\eta_j \zeta_j = \zeta_j$. Moreover, we suppose that

$$|\mathcal{D}_x^{\boldsymbol{k}} \zeta_j(x)| + |\mathcal{D}_x^{\boldsymbol{k}} \eta_j(x)| \leqslant c_2(|\boldsymbol{k}|)\lambda^{-|\boldsymbol{k}|}, \ |\boldsymbol{k}| \geqslant 0.$$

We assume that the radius $\lambda < \rho/2$ of the partition is so small that in the ρ -neighborhood of each point ξ_j, the surface Γ is given in the local Cartesian coordinates (y_1, y_2, y_3) by the equation

$$y_3 = \varphi_j(y_1, \ y_2),$$

where the direction of the axis y_3 is opposite to the outward normal vector $\boldsymbol{n}(\xi_j)$ of the surface Γ at the point ξ_j. The functions φ_j, defined on $K_{j,\rho}$, belong to $C^{2+\alpha}(K_{j,\rho})$; moreover, there is a uniform estimate for all $j \leqslant N_\rho$:

$$|\varphi_j|^{(2+\alpha)}_{K_{j,\rho}} \leqslant M.$$

In addition, $\varphi_j(0) = 0, \nabla\varphi_j(0) = 0$, and

$$|\varphi_j(y')| \leqslant M|y'|, \qquad |\nabla\varphi_j(y^l)| \leqslant M|y'|, \qquad y' = (y_1, y_2).$$

In these coordinates, the outward normal vector to Γ is given by

$$\boldsymbol{n} = \frac{1}{\sqrt{1 + |\nabla\varphi_j|^2}}\Big(\frac{\partial\varphi}{\partial y_1}, \frac{\partial\varphi}{\partial y_2}, -1\Big).$$

By $Z_j(x)$, we denote local boundary straightening transformation near the point ξ_j such that its Jacobian matrix $\mathcal{J}_j(x)$ has unit determinant, $\det \mathcal{J}_j = 1$, and coincides with the unit matrix at ξ_j. Next, let $\mathbb{R}^2_j$ be the plane $\{(Z_j)_3 = 0\}$, and $R_j = \mathbb{R}^3 \backslash \mathbb{R}^2_j$. With this mapping, the neighborhood $\Gamma_j \subset \Gamma$ of the point ξ_j on Γ is transformed into a part of the plane $\mathbb{R}^2_j$.

We put

$$\boldsymbol{v}'(x,\ t) = \sum_{J=1}^{N_\lambda} \boldsymbol{v}_j(x,t)\eta_j(x),$$

$$p'(x.t) = \sum_{j=1}^{N_\lambda} p_j(x,t)\eta_j(x),$$

where $\boldsymbol{v}_j(x,t) = \boldsymbol{w}_j(Z_j(x),t)$, $p_j(x,t) = s_j(Z_j(x),t)$, and $(\boldsymbol{w}_j, s_j)$ are solutions to the following problems with the flat boundaries $\mathbb{R}^2_j$, $j = 1,\ldots,N_\lambda$:

$$\begin{aligned}
&\mathcal{D}_t\boldsymbol{w}_j - \nu^\pm\Delta\boldsymbol{w}_j + \frac{1}{\rho^\pm}\nabla s_j = 0, \qquad \nabla\cdot\boldsymbol{w}_j = 0 \quad \text{in } R_j \times (0,T),\\
&\boldsymbol{w}_j|_{t=0} = 0, \qquad [\boldsymbol{w}_j]\big|_{\mathbb{R}^2_j} = 0,\\
&\big[\Pi_j\mathbb{T}(\boldsymbol{w}_j)\boldsymbol{n}_j\big]\big|_{\mathbb{R}^2_j} = \boldsymbol{b}_j, \qquad\qquad (4.3.1)\\
&\big[\boldsymbol{n}_j\cdot\mathbb{T}(\boldsymbol{w}_j,s_j)\boldsymbol{n}_{\rm j}\big]\big|_{\mathbb{R}^2_j} - \sigma\boldsymbol{n}_j\cdot\Delta_j\int_0^t \boldsymbol{w}_j\mathrm{d}t' = b'_j + \sigma\int_0^t B_j\mathrm{d}t'.
\end{aligned}$$

Here Π_j is the projection onto the plane $\mathbb{R}^2_j$, $\boldsymbol{n}_j = \boldsymbol{n}(\xi_j)$, Δ_j is the Laplacian on $\mathbb{R}^2_j$, $\boldsymbol{b}_j(z,t) = \Pi_j \boldsymbol{b}(x,t)\zeta_j(x)|_{x=Z_j^{-1}(z)}$, and $b'_j(z,t) = b'(x,t)\zeta_j(x)|_{x=Z_j^{-1}(z)}$; B_j is similarly defined.

In the coordinates $x = Z_j^{-1}(z)$, $(\boldsymbol{v}_j, p_j)$ satisfy the system

$$
\begin{aligned}
&\mathcal{D}_t \boldsymbol{v}_j - \nu^{\pm}\nabla_j^2 \boldsymbol{v}_j + \frac{1}{\rho^{\pm}}\nabla_j p_j = 0, \qquad \nabla_j \cdot \boldsymbol{v}_j = 0,\\
&\boldsymbol{v}_j|_{t=0} = 0, \quad [\boldsymbol{v}_j]|_{\Gamma_j} = 0,\\
&\big[\Pi_j \mathbb{T}^{(j)}(\boldsymbol{v}_j)\boldsymbol{n}_j\big]\big|_{\Gamma_j} = \Pi_j \boldsymbol{b}\zeta_j, \qquad (4.3.2)\\
&\big[\boldsymbol{n}_j \cdot \mathbb{T}^{(j)}(\boldsymbol{v}_j, p_j)\boldsymbol{n}_j\big]\big|_{\Gamma_j} - \sigma \boldsymbol{n}_j \cdot \widetilde{\Delta}_j - \int_0^t \boldsymbol{v}_j \mathrm{d}t' = b'\zeta_j + \sigma \int B\zeta_j \mathrm{d}t'.
\end{aligned}
$$

where $\nabla_j = (\mathcal{J}_j^{-1})^T \nabla$, $\mathcal{J}_j^{-1}$ is the Jacobi matrix of the transformation Z_j^{-1} with the elements J_j^{mk}; $\mathbb{T}^{(j)}(\boldsymbol{v}, p)$ is the tensor with the components $T_{ik}^{(j)} = -p\delta_i^k + \mu^{\pm}\sum_{m=1}^3 (J_j^{mk}\partial v_i/\partial x_m + J_j^{mi}\partial v_k/\partial x_m)$; $\Gamma_j = Z_j^{-1}(\mathbb{R}^2_j)$, $\widetilde{\Delta}_j = Z_j^{-1}(\Delta_j)$.

Now we define $\boldsymbol{v}''$ as a solution to the initial-boundary value problem for the heat equation

$$
\mathcal{D}_t \boldsymbol{v}'' - \nu^{\pm}\Delta \boldsymbol{v}'' = \boldsymbol{f}, \qquad [\boldsymbol{v}'']|_{\Gamma} = 0, \quad \boldsymbol{v}''|_{t=0} = 0, \qquad (4.3.3)
$$

with

$$
\begin{aligned}
\boldsymbol{f} = &\sum_{j=1}^{N_\lambda} \eta_j \{\nu^{\pm}(\nabla^2 - \nabla_j^2)\boldsymbol{v}_j + \frac{1}{\rho^{\pm}}(\nabla_j - \nabla)p_j\}\\
&+ \sum_{j=1}^{N_\lambda} \{2\nu^{\pm}(\nabla\eta_j \cdot \nabla)\boldsymbol{v}_j + \nu^{\pm}\boldsymbol{v}_j \nabla^2 \eta_j - \frac{1}{\rho^{\pm}} p_j \cdot \nabla \eta_j\};
\end{aligned}
$$

$\boldsymbol{v}'''$ is the gradient of the solution Φ to the Poisson equation

$$
\Delta\Phi = r - \nabla \cdot \boldsymbol{v}'' \equiv r^l \qquad (4.3.4)
$$

with $r = -\sum_{j=1}^{N_\lambda} \nabla \cdot (\eta_j \boldsymbol{v}_j) = -\sum_j \nabla\eta_j \cdot \boldsymbol{v}_j + \sum_j \eta_j (\nabla_j - \nabla) \cdot \boldsymbol{v}_j$.

We set, by definition,

$$
\mathcal{R}\boldsymbol{D} = (\boldsymbol{v}, p) \equiv (\boldsymbol{v}' + \boldsymbol{v}'' + \boldsymbol{v}''', p' + \mu^{\pm} r' - \rho^{\pm}\mathcal{D}_t \Phi).
$$

We multiply systems (4.3.2) by η_j, combining the last two conditions in them into a single vector one, and sum them up over $j = 1, \dots, N_\lambda$. Now it is easily seen that the pair $(\boldsymbol{v}, p)$ is a solution of problem (1.1.7) with the homogeneous equations and initial condition; moreover, on Γ, the boundary conditions

$$
\begin{aligned}
&[\boldsymbol{v}]|_\Gamma = 0,\\
&[\Pi\mathbb{T}(\boldsymbol{v})\boldsymbol{n}]|_\Gamma = \boldsymbol{b} + \Pi\boldsymbol{b}_1 + \Pi\boldsymbol{d} + \boldsymbol{a},\\
&[\boldsymbol{n}\cdot\mathbb{T}(\boldsymbol{v},p)\boldsymbol{n}]|_\Gamma - \sigma\boldsymbol{n}\cdot\Delta_\Gamma\int_0^t \boldsymbol{v}\mathrm{d}t' = b' + \sigma\int_0^t B\mathrm{d}t' + \boldsymbol{n}\cdot\boldsymbol{d} + \boldsymbol{n}\cdot\boldsymbol{b}_1\\
&\qquad\qquad + a' + \sigma\int_0^t (A + A_1 + A_2)\mathrm{d}t'.
\end{aligned}
$$

hold. Here

$$
\begin{aligned}
\boldsymbol{b}_1 &= \sum_{j=1}^{N_\lambda}(\Pi_j\boldsymbol{b} - \boldsymbol{b})\eta_j,\\
\boldsymbol{d} &= \sum_{j=1}^{N_\lambda}\Big\{\big[\mu^\pm((\boldsymbol{v}_j\cdot\boldsymbol{n})\nabla\eta_j + \frac{\partial\eta_j}{\partial n}\boldsymbol{v}_j)\big]\big|_\Gamma + \eta_j\big[\mathbb{T}(\boldsymbol{v}_j,p_j)\boldsymbol{n} - \mathbb{T}^{(j)}(\boldsymbol{v}_j,p_j)\boldsymbol{n}_j\big]\big|_\Gamma\\
&\quad + \eta_j(\boldsymbol{n}_j - \boldsymbol{n})\big[\boldsymbol{n}_j\cdot\mathbb{T}^{(j)}(\boldsymbol{v}_j,p_j)\boldsymbol{n}_j\big]\big|_\Gamma\Big\},\\
\boldsymbol{a} &= \big[\Pi\mathbb{T}(\boldsymbol{v}'' + \boldsymbol{v}''')\boldsymbol{n}\big]\big|_\Gamma, \qquad a' = \big[\boldsymbol{n}\cdot\mathbb{T}(\boldsymbol{v}'' + \boldsymbol{v}''', \mu^\pm r' - \rho^\pm\mathcal{D}_t\Phi)\boldsymbol{n}\big]\big|_\Gamma,\\
A &= -\boldsymbol{n}\cdot\Delta_\Gamma(\boldsymbol{v}'' + \boldsymbol{v}'''),\\
A_1 &= \boldsymbol{n}\cdot\sum_{j=1}^{N_\lambda}\big(\Delta_\Gamma(\eta_j\boldsymbol{v}_j) - \eta_j\Delta_\Gamma\boldsymbol{v}_j\big), \qquad A_2 = \sum_j \eta_j(\boldsymbol{n}\cdot\Delta_\Gamma\boldsymbol{v}_j - \boldsymbol{n}_j\cdot\widetilde{\Delta}_j\boldsymbol{v}_j).
\end{aligned}
$$

We define the operator $\mathcal{M}$ by the relation

$$
\mathcal{M}\boldsymbol{D} = (\Pi\boldsymbol{b}_1 + \Pi\boldsymbol{d} + \boldsymbol{a},\ \boldsymbol{n}\cdot\boldsymbol{b}_1 + \boldsymbol{n}\cdot\boldsymbol{d} + a',\ A + A_1 + A_2).
$$

(b) Estimating the norm of $\mathcal{M}\boldsymbol{D}$. We need to prove the inequality

$$
\begin{aligned}
&|\boldsymbol{d} + \boldsymbol{a}|_{G_t}^{(1+\alpha,\frac{1+\alpha}{2})} + |a'|_{G_t} + |\nabla_\tau a'|_{G_t}^{(\alpha,\alpha/2)}\\
&\quad + |a'|_{G_t}^{(\gamma,1+\alpha)} + |\boldsymbol{b}_1|_{G_t}^{(1+\alpha,\frac{1+\alpha}{2})} + |A + A_1 + A_2|_{G_t}^{(\alpha,\alpha/2)}\\
&\qquad \leqslant \varepsilon Q(t) + c_3(\varepsilon)\int_0^t Q(t')\mathrm{d}t', \qquad (4.3.5)
\end{aligned}
$$

where $Q(t) = |\boldsymbol{b}|_{G_t}^{(1+\alpha,\frac{1+\alpha}{2})} + |b'|_{G_t} + |\nabla_\tau b'|_{G_t}^{(\alpha,\alpha/2)} + |b'|_{G_t}^{(\gamma,1+\alpha)} + \sigma|B|_{G_t}^{(\alpha,\alpha/2)}$.

It is not difficult to see that pressure functions p_j are cancelled in the expression for the vector $\boldsymbol{d}$; therefore, only the norms of $\boldsymbol{v}_j$ are included in the estimate of the norm of $\boldsymbol{d}$. In the case of single-domain Ω with boundary Γ, the estimates of $\boldsymbol{d}$, A_1, A_2 were obtained in [54]. In our case, the only difference is that Γ is an interface between two domains, and the vector $\boldsymbol{d}$ is given in the form of jumps of the same functions, as in [54], tending to the boundary Γ from inside and from outside. Using the results of [54], we arrive at the inequality

$$|A_1|_{G_t}^{(\alpha,\alpha/2)} + |A_2|_{G_t}^{(\alpha,\alpha/2)} + |\boldsymbol{d}|_{G_t}^{(1+\alpha,\frac{1+\alpha}{2})}$$
$$\leqslant (\varepsilon_1 + \lambda^\alpha c_4) \max_{j\leqslant N_\lambda} |\boldsymbol{v}_j|_{D_t}^{(2+\alpha,1+\frac{\alpha}{2})} + c_6(\lambda,\varepsilon_1) \max_{j\leqslant N_\lambda} \int_0^t |\boldsymbol{v}_j|_{D_{t'}}^{(2+\alpha,1+\frac{\alpha}{2})}\, dt' \tag{4.3.6}$$

with arbitrary $\varepsilon_1 > 0$.

The norms of functions $\boldsymbol{v}_j$ and p_j can be estimated in terms of the norms of solutions to problems (4.3.1) for which, by Theorems 2.5.1 and 3.1.1, we obtain

$$|\boldsymbol{w}_j|_{R_{j,t}}^{(2+\alpha,1+\alpha/2)} + |\nabla s_j|_{R_{j,t}}^{(\alpha,\alpha/2)} + \langle s\rangle_{R_{j,t}}^{(\gamma,1+\alpha)}$$
$$\leqslant c_7(T)\{|\boldsymbol{b}_j|_{\mathbb{R}^2_{j,t}}^{(1+\alpha,\frac{1+a}{2})} + |\nabla_\tau b_j'|_{z',\mathbb{R}^2_{j,t}}^{(\alpha)} + |b_j'|_{\mathbb{R}^2_{j,t}}^{(\gamma,1+\alpha)} + \sigma|B_j|_{\mathbb{R}^2_{j,t}}^{(\alpha,\alpha/2)}\}$$

$(R_{j,t} = R_j \times (0,t), \mathbb{R}^2_{j,t} = \mathbb{R}^2_j \times (0,t), z' = (z_1, z_2))$.

It has been established in Chap. 2 (see inequalities (2.4.20)) that the semi-norm $\langle s_j\rangle_{t',\Omega\times(0,t)}^{(\frac{1+\alpha-\gamma}{2})}$ is bounded above by $c'(\Omega)\langle s_j\rangle_{R_{j,t}}^{(\gamma,1+\alpha)}$ for any bounded domain Ω. Therefore

$$\max_{j\leqslant N_\lambda} \{|\boldsymbol{v}_j|_{D_t}^{(2+\alpha,1+\alpha/2)} + |\nabla p_j|_{D_t}^{(\alpha,\alpha/2)} + |p_j|_{K_{j,2\lambda,t}}^{(\gamma,1+\alpha)}\} \leqslant c_8(T)Q(t) \tag{4.3.7}$$

$(K_{j,2\lambda,t} = K_{j,2\lambda} \times (0,t))$.

Next, it is obvious that

$$|\boldsymbol{a}|_{G_t}^{(1+\alpha,\frac{1+\alpha}{2})} + |a'|_{G_t} + |\nabla_\tau a'|_{G_t}^{(\alpha,\alpha/2)} + |a'|_{G_t}^{(\gamma,1+\alpha)} + |A|_{G_t}^{(\alpha,\alpha/2)} \tag{4.3.8}$$
$$\leqslant c_9\big(|\boldsymbol{v}''|_{D_t}^{(2+\alpha,1+\frac{\alpha}{2})} + |\boldsymbol{v}'''|_{\mathrm{D}_t}^{(2+\alpha,1+\frac{\alpha}{2})} + |r^l|_{D_t}^{(1+\alpha,\frac{1+\alpha}{2})} + |[\rho^\pm \mathcal{D}_t\Phi]|_\Gamma|_{G_t}^{(\gamma,1+\alpha)}\big).$$

Now we estimate the vectors $\boldsymbol{v}''$ and $\boldsymbol{v}'''$. Since the functions in the right-hand sides of system (4.3.2) vanish at $t = 0$, the function $p_j|_{t=0}$ is a solution to homogeneous problem (4.1.9) with $\Gamma = \Gamma_j$, $j = 1, \dots, N_\lambda$. By Theorem 2.5.1 and 3.1.1, in the class of weak power growth functions, the pressure p_j is determined uniquely up to a bounded time function. If we choose as p_j the function that is zero solution to (4.1.9) at $t = 0$, then $\boldsymbol{f}$ in system (4.3.3) also vanishes at $t = 0$. Consequently, compatibility conditions for this problem are satisfied which are necessary for the estimate

$$|\boldsymbol{v}''|_{D_t}^{(2+\alpha,1+\alpha/2)} \leqslant c_{10}|\boldsymbol{f}|_{D_t}^{(\alpha,\alpha/2)}. \tag{4.3.9}$$

For the vector $\boldsymbol{v}'''$, as well as for the gradient of a solution of problem (4.2.12), inequality (4.2.30) holds, i.e.,

$$\langle\nabla\nabla\boldsymbol{v}'''\rangle_{x,D_t}^{(\alpha)} \leqslant c_{11}\big(|r|_{D_t}^{(1+\alpha,\frac{1+\alpha}{2})} + |\boldsymbol{v}''|_{D_t}^{(2+\alpha,1+\alpha/2)}\big). \tag{4.3.10}$$

The estimates obtained in [54] for the norms of the functions $\boldsymbol{f}$ and r lead, in view of (4.3.7), to the inequality

$$|\boldsymbol{f}|_{D_t}^{(\alpha,\alpha/2)} + |r|_{D_t}^{(1+\alpha,\frac{1+\alpha}{2})} \leqslant (c_{12}\lambda^\alpha + \varepsilon_2)Q(t) + c_{13}(\lambda,\varepsilon_2)\int_0^t Q(t')\mathrm{d}t' \tag{4.3.11}$$

with any $\varepsilon_2 > 0$.

Now we turn to time derivative of the solution of problem (4.3.4). It is expressed in terms of the volume potential:

$$\mathcal{D}_t\Phi(x,t) = \int_{\mathbb{R}^3} \mathcal{E}(x,y)\mathcal{D}_t r'(y,t)\mathrm{d}y. \tag{4.3.12}$$

The derivative $\mathcal{D}_t r'$ admits the representation $\mathcal{D}_t r' = \nabla\cdot\boldsymbol{h} + H$, where

$$\boldsymbol{h} = \boldsymbol{h}_k + \boldsymbol{h}_2 + \boldsymbol{h}_3, \quad \boldsymbol{h}_1 = -\mathcal{D}_t\boldsymbol{v}'', \quad \boldsymbol{h}_2 = \sum_{j=1}^{N_\lambda}\eta_j(\mathcal{J}_j^{-1} - \mathbb{I})\mathcal{D}_t\boldsymbol{v}_j,$$

$$\boldsymbol{h}_3 = \sum_{j=1}^{N_\lambda}\mathcal{J}_j^{-1}\Big(\frac{1}{\rho^\pm}p_j\nabla_j\eta_j - \nu^\pm\sum_{k=1}^{3}(\nabla_j\eta_j)_k\nabla_j v_{jk}\Big),$$

$$H = \sum_j\Big\{\frac{1}{\rho^\pm}p_j\nabla_j^2\eta_j - \nu^\pm\sum_k\nabla_j(\nabla_j\eta_j)_k\cdot\nabla_j v_{jk}\Big\},$$

$\mathbb{I}$ being the identity matrix.

After integrating in (4.3.12) by parts, we obtain

$$\mathcal{D}_t\Phi = -\int_{\Omega^-\cup\Omega^+}\nabla_y\mathcal{E}\cdot\boldsymbol{h}\mathrm{d}y + \int_\Gamma \mathcal{E}[\boldsymbol{h}_3\cdot\boldsymbol{n}]\big|_\Gamma\mathrm{d}\Gamma + \int_{R^3}\mathcal{E}H\mathrm{d}y. \tag{4.3.13}$$

Since the function H has a compact support, the estimate of the Newtonian potential (Lemma 4.1.1), together with (4.3.7), implies the inequalities

$$
\begin{aligned}
&|\int_{\mathbb{R}^3} \mathcal{E} H \mathrm{d}y|_{D_t}^{(\gamma,1+\alpha)} + \langle \nabla \int_{\mathbb{R}^3} \mathcal{E} H \mathrm{d}y\rangle_{t,D_t}^{(\alpha/2)} \\
&\quad \leqslant c_{14}\langle H\rangle_{t,D_t}^{(\frac{1+\alpha-\gamma}{2})} \leqslant \varepsilon_3\langle H\rangle_{D_t}^{(\gamma,1+\alpha)} + c_{15}(\varepsilon_3)\int_0^t |H|_{D_t}^{(\gamma,1+\alpha)}\mathrm{d}t' \\
&\quad \leqslant c_{16}\varepsilon_3 Q(t) + c_{17}(\varepsilon_3)\int_0^t Q(t')\mathrm{d}t' \qquad (4.3.14)
\end{aligned}
$$

with any $\varepsilon_3 > 0$. We have also used an estimate similar to that from Lemmas 6.4.1 and 6.4.2. In addition, we apply again similar inequalities below, in estimate (4.3.16).

The terms containing $\boldsymbol{h}_3$ can be estimated in an analogues way:

$$
\begin{aligned}
&|\int_{\Omega^-\cup\Omega^+} \nabla_y \mathcal{E}\cdot h_3 \mathrm{d}y|_{D_t}^{(\gamma,1+\alpha)} + |\int_\Gamma \mathcal{E}[\boldsymbol{h}_3\cdot\boldsymbol{n}]|_\Gamma \mathrm{d}\Gamma|_{D_t}^{(\gamma,1+\alpha)} \\
&\quad \leqslant c_{18}\langle \boldsymbol{h}_3\rangle_{t,D_t}^{(\frac{1+\alpha-\gamma}{2})} \leqslant \varepsilon_4 Q(t) + c_{19}(\varepsilon_4)\int_0^t Q(t')\mathrm{d}t',\ \forall\varepsilon_4 > 0. (4.3.15)
\end{aligned}
$$

In view of (4.1.6), we deduce

$$
\begin{aligned}
&\langle \nabla_x \int_{\Omega^-\cup\Omega^+} \nabla_y\mathcal{E}\cdot\boldsymbol{h}_3\ \mathrm{d}y\rangle_{t',D_t}^{(\alpha/2)} + \langle \nabla_x\int_\Gamma \mathcal{E}[\boldsymbol{h}_3\cdot\boldsymbol{n}]|_\Gamma\ \mathrm{d}\Gamma\rangle_{t',D_t}^{(\alpha/2)} \\
&\quad \leqslant c_{20}|\langle \boldsymbol{h}_3\rangle_{(0,t)}^{(\alpha/2)}|_{\cup\Omega^\pm}^{(\beta)} \leqslant \varepsilon_5|\boldsymbol{h}_3|_{D_t}^{(\gamma,1+\alpha)} + c_{21}(\varepsilon_5)\int_0^t |\boldsymbol{h}_3|_{D_{t'}}^{(\gamma,1+\alpha)}\ \mathrm{d}t',\ \beta<\gamma. \\
&\qquad (4.3.16)
\end{aligned}
$$

Next, by using equations for $(\boldsymbol{v}_j, p_j)$ and $\boldsymbol{v}''$, we can write the sum $\boldsymbol{h}_1+\boldsymbol{h}_2$ as follows: $\boldsymbol{h}_1+\boldsymbol{h}_2 = \boldsymbol{h}_4 - \nu^\pm\nabla\cdot\nabla\boldsymbol{v}'' + \boldsymbol{h}_5$, where

$$
\boldsymbol{h}_4 = \sum_j \eta_j\Big\{(\mathcal{J}_j^{-1}-\mathbb{I})\Big(\nu^\pm\nabla_j^2\boldsymbol{v}_j - \frac{1}{\rho^\pm}\nabla_j p_j\Big) - \nu^\pm(\nabla^2-\nabla_j^2)\boldsymbol{v}_j - \frac{1}{\rho^\pm}(\nabla_j-\nabla)p_j\Big\},
$$

$$
\boldsymbol{h}_5 = \sum_j \Big\{2\nu^\pm(\nabla\eta_j\cdot\nabla)\boldsymbol{v}_j + \nu^\pm\boldsymbol{v}_j\nabla^2\eta_j - \frac{1}{\rho^\pm}\nabla\eta_j p_j\Big\}.
$$

For $\boldsymbol{h}_4$, it was obtained the representation (Chap. 4, [54])

$$
\boldsymbol{h}_4 = \sum_{m=1}^3 \partial\boldsymbol{h}'_m/\partial x_m + \boldsymbol{h}''',
$$

where $\{\boldsymbol{h}'_m\}_{m=1}^3$ and $\boldsymbol{h}'''$ are compactly supported functions which do not contain higher-order derivatives of $\boldsymbol{v}_j$ and p_j. Therefore

$$\int_{\Omega^-\cup\Omega^+} \nabla_y \mathcal{E}\cdot(\boldsymbol{h}_1+\boldsymbol{h}_2)\mathrm{d}y$$

$$=-\sum_{m=1}^{3}\int_{\Omega^-\cup\Omega^+} \nabla_y \partial\mathcal{E}/\partial y_m(\boldsymbol{h}'_m-\nu^{\pm}\partial\boldsymbol{v}''/\partial y_m)\mathrm{d}y$$

$$+\int_\Gamma \nabla_y\mathcal{E}\cdot[\boldsymbol{h}'_m n_m-\nu^{\pm}\partial\boldsymbol{v}''/\partial\boldsymbol{n}]|_\Gamma \mathrm{d}\Gamma+\int_{\Omega^-\cup\Omega^+}\nabla_y\mathcal{E}\cdot(\boldsymbol{h}_5+\boldsymbol{h}''')\mathrm{d}y.$$

The vectors $\boldsymbol{h}_5$ and $\boldsymbol{h}'''$ are estimated in the same way as $\boldsymbol{h}_3$, and for the last term, an estimate like (4.3.15) holds. The first two integrals are evaluated by means of inequalities (4.1.5). For the semi-norm $|\boldsymbol{h}'_m|_{D_t}^{(\gamma,1+\alpha)}$ there is an estimate in terms of the norms $\boldsymbol{v}_j$ and p_j which contains the higher-order semi-norms only with the multiplier λ. By virtue of the above as well as of inequalities (4.3.14), (4.3.15), (4.3.9), and (4.3.11), we finally obtain

$$|\mathcal{D}_{t'}\Phi|_{D_t}^{(\gamma,1+\alpha)} \leqslant (c_{22}\lambda^\alpha+\varepsilon_6)Q(t)+c_{23}(\lambda,\varepsilon_6)\int_0^t Q(t')\mathrm{d}t'. \tag{4.3.17}$$

In order to estimate $\langle\mathcal{D}_{t'}\boldsymbol{v}'''\rangle_{t',D_t}^{(\alpha/2)}=\langle\nabla\mathcal{D}_{t'}\Phi\rangle_{t',D_t}^{(\alpha/2)}$, we differentiate (4.3.13) with respect to x. Due to inequalities (4.3.14) and (4.3.16), we need to consider only the integral containing $\boldsymbol{h}_1$ and $\boldsymbol{h}_2$. By Lemma 4.1.3, we have

$$\langle\nabla_x\int_{\Omega^-\cup\Omega^+}\nabla_y\mathcal{E}\cdot(\boldsymbol{h}_1+\boldsymbol{h}_2)\mathrm{d}y\rangle_{t',D_t}^{(\alpha/2)}$$

$$\leqslant c_{24}|\boldsymbol{h}_1|_{D_t}^{(\alpha,\frac{\alpha}{2})}+|\boldsymbol{h}_2|_{D_t}^{(\alpha,\frac{\alpha}{2})}+|\boldsymbol{h}'_m|_{D_t}^{(\gamma,1+\alpha)}+|\nabla\boldsymbol{v}''|_{D_t}^{(\gamma,1+\alpha)}+|\langle\boldsymbol{h}_5+\boldsymbol{h}'''\rangle_{(0,t)}^{(\frac{\alpha}{2})}|_{\cup\Omega^\pm}^{(\beta)}$$

$$\leqslant(c_{25}\lambda+\varepsilon_7)Q(t)+c_{26}(\lambda,\varepsilon_7)\int_0^t Q(t')\mathrm{d}t'.$$

Consequently

$$\langle\mathcal{D}_{t'}\boldsymbol{v}'''\rangle_{t',D_t}^{(\alpha/2)}\leqslant(c_{25}\lambda+\varepsilon_8)Q(t)+c_{27}(\lambda,\varepsilon_8)\int_0^t Q(t')\mathrm{d}t'. \tag{4.3.18}$$

This inequality, together with (4.3.10) and (4.3.11), implies the desirable estimate for $|\boldsymbol{v}'''|_{D_t}^{(2+\alpha,1+\frac{\alpha}{2})}$ since $\boldsymbol{v}'''|_{t'=0}=\mathcal{D}_{t'}\boldsymbol{v}'''|_{t'=0}=0$, and the complete norm of $\boldsymbol{v}'''$ on the subspace $\overset{\circ}{\boldsymbol{C}}{}^{2+\alpha,1+\alpha/2}(D_t)$ is equivalent to the sum of $\langle\nabla\nabla\boldsymbol{v}'''\rangle_{x,D_t}^{(\alpha)}$ and $\langle\mathcal{D}_{t'}\boldsymbol{v}'''\rangle_{t',D_t}^{(\alpha/2)}$ (see Lemma 4.3 in [77]).

It remains to estimate only the norm $|\boldsymbol{b}_1|_{G_t}^{(1+\alpha,\frac{1+a}{2})}$. Since $\boldsymbol{b}\cdot\boldsymbol{n}=0$, we have $\Pi_j\boldsymbol{b}-\boldsymbol{b}=\boldsymbol{n}_j(\boldsymbol{b}\cdot(\boldsymbol{n}-\boldsymbol{n}_j))$ and, obviously,

$$\Big|\sum_j \eta_j(\Pi_j\boldsymbol{b}-\boldsymbol{b})\Big|_{G_t}^{(1+\alpha,\frac{1+a}{2})} \leqslant c_{28}\lambda|\boldsymbol{b}|_{G_t}^{(1+\alpha,\frac{1+\alpha}{2})}+c_{29}(\lambda)\int_0^t |\boldsymbol{b}|_{G_{t'}}^{(1+\alpha\frac{1+\alpha}{2})}dt'. \tag{4.3.19}$$

Thus, for sufficiently small λ and $\varepsilon_1,\varepsilon_2,\varepsilon_6,\varepsilon_8$, inequalities (4.3.6)–(4.3.11) and (4.3.17)–(4.3.19) imply (4.3.5) with $\varepsilon<1$.

2°. The general case reduces to the one considered above by constructing, firstly, a vector field $\boldsymbol{U}\in \boldsymbol{C}^{2+\alpha,1+\alpha/2}(D_T)$ such that

$$\boldsymbol{U}|_{t=0}=\boldsymbol{v}_0,\quad [\boldsymbol{U}]|_\Gamma=0,\quad \mathcal{D}_t\boldsymbol{U}|_{t=0}=\nu^\pm\Delta\boldsymbol{v}_0-\frac{1}{\rho^\pm}\nabla p_0+\boldsymbol{f}|_{t=0},$$

where p_0 is the solution of problem (4.1.9) studied in Sect. 4.2. In addition, $\boldsymbol{U}$ must satisfy the estimate

$$|\boldsymbol{U}|_{D_T}^{(2+\alpha,1+\alpha/2)}\leqslant c_{30}\Big(|\boldsymbol{v}_0|_{\cup\Omega^\pm}^{(2+\alpha)}+|\nabla p_0|_{\cup\Omega^\pm}^{(\alpha)}+|\boldsymbol{f}|_{D_T}^{(\alpha,\alpha/2)}\Big).$$

Due to compatibility condition (4.1.8), such a vector field $\boldsymbol{U}$ does exist. The proof of this is carried out in the same way as in the case of a plane boundary Γ (Chap. 2).

Secondly, a solution to the system with the new velocity vector field $\boldsymbol{V}=\boldsymbol{v}-\boldsymbol{U}$ can be represented in the form $(\boldsymbol{V},p)=\big(\boldsymbol{w}+\boldsymbol{w}'+\boldsymbol{u},P_0-P_0(x^\pm)+q+\mu^\pm r'-\rho^\pm(\mathcal{D}_t\Phi-\mathcal{D}_t\Phi(x^\pm,t))\big)$, where $\boldsymbol{w}'=\nabla\Phi$, while $\boldsymbol{w}$ and Φ solve problems (4.2.10) and (4.2.12), respectively, $P_0(x,t)\equiv p_0(x)$, and $(\boldsymbol{u},q)$ is the solution of the initial-boundary value problem for the Stokes system from 1°. The vector $\boldsymbol{w}$ is subjected to inequality (4.2.11). The semi-norm $\langle\mathcal{D}_t\boldsymbol{w}'\rangle_{t,D_T}^{(\alpha/2)}$ can be estimated by Lemma 4.1.3, and $\langle\nabla\nabla\boldsymbol{w}'\rangle_{x,D_T}^{(\alpha)}$ can be done by inequality (4.2.30). Therefore, since $\boldsymbol{w}'$ belongs to the space $\overset{\circ}{\boldsymbol{C}}{}^{2+\alpha,1+\alpha/2}$, the norm $\langle\boldsymbol{w}'\rangle_{D_T}^{(2+\alpha,1+\alpha/2)}$can be also estimated. The boundedness of the norms of $\boldsymbol{u}$ and q is proved in the same way as in the case of $\Gamma=\{x_3=0\}$ (see Chap. 2).

The uniqueness of a solution to problem (1.1.7) follows from Theorem 4.1.1.

The proof of Theorem 4.1.2 is complete. □

Chapter 5
Local Solvability of the Problem in Weighted Hölder Spaces

Abstract Here, we analyze nonlinear problem (1.1.5). To this end, we modify the technique developed for studying the problem on the motion of a finite fluid volume in a vacuum [54]. The significant difference is that we work in the weighted Hölder spaces since liquids occupy the entire space, and the nonlinearity of the problem requires the decrease of unknown functions at infinity with respect to the space variables. Under the assumption of a sufficiently rapid decay of mass forces and the initial distribution of velocities at infinity, we prove local-in-time existence of a classical solution to the problem which has a power-law decrease at infinity.

These results were firstly published in our articles [29] and [18].

5.1 Weighted Hölder Spaces: Formulation of the Local Existence Theorem for the Nonlinear Problem

For the convenience of the reader, we state the problem (1.1.5) again:

$$\begin{aligned}
&\mathcal{D}_t \boldsymbol{u} - \nu^{\pm}\nabla_{\boldsymbol{u}}^2 \boldsymbol{u} + \frac{1}{\rho^{\pm}}\nabla_{\boldsymbol{u}} q = \boldsymbol{f}(X_{\boldsymbol{u}}, t),\\
&\nabla_{\boldsymbol{u}} \cdot \boldsymbol{u} = 0 \quad \text{in} \quad Q_T^{\pm} = \Omega_0^{\pm} \times (0,T),\\
&\boldsymbol{u}|_{t=0} = \boldsymbol{v}_0 \ \text{in} \ \Omega^- \cup \Omega^+, \quad \boldsymbol{u} \xrightarrow[|\xi|\to\infty]{} 0, \qquad\qquad (5.1.1)\\
&[\boldsymbol{u}]|_{G_t} = 0, \ \ [\mu^{\pm}\Pi_0\Pi\mathbb{S}_{\boldsymbol{u}}(\boldsymbol{u})\boldsymbol{n}]|_{G_t} = 0 \quad (G_t \equiv \Gamma \times (0,T)),\\
&[\boldsymbol{n}_0 \cdot \mathbb{T}_{\boldsymbol{u}}(\boldsymbol{u}, q)\boldsymbol{u}]|_{G_t} - \sigma \boldsymbol{n}_0 \cdot \Delta(t) X_{\boldsymbol{u}}|_{G_t} = 0.
\end{aligned}$$

© The Author(s), under exclusive license to Springer Nature Switzerland AG 2021

I. V. Denisova, V. A. Solonnikov, *Motion of a Drop in an Incompressible Fluid*, Advances in Mathematical Fluid Mechanics,
https://doi.org/10.1007/978-3-030-70053-9_5

We recall that the problem (5.1.1) was studied in the Sobolev–Slobodetskiĭ spaces in [15, 17, 27] where its solvability was obtained on a finite time interval. In this book, we present this material in Chap. 10.

Let us give the definition of the Hölder spaces with power-law weight.

Let Ω be a domain in $\mathbb{R}^n$, $n \in \mathbb{N}$; for $T > 0$, we set $Q_T = \Omega \times (0, T)$. The space $C_\beta^{\alpha,\alpha/2}(Q_T)$, $\alpha \in (0,1)$, $\beta \geqslant 0$, is a set of functions f given in Q_T with the norm

$$|f|_{\beta,Q_T}^{(\alpha,\alpha/2)} = |f|_{\beta,Q_T} + \langle f\rangle_{\beta,Q_T}^{(\alpha,\alpha/2)},$$

where

$$\begin{aligned}
|f|_{\beta,Q_T} &= \sup_{t\in(0,T)} \sup_{x\in\Omega} (1+|x|)^\beta |f(x,t)|,\\
\langle f\rangle_{\beta,Q_T}^{(\alpha,\alpha/2)} &= \langle f\rangle_{x,\beta,Q_T}^{(\alpha)} + \langle f\rangle_{t,\beta,Q_T}^{(\alpha/2)},\\
\langle f\rangle_{x,\beta,Q_T}^{(\alpha)} &= \sup_{t\in(0,T)} \sup_{x\in\Omega} (1+|x|)^\beta \sup_{y\in\Omega} |f(x,t)-f(y,t)||x-y|^{-\alpha},\\
\langle f\rangle_{t,\beta,Q_T}^{(\mu)} &= \sup_{x\in\Omega} (1+|x|)^\beta \sup_{t,\tau\in(0,T)} |f(x,t)-f(x,\tau)||t-\tau|^{-\mu},
\end{aligned}$$

$$\mu \in (0,1).$$

We set

$$\begin{aligned}
|f|_{x,\beta,Q_T}^{(\alpha)} &= |f|_{\beta,Q_T} + \langle f\rangle_{x,\beta,Q_T}^{(\alpha)},\\
|f|_{t,\beta,Q_T}^{(\mu)} &= |f|_{\beta,Q_T} + \langle f\rangle_{t,\beta,Q_T}^{(\mu)}.
\end{aligned}$$

The space $C_\beta^{k+\alpha,(k+\alpha)/2}(Q_T)$, $k \in \mathbb{N}$, consists, by definition, of functions f with finite norm

$$\begin{aligned}
|f|_{\beta,Q_T}^{(k+\alpha,\frac{k+\alpha}{2})} &= \sum_{|\boldsymbol{r}|+2s\leqslant k} |\mathcal{D}_x^{\boldsymbol{r}}\mathcal{D}_t^s f|_{\beta,Q_T} + \langle f\rangle_{\beta,Q_T}^{(k+\alpha,\frac{k+\alpha}{2})},\\
\langle f\rangle_{\beta,Q_T}^{(k+\alpha,\frac{k+\alpha}{2})} &= \sum_{|\boldsymbol{r}|+2s=k} \langle \mathcal{D}_x^{\boldsymbol{r}}\mathcal{D}_t^s f\rangle_{\beta,Q_T}^{(\alpha,\alpha/2)} + \sum_{|\boldsymbol{r}|+2s=k-1} \langle \mathcal{D}_x^{\boldsymbol{r}}\mathcal{D}_t^s f\rangle_{t,\beta,Q_T}^{(\frac{1+\alpha}{2})}.
\end{aligned}$$

We define $C_\beta^{k+\alpha}(\Omega)$, $k \in \mathbb{N}\cup\{0\}$, as a space of functions $f(x)$, $x \in \Omega$, having finite norm

$$|f|_{\beta,\Omega}^{(k+\alpha)} = |f|_{\beta,\Omega}^{(k)} + \langle f\rangle_{\beta,\Omega}^{(k+\alpha)},$$

where

$$|f|^{(k)}_{\beta,\Omega} = \sum_{|\boldsymbol{r}|\leqslant k} |\mathcal{D}^{\boldsymbol{r}}_x f|_{\beta,\Omega},$$

$$\langle f\rangle^{(k+\alpha)}_{\beta,\Omega} = \sum_{|\boldsymbol{r}|=k} \langle \mathcal{D}^{\boldsymbol{r}}_x f\rangle^{(\alpha)}_{\beta,\Omega} = \sum_{|\boldsymbol{r}|=k} \sup_{x\in\Omega}(1+|x|)^{\beta} \sup_{y\in\Omega} \big|\mathcal{D}^{\boldsymbol{r}}_x f(x) - \mathcal{D}^{\boldsymbol{r}}_y f(y)\big||x-y|^{-\alpha}.$$

We will need below the weighted semi-norms (with $\alpha,\gamma\in(0,1)$)

$$|f|^{(\gamma,1+\alpha)}_{1,\gamma,Q_T} \equiv \langle f\rangle^{(\gamma,1+\alpha)}_{1+\gamma,Q_T} + \langle f\rangle^{(\frac{1+\alpha-\gamma}{2})}_{t,1,Q_T},$$

$$|f|^{(\gamma,1+\alpha)}_{1+\gamma,Q_T} \equiv \langle f\rangle^{(\gamma,1+\alpha)}_{1+\gamma,Q_T} + \langle f\rangle^{(\frac{1+\alpha-\gamma}{2})}_{t,1+\gamma,Q_T},$$

$$\langle f\rangle^{(\gamma,1+\alpha)}_{1+\gamma,Q_T} \equiv \sup_{t,\tau\in(0,T)} \sup_{x,y\in\Omega} (1+|x|)^{1+\gamma}\frac{|f(x,t)-f(y,t)-f(x,\tau)+f(y,\tau)|}{|x-y|^{\gamma}|t-\tau|^{(1+\alpha-\gamma)/2}}.$$

The estimate

$$\langle f\rangle^{(\gamma,1+\alpha)}_{1+\gamma,Q_T} \leqslant c_1\langle f\rangle^{(1+\alpha,\frac{1+\alpha}{2})}_{1+\gamma,Q_T}$$

can be proved just in the same way as a similar inequality without weight (see [77]).

We will write $f\in C^{(\gamma,1+\alpha)}_{1,\gamma}(Q_T)$ if

$$|f|_{1,Q_T} + |f|^{(1+\alpha,\gamma)}_{1,\gamma,Q_T} < \infty.$$

If the norm

$$|f|^{(\gamma,\mu)}_{\beta,\gamma,Q_T} \equiv |f|^{(\mu)}_{t,\beta,Q_T} + \langle f\rangle^{(\gamma)}_{x,\beta+\gamma,Q_T}, \quad \gamma\in(0,1),\ \ \mu\in[0,1),\ \ \beta\geqslant 0,$$

of functions f is finite, we will say that f belongs to the Hölder space $C^{\gamma,\mu}_{\beta,\gamma}(Q_T)$.

Analogously, we define the norm in the space $C^{\nu}_{\beta,\nu}(\Omega)$ by

$$|\cdot|^{(\nu)}_{\beta,\nu,\Omega} \equiv |\cdot|_{\beta,\Omega} + \langle\cdot\rangle^{(\nu)}_{\beta+\nu,\Omega}.$$

By using local maps and a partition of unity, we can introduce these spaces on any smooth manifold, in particular, on Γ and on $G_T=\Gamma\times(0,T)$.

We recall that $\mathbb{R}^3_T=\mathbb{R}^3\times(0,T)$, $Q^{\pm}_T=\Omega^{\pm}\times(0,T)$, $D_T=Q^-_T\cup Q^+_T$, and

$$|f|^{(k,k/2)}_{\beta,D_T} = |f|^{(k,k/2)}_{\beta,Q^-_T} + |f|^{(k,k/2)}_{\beta,Q^+_T},$$

$$|f|^{(k)}_{\beta,\cup\Omega^{\pm}} = |f|^{(k)}_{\beta,\Omega^-} + |f|^{(k)}_{\beta,\Omega^+},$$

$$|f|_{\beta,D_T}^{(k+\alpha,\frac{k+\alpha}{2})} = |f|_{\beta,Q_T^-}^{(k+\alpha,\frac{k+\alpha}{2})} + |f|_{\beta,Q_T^+}^{(k+\alpha,\frac{k+\alpha}{2})},$$

$$|f|_{\beta,\cup\Omega^\pm}^{(k+\alpha)} = |f|_{\beta,\Omega^-}^{(k+\alpha)} + |f|_{\beta,\Omega^+}^{(k+\alpha)}.$$

If $\beta = 0$, all the introduced weighted spaces coincide with ordinary Hölder spaces. In addition, for functions above defined on a compact manifold (e.g., on Γ or on G_T), the weighted norms are equivalent to the ordinary Hölder ones.

We state now the main result of this chapter.

Theorem 5.1.1 *We assume that for some* $\alpha, \gamma \in (0,1)$, $\gamma \leqslant \alpha$, *and* $T < \infty$ $\Gamma \in C^{3+\alpha}$, $\boldsymbol{f}, \mathcal{D}_x\boldsymbol{f} \in \boldsymbol{C}_2^{\alpha,\frac{1+\alpha-\gamma}{2}}(\mathbb{R}_T^3)$, $\boldsymbol{v}_0 \in \boldsymbol{C}_{1+\gamma}^{2+\alpha}(\Omega^+ \cup \Omega^-)$, *and* $\sigma \in C^{1+\alpha}(\Gamma)$, $\sigma(x) \geqslant 0$ *for* $x \in \Gamma$. *Let, in addition, the compatibility conditions be satisfied:*

$$\nabla \cdot \boldsymbol{v}_0 = 0,$$

$$[\boldsymbol{v}_0]|_\Gamma = 0, \quad [\mu^\pm \Pi_0 \mathbb{S}(\boldsymbol{v}_0)\boldsymbol{n}_0]|_\Gamma = 0, \quad \Big[\Pi_0\Big(\nu^\pm \nabla^2 \boldsymbol{v}_0 - \frac{1}{\rho^\pm}\nabla q_0\Big)\Big]\Big|_\Gamma = 0,$$

where the function $q_0(\xi) \equiv q(\xi, 0)$ *is a solution to the diffraction problem*

$$\frac{1}{\rho^\pm}\nabla^2 q_0(\xi) = \nabla \cdot (\boldsymbol{f}(\xi,0) + \mathcal{D}_t \mathbb{B}^*(\boldsymbol{v}_0)\boldsymbol{v}_0(\xi)), \quad \xi \in \Omega^+ \cup \Omega^-, \tag{5.1.2}$$

$$[q_0]|_\Gamma = \Big[2\mu^\pm \frac{\partial \boldsymbol{v}_0}{\partial \boldsymbol{n}_0} \cdot \boldsymbol{n}_0\Big]\Big|_\Gamma - \sigma H_0, \qquad \Big[\frac{1}{\rho^\pm}\partial q_0/\partial \boldsymbol{n}_0\Big]\Big|_\Gamma = [\nu^\pm \boldsymbol{n}_0 \cdot \nabla^2 \boldsymbol{v}_0]|_\Gamma.$$

(Here $\mathbb{B}^*$ *is the transpose of* $\mathbb{B} = \mathbb{A} - \mathbb{I}$, $\mathbb{I}$ *is the identity matrix, and* $H_0(\xi) = \boldsymbol{n}_0 \cdot \Delta(0)\boldsymbol{\xi}$ *is the doubled mean curvature of the surface* Γ.)

Under these conditions, problem (5.1.1) *is uniquely solvable on some finite time interval* $(0, T_0)$, $T_0 \leqslant T$, *whose length depends on the norms of* $\boldsymbol{f}$, *and* $\boldsymbol{v}_0$ *and on the curvature of* Γ. *The solution* $(\boldsymbol{u}, q)$ *is such that* $\boldsymbol{u} \in \boldsymbol{C}_{1+\gamma}^{2+\alpha,1+\alpha/2}(D_{T_0})$, $q \in C_{1,\gamma}^{(\gamma,1+\alpha)}(D_{T_0})$, $\nabla q \in \boldsymbol{C}_{1+\gamma}^{\alpha,\alpha/2}(D_{T_0})$, *and the inequality*

$$\begin{aligned} &|\boldsymbol{u}|_{1+\gamma,D_{T_0}}^{(2+\alpha,1+\alpha/2)} + |\nabla q|_{1+\gamma,D_{T_0}}^{(\alpha,\alpha/2)} + |q|_{1,\gamma,D_{T_0}}^{(\gamma,1+\alpha)} \\ &\qquad \leqslant c_1\Big\{|\boldsymbol{f}|_{2,\mathbb{R}_{T_0}^3}^{(\alpha,\frac{1+\alpha-\gamma}{2})} + |\boldsymbol{v}_0|_{1+\gamma,\cup\Omega^\pm}^{(2+\alpha)} + \sigma|H_0|_\Gamma^{(1+\alpha)}\Big\} \end{aligned} \tag{5.1.3}$$

holds with the constant c_1 *depending on* T *and on the norms of problem data.*

The proof is based on the weighted estimates of the solution of problem (1.1.7) which has been obtained in Chap. 4 (Theorem 4.1.1).

Remark 5.1.1 *The complete norm* $|\boldsymbol{R}|_{D_T}^{(2+\alpha,1+\alpha/2)}$ *in the right-hand side of* (4.1.10) *in Theorem* 4.1.1 *can be replaced with the sum of the semi-norms* $\langle \mathcal{D}_t \boldsymbol{R}\rangle_{D_T}^{(\alpha)}$ *and* $|\boldsymbol{R}|_{D_T}^{(\gamma,1+\alpha)}$ *with some* $\gamma \in (0,1)$ *is required.*

In what follows, we shall also need an existence theorem for the Cauchy problem (4.2.1) in the ordinary Hölder spaces, as well as estimates of a solution (Theorem 4.2.1, see [77]).

Remark 5.1.2 *It is easy to get an estimate for lower-order parts of the norms* $|\boldsymbol{w}|_{\mathbb{R}_T^3}^{(1+\alpha)}$ *and* $|\nabla s|_{\mathbb{R}_T^3}^{(\alpha)}$ *since they are estimated in terms of the semi-norms in* (4.2.3) *and the maxima of moduli of* $|\boldsymbol{w}|_{\mathbb{R}_T^3}$ *and* $|\nabla s|_{\mathbb{R}_T^3}$. *The boundedness of* $\boldsymbol{w}$, *for example, can be proved on the basis of the identity*

$$\boldsymbol{w}(x,t) = \boldsymbol{w}_0(x) + \int_0^t \mathcal{D}_t \boldsymbol{w}(x,t')\,\mathrm{d}t'$$

and the inequality

$$|\mathcal{D}_t \boldsymbol{w}|_{\mathbb{R}_{T_1}^3} \leqslant \varepsilon \langle \mathcal{D}_t \boldsymbol{w}\rangle_{t,\mathbb{R}_{T_1}^3}^{(\alpha/2)} + c_4(\varepsilon, T_1)\big(|\boldsymbol{w}|_{\mathbb{R}_{T_1}^3} + |\mathcal{D}_t \boldsymbol{w}(\cdot,0)|_{\mathbb{R}^3}\big),$$

valid for any $T_1 \leqslant T$ *(see* [54]*). The boundedness of* $|\mathcal{D}_t \boldsymbol{w}(\cdot,0)|_{\mathbb{R}^3}$ *follows from the first equation in* (4.2.1) *at* $t=0$ *in view of the fact that* $s_0(x) \equiv s(x,0)$ *satisfies the Poisson equation*

$$\frac{1}{\rho}\nabla^2 s_0(x) = \nabla \cdot (\nu \nabla r(x,0) - \boldsymbol{g}(x,0)) \equiv \nabla \cdot \boldsymbol{g}_0,$$

its gradient being expressed in term of the singular integral

$$\nabla s_0(x) = \rho \nabla_x \int_{\mathbb{R}^3} \nabla_y \mathcal{E}(x,y) \cdot \boldsymbol{g}_0(y)\,\mathrm{d}y.$$

For $|\nabla s_0|$, *the following estimate*

$$|\nabla s_0|_{\mathbb{R}^3} \leqslant c_5\big(\langle \boldsymbol{g}_0\rangle_{\mathbb{R}^3}^{(\alpha)} + \langle \mathbb{G}^0\rangle_{\mathbb{R}^3}^{(\gamma)}\big), \quad \alpha, \gamma \in (0,1), \tag{5.1.4}$$

is well-known [76], *where the tensor* $\mathbb{G}^0$ *has the components* $G_{ij}^{(0)}(x) = \nu \delta_i^j r(x,0) - G_{ij}(x,0)$, $i,j = 1,2,3$.

Thus, we arrive at the inequality

$$|\boldsymbol{w}|_{\mathbb{R}_t^3} \leqslant c_6(t)\Big(F(t) + \int_0^t |\boldsymbol{w}|_{\mathbb{R}_{t'}^3}\,dt'\Big),$$

where

$$F(t) = |\boldsymbol{f}|_{\mathbb{R}^3_t}^{(\alpha,\alpha/2)} + |r|_{\mathbb{R}^3_t}^{(1+\alpha,\frac{1+\alpha}{2})} + |\boldsymbol{w}_0|_{\mathbb{R}^3}^{(2+\alpha)} + \langle \boldsymbol{g}\rangle_{\mathbb{R}^3_t}^{(\alpha,\alpha/2)} + \langle \mathbb{G}\rangle_{x,\mathbb{R}^3_t}^{(\gamma)} + \langle \mathbb{G}\rangle_{\mathbb{R}^3_t}^{(1+\alpha,\gamma)}. \tag{5.1.5}$$

By applying the Gronwall lemma, we obtain

$$|\boldsymbol{w}|_{\mathbb{R}^3_T} \leqslant c_7(T)F(T).$$

The boundedness of $|\nabla s|_{\mathbb{R}^3_T}$ *follows from* (5.1.4) *and the inequality*

$$|\nabla s|_{\mathbb{R}^3_T} \leqslant T^{\alpha/2}\langle \nabla s\rangle_{t,\mathbb{R}^3_T}^{(\alpha/2)} + |\nabla s_0|_{\mathbb{R}^3}.$$

Thus, we see that to estimate the complete norms of $\boldsymbol{w}$ *and* ∇s, *it is sufficient to add boundedness requirement of* $\langle \mathbb{G}\rangle_{x,\mathbb{R}^3_T}^{(\gamma)}$ *to the assumptions of Theorem* 4.2.1.

5.2 Weighted Estimates for Linear Problem (1.1.7)

Now we prove a theorem on the solvability of problem (1.1.7) in weighted Hölder spaces.

Remark 5.2.1 *Concerning* r *and* $\boldsymbol{f}$, *it will be more convenient to assume the validity of relations* (4.2.2) (*instead of* (4.1.7)). *If* $r|_{t=0} = 0$, *which will always be the case below, then we can set*

$$\boldsymbol{R} = \int_0^t (\boldsymbol{g} + \boldsymbol{f})\,\mathrm{d}t'$$

in (4.1.7). *By Remark* 5.1.1, *the differential properties of this vector-valued function ensure that the corresponding assumptions of Theorem* 4.1.1 *are fulfilled, because the semi-norms* $\langle \mathcal{D}_t\boldsymbol{R}\rangle_{D_T}^{(\alpha)}$ *and* $\langle \boldsymbol{R}\rangle_{D_T}^{(\nu,1+\alpha)}$, $\nu \leqslant \alpha$, *are bounded by the sum* $|\boldsymbol{g}|_{D_T}^{(\alpha,\alpha/2)} + |\boldsymbol{f}|_{D_T}^{(\alpha,\alpha/2)}$. *The normal component of* $\boldsymbol{R}$ *is continuous when passing across the surface* Γ *if we assume that*

$$[(\boldsymbol{g} + \boldsymbol{f})\cdot \boldsymbol{n}]|_{\Gamma} = 0. \tag{5.2.1}$$

Theorem 5.2.1 *Let assumptions* (4.1.8), (4.1.9), (4.2.2), and (5.2.1) *be fulfilled, and let* $\alpha, \gamma \in (0,1)$, $\gamma \leqslant \alpha$, *and* $T < \infty$. *Assume that* $\Gamma \in C^{2+\alpha}$, $\sigma \in C^{1+\alpha}(\Gamma)$, $\sigma \geqslant 0$, $\boldsymbol{f} \in \boldsymbol{C}_{1+\gamma}^{\alpha,\alpha/2}(D_T)$, $r \in C_{1+\gamma}^{1+\alpha,\frac{1+\alpha}{2}}(D_T)$, $r|_{t=0} = 0$, $\boldsymbol{g} \in \boldsymbol{C}_{1+\gamma}^{\alpha,\alpha/2}(D_T)$, $\boldsymbol{v}_0 \in \boldsymbol{C}_{1+\gamma}^{2+\alpha}(\Omega^- \cup \Omega^+)$, $\boldsymbol{b} \in \boldsymbol{C}^{1+\alpha,\frac{1+\alpha}{2}}(G_T)$, $\boldsymbol{b}\cdot\boldsymbol{n} = 0$, $b' \in C^{(\gamma,1+\alpha)}(G_T)$, $B \in C^{\alpha,\alpha/2}(G_T)$, *and the tensor* $\mathbb{G} \in \boldsymbol{C}_{1,\gamma}^{(\gamma,1+\alpha)}(D_T) \cap$

$\boldsymbol{C}^{\gamma,0}_{1,\gamma}(D_T)$. *Then problem* (1.1.7) *has a unique solution* $(\boldsymbol{v},p)$ *such that* $\boldsymbol{v}\in\boldsymbol{C}^{2+\alpha,1+\alpha/2}_{1+\gamma}(D_T)$, $p\in C^{(\gamma,1+\alpha)}_{1,\gamma}(D_T)$, $\nabla p\in\boldsymbol{C}^{\alpha,\alpha/2}_{1+\gamma}(D_T)$, *and it holds the estimate*

$$\begin{aligned}
&|\boldsymbol{v}|^{(2+\alpha,1+\alpha/2)}_{1+\gamma,D_T}+|\nabla p|^{(\alpha,\alpha/2)}_{1+\gamma,D_T}+|p|^{\frac{(1+\alpha-\gamma)}{2}}_{t,1,D_T}+\langle p\rangle^{(\gamma,1+\alpha)}_{1+\gamma,\boldsymbol{D}_T}\\
&\leqslant c_1(T)\{|\boldsymbol{f}|^{(\alpha,\alpha/2)}_{1+\gamma,D_T}+|r|^{(1+\alpha,\frac{1+\alpha}{2})}_{1+\gamma,D_T}+|\boldsymbol{g}|^{(\alpha,\alpha/2)}_{1+\gamma,D_T}+|\boldsymbol{v}_0|^{(2+\alpha)}_{1+\gamma,\cup\Omega^\pm}+|\mathbb{G}|^{(\gamma,1+\alpha)}_{1,\gamma,D_T}\\
&\quad+|\mathbb{G}|^{(\gamma,0)}_{1,\gamma,D_T}+|\boldsymbol{b}|^{(1+\alpha,\frac{1+\alpha}{2})}_{G_T}+|b'|_{G_T}+|b'|^{(\gamma,1+\alpha)}_{G_T}\\
&\quad+|\nabla_\Gamma b'|^{(\alpha,\alpha/2)}_{G_T}+|B|^{(\alpha,\alpha/2)}_{G_T}\}\equiv c_1(T)Q(T).
\end{aligned}\tag{5.2.2}$$

Proof We suppose that the coordinates $\{x\}$ are chosen so that $\overline{\Omega}^+\subset B_1$, while $B_\rho\equiv\{|x|<\rho\}$. The existence of a solution $(\boldsymbol{v},p)$, $\boldsymbol{v}\in\boldsymbol{C}^{2+\alpha,1+\alpha/2}(D_T)$, $p\in C^{(\gamma,1+\alpha)}(B_T)$, and $\nabla p\in\boldsymbol{C}^{\alpha,\alpha/2}(D_T)$ follows from Theorem 4.1.2; hence, all that we need to prove is estimate (5.2.2).

We fix a point $x_0\in\mathbb{R}^3\setminus B_4$ and a smooth cutoff function $\eta(y)$

$$\eta(y)=\begin{cases}1 & \text{if } |x_0-y|\leqslant\frac{|x_0|}{2}\equiv d,\\ 0 & \text{if } |x_0-y|>\frac{3}{4}|x_0|.\end{cases}$$

It is assumed that the derivatives and the Hölder constants of the function η satisfy the inequalities

$$\begin{aligned}
|\nabla^k\eta|_{\mathbb{R}^3}&\leqslant c_2d^{-k},\quad k\in\mathbb{N},\\
\langle\eta\rangle^{(\beta)}_{\mathbb{R}^3}&\leqslant c_3d^{-\beta},\quad \beta\in(0,1).
\end{aligned}$$

We introduce new unknown functions $\boldsymbol{u}=\boldsymbol{v}\eta$, $s=p\eta$. They satisfy the Cauchy problem

$$\begin{aligned}
&\mathcal{D}_t\boldsymbol{u}-\nu^-\nabla^2\boldsymbol{u}+\frac{1}{\rho^-}\nabla s=\boldsymbol{f}\eta-\nu^-(\boldsymbol{v}\nabla^2\eta+2(\nabla\eta\cdot\nabla)\boldsymbol{v})+\frac{1}{\rho^-}p\nabla\eta=\boldsymbol{f}^{(1)},\\
&\nabla\cdot\boldsymbol{u}=r\eta+\boldsymbol{v}\cdot\nabla\eta\equiv r^{(1)}\quad\text{in}\quad\mathbb{R}^3_T,\\
&\boldsymbol{u}|_{t=0}=\boldsymbol{v}_0\eta.
\end{aligned}\tag{5.2.3}$$

The functions in the right-hand sides are compactly supported and admit the representations

$$\mathcal{D}_tr^{(1)}-\nabla\cdot\boldsymbol{f}^{(1)}=\nabla\cdot\boldsymbol{g}^{(1)},\quad \boldsymbol{g}^{(1)}=\nabla\cdot\mathbb{G}^{(1)},$$

where

$$\boldsymbol{g}^{(1)}(x) = \boldsymbol{g}\eta + \nu^-(\boldsymbol{v}\nabla^2\eta + 2(\nabla\eta\cdot\nabla)\boldsymbol{v} + r\nabla\eta + \nabla\eta\times\nabla\times\boldsymbol{v})$$
$$-\frac{2}{\rho^-}p\nabla\eta + \nabla\int_{\mathbb{R}^3}\mathcal{E}(x,y)\Big\{-\boldsymbol{g}\cdot\nabla_y\eta + \Big(\frac{1}{\rho^-}p - \nu^- r\Big)\nabla_y^2\eta\Big\}\,\mathrm{d}y,$$
$$G_{ij}^{(1)}(x) = G_{ij}(x)\eta(x) - \delta_j^i\int_{\mathbb{R}^3}\mathcal{E}(x,y)\Big\{(\nabla\cdot\mathbb{G})\cdot\nabla\eta - \Big(\frac{1}{\rho^-}p - \nu^- r\Big)\nabla^2\eta\Big\}\,\mathrm{d}y$$
$$-\frac{\partial}{\partial x_j}\int_{\mathbb{R}^3}\mathcal{E}(x,y)\Big\{G_{ik}\frac{\partial\eta}{\partial y_k} + \frac{2}{\rho^-}p\frac{\partial\eta}{\partial y_i} - \nu^-\Big[v_i\nabla^2\eta + 2(\nabla\eta\cdot\nabla)v_i$$
$$+ r\frac{\partial\eta}{\partial y_i} - (\nabla\eta\times\nabla\times\boldsymbol{v})_i\Big]\Big\}dy.$$

(The sign $\times$ denotes vector product.)

In order to apply Theorem 4.2.1 to the problem (5.2.3), it only remains to check the boundedness of the norms in (5.1.5):

$$|\boldsymbol{f}^{(1)}|_{B_{d,T}}^{(\alpha,\frac{\alpha}{2})} \leqslant c_4\big(|\boldsymbol{f}|_{B_{d,T}}^{(\alpha,\frac{\alpha}{2})} + d^{-2}|\boldsymbol{v}|_{B_{d,T}}^{(\alpha,\frac{\alpha}{2})} + d^{-1}|\nabla\boldsymbol{v}|_{B_{d,T}}^{(\alpha,\frac{\alpha}{2})} + d^{-1}|p|_{B_{d,T}}^{(\alpha,\frac{\alpha}{2})}\big),$$
$$|r^{(1)}|_{B_{d,T}}^{(1+\alpha,\frac{1+\alpha}{2})} \leqslant c_5\big(|r|_{B_{d,T}}^{(1+\alpha,\frac{1+\alpha}{2})} + d^{-1}|\boldsymbol{v}|_{B_{d,T}}^{(1+\alpha,\frac{1+\alpha}{2})}\big),$$
$$\langle\boldsymbol{g}^{(1)}\rangle_{B_{d,T}}^{(\alpha,\frac{\alpha}{2})} \leqslant c_6\big\{|\boldsymbol{g}|_{B_{d,T}}^{(\alpha,\frac{\alpha}{2})} + |\boldsymbol{f}|_{B_{d,T}}^{(\alpha,\frac{\alpha}{2})} + d^{-2}|\boldsymbol{v}|_{B_{d,T}}^{(\alpha,\frac{\alpha}{2})} +$$
$$+ d^{-1}\big(|\nabla\boldsymbol{v}|_{B_{d,T}}^{(\alpha,\frac{\alpha}{2})} + |r|_{B_{d,T}}^{(\alpha,\frac{\alpha}{2})} + |p|_{B_{d,T}}^{(\alpha,\frac{\alpha}{2})}\big)\big\}, \tag{5.2.4}$$
$$\langle\mathbb{G}^{(1)}\rangle_{B_{d,T}}^{(\gamma,1+\alpha)} \leqslant c_7\big\{\langle\mathbb{G}\rangle_{B_{d,T}}^{(\gamma,1+\alpha)} + d^{-\gamma}\big(\langle\mathbb{G}\rangle_{t,B_{d,T}}^{(\frac{1+\alpha-\gamma}{2})} + \langle p\rangle_{t,B_{d,T}}^{(\frac{1+\alpha-\gamma}{2})} + \langle r\rangle_{t,B_{d,T}}^{(\frac{1+\alpha-\gamma}{2})}$$
$$\langle\mathbb{G}\rangle_{B_{d,T}}^{(\gamma,1+\alpha)} + d^{-\gamma} \quad + d^{-1}\langle\boldsymbol{v}\rangle^{(\frac{1+\alpha-\gamma}{2})} + \langle\nabla\boldsymbol{v}\rangle_{t,B_{d,T}}^{(\frac{1+\alpha-\gamma}{2})}\big)\big\},$$
$$\langle\mathbb{G}^{(1)}\rangle_{x,B_{d,T}}^{(\gamma)} \leqslant c_8\big\{\langle\mathbb{G}\rangle_{x,B_{d,T}}^{(\gamma)} + d^{-\gamma}\big(|\mathbb{G}|_{B_{d,T}} + |p|_{B_{d,T}}$$
$$+ |r|_{B_{d,T}} + |\nabla\boldsymbol{v}|_{B_{d,T}} + d^{-1}|\boldsymbol{v}|_{B_{d,T}}\big)\big\}$$
$$(B_{d,T} \equiv B_d(x_0)\times(0,T);\ B_d(x_0) \equiv \{x : |x-x_0| < d\}).$$

In inequalities (5.2.4), we have used the properties of our cutoff function and Lemma 4.1.1 concerning the estimate of the Newtonian potential (4.1.1).

Combining Theorem 4.2.1, Remark 5.1.2, and estimates (5.2.4), we obtain

$$|\boldsymbol{v}|_{B_{d/2,T}}^{(2+\alpha,1+\frac{\alpha}{2})} + |\nabla p|_{B_{d/2,T}}^{(\alpha,\frac{\alpha}{2})} + \langle p\rangle_{B_{d/2,T}}^{(\gamma,1+\alpha)} \leqslant |\boldsymbol{u}|_{B_{d,T}}^{(2+\alpha,1+\frac{\alpha}{2})} + |\nabla s|_{B_{d,T}}^{(\alpha,\frac{\alpha}{2})} + \langle s\rangle_{B_{d,T}}^{(\gamma,1+\alpha)}$$
$$\leqslant c_{12}\big\{|\boldsymbol{f}|_{B_{d,T}}^{(\alpha,\alpha/2)} + |r|_{B_{d,T}}^{(1+\alpha,\frac{1+\alpha}{2})} + |\boldsymbol{g}|_{B_{d,T}}^{(\alpha,\alpha/2)} + |\boldsymbol{v}_0|_{B_d(x_0)}^{(2+\alpha)} + \langle\mathbb{G}\rangle_{B_{d,T}}^{(\gamma,1+\alpha)} + \langle\mathbb{G}\rangle_{x,B_{d,T}}^{(\gamma)}$$
$$+ d^{-\gamma}\big(|\mathbb{G}|_{t,B_{d,T}}^{(\frac{1+\alpha-\gamma}{2})} + |\boldsymbol{v}|_{B_{d,T}}^{(2+\alpha-\gamma,1+\frac{\alpha-\gamma}{2})} + |p|_{B_{d,T}}^{(\alpha,\alpha/2)} + \langle p\rangle_{t,B_{d,T}}^{(\frac{1+\alpha-\gamma}{2})}\big)\big\}. \tag{5.2.5}$$

We multiply inequality (5.2.5) by $(1+d)^{1+\gamma}$ and observe that $3d/2 \leqslant |x| \leqslant 5d/2$ for any $x \in B_{d/2}(x_0)$. Then, clearly

$$\Big(\frac{2}{5}\Big)^{1+\gamma}\big(|\boldsymbol{v}|^{(2+\alpha,1+\frac{\alpha}{2})}_{1+\gamma,B_{d/2,T}} + |\nabla p|^{(\alpha,\alpha/2)}_{1+\gamma,B_{d/2,T}} + \langle p\rangle^{(\gamma,1+\alpha)}_{1+\gamma,B_{d/2,T}}\big)$$
$$\leqslant c_{12}\big\{|\boldsymbol{f}|^{(\alpha)}_{1+\gamma,Q_T^-} + |r|^{(1+\alpha,\frac{1+\alpha}{2})}_{1+\gamma,Q_T^-} + |\boldsymbol{g}|^{(\alpha,\alpha/2)}_{1+\gamma,Q_T^-}$$
$$+ |\boldsymbol{v}_0|^{(2+\alpha)}_{1+\gamma,\Omega^-} + \langle \mathbb{G}\rangle^{(\gamma,1+\alpha)}_{1+\gamma,Q_T^-} + \langle \mathbb{G}\rangle^{(\gamma)}_{x,1+\gamma,Q_T^-}$$
$$+ 2^{1+\gamma}\big(|\mathbb{G}|^{(\frac{1+\alpha-\gamma}{2})}_{t,1,Q_T^-} + |\boldsymbol{v}|^{(2+\alpha-\gamma,1+\frac{\alpha-\gamma}{2})}_{1,Q_T^-} + |p|^{(\alpha,\alpha/2)}_{1,Q_T^-} + \langle p\rangle^{(\frac{1+\alpha-\gamma}{2})}_{t,1,Q_T^-}\big)\big\}.$$

Now, we take the supremum over all $x_0 \in \mathbb{R}^3 \setminus B_4$. Since the estimate for $(\boldsymbol{v}, p)$ in the ball B_5 follows from Theorem 4.1.1, we conclude

$$|\boldsymbol{v}|^{(2+\alpha,1+\frac{\alpha}{2})}_{1+\gamma,D_T} + |\nabla p|^{(\alpha,\alpha/2)}_{1+\gamma,D_T} + \langle p\rangle^{(\gamma,1+\alpha)}_{1+\gamma,D_T}$$
$$\leqslant c_{13}(T)\big\{Q(T) + |\boldsymbol{v}|^{(2+\alpha,1+\frac{\alpha}{2})}_{1,D_T} + |p|^{(\alpha,\alpha/2)}_{1,D_T} + \langle p\rangle^{(\frac{1+\alpha-\gamma}{2})}_{t,1,D_T}\big\}. \quad (5.2.6)$$

The norms of velocity vector field $\boldsymbol{v}$ and the derivatives of it in the right-hand side of (5.2.6) are estimated with the help of the Young inequality

$$(1+|x|)|u(x)| \equiv (1+|x|)|u(x)|^{p^*}|u(x)|^{q^*} \leqslant \varepsilon(1+|x|)^{1+\gamma}|u(x)| + c(\varepsilon)|u(x)| \quad (5.2.7)$$

with $p^* = \frac{1}{1+\gamma}$, $q^* = \frac{\gamma}{1+\gamma}$. This yields

$$|\boldsymbol{v}|^{(2+\alpha,1+\frac{\alpha}{2})}_{1,D_T} \leqslant \varepsilon|\boldsymbol{v}|^{(2+\alpha,1+\frac{\alpha}{2})}_{1+\gamma,D_T} + c(\varepsilon)|\boldsymbol{v}|^{(2+\alpha,1+\frac{\alpha}{2})}_{D_T}.$$

As for the weightless norm of $\boldsymbol{v}$, it can be estimated with the help of (4.1.10).

Thus, to complete the proof, it only remains to estimate the lower-order norms of the pressure $|p|^{(\alpha,\frac{\alpha}{2})}_{1,D_T}$ and $\langle p\rangle^{(\frac{1+\alpha-\gamma}{2})}_{t,1,D_T}$. Using the interpolation inequality

$$\langle p\rangle^{(\alpha)}_{x,\beta,D_T} \leqslant c_{14}(|\nabla p|_{\beta,D_T} + |p|_{\beta,D_T}) \quad (5.2.8)$$

with $\beta = 1$, and applying estimate (5.2.7) to $|\nabla p|_{1,D_T}$, we see that we need only to prove the boundedness of $|p|^{(\frac{1+\alpha-\gamma}{2})}_{t,1,D_T}$.

To this end, we consider the representation of the pressure given in Section 4.2:

$$p = P_0 + \mu^{\pm} r' - \rho^{\pm}\mathcal{D}_t\Phi + q, \quad (5.2.9)$$

where $P_0(x,t) = p_0(x)$ is a solution of (4.1.9), the function Φ satisfies the Poisson equation, (4.2.12), and vector $\boldsymbol{w}$ solves problem (4.2.10) for the heat equation in D_T. The summand q in (5.2.9) is a solution of diffraction problem (4.2.13), and $\boldsymbol{u} \equiv \boldsymbol{v} - \boldsymbol{w} - \nabla\Phi$; moreover, $\nabla \cdot \boldsymbol{u} = 0$. In addition, q is representable in the form $q = \rho^{\pm} s_1 + s_2$, where s_1 and s_2 are, respectively, a double-layer potential and a single-layer one, namely,

$$s_1(x,t) = -\int_\Gamma \frac{\partial \mathcal{E}(x,y)}{\partial \boldsymbol{n}_y} \mu(y,t)\, d\Gamma(y), \quad x \in \mathbb{R}^3,$$

$$s_2(x,t) = \int_\Gamma \mathcal{E}(x,y)\omega(y,t)\, d\Gamma(y), \quad x \in \mathbb{R}^3, \tag{5.2.10}$$

($\partial/\partial \boldsymbol{n}_y$ is the derivative along the outward normal to the surface $\Gamma(y)$ depending on the variable y). For s_2, formula (4.2.22) has been proved in Chap. 4. Since the integration in (4.2.22) is always carried out over a bounded surface Γ, the decay of the functions s_1 and s_2 as $|x| \to \infty$ is determined by the decay of the kernels in the integrands, hence

$$|s_1(x,t)| \leqslant c_{15} |\mu(\cdot,t)|_\Gamma |x|^{-2}, \quad |x| > 1,$$

$$|s_2(x,t)| \leqslant c_{16} \big(|\boldsymbol{a}(\cdot,t)|_\Gamma^{(\beta)} |x|^{-2} + |\theta(\cdot,t)|_\Gamma |x|^{-1} \big), \quad \beta \in (0,1).$$

In the same way, we estimate the increments of s_1 and s_2 with respect to t. As a result, we have

$$|q|_{t,1,D_T \setminus B_T}^{(\frac{1+\alpha-\gamma}{2})} \leqslant c_{17} \big(|\mu|_{t,G_T}^{(\frac{1+\alpha-\gamma}{2})} + |\boldsymbol{a}|_{G_T}^{(\gamma,1+\alpha)} + |\theta|_{t,G_T}^{(\frac{1+\alpha-\gamma}{2})} \big).$$

By analogy with (4.2.19) and (4.2.21), one can obtain

$$|\mu|_{t,1,D_T}^{(\frac{1+\alpha-\gamma}{2})} \leqslant c |q_0|_{t,G_T}^{(\frac{1+\alpha-\gamma}{2})},$$

$$|\theta|_{t,1,D_T}^{(\frac{1+\alpha-\gamma}{2})} \leqslant c |Q_1|_{t,G_T}^{(\frac{1+\alpha-\gamma}{2})} \leqslant \langle \nabla \boldsymbol{u} \rangle_{G_T}^{(\gamma,1+\alpha)}.$$

Using the estimates of $|q|_{t,D_T}^{\frac{(1+\alpha-\gamma)}{2}}$ and (4.2.23) and (4.2.24), we deduce the inequality

$$|q|_{t,1,D_T}^{(\frac{1+\alpha-\gamma}{2})} \leqslant c_{18} \big(|q_0|_{t,G_T}^{(\frac{1+\alpha-\gamma}{2})} + |\boldsymbol{u}|_{D_T}^{(2+\alpha,1+\alpha/2)} \big) \leqslant c_{19} Q(T). \tag{5.2.11}$$

Next, we consider the term $\rho^{\pm} \mathcal{D}_t \Phi$ in (5.2.9). In view of (5.2.1), the derivative of the solution $\mathcal{D}_t \Phi$ to Eq. (4.2.12) may be expressed as follows:

$$\begin{aligned}\mathcal{D}_t\Phi(x,t)=&\int\limits_{\Omega^+\cup\Omega^-}\mathcal{E}(x,y)\mathcal{D}_t r'(y,t)\,\mathrm{d}y=-\int\limits_{\Omega^-\cup\Omega^+}\nabla_y\mathcal{E}(x,y)\big(\boldsymbol{g}-\nu^{\pm}\Delta\boldsymbol{w}+\frac{1}{\rho^{\pm}}\nabla P_0\big)\,\mathrm{d}y\\ =&-\nabla_x\int\limits_{\Omega^-\cup\Omega^+}\nabla_y\mathcal{E}(x,y):\mathbb{D}(y,t)\,\mathrm{d}y+\nabla_x\int\limits_{\Gamma}\mathcal{E}(x,y)\cdot[\mathbb{D}(y,t)\boldsymbol{n}]|_{\Gamma}\,\mathrm{d}\Gamma,\end{aligned} \tag{5.2.12}$$

where $\mathbb{D}=\mathbb{G}-\nu^{\pm}\nabla\boldsymbol{w}+\frac{1}{\rho^{\pm}}P_0\mathbb{I}$ and $\nabla_x\nabla_y\mathcal{E}:\mathbb{D}\equiv\frac{\partial^2\mathcal{E}}{\partial x_i\partial y_j}D_{ij}$.

In order to estimate the volume integral in (5.2.12), we apply Lemma 5.2.1 (see the end of this section).

A similar estimate for the boundary integral can be proved by the Schauder localization method due to the boundedness of the surface Γ. Since

$$\int\limits_{\Omega^+\cup\Omega^-}\nabla_y\nabla_y\mathcal{E}(x,y):\mathbb{I}P_0(y)dy=P_0(x),$$

we arrive at the estimate

$$|\mathcal{D}_t\Phi|_{1,D_T}\leqslant c_{20}\Big(|\mathbb{G}|^{(\nu)}_{x,1,\nu,D_T}+|\nabla\boldsymbol{w}|^{(\nu)}_{x,1,\nu,D_T}+|P_0|_{1,D_T}+\big\langle[\frac{1}{\rho^{\pm}}P_0]\big|_{\Gamma}\big\rangle^{(\nu)}_{x,G_T}\Big),$$

valid for all $\nu\in(0,1)$. By considering increments with respect to t, we obtain

$$|\mathcal{D}_t\Phi|^{(\frac{1+\alpha-\gamma}{2})}_{t,1,D_T}\leqslant c_{20}\Big(|\mathbb{G}|^{(\gamma,1+\alpha)}_{1,\gamma,D_T}+|\nabla\boldsymbol{w}|^{(\frac{1+\alpha-\gamma}{2})}_{(0,T)}|^{(\nu)}_{1,\nu,\cup\Omega^{\pm}}+|P_0|_{1,D_T}+\big\langle[\frac{1}{\rho^{\pm}}P_0]\big|_{\Gamma}\big\rangle^{(\nu)}_{x,G_T}\Big) \tag{5.2.13}$$

with $\gamma,\nu\in(0,1)$.

Being a solution of (4.1.9), the function P_0 can be written as the sum of a Newtonian and surface potentials, as in (5.2.10): $P_0=s_0+\rho^{\pm}s_1'+s_2'$, where

$$s_0=-\int\limits_{\Omega^+\cup\Omega^-}\rho^{\pm}\nabla_y\mathcal{E}\cdot\boldsymbol{h}\,\mathrm{d}y+\int\limits_{\Gamma}\mathcal{E}[\rho^{\pm}\boldsymbol{h}\cdot\boldsymbol{n}]|_{\Gamma}\,\mathrm{d}\Gamma(y)$$

(see Chap. 4 for details). Taking into consideration the divergence form of $\boldsymbol{h}=\nabla\cdot\mathbb{H}$, $\mathbb{H}(x)=-\mathbb{G}(x,0)+\nu^{\pm}r(x,0)\mathbb{I}$, and calculating as before, we conclude that

$$|P_0|_{1,D_T}+\big\langle[\frac{1}{\rho^{\pm}}P_0]\big|_{\Gamma}\big\rangle^{(\nu)}_{x,G_T}\leqslant c_{21}\big(|\mathbb{H}|^{(\nu)}_{1,\nu,\cup\Omega^{\pm}}+Q(T)\big),\quad \nu\in(0,1). \tag{5.2.14}$$

To complete the proof of Theorem 5.2.1, we estimate weighted norms of the vector field $\boldsymbol{w}$ satisfying relations (4.2.10). In Q_T^-, the field $\boldsymbol{w}$ may be defined as a solution of the Cauchy problem

$$\mathcal{D}_t\boldsymbol{w}^- -\nu^-\nabla^2\boldsymbol{w}^-=\boldsymbol{f}_1,\quad \boldsymbol{w}^-|_{t=0}=\boldsymbol{v}_1,$$

where $\boldsymbol{f}_1 \in C^{\alpha,\alpha/2}_{1+\gamma}(\mathbb{R}^3_T)$ and $\boldsymbol{v}_1 \in C^{2+\alpha}_{1+\gamma}(\mathbb{R}^3)$ are vector fields equal, respectively, to $\boldsymbol{f} - \frac{1}{\rho^-}\nabla P_0$ and $\boldsymbol{v}_0$ in Ω^- and extended into Ω^+ with preservation of class. In the domain Q^+_T, we define $\boldsymbol{w}$ as a solution $\boldsymbol{w}^+$ to the initial-boundary value problem

$$\mathcal{D}_t\boldsymbol{w}^+ - \nu^+\nabla^2\boldsymbol{w}^+ = \boldsymbol{f} - \frac{1}{\rho^+}\nabla P_0,$$

$$\boldsymbol{w}^+|_{t=0} = \boldsymbol{v}_0, \quad \boldsymbol{w}^+|_{G_T} = \lim_{\substack{x\to x_0\in\Gamma_t,\\ x\in\Omega^-_t}} \boldsymbol{w}^-(x,t).$$

It is easily verified that, by (4.1.8) and (4.1.9), $\boldsymbol{w}^+$ satisfies the compatibility condition

$$\mathcal{D}_t\boldsymbol{w}^+|_{\Gamma,t=0} = \nu^-\nabla^2\boldsymbol{v}_0 - \frac{1}{\rho^-}\nabla p_0 + \boldsymbol{f}|_{t=0},$$

therefore

$$|\boldsymbol{w}^+|^{(2+\alpha,1+\alpha/2)}_{Q^+_T} \leqslant c_{22}\big(|\boldsymbol{f}|^{(\alpha,\alpha/2)}_{Q^+_T} + |\nabla p_0|^{(\alpha)}_{\Omega^+} + |\boldsymbol{v}_0|^{(2+\alpha)}_{\Omega^+} + |\boldsymbol{w}^-|^{(2+\alpha,1+\alpha/2)}_{G_T}\big).$$

In all the norms here, one can add the weight due to the boundedness of Ω^+.

Using a similar inequality for the solution of the Cauchy problem and, as above, applying localization procedure, we show that

$$|\boldsymbol{w}^-|^{(2+\alpha,1+\alpha/2)}_{1+\nu,Q^-_T} \leqslant c_{23}\big(|\boldsymbol{f}|^{(\alpha,\alpha/2)}_{1+\nu,Q^-_T} + |\nabla p_0|^{(\alpha)}_{1+\nu,\Omega^-} + |\boldsymbol{v}_0|^{(2+\alpha)}_{1+\nu,\Omega^-}\big)$$

for any $\nu \in (0,1]$. Consequently

$$\begin{aligned} |\nabla\boldsymbol{w}|^{(\frac{1+\alpha-\gamma}{2})}_{(0,T)}|^{(\nu)}_{1,\nu,\cup\Omega^\pm} &\leqslant |\boldsymbol{w}|^{(2+\alpha,1+\alpha/2)}_{1+\nu,D_T} \\ &\leqslant c_{24}\big(|\boldsymbol{f}|^{(\alpha,\alpha/2)}_{1+\nu,D_T} + |\nabla p_0|^{(\alpha)}_{1+\nu,\cup\Omega^\pm} + |\boldsymbol{v}_0|^{(2+\alpha)}_{1+\nu,\cup\Omega^\pm}\big) \end{aligned} \tag{5.2.15}$$

for any $\nu \leqslant \gamma$. Setting $\nu < \gamma$ and estimating ∇p_0 by an inequality similar to (5.2.7) with $p^* = \frac{1+\nu}{1+\gamma}$, $q^* = \frac{\gamma-\nu}{1+\gamma}$, we deduce the relation

$$|\nabla p_0|^{(\alpha)}_{1+\nu,\cup\Omega^\pm} \leqslant \varepsilon|\nabla p|^{(\alpha,\alpha/2)}_{1+\gamma,D_T} + c_{25}(\varepsilon)|\nabla p|^{(\alpha,\alpha/2)}_{D_T}. \tag{5.2.16}$$

Since the weightless norms are weaker, the combination of inequalities (5.2.11) and (5.2.13)–(5.2.16) and (4.1.10) gives us the necessary estimate of the semi-norm $|p|^{(\frac{1+\alpha-\gamma}{2})}_{t,1,D_T}$ in terms of $Q(T)$ and $\varepsilon|\nabla p|^{(\alpha)}_{1+\gamma,D_T}$. Theorem 5.2.1 is proved. □

Lemma 5.2.1 *We assume that the kernel $\mathcal{K}(x,y)$ satisfies the conditions*

$$\Big|\int_{B_\rho(x)} \mathcal{K}(x,y)\,\mathrm{d}y\Big| = \Big|\lim_{\varepsilon\to 0}\int_{B_\rho(x)\setminus B_\varepsilon(x)} \mathcal{K}(x,y)\,\mathrm{d}y\Big| \leqslant c_{26}, \quad x\in\mathbb{R}^3,$$

$$\int_{\mathbb{R}^3} \big(\mathcal{K}(x,y)-\mathcal{K}(x',y)\big)\,\mathrm{d}y = \lim_{\varepsilon\to 0} \int\limits_{\mathbb{R}^3\setminus B_\varepsilon(x)\setminus B_\varepsilon(x')} \big(\mathcal{K}(x,y)-\mathcal{K}(x',y)\big)\,\mathrm{d}y = 0,$$

$$|\mathcal{K}(x,y)| \leqslant \frac{c_{27}}{|x-y|^3}, \quad |\mathcal{K}(x,y)-\mathcal{K}(z,y)| \leqslant c_{28}|x-z|\Big(\frac{1}{|x-y|^4}+\frac{1}{|z-y|^4}\Big). \tag{5.2.17}$$

Then for the integral

$$u(x) = \int_{\mathbb{R}^3} \mathcal{K}(x,y)g(y)\,\mathrm{d}y$$

with $g\in C^{\nu}_{\beta,\nu}(\cup\Omega^{\pm})$, $\nu\in(0,1)$, $\beta>0$, the estimate

$$|u|^{(\nu)}_{\beta,\nu,\mathbb{R}^3} \leqslant c_{29}|g|^{(\nu)}_{\beta,\nu,\mathbb{R}^3} \tag{5.2.18}$$

holds.

If, in addition, a function $g\in C^{\nu}_{\beta+\gamma}(\cup\Omega^{\pm})$, $\gamma\in(0,1)$, is representable in divergence form $g=\nabla\cdot\boldsymbol{G}$, $\boldsymbol{G}\in\boldsymbol{C}^{\gamma}_{\beta+\gamma}(\mathbb{R}^3)$, then the integral u satisfies the inequality

$$|u|^{(\nu)}_{\beta+\gamma,\mathbb{R}^3} \leqslant c\big(|g|^{(\nu)}_{\beta+\gamma,\mathbb{R}^3} + \langle\boldsymbol{G}\rangle^{(\gamma)}_{\beta+\gamma,\mathbb{R}^3}\big), \qquad \beta+\gamma>\nu. \tag{5.2.19}$$

Proof For arbitrary $\rho>0$, we can write

$$u(x) = \int_{|x-y|<\rho} \mathcal{K}(x,y)(g(y)-g(x))\,\mathrm{d}y + g(x)\int_{|x-y|<\rho}\mathcal{K}(x,y)\,\mathrm{d}y - \int_{|x-y|\geqslant\rho}\mathcal{K}(x,y)g(y)\,\mathrm{d}y. \tag{5.2.20}$$

First, let $|x|\leqslant 1/2$, $\rho=1+|x|$. Then, taking into account that $|y|>1>|x|$ for $|x-y|\geqslant\rho$, we have

$$|u(x)| \leqslant c\Big\{\frac{\langle g\rangle^{(\nu)}_{\beta+\nu,\mathbb{R}^3}}{(1+|x|)^{\beta+\nu}} \int\limits_{|x-y|<\rho} \frac{\mathrm{d}y}{|x-y|^{3-\nu}} + \frac{|g|_{\beta,\mathbb{R}^3}}{(1+|x|)^{\beta}} + |g|_{\beta,\mathbb{R}^3}\int\limits_{|x-y|\geqslant\rho}\frac{\mathrm{d}y}{|x-y|^3(1+|y|)^{\beta}}\Big\}$$

$$\leqslant c\Big\{ |g|^{(\nu)}_{\beta,\nu,\mathbb{R}^3}(1+|x|)^{-\beta} + |g|_{\beta,\mathbb{R}^3} \int\limits_{|y|\geqslant 1/2+|x|} \frac{dy}{|y|^{3+\beta}} \Big\}$$

$$\leqslant c_{30}|g|^{(\nu)}_{\beta,\nu,\mathbb{R}^3}(1+|x|)^{-\beta}.$$

Next, for $|x| > 1/2$, we put $\rho = 2|x|$. Now $1+|y| \geqslant |x-y|/2$ for $|x-y| \geqslant \rho$, therefore, (5.2.20) can be estimated as follows:

$$|u(x)| \leqslant c\Big\{ \langle g\rangle^{(\nu)}_{\beta+\nu,\mathbb{R}^3} \frac{\rho^{\nu}}{(1+|x|)^{\beta+\nu}} + \frac{|g|_{\beta,\mathbb{R}^3}}{(1+|x|)^{\beta}} + |g|_{\beta,\mathbb{R}^3} \int\limits_{|x-y|\geqslant\rho} \frac{2^{\beta}\,dy}{|x-y|^{3+\beta}} \Big\}$$

$$\leqslant c_{31}|g|^{(\nu)}_{\beta,\nu,\mathbb{R}^3}(1+|x|)^{-\beta}.$$

Taking the supremum over $x \in \mathbb{R}^3$, we arrive at the estimate of $|u|_{\beta,\mathbb{R}^3}$.

Now, consider two arbitrary points x and z. We set $\rho = |x - z|$. Let, for definiteness, $|x| \leqslant |z|$.

First, we study the case where $\rho \leqslant (1+|x|)/6$. Using (5.2.17), we represent $u(x) - u(z)$ in the form of a sum of four integrals:

$$u(x) - u(z) =$$

$$= \int\limits_{|x-y|\leqslant 2\rho} \mathcal{K}(x,y)[g(y)-g(x)]\,dy - \int\limits_{|x-y|\leqslant 2\rho} \mathcal{K}(z,y)[g(y)-g(z)]\,dy$$

$$+[g(x)-g(z)] \int\limits_{|x-y|\leqslant 2\rho} \mathcal{K}(x,y)\,dy + \int\limits_{|x-y|>2\rho} (\mathcal{K}(x,y)-\mathcal{K}(z,y))[g(y)-g(z)]\,dy$$

$$\equiv I_1 + I_2 + I_3 + I_4.$$

Since $|x| \leqslant |z|$, we have

$$|I_1| + |I_2| + |I_3| \leqslant c_{32}\langle g\rangle^{(\nu)}_{\beta+\nu,\mathbb{R}^3}\rho^{\nu}(1+|x|)^{-\beta-\nu}.$$

The last integral is estimated by using (5.2.17)

$$|I_4| \leqslant c_{33}\langle g\rangle^{(\nu)}_{\beta+\nu,\mathbb{R}^3}|x-z| \int\limits_{|x-y|>2\rho} \frac{|z-y|^{\nu}}{(1+|z|)^{\beta+\nu}}\Big(\frac{1}{|x-y|^4} + \frac{1}{|z-y|^4}\Big)\,dy$$

$$\leqslant c\langle g\rangle^{(\nu)}_{\beta+\nu,\mathbb{R}^3}\rho^{\nu}(1+|x|)^{-\beta-\nu}.$$

If $\rho > (1+|x|)/6$, then

$$\frac{|u(x)-u(z)|}{|z-x|^\nu}(1+|x|)^{\beta+\nu} \leqslant 6^\nu(|u(x)|+|u(z)|)(1+|x|)^\beta \leqslant c_{35}|u|_{\beta,\mathbb{R}^3}$$
$$\leqslant c_{35}c_{31}|g|^{(\nu)}_{\beta,\nu,\mathbb{R}^3}.$$

Taking the supremum first over $z \in \mathbb{R}^3$, and then over $x \in \mathbb{R}^3$, we obtain an estimate of the weighted Hölder constant for function u.

Thus, (5.2.18) is completely proved.

We prove now (5.2.19). For arbitrary $\rho > 0$, one can write

$$u(x) = \int_{|x-y|<\rho} \mathcal{K}(x,y)(g(y)-g(x))\,\mathrm{d}y + g(x)\int_{|x-y|<\rho} \mathcal{K}(x,y)\,\mathrm{d}y$$
$$-\int_{|x-y|\geqslant\rho} \nabla_y\mathcal{K}(x,y)\cdot(\boldsymbol{G}(y)-\boldsymbol{G}(x))\,\mathrm{d}y$$
$$+\int_{|x-y|=\rho} \mathcal{K}(x,y)\boldsymbol{n}_S\cdot(\boldsymbol{G}(y)-\boldsymbol{G}(x))\,\mathrm{d}S(y)$$

($\boldsymbol{n}_S$ is the outward normal to the sphere $S = \{|x-y| = \rho\}$), hence

$$|u(x)| \leqslant c_{30}\Big\{\langle g\rangle^{(\nu)}_{\beta+\gamma,\mathbb{R}^3}\int_{|x-y|<\rho}\frac{\mathrm{d}y}{|x-y|^{3-\nu}(1+|x|)^{\beta+\gamma}} + \frac{|g|_{\beta+\gamma,\mathbb{R}^3}}{(1+|x|)^{\beta+\gamma}}$$
$$+\langle\boldsymbol{G}\rangle^{(\gamma)}_{\beta+\gamma,\mathbb{R}^3}\Big(\int_{|x-y|\geqslant\rho}\frac{\mathrm{d}y}{|x-y|^{4-\gamma}(1+|x|)^{\beta+\gamma}} + \frac{\rho^\gamma}{(1+|x|)^{\beta+\gamma}}\Big)\Big\}$$
$$\leqslant c_{31}(\rho)\big(|g|^{(\nu)}_{\beta+\gamma,\mathbb{R}^3} + \langle\boldsymbol{G}\rangle^{(\gamma)}_{\beta+\gamma,\mathbb{R}^3}\big)(1+|x|)^{-\beta-\gamma}.$$

From here we obtain the estimate of $|u|_{\beta+\gamma,\mathbb{R}^3}$.

In order to estimate $\langle u\rangle^{(\nu)}_{\beta+\gamma,\mathbb{R}^3}$, we replace β in (5.2.18) by $\beta+\gamma-\nu$, which is a positive constant for $\beta+\gamma > \nu$. □

Remark 5.2.2 *The proof of this lemma can easily be modified to embrace the case where the integral is taken over* $\Omega^+ \cup \Omega^-$.

5.3 Solvability of a Linearized Problem on a Finite Time Interval

In this section, we study the following linearization of problem (5.1.1):

$$\mathcal{D}_t\boldsymbol{w} - \nu^\pm\nabla^2_{\boldsymbol{u}}\boldsymbol{w} + \frac{1}{\rho^\pm}\nabla_{\boldsymbol{u}}s = \boldsymbol{f}, \qquad \nabla_{\boldsymbol{u}}\cdot\boldsymbol{w} = r \quad \text{in } D_T,$$

$$\boldsymbol{w}|_{t=0} = \boldsymbol{w}_0 \quad \text{in } \cup\Omega^\pm, \qquad \boldsymbol{w} \xrightarrow[|\xi|\to\infty]{} 0, \tag{5.3.1}$$

$$[\boldsymbol{w}]|_{G_T} = 0, \quad [\mu^\pm \Pi_0 \Pi \mathbb{S}_{\boldsymbol{u}}(\boldsymbol{w})\boldsymbol{n}]|_{G_T} = \Pi_0 \boldsymbol{a},$$

$$[\boldsymbol{n}_0 \cdot \mathbb{T}_{\boldsymbol{u}}(\boldsymbol{w}, s)\boldsymbol{n}]|_{G_T} - \sigma \boldsymbol{n}_0 \cdot \Delta(t) \int_0^t \boldsymbol{w}\, \mathrm{d}t'|_{G_T} = \Big(b + \int_0^t B\, \mathrm{d}t'\Big)\Big|_{G_T},$$

where $\boldsymbol{u}$ is given and

$$(\mathbb{S}_{\boldsymbol{u}}(\boldsymbol{w}))_{ij} = A_{jk}\partial w_i/\partial \xi_k + A_{ik}\partial w_j/\partial \xi_k, \quad i, j, k = 1, 2, 3.$$

Theorem 5.3.1 *We assume that for some $\alpha, \gamma \in (0,1)$, $\gamma \leqslant \alpha$, $0 < T < \infty$, the surface $\Gamma \in C^{2+\alpha}$, the function $\sigma \in C^{1+\alpha}(\Gamma)$, $\sigma \geqslant \sigma_0 > 0$, and vector field $\boldsymbol{u} \in \boldsymbol{C}^{2+\alpha,1+\alpha/2}_{1+\gamma}(D_T)$, $[\boldsymbol{u}]|_{G_T} = 0$, are subjected to the inequality*

$$(T + T^{\gamma/2})|\boldsymbol{u}|^{(2+\alpha,1+\alpha/2)}_{1+\gamma,D_T} \leqslant \delta \tag{5.3.2}$$

with some sufficiently small $\delta > 0$.

Moreover, let the following four groups of assumptions be fulfilled:

1. $\boldsymbol{f} \in \boldsymbol{C}^{\alpha,\alpha/2}_{1+\gamma}(D_T)$, $r \in C^{1+\alpha,\frac{1+\alpha}{2}}_{1+\gamma}(D_T)$, $\boldsymbol{w}_0 \in \boldsymbol{C}^{2+\alpha}_{1+\gamma}(\cup\Omega^\pm)$, $\boldsymbol{a} \in \boldsymbol{C}^{1+\alpha,\frac{1+\alpha}{2}}(G_T)$, $b \in C^{(\gamma,1+\alpha)}(G_T)$, $B \in C^{\alpha,\alpha/2}(G_T)$*;*
2.

$$\nabla \cdot \boldsymbol{w}_0(\xi) = r(\xi, 0) = 0, \quad [\boldsymbol{w}_0]|_\Gamma = 0,$$

$$[\mu^\pm \Pi_0 \mathbb{S}(\boldsymbol{w}_0(\xi))\boldsymbol{n}_0]|_{\xi\in\Gamma} = \Pi_0 \boldsymbol{a}(\xi, 0), \quad \xi \in \Gamma, \tag{5.3.3}$$

$$\Big[\Pi_0\Big(\boldsymbol{f}(\xi,0) - \frac{1}{\rho^\pm}\nabla s(\xi,0) + \nu^\pm \nabla^2 \boldsymbol{w}_0(\xi)\Big)\Big]\Big|_{\xi\in\Gamma} = 0;$$

3. *there exist a vector $\boldsymbol{g} \in \boldsymbol{C}^{\alpha,\alpha/2}_{1+\gamma}(D_T)$ and tensor $\mathbb{G} = \{G_{ik}\}^3_{i,k=1}$, $G_{ik} \in C^{(\gamma,1+\alpha)}_{1,\gamma}(D_T) \cap C^{\gamma,0}_{1,\gamma}(D_T)$ such that the representations take place:*

$$\mathcal{D}_t r - \nabla_{\boldsymbol{u}} \cdot \boldsymbol{f} = \nabla \cdot \boldsymbol{g}, \quad \boldsymbol{g} = \nabla \cdot \mathbb{G} \quad (g_i = \partial G_{ik}/\partial \xi_k, \quad i = 1, 2, 3), \tag{5.3.4}$$

(these equalities are understood in a weak sense) and, in addition,

$$[(\boldsymbol{g} + \mathbb{A}^* \boldsymbol{f}) \cdot \boldsymbol{n}_0]\big|_{G_T} = 0; \tag{5.3.5}$$

4. $s_0(\xi) = s(\xi, 0)$ *is a solution of the problem*

$$\frac{1}{\rho^\pm}\nabla^2 s_0(\xi) = \nabla\cdot\big(\mathcal{D}_t\mathbb{B}^*|_{t=0}\boldsymbol{w}_0(\xi) - \boldsymbol{g}(\xi,0)\big) \equiv \nabla\cdot\boldsymbol{d} \quad in \quad \Omega^+\cup\Omega^-,$$

$$[s_0]|_\Gamma = \Big[2\mu^\pm\frac{\partial \boldsymbol{w}_0}{\partial \boldsymbol{n}_0}\cdot\boldsymbol{n}_0\Big]\Big|_\Gamma - b|_{t=0} \equiv p_{00}, \tag{5.3.6}$$

$$\Big[\frac{1}{\rho^\pm}\frac{\partial s_0}{\partial \boldsymbol{n}_0}\Big]\Big|_\Gamma = \big[\boldsymbol{n}_0\cdot(\boldsymbol{f}|_{t=0} + \nu^\pm\nabla^2\boldsymbol{w}_0)\big]\big|_\Gamma \equiv p_{01}.$$

Under these assumptions, problem (5.3.1) *has a unique solution* $(\boldsymbol{w}, s)$, $\boldsymbol{w}\in \boldsymbol{C}^{2+\alpha,1+\alpha/2}_{1+\gamma}(D_T)$, $s\in C^{(1+\alpha,\gamma)}_{1,\gamma}(D_T)$, $\nabla s\in \boldsymbol{C}^{\alpha,\alpha/2}_{1+\gamma}(D_T)$. *For it the inequality*

$$\begin{aligned}
N_{t'}[\boldsymbol{w}, s] &\equiv |\boldsymbol{w}|^{(2+\alpha,1+\alpha/2)}_{1+\gamma,D_{t'}} + |\nabla s|^{(\alpha,\alpha/2)}_{1+\gamma,D_{t'}} + |s|^{(\frac{1+\alpha-\gamma}{2})}_{t,1,D_{t'}} + \langle s\rangle^{(\gamma,1+\alpha)}_{1+\gamma,D_{t'}}\\
&\leqslant c_1(t')\big\{|\boldsymbol{f}|^{(\alpha,\alpha/2)}_{1+\gamma,D_{t'}} + |r|^{(1+\alpha,\frac{1+\alpha}{2})}_{1+\gamma,D_{t'}} + |\boldsymbol{w}_0|^{(2+\alpha)}_{1+\gamma,\cup\Omega^\pm} + |\boldsymbol{g}|^{(\alpha,\alpha/2)}_{1+\gamma,D_{t'}}\\
&\quad + |\mathbb{G}|^{(\gamma,1+\alpha)}_{1,\gamma,D_{t'}} + |\mathbb{G}|^{(\gamma,0)}_{1,\gamma,D_{t'}} + |\boldsymbol{a}|^{(1+\alpha,\frac{1+\alpha}{2})}_{G_{t'}} + |b|_{G_{t'}}\\
&\quad + |b|^{(\gamma,1+\alpha)}_{G_{t'}} + |\nabla_\Gamma b|^{(\alpha,\alpha/2)}_{G_{t'}} + |B|^{(\alpha,\alpha/2)}_{G_{t'}} + P_{t'}[\boldsymbol{u}]|\boldsymbol{w}_0|^{(1)}_{1+\gamma,\cup\Omega^\pm}\big\}\\
&\equiv c_1(t')\big\{Q_1(t') + P_{t'}[\boldsymbol{u}]|\boldsymbol{w}_0|^{(1)}_{1+\gamma,\cup\Omega^\pm}\big\}.
\end{aligned} \tag{5.3.7}$$

holds, where $c_1(t')$ *is a nondecreasing function of* $t'\leqslant T$, $\nabla_\Gamma = \Pi_0\nabla$, *and*

$$P_t[\boldsymbol{u}] = t^{\frac{1-\alpha}{2}}|\nabla\boldsymbol{u}|_{1+\gamma,D_t} + |\nabla\boldsymbol{u}|^{(\alpha,\alpha/2)}_{1+\gamma,D_t}. \tag{5.3.8}$$

The technique of the proof of Theorem 5.3.1 is similar to that used in [54], §5, and is based on successive approximations and on some coercive estimates for linear problem (1.1.7).

We rewrite system (5.3.1) in the form

$$\begin{aligned}
&\mathcal{D}_t\boldsymbol{w} - \nu^\pm\nabla^2\boldsymbol{w} + \frac{1}{\rho^\pm}\nabla s = \boldsymbol{f} + \boldsymbol{l}_1(\boldsymbol{w}, s) \equiv \boldsymbol{f}_1,\\
&\nabla\cdot\boldsymbol{w} = r + l_2(\boldsymbol{w}) = r_1 \quad \text{in} \quad D_T,\\
&\boldsymbol{w}|_{t=0} = \boldsymbol{w}_0, \quad [\boldsymbol{w}]|_{G_T} = 0, \quad \boldsymbol{w}\xrightarrow[|\xi|\to\infty]{} 0,\\
&[\mu^\pm\Pi_0\mathbb{S}(\boldsymbol{w})\boldsymbol{n}_0]|_{G_T} = \boldsymbol{l}_3(\boldsymbol{w}) + \Pi_0\boldsymbol{a},\\
&[\boldsymbol{n}_0\cdot\mathbb{T}(\boldsymbol{w}, s)\boldsymbol{n}_0]|_{G_T} - \sigma\boldsymbol{n}_0\cdot\Delta(0)\int_0^t \boldsymbol{w}\,\mathrm{d}t'|_{G_T}\\
&\quad = b + \int_0^t B\,\mathrm{d}t' + l_4(\boldsymbol{w}, s) + \sigma\int_0^t l_5(\boldsymbol{w})\,\mathrm{d}t', \quad t\in(0,T).
\end{aligned} \tag{5.3.9}$$

Here we have used the following notation:

$$\begin{aligned}
\boldsymbol{l}_1(\boldsymbol{w}, s) &= \nu^{\pm}(\nabla_{\boldsymbol{u}}^2 - \nabla^2)\boldsymbol{w} + \frac{1}{\rho^{\pm}}(\nabla - \nabla_{\boldsymbol{u}})s \\
&= \nu^{\pm}\frac{\partial}{\partial \xi_j}\Big\{(A_{ij}A_{im} - \delta_j^m)\frac{\partial \boldsymbol{w}}{\partial \xi_m}\Big\} - \mathbb{B}\frac{\nabla s}{\rho^{\pm}}, \\
l_2(\boldsymbol{w}) &= (\nabla - \nabla_{\boldsymbol{u}})\cdot \boldsymbol{w} = -\mathbb{B}\nabla\cdot\boldsymbol{w}, \\
\boldsymbol{l}_3(\boldsymbol{w}) &= \big[\mu^{\pm}\Pi_0\big(\mathbb{S}(\boldsymbol{w})\boldsymbol{n}_0 - \Pi\mathbb{S}_{\boldsymbol{u}}(\boldsymbol{w})\boldsymbol{n}\big)\big]|_{\Gamma}, \\
l_4(\boldsymbol{w}, s) &= \big[\boldsymbol{n}_0\cdot\big(\mathbb{T}(\boldsymbol{w}, s)\boldsymbol{n}_0 - \mathbb{T}_{\boldsymbol{u}}(\boldsymbol{w}, s)\boldsymbol{n}\big)\big]|_{\Gamma} \qquad (5.3.10)\\
&= \big[s\boldsymbol{n}_0\cdot(\boldsymbol{n} - \boldsymbol{n}_0) + \mu^{\pm}\boldsymbol{n}_0\cdot\big(\mathbb{S}(\boldsymbol{w})\boldsymbol{n}_0 - \mathbb{S}_{\boldsymbol{u}}(\boldsymbol{w})\boldsymbol{n}\big)\big]|_{\Gamma}, \\
l_5(\boldsymbol{w}) &= \boldsymbol{n}_0\cdot\mathcal{D}_t\Big\{(\Delta(t) - \Delta(0))\int_0^t \boldsymbol{w}|_{\Gamma}\,\mathrm{d}t'\Big\} \\
&= \boldsymbol{n}_0\cdot\Big\{(\Delta(t) - \Delta(0))\boldsymbol{w} + \dot{\Delta}(t)\int_0^t \boldsymbol{w}|_{\Gamma}\,\mathrm{d}t'\Big\},
\end{aligned}$$

$\dot{\Delta}(t)$ is the operator obtained from the Beltrami–Laplace operator (1.1.4) by differentiating its coefficients with respect to t:

$$\dot{\Delta}(t) = \dot{g}^{\alpha\beta}\frac{\partial^2}{\partial s_\alpha \partial s_\beta} + \dot{h}^{\beta}\frac{\partial}{\partial s_\beta},$$

where the dot means the differentiation with respect to t (see [54] for details).

We observe that the operators $\boldsymbol{l}_1$ and l_2 have divergence form

$$\begin{aligned}
\boldsymbol{l}_1(\boldsymbol{w}, s) &= \partial \boldsymbol{L}_{1j}(\boldsymbol{w}, s)/\partial \xi_j, \\
l_2(\boldsymbol{w}) &= \nabla\cdot\boldsymbol{L}_2(\boldsymbol{w}); \\
\boldsymbol{L}_{1j}(\boldsymbol{w}, s) &= \nu^{\pm}(A_{ij}A_{im} - \delta_j^m)\partial\boldsymbol{w}/\partial\xi_m - \mathbb{B}\boldsymbol{e}_j s/\rho^{\pm} \\
&= \nu^{\pm}(B_{ij}A_{im} + B_{jm})\partial\boldsymbol{w}/\partial\xi_m - \mathbb{B}\boldsymbol{e}_j s/\rho^{\pm}, \\
\boldsymbol{L}_2(\boldsymbol{w}) &= (\mathbb{I} - \mathbb{A}^*)\boldsymbol{w} = -\mathbb{B}^*\boldsymbol{w}
\end{aligned}$$

($\boldsymbol{e}_j$ is the unit vector in the direction of ξ_j). The equality $\mathbb{A}\nabla\cdot\boldsymbol{w} = \nabla\cdot\mathbb{A}^*\boldsymbol{w}$ follows from the identity

$$\frac{\partial A_{ij}}{\partial \xi_j} = 0,$$

which is valid for the cofactors of the Jacobi matrix of any transformation.

Furthermore, the expression $\mathcal{D}_t r_1 - \nabla\cdot\boldsymbol{f}_1$ is also representable in the divergence form

$$\begin{aligned}\mathcal{D}_t r_1 - \nabla \cdot \boldsymbol{f}_1 &= \mathcal{D}_t r - \nabla \cdot \boldsymbol{f} + \mathcal{D}_t l_2(\boldsymbol{w}) - \nabla \cdot \boldsymbol{l}_1(\boldsymbol{w}, s)\\ &= \mathcal{D}_t r - \nabla_{\boldsymbol{u}} \cdot \boldsymbol{f} + \nabla \cdot [\mathbb{B}^*(\boldsymbol{f} - \mathcal{D}_t \boldsymbol{w}) - (\mathcal{D}_t \mathbb{B}^*)\boldsymbol{w} - \boldsymbol{l}_1(\boldsymbol{w}, s)]\\ &\equiv \nabla \cdot (\boldsymbol{g} + \boldsymbol{l}_6(\boldsymbol{w}, s))\end{aligned}$$

with

$$\begin{aligned}\boldsymbol{l}_6(\boldsymbol{w}, s) &= \mathbb{B}^*(\boldsymbol{f} - \mathcal{D}_t \boldsymbol{w}) - (\mathcal{D}_t \mathbb{B}^*)\boldsymbol{w} - \boldsymbol{l}_1(\boldsymbol{w}, s)\\ &= -\mathbb{B}^*\Big(\nu^{\pm}\nabla^2 \boldsymbol{w} - \frac{1}{\rho^{\pm}}\nabla s\Big) - (\mathcal{D}_t \mathbb{B}^*)\boldsymbol{w} - \mathbb{A}^* \boldsymbol{l}_1(\boldsymbol{w}, s)\\ &\equiv \partial \boldsymbol{L}_{6j}(\boldsymbol{w}, s)/\partial \xi_j,\\ \boldsymbol{L}_{6j}(\boldsymbol{w}, s) &= -\nu^{\pm}\mathbb{B}^* \partial \boldsymbol{w}/\partial \xi_j + \mathbb{B}^* \boldsymbol{e}_j s/\rho^{\pm} - \mathbb{A}^* \boldsymbol{L}_{1j} + \partial \boldsymbol{V}/\partial \xi_j,\\ \boldsymbol{V}(\xi, t) &= \int\limits_{\Omega^+ \cup \Omega^-} \mathcal{E}(\xi, \eta)\Big\{\frac{\partial \mathbb{B}^*}{\partial \eta_m}\Big(\nu^{\pm}\frac{\partial \boldsymbol{w}}{\partial \eta_m} - \frac{1}{\rho^{\pm}}\boldsymbol{e}_m s + \boldsymbol{L}_{1m}\Big) - (\mathcal{D}_t \mathbb{B}^*)\boldsymbol{w}\Big\}\, d\eta.\end{aligned}$$

Now, we formulate several auxiliary statements.

It is easily seen that if functions $f, g \in C_{\beta}^{\alpha,\alpha/2}(D_T)$, $\alpha \in (0,1)$, $\beta > 0$, then its product $fg \in C_{2\beta}^{\alpha,\alpha/2}(D_T)$ and the following inequalities, similar to the weightless estimates, are fulfilled:

$$\begin{aligned}\langle fg\rangle_{x,\beta,D_T}^{(\alpha)} &\leqslant \langle f\rangle_{x,\beta,D_T}^{(\alpha)}|g|_{D_T} + |f|_{D_T}\langle g\rangle_{x,\beta D_T}^{(\alpha)}\\ &\leqslant |f|_{\nu,\alpha,D_T}^{(\alpha,0)}|g|_{\nu,\alpha,D_T}^{(\alpha,0)}, \quad \nu = \max(0, \beta - \alpha),\\ |fg|_{\beta+\gamma,\mathbb{D}_T}^{(\alpha,\alpha/2)} &\leqslant |f|_{\beta,D_T}^{(\alpha,\alpha/2)}|g|_{\gamma,D_T}^{(\alpha,\alpha/2)}, \quad \gamma \leqslant \beta.\end{aligned} \tag{5.3.11}$$

Also, if $f, g \in C_{1,\gamma}^{(1+\alpha,\gamma)}(D_T)$, then

$$|fg|_{1,\gamma,D_T}^{(\gamma,1+\alpha)} \leqslant |f|_{1,\gamma,D_T}^{(\gamma,1+\alpha)}|g|_{x,D_T}^{(\gamma)} + |f|_{x,D_T}^{(\gamma)}|g|_{1,\gamma,D_T}^{(\gamma,1+\alpha)}. \tag{5.3.12}$$

Next, we consider the matrix $\mathbb{B}$. Its elements $B_{ij} = A_{ij} - \delta_j^i$ are given by the formulas

$$\begin{aligned}B_{ii} &= b_{jj} + b_{kk} + b_{jj}b_{kk} - b_{jk}b_{kj},\\ B_{ij} &= -b_{ji} - b_{ji}b_{kk} + b_{kj}b_{ik},\end{aligned}$$

where $i \neq j$, $j \neq k$, $k \neq i$, $b_{ij}(\xi, t) = \int_0^t \partial u_i(\xi, t')/\partial \xi_j \,\mathrm{d}t'$.

If vector $\boldsymbol{u}$ satisfies condition (5.3.2), then

$$|b_{km}(t)|_{1+\gamma,\cup\Omega^{\pm}}^{(1+\alpha)} \leqslant \int\limits_0^t |\partial u_k(t')/\partial \xi_m|_{1+\gamma,\cup\Omega^{\pm}}^{(1+\alpha)}\,\mathrm{d}t' \leqslant \delta, \quad t \in (0, T),$$

and

$$|b_{km}|^{(\gamma,1+\alpha)}_{1+\gamma,D_T} \leqslant T^{1-\frac{1+\alpha-\gamma}{2}} |\partial u_k/\partial \xi_m|^{(\gamma)}_{x,1+\gamma,D_T} \leqslant \delta, \quad k,m = 1,2,3.$$

Consequently, for coefficients B_{ij}, the estimate

$$\begin{aligned}|B_{ij}|^{(\gamma,1+\alpha)}_{1+\gamma,D_T} &\leqslant \max_{k,m} |b_{k,m}|^{(\gamma,1+\alpha)}_{1+\gamma,D_T} \big(2 + 8 \max_{k,m} |b_{km}|^{(\gamma)}_{x,1+\gamma,D_T}\big)\\ &\leqslant (2+8\delta) \max_{k,m} |b_{km}|^{(\gamma,1+\alpha)}_{1+\gamma,D_T}\end{aligned}$$

holds. The other inequalities collected in the following lemma can be proved in a similar way.

Lemma 5.3.1 *If* $\boldsymbol{u}$ *satisfies condition* (5.3.2), *then*

$$\begin{aligned}
|B_{ij}|^{(\alpha)}_{x,1+\gamma,D_T} &\leqslant c_2 \max_{k,m} |b_{km}|^{(\alpha)}_{x,1+\gamma,D_T},\\
|\nabla B_{ij}|^{(\alpha)}_{x,1+\gamma,D_T} &\leqslant c_3 \max_{k,m} |\nabla b_{km}|^{(\alpha)}_{x,1+\gamma,D_T},\\
\langle B_{ij}\rangle^{(\beta)}_{t,1+\gamma,D_T} &\leqslant c_4 \max_{k,m} \langle b_{km}\rangle^{(\beta)}_{t,1+\gamma,D_T}, \quad \beta \in (0,1),\\
|B_{ij}|^{(\gamma,1+\alpha)}_{1+\gamma,D_T} &\leqslant c_5 \max_{k,m} |b_{km}|^{(\gamma,1+\alpha)}_{1+\gamma,D_T},\\
\langle \nabla B_{ij}\rangle^{(\frac{1+\alpha-\gamma}{2})}_{t,1+\gamma,D_T} &\leqslant c_6 \max_{k,m} \langle \nabla b_{km}\rangle^{(\frac{1+\alpha-\gamma}{2})}_{t,1+\gamma,D_T}\\
&\leqslant c_7 T^{\frac{1-\alpha+\gamma}{2}} |\nabla\nabla \boldsymbol{u}|_{1+\gamma,D_T},\\
|\mathcal{D}_t B_{ij}|^{(\alpha,\alpha/2)}_{1+\gamma,D_T} &\leqslant c_8 \max_{k,m} |\mathcal{D}_t b_{km}|^{(\alpha,\alpha/2)}_{1+\gamma,D_T} \leqslant c_8 |\nabla \boldsymbol{u}|^{(\alpha,\alpha/2)}_{1+\gamma,D_T}, \qquad (5.3.13)\\
\langle \mathcal{D}_t B_{ij}\rangle^{(\gamma,1+\alpha)}_{1+\gamma,D_T} &\leqslant c_9 \max_{k,m} \big(|\mathcal{D}_t b_{km}|^{(\gamma,1+\alpha)}_{1+\gamma,D_t} + |\mathcal{D}_t b_{km}|^{(\gamma)}_{x,1+\gamma,D_T}\big)\\
&\leqslant c_9 \big(|\nabla \boldsymbol{u}|^{(\gamma,1+\alpha)}_{1+\gamma,D_T} + |\nabla \boldsymbol{u}|^{(\gamma)}_{x,1+\gamma,D_T}\big);\\
\langle B_{ij}\rangle^{(\frac{1+\alpha}{2})}_{t,1+\gamma,D_T} &\leqslant c_{10} T^{\frac{1-\alpha}{2}} |\nabla \boldsymbol{u}|_{1+\gamma,D_T}.\\
\langle \mathcal{D}_t B_{ij}\rangle^{(\frac{1+\alpha-\gamma}{2})}_{t,1+\gamma,D_T} &\leqslant c'_{10} |\nabla \boldsymbol{u}|^{(\frac{1+\alpha-\gamma}{2})}_{t,1+\gamma,D_T}.
\end{aligned}$$

Corollary 5.3.1 *If the assumptions of Lemma* 5.3.1 *are fulfilled, then*

$$|B_{ij}|^{(\alpha,\alpha/2)}_{1+\gamma,D_T} + |\nabla B_{ij}|^{(\alpha,\alpha/2)}_{1+\gamma,D_T} + \langle \nabla B_{ij}\rangle^{(\frac{1+\alpha-\gamma}{2})}_{t,1+\gamma,D_T} + |B_{ij}|^{(\gamma,1+\alpha)}_{1+\gamma,D_T} \leqslant c_{11}\delta, \quad i,j = 1,2,3. \qquad (5.3.14)$$

Remark 5.3.1 *The vector* $\mathbb{B}\boldsymbol{n}_0$ *is continuous as it passes across the surface* Γ: $[\mathbb{B}\boldsymbol{n}_0]|_\Gamma = [\mathbb{A}\boldsymbol{n}_0]|_\Gamma = 0$. *This follows immediately from the formula, e.g., for* A_{1j}, *which is the cofactor of* $a_{1j} = \partial x_1/\partial \xi_j$:

$$A_{1j}n_{0j} = \boldsymbol{n}_0 \cdot (\nabla x_2 \times \nabla x_3) = \boldsymbol{n}_0 \cdot (\nabla_\Gamma x_2 \times \nabla_\Gamma x_3).$$

The jump of this expression at the points of Γ *is equal to zero since* $\boldsymbol{x}$ *and its tangential derivatives are continuous.*

Lemma 5.3.2 *We suppose that* $\boldsymbol{u} \in \boldsymbol{C}_{1+\gamma}^{2+\alpha,1+\alpha/2}(D_T)$, $[\boldsymbol{u}]|_\Gamma = 0$, *and it is subjected to inequality* (5.3.2). *Then the operators* l_i *satisfy the estimates*

$$\begin{aligned}
&|\boldsymbol{l}_1(\boldsymbol{w},s)|_{1+\gamma,D_T}^{(\alpha,\alpha/2)} + |l_2(\boldsymbol{w})|_{1+\gamma,D_T} + |\nabla l_2(\boldsymbol{w})|_{1+\gamma,D_T}^{(\alpha,\alpha/2)} + |\boldsymbol{l}_3(\boldsymbol{w})|_{x,G_T}^{(1+\alpha)} + |l_4(\boldsymbol{w},s)|_{G_T}\\
&\quad + |\nabla_\Gamma l_4(\boldsymbol{w},s)|_{G_T}^{(\alpha,\alpha/2)} + |l_4(\boldsymbol{w},s)|_{G_T}^{(\gamma,1+\alpha)} + |l_5(\boldsymbol{w})|_{G_T}^{(\alpha,\alpha/2)}\\
&\quad + \max_j \big(|\boldsymbol{L}_{1j}(\boldsymbol{w},s)|_{1,\gamma,D_T}^{(1+\alpha,\gamma)} + |\boldsymbol{L}_{1j}(\boldsymbol{w},s)|_{1,\gamma,D_T}^{(\gamma,0)}\big)\\
&\quad \leqslant c_{11}'\delta\big(|\boldsymbol{w}|_{1+\gamma,D_T}^{(2+\alpha,1+\alpha/2)} + |\nabla s|_{1+\gamma,D_T}^{(\alpha,\alpha/2)} + |s|_{1,\gamma,D_T}^{(\gamma,1+\alpha)} + |s|_{x,1,\gamma,D_T}^{(\gamma)}\big), \qquad (5.3.15)
\end{aligned}$$

$$\begin{aligned}
&\langle l_2(\boldsymbol{w})\rangle_{t,1+\gamma,D_T}^{(\frac{1+\alpha}{2})} + \langle l_3(\boldsymbol{w})\rangle_{t,G_T}^{(\frac{1+\alpha}{2})} + |\boldsymbol{l}_6(\boldsymbol{w},s)|_{1+\gamma,D_T}^{(\alpha,\alpha/2)} \leqslant c_{12}\delta\big(|\boldsymbol{w}|_{1+\gamma,D_T}^{(2+\alpha,1+\alpha/2)} + |\nabla s|_{1+\gamma,D_T}^{(\alpha,\alpha/2)}\big)\\
&\quad + c_{13}\big(T^{\frac{1-\alpha}{2}}|\nabla \boldsymbol{u}|_{1+\gamma,D_T}|\nabla \boldsymbol{w}|_{1+\gamma,D_T} + |\nabla \boldsymbol{u}|_{1+\gamma,D_T}^{(\alpha,\alpha/2)}|\boldsymbol{w}|_{1+\gamma,D_T}^{(\alpha,\alpha/2)}\big). \qquad (5.3.16)
\end{aligned}$$

Proof With the help of (5.3.11), (5.3.13), and (5.3.14), it is not hard to show that

$$\begin{aligned}
&|\boldsymbol{l}_1(\boldsymbol{w},s)|_{1+\gamma,D_T}^{(\alpha,\alpha/2)} + |l_2(\boldsymbol{w})|_{1+\gamma,D_T} + |\nabla l_2(w)|_{1+\gamma,D_T}^{(\alpha,\alpha/2)}\\
&\qquad \leqslant c_{14}\big(|\mathbb{B}|_{1+\gamma,D_T}^{(\alpha,\alpha/2)} + |\nabla\mathbb{B}|_{1+\gamma,D_T}^{(\alpha,\alpha/2)}\big)\big(|\boldsymbol{w}|_{1+\gamma,D_T}^{(2+\alpha,1+\alpha/2)} + |\nabla s|_{1+\gamma,D_T}^{(\alpha,\alpha/2)}\big)\\
&\qquad \leqslant c_{15}\delta\big(|\boldsymbol{w}|_{1+\gamma,D_T}^{(2+\alpha,1+\alpha/2)} + |\nabla s|_{1+\gamma,D_T}^{(\alpha,\alpha/2)}\big),\\
&\langle l_2\rangle_{t,1+\gamma,D_T}^{\frac{(1+\alpha)}{2}} \leqslant c_{16}\big(\langle\mathbb{B}\rangle_{t,1+\gamma,D_T}^{(\frac{1+\alpha}{2})}|\nabla \boldsymbol{w}|_{1+\gamma,D_T} + |\mathbb{B}|_{1+\gamma,D_T}\langle\nabla \boldsymbol{w}\rangle_{t,1+\gamma,D_T}^{(\frac{1+\alpha}{2})}\big)\\
&\qquad \leqslant c_{17}\big(T^{\frac{1-\alpha}{2}}|\nabla \boldsymbol{u}|_{1+\gamma,D_T}|\nabla \boldsymbol{w}|_{1+\gamma,D_T} + \delta|\boldsymbol{w}|_{1+\gamma,D_T}^{(2+\alpha,1+\alpha/2)}\big), \qquad (5.3.17)\\
&|\boldsymbol{l}_6(\boldsymbol{w},s)|_{1+\gamma,D_T}^{(\alpha,\alpha/2)} \leqslant c_{18}|\nabla \boldsymbol{u}|_{1+\gamma,D_T}^{(\alpha,\alpha/2)}|\boldsymbol{w}|_{1+\gamma,D_T}^{(\alpha,\alpha/2)} + c_{19}\delta\big(|\boldsymbol{w}|_{1+\gamma,D_T}^{(2+\alpha,1+\alpha/2)} + |\nabla s|_{1+\gamma,D_T}^{(\alpha,\alpha/2)}\big),\\
&\max_{j=1,2,3}\big(|\boldsymbol{L}_{1j}(\boldsymbol{w},s)|_{1,\gamma,D_T}^{(\gamma,1+\alpha)} + |\boldsymbol{L}_{1j}(\boldsymbol{w},s)|_{1,\gamma,D_T}^{(\gamma,0)}\big)\\
&\qquad \leqslant c_{20}\delta\big(|\boldsymbol{w}|_{1+\gamma,D_T}^{(2+\alpha,1+\alpha/2)} + |s|_{1,\gamma,D_T}^{(\gamma,1+\alpha)} + |s|_{x,1,\gamma,D_T}^{(\gamma)}\big).
\end{aligned}$$

Boundary operators can be estimated similarly. (Their analogs in the case of a single fluid were analyzed in [54].) For instance, we consider

$$l_4(\boldsymbol{w},s) = [s\boldsymbol{n}_0 \cdot (\boldsymbol{n} - \boldsymbol{n}_0) + \mu^\pm \boldsymbol{n}_0 \cdot \{(\mathbb{S}(\boldsymbol{w}) - \mathbb{S}_{\boldsymbol{u}}(\boldsymbol{w}))\boldsymbol{n}_0 + \mathbb{S}_{\boldsymbol{u}}(\boldsymbol{w})(\boldsymbol{n}_0 - \boldsymbol{n})\}]|_\Gamma,$$

where

$$\boldsymbol{n}-\boldsymbol{n}_0=\frac{\mathbb{B}\boldsymbol{n}_0}{|\mathbb{A}\boldsymbol{n}_0|}-\frac{\boldsymbol{n}_0(\mathbb{A}\boldsymbol{n}_0\cdot\mathbb{A}\boldsymbol{n}_0-\boldsymbol{n}_0\cdot\boldsymbol{n}_0)}{|\mathbb{A}\boldsymbol{n}_0|(|\mathbb{A}\boldsymbol{n}_0|+1)},$$

$$\mathbb{A}\boldsymbol{n}_0\cdot\mathbb{A}\boldsymbol{n}_0-\boldsymbol{n}_0\cdot\boldsymbol{n}_0=2\sum_{i=1}^{3}B_{ij}n_{0j}+B_{ij}n_{0j}B_{im}n_{0m}.$$

By Lemma 5.3.1, we have

$$\begin{aligned}|l_4(\boldsymbol{w},s)|_{G_T}^{(\gamma,1+\alpha)}&\leqslant c_{21}\big\{|\mathbb{B}|_{G_T}^{(1+\alpha,\gamma)}\big(|[s]|_\Gamma|_{x,G_T}^{(\gamma)}+|\nabla\boldsymbol{w}|_{x,G_T}^{(\gamma)}\big)\\&\qquad+|\mathbb{B}|_{x,G_T}^{(\gamma)}\big(|[s]|_\Gamma|_{G_T}^{(\gamma,1+\alpha)}+|\nabla\boldsymbol{w}|_{G_T}^{(\gamma,1+\alpha)}\big)\big\}\\&\leqslant c_{22}\delta\big(|\boldsymbol{w}|_{1+\gamma,D_T}^{(2+\alpha,1+\alpha/2)}+|s|_{t,1,D_T}^{(\frac{1+\alpha-\gamma}{2})}+\langle s\rangle_{1+\gamma,D_T}^{(1+\alpha,\gamma)}+|\nabla s|_{1+\gamma,D_T}\big),\end{aligned}$$

$$\begin{aligned}&|l_4(\boldsymbol{w},s)|_{G_T}+|\nabla_\Gamma l_4(\boldsymbol{w},s)|_{G_T}^{(\alpha,\alpha/2)}\\&\quad\leqslant c_{23}\delta\big(|\boldsymbol{w}|_{1+\gamma,D_T}^{(2+\alpha,1+\alpha/2)}+|\nabla s|_{1+\gamma,D_T}^{(\alpha,\alpha/2)}+|s|_{t,1,D_T}^{(\frac{1+\alpha-\gamma}{2})}\big).\end{aligned}$$

From [54] we know that the operators $\boldsymbol{l}_3$ and l_5 satisfy the estimates

$$|\boldsymbol{l}_3(\boldsymbol{w})|_{G_T}+|\nabla\boldsymbol{l}_3(\boldsymbol{w})|_{x,G_T}^{(\alpha)}+|l_5(\boldsymbol{w})|_{G_T}^{(\alpha,\alpha/2)}\leqslant c_{24}\delta|\boldsymbol{w}|_{G_T}^{(2+\alpha,1+\alpha/2)},$$

$$\langle\boldsymbol{l}_3(\boldsymbol{w})\rangle_{t,G_T}^{(\frac{1+\alpha}{2})}\leqslant c_{25}\big(\langle\mathbb{B}\rangle_{t,G_T}^{(\frac{1+\alpha}{2})}|\nabla\boldsymbol{w}|_{G_T}+|\mathbb{B}|_{G_T}\langle\nabla\boldsymbol{w}\rangle_{t,G_T}^{(\frac{1+\alpha}{2})}\big).$$

Summing the inequalities obtained, we arrive at (5.3.15) and (5.3.16). □

Lemma 5.3.3 *If* $F\in C_2^{(0,\frac{1+\alpha-\gamma}{2})}(D_T)$, *then the gradient of the Newtonian potential*

$$\nabla_x V(x,t)=\nabla_x\int_{\mathbb{R}^3}\mathcal{E}(x,y)F(y,t)\,\mathrm{d}y$$

satisfies the estimates

$$|\nabla V|_{1,\gamma,D_T}^{(\gamma,0)}\leqslant c_{26}|F|_{2,D_T},\tag{5.3.18}$$

$$|\nabla V|_{1,\gamma,D_T}^{(1+\alpha,\gamma)}\leqslant c_{27}|F|_{t,2,D_T}^{(\frac{1+\alpha-\gamma}{2})}.\tag{5.3.19}$$

Proof Let $x\in\Omega^+\cup\Omega^-$. We set $\rho=1+|x|$. Obviously

$$|\nabla V(x,t)|\leqslant c_{28}|F|_{2,D_T}\int_{\mathbb{R}^3}\frac{dy}{|x-y|^2(1+|y|)^2}$$

$$\leqslant c_{28}|F|_{2,D_T}\left\{\left(\frac{\rho}{2}\right)^{-2}\int_{|x-y|\leqslant\rho/2}\frac{dy}{|x-y|^2}+\int_{|x-y|>\rho/2}\frac{dy}{|x-y|^2|y|^2}\right\}$$
$$\leqslant c_{29}|F|_{2,D_T}\rho^{-1}.$$

We consider now two points $x, z \in Q_T^+$ and $x, z \in Q_T^-$. If $|x-z| \leqslant \rho/4$, then

$$|\nabla_x V(x,t) - \nabla_z V(z,t)| = \left|\int_{\mathbb{R}^3}(\nabla_x\mathcal{E}(x,y) - \nabla_z\mathcal{E}(z,y))F(y,t)\,\mathrm{d}y\right|$$
$$\leqslant c_{30}|F|_{2,D_T}|x-z|^\gamma\int_{\mathbb{R}^3}\left(\frac{1}{|x-y|^{2+\gamma}}+\frac{1}{|z-y|^{2+\gamma}}\right)\frac{dy}{(1+|y|)^2}$$
$$\leqslant c_{31}|F|_{2,D_T}|x-z|^\gamma\rho^{-(1+\gamma)};$$

otherwise

$$|\nabla_x V(x,t) - \nabla_z V(z,t)| \leqslant 2c_{29}|F|_{2,D_T}\rho^{-1} \leqslant c_{32}|F|_{2,D_T}|x-z|^\gamma\rho^{-(1+\gamma)}.$$

Taking the maximum in these inequalities (first, with respect to x and z and, then, with respect to $t \in (0,T)$), we get (5.3.18). Application of the above arguments to the increment with respect to t leads us to (5.3.19). □

Lemma 5.3.4 *Under the assumptions of Lemma* 5.3.2, *we have the inequality*

$$\max_{j=1,2,3}\big(|\boldsymbol{L}_{6j}|^{(\gamma,1+\alpha)}_{1,\gamma,D_T}+|\boldsymbol{L}_{6j}|^{(\gamma,0)}_{1,\gamma,D_T}\big) \leqslant c_{33}\delta\big(|\boldsymbol{w}|^{(2+\alpha,1+\alpha/2)}_{1+\gamma,D_T}+|s|^{(1+\alpha,\gamma)}_{1,\gamma,D_T}+|s|^{(\gamma)}_{x,1,\gamma,D_T}\big)$$
$$+c_{34}|\nabla\boldsymbol{u}|^{(\frac{1+\alpha-\gamma}{2})}_{t,1+\gamma,D_T}|\boldsymbol{w}|^{(\frac{1+\alpha-\gamma}{2})}_{t,1+\gamma,D_T}. \qquad (5.3.20)$$

Proof Using (5.3.11) and (5.3.12), as well as Lemma 5.3.1 and 5.3.2 we conclude that

$$\max_{j=1,2,3}|\boldsymbol{L}_{6j}|^{(\gamma,0)}_{1,\gamma,D_T} \leqslant c_{35}\delta\big(|\boldsymbol{w}|^{(2+\alpha,1+\alpha/2)}_{1+\gamma,D_T}+|s|^{(\gamma)}_{x,1,\gamma,D_T}\big)+|\nabla\boldsymbol{V}|^{(\gamma,0)}_{1,\gamma,D_T},$$
$$\max_{j=1,2,3}|\boldsymbol{L}_{6j}|^{(\gamma,1+\alpha)}_{1,\gamma,D_T} \leqslant c_{36}\delta\big(|\boldsymbol{w}|^{(2+\alpha,1+\alpha/2)}_{1+\gamma,D_T}+|s|^{(\gamma,1+\alpha)}_{1,\gamma,D_T}\big)+|\nabla\boldsymbol{V}|^{(\gamma,1+\alpha)}_{1,\gamma,D_T}. \qquad (5.3.21)$$

The inequality

$$|fg|_{\nu+\beta,D_T} \leqslant |f|_{\nu,D_T}|g|_{\beta,D_T},$$

makes it possible to estimate the gradient of the potential $\boldsymbol{V}$ by Lemma 5.3.3:

$$|\nabla\boldsymbol{V}|^{(\gamma,0)}_{1,\gamma,D_T}+|\nabla\boldsymbol{V}|^{(\gamma,1+\alpha)}_{1,\gamma,D_T}$$

$$\leqslant c_{37}\Big\{|\nabla\mathbb{B}|^{(\frac{1+\alpha-\gamma}{2})}_{t,1,D_T}\Big(|\nabla \boldsymbol{w}|^{(\frac{1+\alpha-\gamma}{2})}_{t,1,D_t}+|s|^{(\frac{1+\alpha-\gamma}{2})}_{t,1,D_T}+\max_m|\boldsymbol{L}_{1m}|^{(\frac{1+\alpha-\gamma}{2})}_{t,1,D_T}\Big)$$
$$+|\mathcal{D}_t\mathbb{B}|^{(\frac{1+\alpha-\gamma}{2})}_{t,1,D_T}|\boldsymbol{w}|^{(\frac{1+\alpha-\gamma}{2})}_{t,1,D_T}\Big\}.$$

Combined with (5.3.13), (5.3.17), and (5.3.21), this yields inequality (5.3.20). □

Lemma 5.3.5 *Lemmas* 5.3.2 *and* 5.3.4 *imply the inequality*

$$|\boldsymbol{l}_1|^{(\alpha,\alpha/2)}_{1+\gamma,D_T}+|l_2|^{(1+\alpha,\frac{1+\alpha}{2})}_{1+\gamma,D_T}+|\boldsymbol{l}_3|^{(1+\alpha,\frac{1+\alpha}{2})}_{G_T}+|l_4|_{G_T}+|\nabla_\Gamma l_4|^{(\alpha,\alpha/2)}_{G_T}$$
$$+|l_4|^{(\gamma,1+\alpha)}_{G_T}+|l_5|^{(\alpha,\alpha/2)}_{G_T}+|\boldsymbol{l}_6|^{(\alpha,\alpha/2)}_{1+\gamma,D_T}+\max_j\big(|\boldsymbol{L}_{6j}|^{(\gamma,1+\alpha)}_{1,\gamma,D_T}+|\boldsymbol{L}_{6j}|^{(\gamma,0)}_{1,\gamma,D_T}\big)$$
$$\leqslant c_{38}\big\{\delta N_T[\boldsymbol{w},s]+P_T[\boldsymbol{u}]|\boldsymbol{w}(\cdot,0)|^{(1)}_{1+\gamma,\cup\Omega^\pm}\big\}, \tag{5.3.22}$$

where $P_T[\boldsymbol{u}]$ *is given by formula* (5.3.8).

Proof Since

$$|\boldsymbol{w}|^{(\frac{1+\alpha-\gamma}{2})}_{t,1+\gamma,D_T}\leqslant|\boldsymbol{w}(\cdot,0)|_{1+\gamma,\cup\Omega^\pm}$$
$$+\int_0^T|\mathcal{D}_t\boldsymbol{w}(\cdot,t)|_{1+\gamma,\cup\Omega^\pm}\,\mathrm{d}t+T^{\frac{1-\alpha+\gamma}{2}}|\mathcal{D}_t\boldsymbol{w}|_{1+\gamma,D_T},$$
$$|\boldsymbol{w}|^{(\alpha,\frac{\alpha}{2})}_{1+\gamma,D_T}\leqslant|\boldsymbol{w}(\cdot,0)|^{(\alpha)}_{1+\gamma,\cup\Omega^\pm}+(T+T^{1-\alpha/2})|\boldsymbol{w}|^{(2+\alpha,1+\frac{\alpha}{2})}_{1+\gamma,D_T},$$
$$|\nabla\boldsymbol{w}|_{1+\gamma,D_T}\leqslant|\nabla\boldsymbol{w}(\cdot,0)|_{1+\gamma,D_T}+T^{\frac{1+\alpha}{2}}\langle\nabla\boldsymbol{w}\rangle^{(\frac{1+\alpha}{2})}_{t,1+\gamma,D_T},$$
$$\langle\nabla\boldsymbol{u}\rangle^{(\frac{1+\alpha-\gamma}{2})}_{t,1+\gamma,D_T}\leqslant T^{\gamma/2}|\boldsymbol{u}|^{(2+\alpha,1+\frac{\alpha}{2})}_{1+\gamma,D_T},$$

we have

$$T^{\frac{1-\alpha}{2}}|\nabla\boldsymbol{u}|_{1+\gamma,D_T}|\nabla\boldsymbol{w}|_{1+\gamma,D_T}+|\nabla\boldsymbol{u}|^{(\alpha,\frac{\alpha}{2})}_{1+\gamma,D_T}|\boldsymbol{w}|^{(\alpha,\frac{\alpha}{2})}_{1+\gamma,D_T}+|\nabla\boldsymbol{u}|^{(\frac{1+\alpha-\gamma}{2})}_{t,1+\gamma,D_T}|\boldsymbol{w}|^{(\frac{1+\alpha-\gamma}{2})}_{t,1+\gamma,D_T}$$
$$\leqslant c_{39}\big\{\delta|\boldsymbol{w}|^{(2+\alpha)}_{1+\gamma,D_T}+P_T[\boldsymbol{u}]|\boldsymbol{w}(\cdot,0)|^{(1)}_{1+\gamma,\cup\Omega^\pm}\big\}.$$

In addition, the weighted Hölder constant of pressure function is estimated by the evident inequality

$$\langle s\rangle^{(\gamma)}_{x,1+\gamma,D_T}\leqslant|\nabla s|_{1+\gamma,D_T}+|s|_{1,D_T}.$$

Thus, (5.3.22) follows from Lemmas 5.3.2 and 5.3.4. □

Proof of Theorem 5.3.1 We prove the solvability of (5.3.1) by successive approximations. As zero approximation we take $(\boldsymbol{w}^{(0)},s^{(0)})$, where $s^{(0)}(\xi,t)=$

$s_0(\xi)$ is a solution of (5.3.6) and $\boldsymbol{w}^{(0)}$ is a solution to the following problem similar to (4.2.10):

$$\mathcal{D}_t\boldsymbol{w}^{(0)} - \nu^{\pm}\nabla^2\boldsymbol{w}^{(0)} = \boldsymbol{f} - \frac{1}{\rho^{\pm}}\nabla s^{(0)}, \qquad \boldsymbol{w}^{(0)}|_{t=0} = \boldsymbol{w}_0, \quad [\boldsymbol{w}^{(0)}]|_{\Gamma} = 0. \tag{5.3.23}$$

Equalities (5.3.3) imply compatibility conditions

$$[\boldsymbol{w}^{(0)}]|_{\Gamma,t=0} = [\boldsymbol{w}_0]|_{\Gamma} = 0, \qquad [\mathcal{D}_t\boldsymbol{w}^{(0)}]|_{\Gamma,t=0} = 0$$

for (5.3.23); therefore, it is possible to find $\boldsymbol{w}^{(0)}$ satisfying estimate (5.2.15) with $\nu = \gamma$, namely,

$$|\boldsymbol{w}^{(0)}|^{(2+\alpha,1+\frac{\alpha}{2})}_{1+\gamma,D_T} \leqslant c_{40}\big(|\boldsymbol{f}|^{(\alpha,\frac{\alpha}{2})}_{1+\gamma,D_T} + |\nabla s_0|^{(\alpha)}_{1+\gamma,\cup\Omega^{\pm}} + |\boldsymbol{w}_0|^{(2+\alpha)}_{1+\gamma,\cup\Omega^{\pm}}\big). \tag{5.3.24}$$

The $(m+1)$th approximations $(\boldsymbol{w}^{(m+1)}, s^{(m+1)})$, $m \geqslant 0$, are defined as solutions of the initial-boundary value problems

$$\begin{aligned}
&\mathcal{D}_t\boldsymbol{w}^{(m+1)} - \nu^{\pm}\nabla^2\boldsymbol{w}^{(m+1)} + \frac{1}{\rho^{\pm}}\nabla s^{(m+1)} = \boldsymbol{f} + \boldsymbol{l}_1(\boldsymbol{w}^{(m)}, s^{(m)}),\\
&\nabla\cdot\boldsymbol{w}^{(m+1)} = r + l_2(\boldsymbol{w}^{(m)}) \quad \text{in} \quad D_T,\\
&\boldsymbol{w}^{(m+1)}|_{t=0} = \boldsymbol{w}_0, \quad [\boldsymbol{w}^{(m+1)}]|_{G_T} = 0,\\
&[\mu^{\pm}\Pi_0\mathbb{S}(\boldsymbol{w}^{(m+1)})\boldsymbol{n}_0]|_{G_T} = \Pi_0\boldsymbol{a} + \boldsymbol{l}_3(\boldsymbol{w}^{(m)}),\\
&[\boldsymbol{n}_0\cdot\mathbb{T}(\boldsymbol{w}^{(m+1)}, s^{(m+1)})\boldsymbol{n}_0]|_{G_T} - \sigma\boldsymbol{n}_0\cdot\Delta(0)\int_0^t \boldsymbol{w}^{(m+1)}\,\mathrm{d}t'|_{G_T}\\
&\quad = b + \int_0^t B\,\mathrm{d}t' + l_4(\boldsymbol{w}^{(m)}, s^{(m)}) + \sigma\int_0^t l_5(\boldsymbol{w}^{(m)})\,\mathrm{d}t', \quad t\in(0,T).
\end{aligned} \tag{5.3.25}$$

We consider the case $m = 0$. In order to apply Theorem 5.2.1 on the solvability of the linear problem, we verify conditions (4.2.2) and (5.2.1) for problem (5.3.25). Due to (5.3.4), we have

$$\mathcal{D}_t r + \mathcal{D}_t l_2(\boldsymbol{w}^{(0)}) - \nabla\cdot\big(\boldsymbol{f} + \boldsymbol{l}_1(\boldsymbol{w}^{(0)}, s^{(0)})\big) = \nabla\cdot\big(\boldsymbol{g} + \boldsymbol{l}_7(\boldsymbol{w}^{(0)}, s^{(0)})\big);$$

and, consequently, representation formulas (4.2.2) are valid with

$$\boldsymbol{g}^{(1)} = \boldsymbol{g} + \boldsymbol{l}_7(\boldsymbol{w}^{(0)}, s^{(0)}), \qquad \mathbb{G}^{(1)} = \mathbb{G} + \mathbb{L}_7(\boldsymbol{w}^{(0)}, s^{(0)}),$$

where $\boldsymbol{l}_7 = \boldsymbol{l}_6 + \mathbb{B}^*\boldsymbol{l}_1$, $\mathbb{L}_7 = \{\boldsymbol{L}_{7k}\}_{k=1}^3$,

$$\boldsymbol{L}_{7k}(\boldsymbol{w}^{(0)}, s^{(0)}) = -\mathbb{B}^*\Big(\nu^{\pm}\frac{\partial\boldsymbol{w}^{(0)}}{\partial\xi_k} - \frac{1}{\rho^{\pm}}\boldsymbol{e}_k s^{(0)}\Big) - \boldsymbol{L}_{1k}(\boldsymbol{w}^{(0)}, s^{(0)}) + \frac{\partial\boldsymbol{V}^{(0)}}{\partial\xi_k},$$

$$\boldsymbol{V}^{(0)}(\xi,t)=\int_{\Omega^+\cup\Omega^-}\mathcal{E}(\xi,\eta)\Big\{\frac{\partial\mathbb{B}^*}{\partial\eta_m}\Big(\nu^{\pm}\frac{\partial\boldsymbol{w}^{(0)}}{\partial\eta_m}-\frac{1}{\rho^{\pm}}\boldsymbol{e}_m s^{(0)}\Big)-\mathcal{D}_t\mathbb{B}^*\boldsymbol{w}^{(0)}\Big\}\,\mathrm{d}\eta.$$

The jump

$$\begin{aligned}&\big[(\boldsymbol{g}+\boldsymbol{l}_7(\boldsymbol{w}^{(0)},s^{(0)})+\boldsymbol{f}+\boldsymbol{l}_1(\boldsymbol{w}^{(0)},s^{(0)}))\cdot\boldsymbol{n}_0\big]\big|_\Gamma\\&\quad=\Big[-\mathbb{B}^*\Big(\boldsymbol{f}+\nu^{\pm}\nabla^2\boldsymbol{w}^{(0)}-\frac{1}{\rho^{\pm}}\nabla s^{(0)}\Big)\cdot\boldsymbol{n}_0-(\mathcal{D}_t\mathbb{B}^*)\boldsymbol{w}^{(0)}\cdot\boldsymbol{n}_0\Big]\Big|_\Gamma\\&\quad=-[\mathbb{B}\boldsymbol{n}_0]|_\Gamma\mathcal{D}_t\boldsymbol{w}^{(0)}-[\mathcal{D}_t\mathbb{B}\boldsymbol{n}_0]|_\Gamma\boldsymbol{w}^{(0)}\end{aligned}$$

is equal to 0 on $[0,T]$ (see Remark 5.3.1).

Next, conditions (4.1.8) and (4.1.9) for $\boldsymbol{w}_0$ and $s_0=s^{(1)}|_{t=0}$ follow from conditions (5.3.3) and (5.3.6), and in (4.1.9), we have

$$\mathcal{D}_t\boldsymbol{R}|_{t=0}=\big(\boldsymbol{f}+\boldsymbol{g}^{(1)}+\boldsymbol{l}_1(\boldsymbol{w}^{(0)},s^{(0)})\big)\big|_{t=0}=(\boldsymbol{f}+\boldsymbol{g})|_{t=0}-\mathcal{D}_t\mathbb{B}^*|_{t=0}\boldsymbol{w}_0.$$

The norms of the functions in the right-hand side of (5.3.25) for $m=0$ are bounded by Lemma 5.3.5. Therefore, Theorem 5.2.1 shows that there exists a unique solution $(\boldsymbol{w}^{(1)},s^{(1)})$ of system (5.3.25), and this solution is subjected to the estimate

$$N_{t'}[\boldsymbol{w}^{(1)},s^{(1)}]\leqslant c_{41}(t')\{Q_1(t')+\delta N_{t'}[\boldsymbol{w}^{(0)},s^{(0)}]+P_{t'}[\boldsymbol{u}]|\boldsymbol{w}_0|^{(1)}_{1+\gamma,\cup\Omega^{\pm}}\},\quad t'\leqslant T.$$

From (5.3.24), we obtain

$$N_{t'}[\boldsymbol{w}^{(1)},s^{(1)}]\leqslant c_{42}(t')\{Q_1(t')+|\nabla s_0|^{(\alpha)}_{1+\gamma,\cup\Omega^{\pm}}+|s_0|_{1,\cup\Omega^{\pm}}+P_{t'}[\boldsymbol{u}]|\boldsymbol{w}_0|^{(1)}_{1+\gamma,\cup\Omega^{\pm}}\}. \tag{5.3.26}$$

The function s_0, as well as the solution p_0 of problem (4.1.9), satisfies inequality (5.2.14):

$$|s_0|_{1,\cup\Omega^{\pm}}=|s^{(0)}|_{1,D_{t'}}\leqslant c_{43}\big(Q_1(t')+|\mathbb{G}^{(1)}(\cdot,0)|^{(\gamma)}_{1,\gamma,\cup\Omega^{\pm}}\big).$$

Since

$$\mathbb{G}^{(1)}\big|_{t=0}=\mathbb{G}\big|_{t=0}+\mathbb{L}_7(\boldsymbol{w}^{(0)},s^{(0)})\big|_{t=0}=\mathbb{G}\big|_{t=0}-\nabla\int_{\Omega^+\cup\Omega^-}\mathcal{E}D_t\mathbb{B}^*\big|_{t=0}\boldsymbol{w}_0\,\mathrm{d}\eta,$$

by Lemma 5.3.3, we obtain

$$|\mathbb{G}^{(1)}(\cdot,0)|^{(\gamma)}_{1,\gamma,\cup\Omega^{\pm}}\leqslant|\mathbb{G}(\cdot,0)|^{(\gamma)}_{1,\gamma,\cup\Omega^{\pm}}+c_{26}|\mathcal{D}_t\mathbb{B}^*\big|_{t=0}\boldsymbol{w}_0|_{2,\cup\Omega^{\pm}}.$$

Differentiating the integral representation for s_0, we estimate the norm of the gradient of s_0 with the help of inequality (5.2.18):

$$|\nabla s_0|^{(\alpha)}_{1+\gamma,\cup\Omega^\pm} \leqslant c_{44}\big(Q_1(t') + |\boldsymbol{g}(\cdot,0)|^{(\alpha)}_{1+\gamma,\cup\Omega^\pm} + |\mathcal{D}_t\mathbb{B}^*\boldsymbol{w}_0|^{(\alpha)}_{1+\gamma,\cup\Omega^\pm}\big).$$

Finally, applying (5.3.11), as well as (5.3.13) for estimating $|\mathcal{D}_t\mathbb{B}^*|_{t=0}|^{(\alpha)}_{1+\gamma,\cup\Omega^\pm}$, we get the inequality

$$|s_0|_{1,\cup\Omega^\pm} + |\nabla s_0|^{(\alpha)}_{1+\gamma,\cup\Omega^\pm} \leqslant c_{45}\big(Q_1(t') + |\nabla \boldsymbol{u}|^{(\alpha,\alpha/2)}_{1+\gamma,D_{t'}}|\boldsymbol{w}_0|^{(\alpha)}_{1+\gamma,\cup\Omega^\pm}\big). \tag{5.3.27}$$

Combined with (5.3.26), this gives estimate (5.3.7) for $(\boldsymbol{w}^{(1)}, s^{(1)})$.

In the same way, using the relations

$$\begin{aligned}&\big[(\boldsymbol{g} + \boldsymbol{l}_6(\boldsymbol{w}^{(m)}, s^{(m)}) + \boldsymbol{f} + \boldsymbol{l}_1(\boldsymbol{w}^{(m)}, s^{(m)})) \cdot \boldsymbol{n}_0]\big|_\Gamma \\ &\quad = [(\boldsymbol{g} + \mathbb{A}^*\boldsymbol{f}) \cdot \boldsymbol{n}_0]\big|_\Gamma - [\mathbb{B}\boldsymbol{n}_0]\big|_\Gamma \cdot \mathcal{D}_t\boldsymbol{w}^{(m)} - [\mathcal{D}_t\mathbb{B}\boldsymbol{n}_0]\big|_\Gamma \cdot \boldsymbol{w}^{(m)} = 0,\end{aligned}$$

we obtain, by induction, the existence and the boundedness of the solutions $(\boldsymbol{w}^{(m)}, s^{(m)})$, $m \geqslant 2$, of systems (5.3.25).

Now we consider the differences

$$\boldsymbol{z}^{(m+1)} = \boldsymbol{w}^{(m+1)} - \boldsymbol{w}^{(m)}, \quad q^{(m+1)} = s^{(m+1)} - s^{(m)}, \quad m \geqslant 1,$$

which are solutions to the problems

$$\begin{aligned}&\mathcal{D}_t\boldsymbol{z}^{(m+1)} - \nu^\pm\nabla^2\boldsymbol{z}^{(m+1)} + \frac{1}{\rho^\pm}\nabla q^{(m+1)} = \boldsymbol{l}_1(\boldsymbol{z}^{(m)}, q^{(m)}),\\ &\nabla\cdot\boldsymbol{z}^{(m+1)} = l_2(\boldsymbol{z}^{(m)}) \quad \text{in} \quad D_T,\\ &\boldsymbol{z}^{(m+1)}\big|_{t=0} = 0, \quad [\boldsymbol{z}^{(m+1)}]\big|_{G_T} = 0, \quad [\mu^\pm\Pi_0\mathbb{S}(\boldsymbol{z}^{(m+1)})\boldsymbol{n}_0]\big|_{G_T} = \boldsymbol{l}_3(\boldsymbol{z}^{(m)}),\\ &\big[\boldsymbol{n}_0\cdot\mathbb{T}(\boldsymbol{z}^{(m+1)}, q^{(m+1)})\boldsymbol{n}_0]\big|_{G_T} - \sigma\boldsymbol{n}_0\cdot\Delta(0)\int_0^t \boldsymbol{z}^{(m+1)}\,d\tau\big|_{G_T} \\ &\quad = l_4(\boldsymbol{z}^{(m)}, q^{(m)}) + \sigma\int_0^t l_5(\boldsymbol{z}^{(m)})\,d\tau\big|_{G_T}.\end{aligned} \tag{5.3.28}$$

We set $\boldsymbol{z}^{(1)} = \boldsymbol{w}^{(1)} - \boldsymbol{w}^{(0)}$, $q^{(1)} = s^{(1)} - s^{(0)}$, $\boldsymbol{z}^{(0)} = \boldsymbol{w}^{(0)}$, $q^{(0)} = s^{(0)}$.

It is easily seen that all the assumptions of Theorem 5.2.1 are fulfilled for the system (5.3.28); in particular, we have representation formulas (4.2.2):

$$\begin{aligned}&\mathcal{D}_t l_2(\boldsymbol{z}^{(m)}) - \nabla\cdot\boldsymbol{l}_1(\boldsymbol{z}^{(m)}, q^{(m)})\\ &\quad = \nabla\cdot\Big\{\mathbb{B}^*\Big(-\nu^\pm\nabla^2\boldsymbol{z}^{(m)} + \frac{1}{\rho^\pm}\nabla q^{(m)} - \boldsymbol{l}_1(\boldsymbol{z}^{(m-1)}, q^{(m-1)})\Big)\\ &\qquad - \mathcal{D}_t\mathbb{B}^*\boldsymbol{z}^{(m)} - \boldsymbol{l}_1(\boldsymbol{z}^{(m)}, q^{(m)})\Big\}\end{aligned}$$

$$\begin{aligned}
&\equiv \nabla \cdot (\partial \boldsymbol{M}_k^{(m)}/\partial \xi_k),\\
\boldsymbol{M}_k^{(m)} &= -\mathbb{B}^*\Big\{\nu^\pm \frac{\partial \boldsymbol{z}^{(m)}}{\partial \xi_k} - \frac{1}{\rho^\pm}\boldsymbol{e}_k q^{(m)} + \boldsymbol{L}_{1k}(\boldsymbol{z}^{(m-1)}, q^{(m-1)})\Big\}\\
&\qquad - \boldsymbol{L}_{1k}(\boldsymbol{z}^{(m)}, q^{(m)}) + \frac{\partial \boldsymbol{W}^{(m)}}{\partial \xi_k},\\
\boldsymbol{W}^{(m)} &= \int\limits_{\Omega^+ \cup \Omega^-} \mathcal{E}(\cdot, \eta)\Big\{\frac{\partial \mathbb{B}^*}{\partial \eta_j}\Big(\nu^\pm \frac{\partial \boldsymbol{z}^{(m)}}{\partial \eta_j} - \boldsymbol{e}_j q^{(m)}/\rho^\pm + \boldsymbol{L}_{1j}(\boldsymbol{z}^{(m-1)}, q^{(m-1)})\Big)\\
&\qquad - (\mathcal{D}_t \mathbb{B}^*)\boldsymbol{z}^{(m)}\Big\}\, d\eta.
\end{aligned}$$

Equality (5.2.1) is true, because

$$[\mathcal{D}_t \boldsymbol{L}_2(\boldsymbol{z}^{(m)}) \cdot \boldsymbol{n}_0]\big|_\Gamma = -[\mathbb{B}\boldsymbol{n}_0]\big|_\Gamma \cdot \mathcal{D}_t \boldsymbol{z}^{(m)} - [\mathcal{D}_t \mathbb{B}\boldsymbol{n}_0]\big|_\Gamma \cdot \boldsymbol{z}^{(m)} = 0.$$

If $m > 1$, then $\boldsymbol{z}^{(m)}\big|_{t=0} = 0$, and, hence, Theorem 5.2.1 and Lemma 5.3.5 provide the following estimate for $(\boldsymbol{z}^{(m+1)}, q^{(m+1)})$:

$$N_T[\boldsymbol{z}^{(m+1)}, q^{(m+1)}] \leqslant c_{46}\delta\big(N_T[\boldsymbol{z}^{(m)}, q^{(m)}] + N_T[\boldsymbol{z}^{(m-1)}, q^{(m-1)}]\big). \tag{5.3.29}$$

If $m = 1$, then we have

$$N_T[\boldsymbol{z}^{(2)}, q^{(2)}] \leqslant c_{46}\Big\{\delta\big(N_T[\boldsymbol{z}^{(1)}, q^{(1)}] + N_T[\boldsymbol{z}^{(0)}, q^{(0)}]\big) + P_T[\boldsymbol{u}]|\boldsymbol{w}_0|^{(1)}_{1+\gamma, \cup\Omega^\pm}\Big\}, \tag{5.3.30}$$

since $\boldsymbol{z}^{(0)}|_{t=0} = \boldsymbol{w}_0$.

We sum inequalities (5.3.29) from $m = 2$ to $M - 1$ and add (5.3.30), then

$$\begin{aligned}
&\sum_{m=2}^{M} N_T[\boldsymbol{z}^{(m)}, q^{(m)}]\\
&\leqslant c_{46}\Big(\delta \sum_{m=1}^{M-1} N_T[\boldsymbol{z}^{(m)}, q^{(m)}] + \delta \sum_{m=0}^{M-2} N_T[\boldsymbol{z}^{(m)}, q^{(m)}] + P_T[\boldsymbol{u}]|\boldsymbol{w}_0|^{(1)}_{1+\gamma, \cup\Omega^\pm}\Big)\\
&\leqslant c_{46}\Big(2\delta \sum_{m=2}^{M} N_T[\boldsymbol{z}^{(m)}, q^{(m)}] + 2\delta N_T[\boldsymbol{w}^{(1)}, s^{(1)}] + 3\delta N_T[\boldsymbol{w}^{(0)}, s^{(0)}]\\
&\qquad + P_T[\boldsymbol{u}]|\boldsymbol{w}_0|^{(1)}_{1+\gamma, \cup\Omega^\pm}\Big).
\end{aligned}$$

If δ is chosen so small that $2c_{46}\delta < 1$, then from (5.3.24), (5.3.26), and (5.3.27), one can deduce the chain of the inequalities

$$N_T[\boldsymbol{w}^{(M)}, s^{(M)}] \leqslant \sum_{m=2}^{M} N_T[\boldsymbol{z}^{(m)}, q^{(m)}] + N_T[\boldsymbol{w}^{(1)}, s^{(1)}] \leqslant N_T[\boldsymbol{w}^{(1)}, s^{(1)}]$$

$$+ \frac{c_{46}}{1 - 2\delta c_{46}} \Big\{ \delta\big(2N_T[\boldsymbol{w}^{(1)}, s^{(1)}] + 3N_T[\boldsymbol{w}^{(0)}, s^{(0)}]\big) + P_T[\boldsymbol{u}] |\boldsymbol{w}_0|^{(1)}_{1+\gamma, \cup\Omega^\pm} \Big\}$$

$$\leqslant c_{47}(T) \big\{ Q_1(T) + P_T[\boldsymbol{u}] |\boldsymbol{w}_0|^{(1)}_{1+\gamma, \cup\Omega^\pm} \big\}. \tag{5.3.31}$$

Now we see that the series $\sum_{m=2}^{\infty} N_T[\boldsymbol{z}^{(m)}, q^{(m)}]$ is convergent and, consequently, the sequence $(\boldsymbol{w}^{(m)}, s^{(m)})$ has a limit $(\boldsymbol{w}, s)$. It is a solution to problem (5.3.9). Passing to the limit in (5.3.31) as $M \to \infty$, we obtain the estimate

$$N_T[\boldsymbol{w}, s] \leqslant c_1(T)\big(Q_1(T) + P_T[\boldsymbol{u}] |\boldsymbol{w}_0|^{(1)}_{1+\gamma, \cup\Omega^\pm}\big),$$

whence it follows inequality (5.3.7), because T has been arbitrary. □

5.4 The Proof of the Solvability of Nonlinear Problem (5.1.1)

In this section, we follow the lines of [54], §6, where the solvability of a similar problem for a single liquid was proved.

Before proving Theorem 5.1.1, we study the differences of the operators l_i and l'_i calculated by formulas (5.3.10) for the vectors $\boldsymbol{u}$ and $\boldsymbol{u}'$, respectively.

We introduce the notation, $\widetilde{l}_i = l_i - l'_i$, $\widetilde{B}_{ij} = B_{ij} - B'_{ij}$, $\widetilde{b}_{km} = b_{km} - b'_{km}$, where B'_{ij} and b'_{km} are also computed for the vector $\boldsymbol{u}'$.

Lemma 5.4.1 *If two vectors $\boldsymbol{u}$, $\boldsymbol{u}'$ satisfy condition* (5.3.2), *then*

$$|\widetilde{B}_{ij}|^{(\alpha,\alpha/2)}_{1+\gamma,D_T} + |\nabla \widetilde{B}_{ij}|^{(\alpha,\alpha/2)}_{1+\gamma,D_T} + \langle \nabla \widetilde{B}_{ij} \rangle^{(\frac{1+\alpha-\gamma}{2})}_{t,1+\gamma,D_T} + |\widetilde{B}_{ij}|^{(\gamma,1+\alpha)}_{1+\gamma,D_T} \leqslant c_1 m[\boldsymbol{u} - \boldsymbol{u}'],$$

where

$$m[\boldsymbol{v}] \equiv \int_0^T |\nabla \boldsymbol{v}|^{(1+\alpha)}_{1+\gamma, \cup\Omega^\pm}\, \mathrm{d}t'$$
$$+ T^{\frac{1-\alpha+\gamma}{2}} \big(|\nabla \boldsymbol{v}|^{(\gamma)}_{x,1+\gamma,D_T} + |\nabla\nabla \boldsymbol{v}|_{1+\gamma,D_T}\big) + T^{\gamma/2} |\nabla \boldsymbol{v}|^{(\frac{1+\alpha}{2})}_{t,1+\gamma,D_T},$$

and

$$\langle \widetilde{B}_{ij} \rangle^{(\frac{1+\alpha}{2})}_{t,1+\gamma,D_T} \leqslant c_2 T^{\frac{1-\alpha}{2}} |\nabla(\boldsymbol{u} - \boldsymbol{u}')|_{1+\gamma,D_T},$$

$$|\mathcal{D}_t \widetilde{B}_{ij}|^{(\alpha,\alpha/2)}_{1+\gamma,D_T} \leqslant c_3 |\nabla(\boldsymbol{u} - \boldsymbol{u}')|^{(\alpha,\alpha/2)}_{1+\gamma,D_T},$$

$$|\mathcal{D}_t\widetilde{B}_{ij}|^{(1+\alpha,\gamma)}_{1+\gamma,D_T} \leqslant c_4(|\nabla(\boldsymbol{u}-\boldsymbol{u}')|^{(\gamma,1+\alpha)}_{1+\gamma,D_T} + |\nabla(\boldsymbol{u}-\boldsymbol{u}')|^{(\gamma)}_{x,1+\gamma,D_T}).$$

This Lemma is proved similarly to Lemma 5.3.1.

From the analogs of Lemmas 5.3.2 and 5.3.4 for the operators $\widetilde{l}_i$, we deduce the following statement.

Lemma 5.4.2 *If two continuous vectors* $\boldsymbol{u}$, $\boldsymbol{u}'$ *satisfy inequality* (5.3.2), *and* $\Gamma \in C^{2+\alpha}$, *then*

$$\begin{aligned}
&|\widetilde{\boldsymbol{l}}_1(\boldsymbol{w},s)|^{(\alpha,\alpha/2)}_{1+\gamma,D_T} + |\widetilde{l}_2(\boldsymbol{w})|^{(1+\alpha,\frac{1+\alpha}{2})}_{1+\gamma,D_T} + |\widetilde{\boldsymbol{l}}_3(\boldsymbol{w})|^{(1+\alpha,\frac{1+\alpha}{2})}_{G_T}\\
&+|\widetilde{l}_4(\boldsymbol{w},s)|_{G_T} + |\nabla_\Gamma \widetilde{l}_4(\boldsymbol{w},s)|^{(\alpha,\alpha/2)}_{G_T} + |\widetilde{l}_4(\boldsymbol{w},s)|^{(\gamma,1+\alpha)}_{G_T} + |\widetilde{l}_5(\boldsymbol{w})|^{(\alpha,\alpha/2)}_{G_T}\\
&+\max_k\big(|\widetilde{\boldsymbol{L}}_{1k}(\boldsymbol{w},s)|^{(\gamma,0)}_{1,\gamma,D_T} + |\widetilde{\boldsymbol{L}}_{1k}(\boldsymbol{w},s)|^{(\gamma,1+\alpha)}_{1,\gamma,D_T}\big) + |\widetilde{\boldsymbol{l}}_6(\boldsymbol{w},s)|^{(\alpha,\alpha/2)}_{1+\gamma,D_T}\\
&+\max_k\big(|\widetilde{\boldsymbol{L}}_{6k}(\boldsymbol{w},s)|^{(\gamma,1+\alpha)}_{1,\gamma,D_T} + |\widetilde{\boldsymbol{L}}_{6k}(\boldsymbol{w},s)|^{(\gamma,0)}_{1,\gamma,D_T}\big)\\
&\qquad \leqslant c_5\big\{m[\boldsymbol{u}-\boldsymbol{u}']N_T[\boldsymbol{w},s] + P_T[\boldsymbol{u}-\boldsymbol{u}']|\boldsymbol{w}(\cdot,0)|^{(1)}_{1+\gamma,\cup\Omega^\pm}\big\},
\end{aligned}$$

where $P_T[\boldsymbol{v}]$ *is calculated by formula* (5.3.8).

The proof of this lemma reduces to that of Lemma 5.3.5 (we must replace δ with $m[\boldsymbol{u}-\boldsymbol{u}']$ and $P_T[\boldsymbol{u}]$ with $P_T[\boldsymbol{u}-\boldsymbol{u}']$, respectively).

Now, we turn to the system (5.1.1). From (1.1.3) it follows that

$$\Delta(t)\boldsymbol{X}_{\boldsymbol{u}} = \Delta(t)\boldsymbol{\xi} + \Delta(t)\int_0^t \boldsymbol{u}\,\mathrm{d}\tau = \Delta(0)\boldsymbol{\xi} + \int_0^t \dot{\Delta}(\tau)\boldsymbol{\xi}\,\mathrm{d}\tau + \Delta(t)\int_0^t \boldsymbol{u}\,\mathrm{d}\tau;$$

therefore, the last boundary condition in (5.1.1) can be rewritten in the form

$$[\boldsymbol{n}_0\cdot\mathbb{T}_{\boldsymbol{u}}(\boldsymbol{u},q)\boldsymbol{n}]|_\Gamma - \sigma\boldsymbol{n}_0\cdot\Delta(t)\int_0^t \boldsymbol{u}\,\mathrm{d}\tau = \sigma\boldsymbol{n}_0\cdot\int_0^t \dot{\Delta}(\tau)\boldsymbol{\xi}\,\mathrm{d}\tau + \sigma H_0(\xi),$$

where $H_0(\xi) = \boldsymbol{n}_0\cdot\Delta(0)\boldsymbol{\xi}$ denotes the doubled mean curvature of the surface Γ.

As for the operator $\dot{\Delta}(t)$, the following assertion was proved in paper [54] (Lemma 6.2):

Lemma 5.4.3 *If two vectors* $\boldsymbol{u}$, $\boldsymbol{u}' \in C^{2+\alpha,1+\alpha/2}(Q_T^+)$ *satisfy inequality* (5.3.2) *with respect to the domain* Q_T^+, *and if the surface* $\Gamma \in C^{2+\alpha}$, *then*

$$|\boldsymbol{n}_0\cdot(\dot{\Delta}(t) - \dot{\Delta}'(t))\boldsymbol{\xi}|^{(\alpha,\alpha/2)}_{G_T} \leqslant c_6|\nabla(\boldsymbol{u}-\boldsymbol{u}')|^{(\alpha,\alpha/2)}_{G_T}, \tag{5.4.1}$$

where $\Delta'(t)$ *is the Laplace–Beltrami operator on the surface* $\Gamma'_t = X_{\boldsymbol{u}'}\Gamma$ *obtained by transformation* (1.1.2) *with the vector* $\boldsymbol{u}'$.

We will also use the following statement.

Lemma 5.4.4 *If* $\boldsymbol{f}$ *satisfies the assumptions of Theorem* 5.1.1, *and* $\boldsymbol{u}$, $\boldsymbol{u}'$ *do condition* (5.3.2), *then*

$$|\boldsymbol{f}(X_{\boldsymbol{u}},t)-\boldsymbol{f}(X_{\boldsymbol{u}'},t)|^{(\alpha)}_{x,2,D_T}+\langle \boldsymbol{f}(X_{\boldsymbol{u}},t)-\boldsymbol{f}(X_{\boldsymbol{u}'},t)\rangle^{(\frac{1+\alpha-\gamma}{2})}_{t,2,D_T}$$
$$\leqslant c_7(T+T^{\gamma/2})|\boldsymbol{u}-\boldsymbol{u}'|^{(\alpha)}_{x,1+\gamma,D_T}. \tag{5.4.2}$$

Proof Consider the continuous transformation $\boldsymbol{u}_s=\boldsymbol{u}'+s\widetilde{\boldsymbol{u}}$ of the vector $\boldsymbol{u}_0\equiv\boldsymbol{u}'$ into $\boldsymbol{u}_1\equiv\boldsymbol{u}$; here $\widetilde{\boldsymbol{u}}\equiv\boldsymbol{u}-\boldsymbol{u}'$. Obviously, we have

$$\boldsymbol{f}(X_{\boldsymbol{u}},t)-\boldsymbol{f}(X_{\boldsymbol{u}'},t)=\sum_{k=1}^{3}\int_0^t(u_k-u_k')\,\mathrm{d}\tau\int_0^1\partial\boldsymbol{f}(X_{\boldsymbol{u}_s},t)/\partial x_k\,\mathrm{d}s.$$

This representation implies inequality (5.4.2) in view of differentiability properties of $\boldsymbol{f}$. □

Proof of Theorem 5.1.1 After elementary transformations, system (5.1.1) acquires the form

$$\begin{aligned}
&\mathcal{D}_t\boldsymbol{u}-\nu^{\pm}\nabla^2_{\boldsymbol{u}}\boldsymbol{u}+\frac{1}{\rho^{\pm}}\nabla_{\boldsymbol{u}}q=\boldsymbol{f}(X_{\boldsymbol{u}},t),\qquad \nabla_{\boldsymbol{u}}\cdot\boldsymbol{u}=0\quad\text{in}\quad D_T,\\
&\boldsymbol{u}|_{t=0}=\boldsymbol{v}_0\ \text{in}\ \Omega^-\cup\Omega^+,\qquad \boldsymbol{u}\xrightarrow[|\xi|\to\infty]{}0,\\
&[\boldsymbol{u}]\big|_{G_T}=0,\qquad [\mu^{\pm}\Pi_0\Pi\mathbb{S}_{\boldsymbol{u}}(\boldsymbol{u})\boldsymbol{n}]\big|_{G_T}=0,\\
&[\boldsymbol{n}_0\cdot\mathbb{T}_{\boldsymbol{u}}(\boldsymbol{u},q)\boldsymbol{n}]\big|_{G_T}-\sigma\boldsymbol{n}_0\cdot\Delta(t)\int_0^t\boldsymbol{u}\,\mathrm{d}\tau\big|_{G_T}=\sigma\Big(H_0(\xi)+\boldsymbol{n}_0\cdot\int_0^t\dot{\Delta}(\tau)\boldsymbol{\xi}\,\mathrm{d}\tau\Big)\big|_{G_T}.
\end{aligned}\tag{5.4.3}$$

We use again the method of successive approximations. As the initial approximation $(\boldsymbol{u}^{(0)},q^{(0)})$, we take $(\boldsymbol{v}_0,q_0)$. The approximations $(\boldsymbol{u}^{(m+1)},q^{(m+1)})$, $m\geqslant 0$, are defined as solutions of the linearized problems

$$\begin{aligned}
&\mathcal{D}_t\boldsymbol{u}^{(m+1)}-\nu^{\pm}\nabla^2_m\boldsymbol{u}^{(m+1)}+\frac{1}{\rho^{\pm}}\nabla_m q^{(m+1)}=\boldsymbol{f}(X_m,t),\quad \nabla_m\cdot\boldsymbol{u}^{(m+1)}=0\ \text{in}\ D_T,\\
&\boldsymbol{u}^{(m+1)}|_{t=0}=\boldsymbol{v}_0\ \text{in}\ \Omega^-\cup\Omega^+,\quad \boldsymbol{u}^{(m+1)}\xrightarrow[|\xi|\to\infty]{}0,\\
&[\boldsymbol{u}^{(m+1)}]|_{G_T}=0,\quad [\mu^{\pm}\Pi_0\Pi_m\mathbb{S}_m(\boldsymbol{u}^{(m+1)})\boldsymbol{n}_m]|_{G_T}=0,\\
&[\boldsymbol{n}_0\cdot\mathbb{T}_m(\boldsymbol{u}^{(m+1)},q^{(m+1)})\boldsymbol{n}_m]|_{G_T}-\sigma\boldsymbol{n}_0\cdot\Delta_m(t)\int_0^t\boldsymbol{u}^{(m+1)}d\tau|_{G_T}\\
&\quad=\sigma H_0(\xi)|_\Gamma+\sigma\int_0^t\boldsymbol{n}_0\cdot\dot{\Delta}_m(\tau)\boldsymbol{\xi}d\tau|_{G_T}.
\end{aligned}\tag{5.4.4}$$

Here we have used the notation: $X_m = X_{\boldsymbol{u}^{(m)}}$; $\nabla_m = \nabla_{\boldsymbol{u}^{(m)}}$; $\boldsymbol{n}_m$ is the normal to the surface $\Gamma_m(t) = \Gamma(X_m, t)$; $\Delta_m(t)$ is the Laplace-Beltrami operator on this surface; $\Pi_m\boldsymbol{\omega} = \boldsymbol{\omega} - \boldsymbol{n}_m(\boldsymbol{n}_m \cdot \boldsymbol{\omega})$; and $\mathbb{S}_m = \mathbb{S}_{\boldsymbol{u}^{(m)}}$, $\mathbb{T}_m = \mathbb{T}_{\boldsymbol{u}^{(m)}}$.

We shall apply Theorem 5.3.1 successively to the problems (5.4.4). In condition (5.3.4), we set $\boldsymbol{g}_m = -\mathbb{A}^{(m)*}\boldsymbol{f}$, where $\mathbb{A}^{(m)} = \mathbb{A}(\boldsymbol{u}^{(m)})$, and

$$\mathbb{G}_m(\xi, t) = \nabla \int_{\mathbb{R}^3} \mathcal{E}(\xi, \eta)\mathbb{A}^{(m)*}\boldsymbol{f}(\eta, t)\,\mathrm{d}\eta.$$

Then the validity of (5.3.5) is obvious, and (5.3.3) and (5.3.6) follow from the assumptions of Theorem 5.1.1. We estimate $\mathbb{G}_m$ by Lemma 5.3.3:

$$\begin{aligned} |\mathbb{G}_m|^{(\gamma,0)}_{1,\gamma,D_T} &\leqslant c_8|\mathbb{A}^{(m)*}\boldsymbol{f}|_{2,\mathbb{D}_T} \leqslant c_9(1 + T|\boldsymbol{u}^{(m)}|^{(2+\alpha,0)}_{1+\gamma,D_T})|\boldsymbol{f}|_{2,\mathbb{R}^3_T}, \\ |\mathbb{G}_m|^{(\gamma,1+\alpha)}_{1,\gamma,D_T} &\leqslant c_{10}(T)(1 + |\boldsymbol{u}^{(m)}|^{(2+\alpha,1+\alpha/2)}_{1+\gamma,D_T})|\boldsymbol{f}|^{(\frac{1+\alpha-\gamma}{2})}_{t,2,\mathbb{R}^3_T}, \end{aligned} \tag{5.4.5}$$

and $\boldsymbol{n}_0 \cdot \dot{\Delta}_m(\tau)\boldsymbol{\xi}$ by Lemma 5.4.3 with $\boldsymbol{u}' = 0$.

In this way, we conclude by induction that for every $m \geqslant 0$ system (5.4.4) has a solution $(\boldsymbol{u}^{(m+1)}, q^{(m+1)})$ on a time interval $(0, T_{m+1}) \subset (0, T)$ such that the preceding approximation $(\boldsymbol{u}^{(m)}, q^{(m)})$ is also defined on this interval and $\boldsymbol{u}^{(m)}$ satisfies inequality (5.3.2) with small δ. We are going to show now that $T_m \geqslant T' > 0$ for every m, the norms $N_{T'}[\boldsymbol{u}^{(m)}, q^{(m)}]$ are uniformly bounded, and the sequence $(\boldsymbol{u}^{(m)}, q^{(m)})$ converges to a solution of system (5.4.3).

We assume that $(\boldsymbol{u}^{(j)}, q^{(j)})$, $j = 1, \dots, m+1$, are defined on $(0, T_{m+1})$. We consider the differences

$$\boldsymbol{w}^{(j+1)} = \boldsymbol{u}^{(j+1)} - \boldsymbol{u}^{(j)}, \qquad s^{(j+1)} = q^{(j+1)} - q^{(j)};$$

they satisfy the system

$$\begin{aligned} &\mathcal{D}_t\boldsymbol{w}^{(j+1)} - \nu^{\pm}\nabla_j^2\boldsymbol{w}^{(j+1)} + \frac{1}{\rho^{\pm}}\nabla_j s^{(j+1)} \\ &\qquad = \boldsymbol{l}_1^{(j)}(\boldsymbol{u}^{(j)}, q^{(j)}) - \boldsymbol{l}_1^{(j-1)}(\boldsymbol{u}^{(j)}, q^{(j)}) + \boldsymbol{f}(X_j, t) - \boldsymbol{f}(X_{j-1}, t) \equiv \boldsymbol{f}^{(j)}, \\ &\nabla_j \cdot \boldsymbol{w}^{(j+1)} = l_2^{(j)}(\boldsymbol{u}^{(j)}) - l_2^{(j-1)}(\boldsymbol{u}^{(j)}) \equiv r^{(j)} \quad \text{in} \quad D_{T_{m+1}}, \\ &\boldsymbol{w}^{(j+1)}(\xi, 0) = 0, \quad \xi \in \Omega^+ \cup \Omega^-, \\ &[\boldsymbol{w}^{(j+1)}]|_\Gamma = 0, \quad \boldsymbol{w}^{(j+1)} \xrightarrow[|\xi|\to\infty]{} 0, \\ &[\mu^{\pm}\Pi_0\Pi_j\mathbb{S}_j(\boldsymbol{w}^{(j+1)})\boldsymbol{n}_j]|_\Gamma = \boldsymbol{l}_3^{(j)}(\boldsymbol{u}^{(j)}) - \boldsymbol{l}_3^{(j-1)}(\boldsymbol{u}^{(j)}), \qquad (5.4.6) \\ &[\boldsymbol{n}_0 \cdot \mathbb{T}_j(\boldsymbol{w}^{(j+1)}, s^{(j+1)})\boldsymbol{n}_j]|_\Gamma - \sigma\boldsymbol{n}_0 \cdot \Delta_j(t)\int_0^t \boldsymbol{w}^{(j+1)}\,\mathrm{d}\tau|_\Gamma \end{aligned}$$

$$= l_4^{(j)}(\boldsymbol{u}^{(j)}, q^{(j)}) - l_4^{(j-1)}(\boldsymbol{u}^{(j)}, q^{(j)}) + \sigma \int_0^t (l_5^{(j)}(\boldsymbol{u}^{(j)}) - l_5^{(j-1)}(\boldsymbol{u}^{(j)}))\, \mathrm{d}\tau$$

$$+\sigma \int_0^t \boldsymbol{n}_0 \cdot (\dot{\Delta}_j(\tau) - \dot{\Delta}_{j-1}(\tau))\boldsymbol{\xi}\, \mathrm{d}\tau|_\Gamma, \quad t \in (0, T_{m+1}),$$

where operators $l_i^{(k)}$ are computed by formulas (5.3.10) with $\boldsymbol{u} = \boldsymbol{u}^{(k)}$, $k \in \mathbb{N}$, and $\boldsymbol{u}^{(0)} = \boldsymbol{v}_0$, $q^{(0)} = q_0$.

In order to apply Theorem 5.3.1 again, we need, in fact, to verify only the third hypothesis of that theorem. So, setting

$$\boldsymbol{g} = \boldsymbol{g}^{(j)} = (\mathbb{B}^{(j-1)*} - \mathbb{B}^{(j)*})\mathcal{D}_t \boldsymbol{u}^{(j)} + \mathcal{D}_t(\mathbb{B}^{(j-1)*} - \mathbb{B}^{(j)*})\boldsymbol{u}^{(j)} - \mathbb{A}^{(j)*}\boldsymbol{f}^{(j)}, \tag{5.4.7}$$

we have

$$\mathcal{D}_t r^{(j)} - \nabla_j \cdot \boldsymbol{f}^{(j)} = \nabla \cdot \boldsymbol{g}^{(j)}, \qquad [(\boldsymbol{g}^{(j)} + \mathbb{A}^{(j)*}\boldsymbol{f}^{(j)}) \cdot \boldsymbol{n}_0]|_{G_t} = 0.$$

Next, since

$$\boldsymbol{f}^{(j)} = \partial\big(\boldsymbol{L}_{1k}^{(j)}(\boldsymbol{u}^{(j)}, q^{(j)}) - \boldsymbol{L}_{1k}^{(j-1)}(\boldsymbol{u}^{(j)}, q^{(j)})\big)/\partial\xi_k + \boldsymbol{f}(X_j, t) - \boldsymbol{f}(X_{j-1}, t)$$

and

$$\mathcal{D}_t \boldsymbol{u}^{(j)} = \partial \boldsymbol{M}_k^{(j)}/\partial\xi_k + \boldsymbol{f}(X_{j-1}, t)$$

with $\boldsymbol{M}_k^{(j)} = \nu^{\pm}(\mathbb{A}^{(j-1)*}\boldsymbol{e}_k \cdot \nabla_{j-1})\boldsymbol{u}^{(j)} - \mathbb{A}^{(j-1)*}\boldsymbol{e}_k q^{(j)}/\rho^{\pm}$, we can write

$$\boldsymbol{g}^{(j)} = \nabla \cdot \mathbb{G}^{(j)} \equiv \partial \boldsymbol{G}_k^{(j)}/\partial\xi_k,$$

where

$$\begin{aligned} \boldsymbol{G}_k^{(j)} &= (\mathbb{B}^{(j-1)*} - \mathbb{B}^{(j)*})\boldsymbol{M}_k^{(j)} \\ &\quad - \mathbb{A}^{(j)*}\big(\boldsymbol{L}_{1k}^{(j)}(\boldsymbol{u}^{(j)}, q^{(j)}) - \boldsymbol{L}_{1k}^{(j-1)}(\boldsymbol{u}^{(j)}, q^{(j)})\big) + \partial \boldsymbol{W}^{(j)}/\partial\xi_k, \qquad (5.4.8) \\ \boldsymbol{W}^{(j)} &= -\int_{\Omega^+\cup\Omega^-} \mathcal{E}(\xi, \eta)\Big\{\frac{\partial}{\partial\eta_i}(\mathbb{B}^{(j-1)*} - \mathbb{B}^{(j)*})\boldsymbol{M}_i^{(j)} \\ &\quad - (\mathbb{B}^{(j-1)*} - \mathbb{B}^{(j)*})\boldsymbol{f}(X_{j-1}, t) - \mathcal{D}_t(\mathbb{B}^{(j-1)*} - \mathbb{B}^{(j)*})\boldsymbol{u}^{(j)} \\ &\quad - \partial\mathbb{A}^{(j)*}/\partial\eta_i\big(\boldsymbol{L}_{1i}^{(j)}(\boldsymbol{u}^{(j)}, q^{(j)}) - \boldsymbol{L}_{1i}^{(j-1)}(\boldsymbol{u}^{(j)}, q^{(j)})\big) \\ &\quad + \mathbb{A}^{(j)*}\big(\boldsymbol{f}(X_j, t) - \boldsymbol{f}(X_{j-1}, t)\big)\Big\}\, \mathrm{d}\eta. \end{aligned}$$

We assume that $(\boldsymbol{u}^{(j)}, q^{(j)})$, $j = 1, \dots, m$ satisfy the inequality

$$(T + T^{\gamma/2}) N_T[\boldsymbol{u}^{(j)}, q^{(j)}] < \delta, \quad T \leqslant T_{m+1}, \tag{5.4.9}$$

which is stronger than (5.3.2).

The norms of the right-hand sides of the system (5.4.6) are estimated with the help of Lemmas 5.4.1–5.4.4:

$$\begin{aligned}
&|\boldsymbol{f}^{(j)}|^{(\alpha,\alpha/2)}_{1+\gamma,D_T} + |r^{(j)}|^{(1+\alpha,\frac{1+\alpha}{2})}_{1+\gamma,D_T} + |\boldsymbol{g}^{(j)}|^{(\alpha,\alpha/2)}_{1+\gamma,D_T} + |\mathbb{G}^{(j)}|^{(\gamma,1+\alpha)}_{1,\gamma,D_T} + |\mathbb{G}^{(j)}|^{(\gamma,0)}_{1,\gamma,D_T}\\
&\quad + |\boldsymbol{l}_3^{(j)} - \boldsymbol{l}_3^{(j-1)}|^{(1+\alpha,\frac{1+\alpha}{2})}_{G_T} + |l_4^{(j)} - l_4^{(j-1)}|_{G_T} + |\nabla_\Gamma(l_4^{(j)} - l_4^{(j-1)})|^{(\alpha,\alpha/2)}_{G_T}\\
&\quad + |l_4^{(j)} - l_4^{(j-1)}|^{(\gamma,1+\alpha)}_{G_T} + |l_5^{(j)} - l_5^{(j-1)}|^{(\alpha,\alpha/2)}_{G_T} + |\boldsymbol{n}_0 \cdot (\dot{\boldsymbol{\Delta}}_j - \dot{\boldsymbol{\Delta}}_{j-1})\boldsymbol{\xi}|^{(\alpha,\alpha/2)}_{G_T}\\
&\leqslant c_{11}\big\{m[\boldsymbol{w}^{(j)}] N_T[\boldsymbol{u}^{(j)}, q^{(j)}] + |\nabla \boldsymbol{w}^{(j)}|^{(\alpha,\alpha/2)}_{1+\gamma,D_T}\\
&\quad + (T + T^{\gamma/2})|\boldsymbol{w}^{(j)}|^{(\alpha,\alpha/2)}_{1+\gamma,D_T} + P_T[\boldsymbol{w}^{(j)}]|\boldsymbol{u}^{(j)}(\cdot,0)|^{(1)}_{1+\gamma,\cup\Omega^\pm}\big\}.
\end{aligned} \tag{5.4.10}$$

The vectors $\boldsymbol{G}_k^{(j)}$ can be estimated in the same way as the operators $\boldsymbol{L}_{6k}$ having a similar structure.

Further, by (5.4.9)

$$\begin{aligned}
m[\boldsymbol{w}^{(j)}] N_T[\boldsymbol{u}^{(j)}, q^{(j)}] &\leqslant c_{12}(T + T^{\gamma/2})|\boldsymbol{w}^{(j)}|^{(2+\alpha,1+\alpha/2)}_{1+\gamma,D_T} N_T[\boldsymbol{u}^{(j)}, q^{(j)}]\\
&\leqslant c_{12}\delta N_T[\boldsymbol{w}^{(j)}, s^{(j)}].
\end{aligned} \tag{5.4.11}$$

Since $\nabla \boldsymbol{w}^{(j)}|_{t=0} = 0$ for $j \geqslant 1$, we have

$$P_T[\boldsymbol{w}^{(j)}]|\boldsymbol{u}^{(j)}(\cdot,0)|^{(1)}_{1+\gamma,\cup\Omega^\pm} \leqslant c_{13}(T + T^{1/2})|\boldsymbol{w}^{(j)}|^{(2+\alpha,1+\alpha/2)}_{1+\gamma,D_T} |\boldsymbol{v}_0|^{(1)}_{1+\gamma,\cup\Omega^\pm}. \tag{5.4.12}$$

Using the interpolation inequalities (it is easily seen that they are valid also for the weighted Hölder spaces), we obtain

$$|\boldsymbol{w}^{(j)}|^{(\alpha)}_{1+\gamma,D_T} + |\nabla \boldsymbol{w}^{(j)}|^{(\alpha)}_{1+\gamma,D_T} \leqslant \varepsilon|\boldsymbol{w}^{(j)}|^{(2+\alpha)}_{1+\gamma,D_T} + c_{16}(\varepsilon)|\boldsymbol{w}^{(j)}|_{1+\gamma,D_T}. \tag{5.4.13}$$

Thus, the following estimate is a consequence of Theorem 5.3.1 and inequalities (5.4.10)–(5.4.13):

$$\begin{aligned}
N_T[\boldsymbol{w}^{(j+1)}, s^{(j+1)}] \leqslant c_{17}(T)\big\{\big((c_{12} + c_{13})\delta + (1 + T + T^{1/2})\varepsilon\big) N_T[\boldsymbol{w}^{(j)}, s^{(j)}]&\\
+ c_{18}(\varepsilon)|\boldsymbol{w}^{(j)}|_{1+\gamma,D_T}\big\}.&
\end{aligned} \tag{5.4.14}$$

We choose δ and ε so that

$$\varkappa \equiv c_{17}(T)((c_{12} + c_{13})\delta + (1 + T + T^{1/2})\varepsilon) < 1,$$

and observe that

$$|\boldsymbol{w}^{(j)}|_{1+\gamma,D_T} \leqslant |\boldsymbol{w}^{(j)}(\cdot,0)|_{1+\gamma,\cup\Omega^\pm} + \int_0^T |\mathcal{D}_t \boldsymbol{w}^{(j)}(\cdot,t)|_{1+\gamma,\cup\Omega^\pm}\,\mathrm{d}t$$
$$\leqslant \int_0^T N_t[\boldsymbol{w}^{(j)}, s^{(j)}]\,\mathrm{d}t.$$

Summing inequalities (5.4.14) from $j=1$ to $j=m$, for the expression

$$\Sigma_{m+1}(T) = \sum_{j=1}^{m+1} N_T[\boldsymbol{w}^{(j)}, s^{(j)}]$$

we get the estimate

$$\Sigma_{m+1}(T) \leqslant \varkappa \Sigma_{m+1}(T) + c_{19}\int_0^T \Sigma_{m+1}(t)\,\mathrm{d}t + N_T[\boldsymbol{w}^{(1)}, s^{(1)}].$$

Finally, from the Gronwall lemma applied to the inequality

$$\Sigma_{m+1}(T) \leqslant c_{20}\int_0^T \Sigma_{m+1}(t)\,\mathrm{d}t + \frac{1}{1-\varkappa} N_T[\boldsymbol{w}^{(1)}, s^{(1)}],$$

it follows that

$$N_T[\boldsymbol{u}^{(m+1)}, q^{(m+1)}] \leqslant \Sigma_{m+1}(T) + N_T[\boldsymbol{u}^{(0)}, q^{(0)}] \leqslant \frac{N_T[\boldsymbol{w}^{(1)}, s^{(1)}]}{1-\varkappa} e^{c_{20}T} + N_T[\boldsymbol{v}_0, q_0]. \tag{5.4.15}$$

Since $(\boldsymbol{w}^{(1)}, s^{(1)}) = (\boldsymbol{u}^{(1)} - \boldsymbol{v}_0, q^{(1)} - q_0)$, we have

$$N_T[\boldsymbol{w}^{(1)}, s^{(1)}] \leqslant N_T[\boldsymbol{u}^{(1)}, q^{(1)}] + N_T[\boldsymbol{v}_0, q_0], \tag{5.4.16}$$

where the norm $N_T[\boldsymbol{u}^{(1)}, q^{(1)}]$ can be evaluated by Theorem 5.3.1. So, by virtue of (5.3.7) with $\boldsymbol{w} = \boldsymbol{u}^{(1)}$, $s = q^{(1)}$, and $\boldsymbol{u} = \boldsymbol{w}_0 = \boldsymbol{v}_0$; (5.4.1) with $\boldsymbol{u} = \boldsymbol{v}_0$ and $\boldsymbol{u}' = 0$; as well as (5.4.5) with $\boldsymbol{u} = \boldsymbol{v}_0$, we obtain

$$N_T[\boldsymbol{u}^{(1)}, q^{(1)}] \leqslant c\big(T, |\boldsymbol{v}_0|^{(2+\alpha)}_{1+\gamma,\cup\Omega^\pm}, |\mathcal{D}_x \boldsymbol{f}|^{(\alpha,\frac{1+\alpha-\gamma}{2})}_{2,\mathbb{R}^3_T}\big)\Big\{|\boldsymbol{f}|^{(\alpha,\frac{1+\alpha-\gamma}{2})}_{2,\mathbb{R}^3_T} + |\boldsymbol{v}_0|^{(2+\alpha)}_{1+\gamma,\cup\Omega^\pm}$$
$$+ \sigma |H_0|^{(1+\alpha)}_{\Gamma}\Big\}. \tag{5.4.17}$$

For pressure function q_0, as a solution to problem (5.1.2), according to Lemma 5.2.1, it holds an inequality similar to (5.3.27), where Q_1 is the sum of the norms of the given functions and $\boldsymbol{v}_0$ is taken as $\boldsymbol{u}$ and $\boldsymbol{w}_0$. Hence, $N_T[\boldsymbol{w}^{(1)}, s^{(1)}]$ is bounded by the norms of the given functions. If, moreover,

$$(T+T^{\gamma/2})\{N_T[\boldsymbol{w}^{(1)},s^{(1)}](1-\varkappa)^{-1}e^{c_{20}T}+N_T[\boldsymbol{v}_0,q_0]\}\leqslant\delta, \tag{5.4.18}$$

then condition (5.4.9) holds also for $(\boldsymbol{u}^{(m+1)},q^{(m+1)})$. It is clear that there is a number T satisfying (5.4.18). We denote it by T_0.

Since the right-hand side of the last inequality in (5.4.15) is independent of m, the functions $(\boldsymbol{u}^{(m)},q^{(m)})$, $m\in\mathbb{N}$, are defined in the interval $(0,T_0)$ and satisfy the uniform estimate (5.4.15). It follows that the series $\sum_{j=1}^{\infty}N_{T_0}[\boldsymbol{w}^{(j)},s^{(j)}]$ is convergent, and therefore we see that the sequence $(\boldsymbol{u}^{(m)},q^{(m)})$ is convergent in the norm N_{T_0}. Passing to the limit as $m\to\infty$ in system (5.4.4) and in the inequalities, we make sure that $(\boldsymbol{u},q)=\lim_{m\to\infty}(\boldsymbol{u}^{(m)},q^{(m)})$ is a solution of problem (5.1.1) and, due to (5.4.15), (5.4.16), and (5.4.17), estimate (5.1.3) holds for it.

Now we prove the uniqueness of the solution. Let us suppose that $(\boldsymbol{u},q)$ and $(\boldsymbol{u}',q')$ are two solutions of (5.1.1) and consider the differences $\boldsymbol{w}=\boldsymbol{u}-\boldsymbol{u}'$, $s=q-q'$. The couple $(\boldsymbol{w},s)$ satisfies a problem of type (5.4.6):

$$\begin{aligned}
&\mathcal{D}_t\boldsymbol{w}-\nu^{\pm}\nabla_{\boldsymbol{u}}^2\boldsymbol{w}+\frac{1}{\rho^{\pm}}\nabla_{\boldsymbol{u}}s=\boldsymbol{l}_1(\boldsymbol{u}',q')-\boldsymbol{l}_1'(\boldsymbol{u}',q')+\boldsymbol{f}(X_{\boldsymbol{u}},t)-\boldsymbol{f}(X_{\boldsymbol{u}'},t),\\
&\nabla_{\boldsymbol{u}}\cdot\boldsymbol{w}=l_2(\boldsymbol{u}')-l_2'(\boldsymbol{u}')\quad\text{in}\quad D_{T_0},\\
&\boldsymbol{w}|_{t=0}=0\ \text{in}\ \Omega^-\cup\Omega^+,\quad \boldsymbol{w}\xrightarrow[|\xi|\to\infty]{}0,\\
&[\boldsymbol{w}]|_{G_{T_0}}=0,\quad [\mu^{\pm}\Pi_0\Pi\mathbb{S}_{\boldsymbol{u}}(\boldsymbol{w})\boldsymbol{n}]|_{G_{T_0}}=\boldsymbol{l}_3(\boldsymbol{u}')-\boldsymbol{l}_3'(\boldsymbol{u}'),\\
&[\boldsymbol{n}_0\cdot\mathbb{T}_{\boldsymbol{u}}(\boldsymbol{w},s)\boldsymbol{n}]|_{G_{T_0}}-\sigma\boldsymbol{n}_0\cdot\Delta(t)\int_0^t\boldsymbol{w}\,\mathrm{d}\tau|_{G_{T_0}}=l_4(\boldsymbol{u}',q')-l_4'(\boldsymbol{u}',q')+\\
&\qquad+\sigma\int_0^t(l_5(\boldsymbol{u}')-l_5'(\boldsymbol{u}'))\,\mathrm{d}\tau+\sigma\int_0^t\boldsymbol{n}_0\cdot(\dot{\Delta}(\tau)-\dot{\Delta}'(\tau))\boldsymbol{\xi}\,\mathrm{d}\tau|_{G_{T_0}}.
\end{aligned}$$

Repeating the above arguments, we arrive at an inequality similar to (5.4.14):

$$\begin{aligned}
N_{T_0}[\boldsymbol{w},s]&\leqslant\varkappa N_{T_0}[\boldsymbol{w},s]+c_{21}|\boldsymbol{w}|_{1+\gamma,D_{T_0}}\\
&\leqslant\varkappa N_{T_0}[\boldsymbol{w},s]+c_{21}\int_0^{T_0}N_t[\boldsymbol{w},s]\,\mathrm{d}t,\quad\varkappa<1,
\end{aligned}$$

which, by the Gronwall lemma, implies that $\boldsymbol{w}=0$ and $s=0$.

Theorem 5.1.1 is completely proved. □

Chapter 6
Global Solvability in the Hölder Spaces for the Nonlinear Problem Without Surface Tension

Abstract In this chapter, we consider that the fluids are located in a container, and they do not fill the whole space as has been before. It means that the outer fluid is bounded by a surface Σ on which the nonslip condition is posed. We suppose that the boundaries Γ and Σ have no intersection. On the interface Γ_t, we do not take the surface tension into account, i.e., $\sigma = 0$. The main result is global-in-time unique solvability of the problem in the ordinary Hölder spaces. It is based on the local existence theorem which is stated in Chap. 5. We use the technique developed by V. A. Solonnikov for the investigation of a problem governing drop motion in a vacuum [88].

6.1 Statement of the Main Result

The material of this chapter was published in papers [21, 24].

At the initial moment $t = 0$, let a fluid with the viscosity $\nu^+ > 0$ and the density $\rho^+ > 0$ occupy a bounded domain $\Omega_0^+ \subset \mathbb{R}^3$, and let a fluid with the viscosity $\nu^- > 0$ and the density $\rho^- > 0$ be in the domain Ω_0^- which surrounds Ω_0^+. We denote $\partial\Omega_0^+$ by Γ_0. The boundary $\Sigma \equiv \partial(\Omega_0^+ \cup \Gamma \cup \Omega_0^-)$ is a given closed surface, $\Sigma \cap \Gamma_0 = \emptyset$ (see Fig. 6.1).

For $t > 0$, velocity vector field $\boldsymbol{v}$ and pressure function p solve system (1.1.1) with homogeneous boundary condition on the jump of the normal stresses. On the outer boundary Σ, adhesion condition for the velocity is specified: $\boldsymbol{v}|_\Sigma = 0$. Condition (1.0.3) completes the system.

We pass from the Eulerian coordinates to the Lagrangian ones by the formula (1.1.2). Now the problem is formulated for the functions $\boldsymbol{u}$ and $q = p(X_{\boldsymbol{u}}, t)$ in the domain with the given interface $\Gamma \equiv \Gamma_0$. If the angle between $\boldsymbol{n}$ and the outward normal $\boldsymbol{n}_0$ to Γ is acute, the problem is equivalent to the system

© The Author(s), under exclusive license to Springer Nature Switzerland AG 2021

I. V. Denisova, V. A. Solonnikov, *Motion of a Drop in an Incompressible Fluid*, Advances in Mathematical Fluid Mechanics,
https://doi.org/10.1007/978-3-030-70053-9_6

$$\mathcal{D}_t \boldsymbol{u} - \nu^{\pm}\nabla_{\boldsymbol{u}}^2 \boldsymbol{u} + \frac{1}{\rho^{\pm}}\nabla_{\boldsymbol{u}} q = \boldsymbol{f}(X_{\boldsymbol{u}}, t), \quad \nabla_{\boldsymbol{u}} \cdot \boldsymbol{u} = 0 \quad \text{in } Q_T^{\pm} = \Omega_0^{\pm} \times (0,T),$$
$$\boldsymbol{u}|_{t=0} = \boldsymbol{v}_0 \qquad \text{in } \Omega_0^- \cup \Omega_0^+, \tag{6.1.1}$$
$$[\boldsymbol{u}]|_{G_T} = 0, \quad \boldsymbol{u}|_{\Upsilon_T} = 0 \quad (\Upsilon_T = \Sigma \times (0,T)),$$
$$[\mu^{\pm}\Pi_0 \Pi \mathbb{S}_{\boldsymbol{u}}(\boldsymbol{u})\boldsymbol{n}]|_{G_T} = 0, \qquad [\boldsymbol{n}_0 \cdot \mathbb{T}_{\boldsymbol{u}}(\boldsymbol{u}, q)\boldsymbol{n}]|_{G_T} = 0 \quad \text{on } G_T = \Gamma \times (0,T).$$

Fig. 6.1 Motion of a Drop in a Container

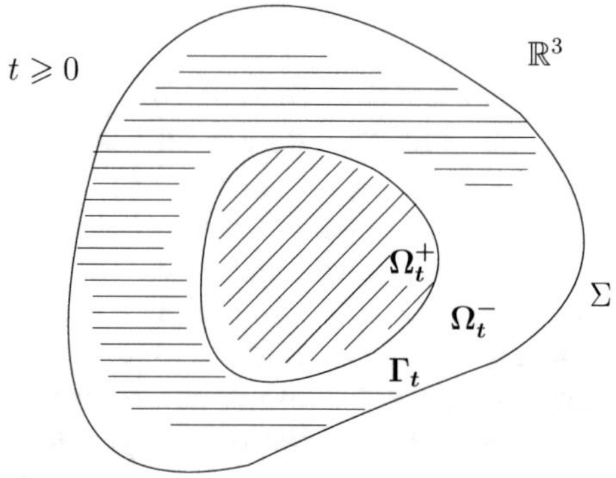

Let $T \in (0,\infty]$, $t, \tau > 0$. We set $D_{\Sigma T} = Q_T^- \cup Q_T^+$, $Q_T^{\pm} = \Omega_0^{\pm} \times (0,T)$, $Q_{(t,t+\tau)}^{\pm} = \Omega_t^{\pm} \times (t, t+\tau)$, $\Omega = \Omega_0^- \cup \overline{\Omega_0^+} \equiv \Omega_t^- \cup \overline{\Omega_t^+}$, $Q_T = \Omega \times (0,T)$. By $\|\cdot\|_{\Omega}$, we denote the norm $\|\cdot\|_{L_2(\Omega)}$.

The following theorem is the main result of this chapter.

Theorem 6.1.1 *Let $\alpha, \gamma \in (0,1)$. We assume that $\Gamma, \Sigma \in C^{2+\alpha}$ and $\boldsymbol{v}_0 \in C^{2+\alpha}(\Omega_0^- \cup \Omega_0^+)$, $\boldsymbol{f}, \nabla \boldsymbol{f} \in C^{\alpha, \frac{1+\alpha-\gamma}{2}}(Q_\infty)$ satisfy compatibility conditions*

$$\nabla \cdot \boldsymbol{v}_0 = 0, \quad [\boldsymbol{v}_0]\big|_{\Gamma} = 0, \quad [\mu^{\pm}\Pi_0 \mathbb{S}(\boldsymbol{v}_0(\xi))\boldsymbol{n}_0]\big|_{\xi\in\Gamma} = 0,$$
$$\Big[\Pi_0\big(\nu^{\pm}\nabla^2 \boldsymbol{v}_0(\xi) - \frac{1}{\rho^{\pm}}\nabla p_0(\xi)\big)\Big]\Big|_{\xi\in\Gamma} = 0, \tag{6.1.2}$$
$$\boldsymbol{v}_0\big|_{\Sigma} = 0, \quad \Pi_{\Sigma}\Big(\boldsymbol{f}(\xi,0) - \frac{1}{\rho^-}\nabla p_0(\xi) + \nu^- \nabla^2 \boldsymbol{v}_0(\xi)\Big)\Big|_{\xi\in\Sigma} = 0,$$

where $\Pi_{\Sigma}\boldsymbol{H} = \boldsymbol{H} - \boldsymbol{n}_{\Sigma}(\boldsymbol{n}_{\Sigma} \cdot \boldsymbol{H})$, $\boldsymbol{n}_{\Sigma}$ is the outward normal to Σ and $p_0 = p(\xi, 0)$ is a solution to the diffraction problem

$$\frac{1}{\rho^{\pm}}\nabla^2 p_0(\xi) = \nabla \cdot (\boldsymbol{f}(\xi,0) - \mathcal{D}_t \mathbb{B}^*(\boldsymbol{v}_0)\boldsymbol{v}_0(\xi)), \quad \xi \in \Omega_0^- \cup \Omega_0^+,$$
$$[p_0]|_{\Gamma} = \Big[2\mu^{\pm}\frac{\partial \boldsymbol{v}_0}{\partial \boldsymbol{n}_0} \cdot \boldsymbol{n}_0\Big]\Big|_{\Gamma}, \quad \Big[\frac{1}{\rho^{\pm}}\frac{\partial p_0}{\partial \boldsymbol{n}_0}\Big]\Big|_{\Gamma} = [\nu^{\pm}\boldsymbol{n}_0 \cdot \nabla^2 \boldsymbol{v}_0]\big|_{\Gamma}, \tag{6.1.3}$$

$$\frac{1}{\rho^-}\frac{\partial p_0}{\partial \boldsymbol{n}_\Sigma}\Big|_\Sigma = \boldsymbol{n}_\Sigma \cdot \big(\nu^- \nabla^2 \boldsymbol{v}_0 + \boldsymbol{f}\big|_{t=0}\big)\Big|_\Sigma \qquad \Big(\frac{\partial}{\partial \boldsymbol{n}_\Sigma} = \boldsymbol{n}_\Sigma \cdot \nabla\Big).$$

In addition, we assume that the initial data are sufficiently small and mass forces decrease exponentially as $t \to \infty$*, i.e.,*

$$|\boldsymbol{v}_0|^{(2+\alpha)}_{\cup\Omega_0^\pm} + |\mathrm{e}^{bt}\boldsymbol{f}|^{(\alpha,\frac{1+\alpha-\gamma}{2})}_{Q_\infty} + \int_0^\infty \|\mathrm{e}^{bt}\boldsymbol{f}\|_\Omega\,\mathrm{d}t \leqslant \varepsilon \ll 1, \tag{6.1.4}$$

where $b = \min\{\nu^+, \nu^-\}/(2c_0)$ *with the constant* c_0 *from inequality* (6.4.3).

Then problem (1.1.1), (1.0.3) *with* $\sigma = 0$ *in a bounded domain is uniquely solvable on infinite time interval* $t > 0$*. The solution* $(\boldsymbol{v}, p)$ *has the following properties:* $\boldsymbol{v} \in \boldsymbol{C}^{2+\alpha,1+\alpha/2}$, $p \in C^{(\gamma,1+\alpha)}$, $\nabla p \in \boldsymbol{C}^{\alpha,\alpha/2}$, Γ_t *being from* $C^{2+\alpha}$*-class. It means that for every* $t_0 \in (0,\infty)$*, the solution* $(\boldsymbol{u}, q)$ *and its derivatives in the Lagrangian coordinates belong to the corresponding spaces over* $\cup Q^\pm_{(t_0-\tau,t_0)}$ *for a small enough time interval* $(t_0-\tau, t_0)$ *and*

$$N_{(t_0-\tau,t_0)}[\boldsymbol{v},p] \leqslant c\big(\tau, |\boldsymbol{v}_0|^{(2+\alpha)}_{\cup\Omega_0^\pm}, |\nabla \boldsymbol{f}|^{(\alpha,\frac{1+\alpha-\gamma}{2})}_{Q_\infty}\big)\mathrm{e}^{-bt_0}$$
$$\Big\{|\boldsymbol{v}_0|^{(2+\alpha)}_{\cup\Omega_0^\pm} + \int_0^\infty \mathrm{e}^{bt}\|\boldsymbol{f}(\cdot,t)\|_\Omega\,\mathrm{d}t + |\mathrm{e}^{bt}\boldsymbol{f}|^{(\alpha,\frac{1+\alpha-\gamma}{2})}_{Q_\infty}\Big\}. \tag{6.1.5}$$

The proof of this theorem is based on the solvability of auxiliary linearized problems.

6.2 A Linear Problem with Closed Interface Between the Fluids

Diffraction problem (3.1.1) for the Stokes system with plane interface has been studied in Chap. 3. There existence Theorem 3.1.1 has been proved for this problem.

Now we consider linear diffraction problem with a closed interface between the fluids:

$$\mathcal{D}_t\boldsymbol{v} - \nu^\pm\nabla^2\boldsymbol{v} + \frac{1}{\rho^\pm}\nabla p = \boldsymbol{f}, \qquad \nabla\cdot\boldsymbol{v} = r \quad \text{in } D_{\Sigma T},$$
$$\boldsymbol{v}|_{t=0} = \boldsymbol{v}_0 \qquad \text{in } \Omega_0^- \cup \Omega_0^+, \tag{6.2.1}$$

$$[\boldsymbol{v}]|_{G_T} = 0, \quad \boldsymbol{v}|_\Sigma = 0, \qquad [\mu^\pm \Pi_0 \mathbb{S}(\boldsymbol{v})\boldsymbol{n}_0]|_{G_T} = \boldsymbol{d}, \quad \boldsymbol{d}\cdot\boldsymbol{n}_0 = 0,$$
$$[\boldsymbol{n}_0 \cdot \mathbb{T}(\boldsymbol{v}, p)\boldsymbol{n}_0]|_{G_T} = b \quad \text{on } G_T.$$

The solvability of problem (6.2.1) can be obtained in the same way as in Chap. 4, the existence of a solution has been proved for a similar problem in an unbounded domain. For estimating near the outer boundary, one needs to apply the theorem on the solvability of the Dirichlet problem for the Stokes system in a half-space [76].

Let two representations take place:

$$r = \nabla\cdot\boldsymbol{R} \quad \text{and} \quad \mathcal{D}_t\boldsymbol{R} - \boldsymbol{f} = \nabla\cdot\mathbb{G}, \tag{6.2.2}$$

and let the pressure $p_0(x) = p(x, 0)$ be a solution to the problem

$$\frac{1}{\rho^\pm}\nabla^2 p_0 = \left(\nabla\cdot\boldsymbol{f} + \nu^\pm\Delta r - \mathcal{D}_t r\right)\big|_{t=0} \quad \text{in} \quad \Omega_0^- \cup \Omega_0^+, \tag{6.2.3}$$

$$[p_0]|_\Gamma = \left[2\mu^\pm\frac{\partial\boldsymbol{v}_0}{\partial\boldsymbol{n}_0}\cdot\boldsymbol{n}_0\right]\Big|_\Gamma - b|_{t=0} \equiv p_{00},$$
$$\left[\frac{1}{\rho^\pm}\frac{\partial p_0}{\partial\boldsymbol{n}_0}\right]\Big|_\Gamma = [\boldsymbol{n}_0\cdot(\boldsymbol{f}|_{t=0} + \nu^\pm\nabla^2\boldsymbol{v}_0)]|_\Gamma \equiv p_{01}, \tag{6.2.4}$$
$$\frac{1}{\rho^-}\frac{\partial p_0}{\partial\boldsymbol{n}_\Sigma}\Big|_\Sigma = \boldsymbol{n}_\Sigma\cdot\left(\nu^-\nabla^2\boldsymbol{v}_0 + \boldsymbol{f}\big|_{t=0}\right)\Big|_\Sigma \equiv p_{02}.$$

We state existence theorem for (6.2.1) (the case of finite fluid mass and $\sigma = 0$).

Theorem 6.2.1 *We suppose that* $\alpha, \gamma \in (0, 1)$, $0 < T < \infty$ *and* $\Gamma, \Sigma \in C^{2+\alpha}$. *In addition, let conditions* (6.2.2), (6.2.3), *and* (6.2.4) *hold and* $\boldsymbol{f} \in \boldsymbol{C}^{\alpha,\alpha/2}(D_{\Sigma T})$, $r \in C^{1+\alpha,\frac{1+\alpha}{2}}(D_{\Sigma T})$, $\boldsymbol{R} \in \boldsymbol{C}^{2+\alpha,1+\alpha/2}(D_{\Sigma T})$, $[\boldsymbol{R}\cdot\boldsymbol{n}_0]|_{G_T} = 0$, $\boldsymbol{v}_0 \in \boldsymbol{C}^{2+\alpha}(\Omega_0^- \cup \Omega_0^+)$, $\boldsymbol{d} \in \boldsymbol{C}^{1+\alpha,\frac{1+\alpha}{2}}(G_T)$, $b \in C^{(\gamma,1+\alpha)}(G_T)$, $\nabla_\Gamma b \in \boldsymbol{C}^{\alpha,\alpha/2}(G_T)$, *and* $\mathbb{G} = \{G_{ik}\}_{i,k=1}^3$, $G_{ik} \in C^{(\gamma,1+\alpha)}(D_{\Sigma T}) \cap C^{\gamma,0}(D_{\Sigma T})$.

We also assume that compatibility conditions hold:

$$\nabla\cdot\boldsymbol{v}_0(x) = r(x, 0) = 0, \quad x \in \Omega^+\cup\Omega^-, \quad [\boldsymbol{v}_0]\big|_\Gamma = 0, \quad \boldsymbol{v}_0\big|_\Sigma = 0,$$
$$[\mu^\pm\Pi_0\mathbb{S}(\boldsymbol{v}_0(x))\boldsymbol{n}_0]|_{x\in\Gamma} = \Pi_0\boldsymbol{d}(x, 0), \quad x \in \Gamma,$$
$$\left[\Pi_0\left(\boldsymbol{f}(x, 0) - \frac{1}{\rho^\pm}\nabla p_0(x) + \nu^\pm\nabla^2\boldsymbol{v}_0(x)\right)\right]\Big|_{x\in\Gamma} = 0, \tag{6.2.5}$$
$$\Pi_\Sigma\left(\boldsymbol{f}(x, 0) - \frac{1}{\rho^-}\nabla p_0(x) + \nu^-\nabla^2\boldsymbol{v}_0(x)\right)\Big|_{x\in\Sigma} = 0.$$

Then problem (6.2.1) *has a unique solution* $(\boldsymbol{v}, p)$ *with the following properties:* $\boldsymbol{v} \in \boldsymbol{C}^{2+\alpha,1+\alpha/2}(D_{\Sigma T})$, $p \in C^{(\gamma,1+\alpha)}(D_{\Sigma T})$, $\nabla p \in \boldsymbol{C}^{\alpha,\alpha/2}(D_{\Sigma T})$. *This solution satisfies the inequality*

$$
\begin{aligned}
&|\boldsymbol{v}|^{(2+\alpha,1+\alpha/2)}_{D_{\Sigma t'}} + |\nabla p|^{(\alpha,\alpha/2)}_{D_{\Sigma t'}} + |p|^{(\gamma,1+\alpha)}_{D_{\Sigma t'}} \\
&\qquad \leqslant c(t')\Big\{ |\boldsymbol{f}|^{(\alpha,\alpha/2)}_{D_{\Sigma t'}} + |r|^{(1+\alpha,\frac{1+\alpha}{2})}_{D_{\Sigma t'}} + |\mathcal{D}_t \boldsymbol{R}|^{(\alpha,\alpha/2)}_{D_{\Sigma t'}} + \langle \boldsymbol{R}\rangle^{(\nu,1+\alpha)}_{D_{\Sigma t'}} \\
&\qquad\quad + |\boldsymbol{v}_0|^{(2+\alpha)}_{\cup\Omega^{\pm}} + \langle \mathbb{G}\rangle^{(\gamma,1+\alpha)}_{D_{\Sigma t'}} + |\mathbb{G}|^{(\gamma,0)}_{D_{\Sigma t'}} + |\boldsymbol{d}|^{(1+\alpha,\frac{1+\alpha}{2})}_{G_{t'}} + |b|_{G_{t'}} \\
&\qquad\quad + \langle b\rangle^{(\gamma,1+\alpha)}_{G_{t'}} + |\nabla_{\Gamma} b|^{(\alpha,\alpha/2)}_{G_{t'}} \Big\}, \qquad \forall t' \in (0,T],
\end{aligned}
$$

where $c(t')$ *is a nondecreasing function of* $t' \leqslant T$, *and* $\nabla_\Gamma = \Pi_0 \nabla$.

6.3 A Linearized Problem

Now let us study the linearized problem

$$
\begin{aligned}
&\mathcal{D}_t \boldsymbol{w} - \nu^{\pm}\nabla^2_{\boldsymbol{u}} \boldsymbol{w} + \frac{1}{\rho^{\pm}}\nabla_{\boldsymbol{u}} s = \boldsymbol{f}, \qquad \nabla_{\boldsymbol{u}} \cdot \boldsymbol{w} = r \quad \text{in } D_{\Sigma T}, \\
&\boldsymbol{w}|_{t=0} = \boldsymbol{w}_0 \text{ in } \Omega^- \cup \Omega^+, \qquad (6.3.1) \\
&[\boldsymbol{w}]|_{G_T} = 0, \qquad \boldsymbol{w}|_{\Sigma} = 0, \qquad [\mu^{\pm}\Pi_0 \Pi \mathbb{S}_{\boldsymbol{u}}(\boldsymbol{w})\boldsymbol{n}]|_{G_T} = \Pi_0 \boldsymbol{d}, \\
&[\boldsymbol{n}_0 \cdot \mathbb{T}_{\boldsymbol{u}}(\boldsymbol{w}, s)\boldsymbol{n}]|_{G_T} = b \text{ on } G_T.
\end{aligned}
$$

An analogue to system (6.3.1) with the surface tension has been investigated in Sect. 5.3. There it has been proved unique solvability for the problem in any finite time interval when Σ is absent and the domain $\overline{\Omega^+ \cup \Omega^-}$ coincides with the whole space $\mathbb{R}^3$ (Theorem 5.3.1). This result has been obtained in the Hölder spaces with power-law weights at infinity, but it is also valid in our case. The proof is based on Theorem 6.2.1, it being only simpler without weights and surface tension; in addition, the weighted spaces are equivalent to the ordinary Hölder spaces in bounded domains. Now we cite existence theorem for system (6.3.1).

Theorem 6.3.1 *Let* $\alpha, \gamma \in (0,1)$ *and* $0 < T < \infty$. *We assume that* $\Gamma, \Sigma \in C^{2+\alpha}$ *and that for* $\boldsymbol{u} \in \boldsymbol{C}^{2+\alpha,1+\alpha/2}(D_{\Sigma T})$, $[\boldsymbol{u}]|_{G_T} = 0$, *the inequality*

$$
(T + T^{\gamma/2})|\boldsymbol{u}|^{(2+\alpha,1+\alpha/2)}_{D_{\Sigma T}} \leqslant \delta \qquad (6.3.2)
$$

holds for sufficiently small $\delta > 0$.

Moreover, we assume that the following four groups of conditions are fulfilled:

(1) $\boldsymbol{f} \in \boldsymbol{C}^{\alpha,\frac{\alpha}{2}}(D_{\Sigma T})$, $r \in C^{1+\alpha,\frac{1+\alpha}{2}}(D_{\Sigma T})$, $\boldsymbol{w}_0 \in \boldsymbol{C}^{2+\alpha}(\cup\Omega^{\pm})$, $\boldsymbol{d} \in \boldsymbol{C}^{1+\alpha,\frac{1+\alpha}{2}}(G_T)$, $b \in C^{(\gamma,1+\alpha)}(G_T)$;

(2) *compatibility conditions* (6.2.5) *with* $\boldsymbol{v}_0 = \boldsymbol{w}_0$ *and* $p_0(\xi) = s(\xi,0)$ *are satisfied;*

(3) *there exist a vector* $\boldsymbol{g} \in \boldsymbol{C}^{\alpha,\alpha/2}(D_{\Sigma T})$ *and a tensor* $\mathbb{G} = \{G_{ik}\}_{i,k=1}^3$ *with* $G_{ik} \in C^{(\gamma,1+\alpha)}(D_{\Sigma T}) \cap C^{\gamma,0}(D_{\Sigma T})$ *such that*

$$\mathcal{D}_t r - \nabla_{\boldsymbol{u}} \cdot \boldsymbol{f} = \nabla \cdot \boldsymbol{g}, \quad g_i = \partial G_{ik}/\partial \xi_k, \quad i = 1,2,3,$$

(*these equalities are understood in a weak sense*) *and, moreover,*

$$[(\boldsymbol{g} + \mathbb{A}^* \boldsymbol{f}) \cdot \boldsymbol{n}_0]|_{G_T} = 0;$$

(4) $s_0(\xi) = s(\xi,0)$ *is a solution of the equation*

$$\frac{1}{\rho^{\pm}} \nabla^2 s_0(\xi) = \nabla \cdot (\mathcal{D}_t \mathbb{B}^*|_{t=0} \boldsymbol{w}_0(\xi) - \boldsymbol{g}(\xi,0)), \quad \xi \in \Omega^+ \cup \Omega^-, \qquad (6.3.3)$$

with boundary conditions (6.2.4), *where* $p_0 = s_0$ *and* $\boldsymbol{v}_0 = \boldsymbol{w}_0$.

Under all these assumptions, problem (6.3.1) *has a unique solution* $(\boldsymbol{w}, s)$ *with the following properties:* $\boldsymbol{w} \in \boldsymbol{C}^{2+\alpha,1+\alpha/2}(D_{\Sigma T})$, $s \in C^{(\gamma,1+\alpha)}(D_{\Sigma T})$, $\nabla s \in \boldsymbol{C}^{\alpha,\alpha/2}(D_{\Sigma T})$; *moreover, this solution satisfies the inequality*

$$\begin{aligned} &|\boldsymbol{w}|_{D_{\Sigma t'}}^{(2+\alpha,1+\alpha/2)} + |\nabla s|_{D_{\Sigma t'}}^{(\alpha,\alpha/2)} + |s|_{D_{\Sigma t'}}^{(\gamma,1+\alpha)} \\ &\quad \leqslant c_1(t') \Big\{ |\boldsymbol{f}|_{D_{\Sigma t'}}^{(\alpha,\alpha/2)} + |r|_{D_{\Sigma t'}}^{(1+\alpha,\frac{1+\alpha}{2})} + |\boldsymbol{w}_0|_{\cup\Omega^{\pm}}^{(2+\alpha)} + |\boldsymbol{g}|_{D_{\Sigma t'}}^{(\alpha,\alpha/2)} \\ &\qquad + \langle \mathbb{G} \rangle_{D_{\Sigma t'}}^{(\gamma,1+\alpha)} + |\mathbb{G}|_{D_{\Sigma t'}}^{(\gamma,0)} + |\boldsymbol{d}|_{G_{t'}}^{(1+\alpha,\frac{1+\alpha}{2})} + |b|_{G_{t'}} + \langle b \rangle_{G_{t'}}^{(\gamma,1+\alpha)} \\ &\qquad + |\nabla_{\Gamma} b|_{G_{t'}}^{(\alpha,\alpha/2)} + P_{t'}[\boldsymbol{u}] |\boldsymbol{w}_0|_{\cup\Omega^{\pm}}^{(1)} \Big\}, \end{aligned} \qquad (6.3.4)$$

where $c_1(t')$ *a nondecreasing function of* $t' \leqslant T$, *and*

$$P_t[\boldsymbol{u}] = t^{\frac{1-\alpha}{2}} |\nabla \boldsymbol{u}|_{D_{\Sigma t}} + |\nabla \boldsymbol{u}|_{D_{\Sigma t}}^{(\alpha,\alpha/2)}.$$

6.4 Global Solvability of Problem (1.1.1) with $\sigma = 0$

Our purpose in this section is to prove an existence theorem for problem (1.1.1), (1.0.3) without taking surface tension into account for all $t > 0$.

Local solvability for this problem transformed into the Lagrangian coordinates $\{\xi\}$ by (1.1.2) is stated in Theorem 5.1.1 for $\sigma = 0$. We reformulate it as follows.

Theorem 6.4.1 (Local Solvability in a Bounded Domain) *We assume that for $\alpha, \gamma \in (0,1)$, $T < \infty$, the surfaces $\Gamma, \Sigma \in C^{2+\alpha}$, and $\boldsymbol{f}, \mathcal{D}_x \boldsymbol{f} \in \boldsymbol{C}^{\alpha, \frac{1+\alpha-\gamma}{2}}(Q_T)$, $\boldsymbol{v}_0 \in \boldsymbol{C}^{2+\alpha}(\Omega_0^- \cup \Omega_0^+)$, compatibility conditions (6.1.2) are satisfied, and the pressure $q(\xi, 0) \equiv p(\xi, 0)$ at the initial moment $t = 0$ is a solution to diffraction problem (6.1.3).*

Then for any $T < \infty$, there is $\varepsilon(T)$ such that under the condition

$$|\boldsymbol{f}|_{Q_T}^{(\alpha, \frac{1+\alpha-\gamma}{2})} + |\mathcal{D}_x \boldsymbol{f}|_{Q_T} + |\boldsymbol{v}_0|_{\cup\Omega_0^\pm}^{(2+\alpha)} \leqslant \varepsilon(T), \tag{6.4.1}$$

system (6.1.1) has a unique solution $(\boldsymbol{u}, q)$ which possesses the following properties: $\boldsymbol{u} \in \boldsymbol{C}^{2+\alpha, 1+\alpha/2}(D_{\Sigma T})$, $q \in C^{(\gamma, 1+\alpha)}(D_{\Sigma T})$, $\nabla q \in \boldsymbol{C}^{\alpha, \alpha/2}(D_{\Sigma T})$. Pressure function is defined uniquely up to a bounded functions of time. The interface Γ_t is of the class $C^{2+\alpha}$. In addition, the estimate

$$|\boldsymbol{u}|_{D_{\Sigma T}}^{(2+\alpha, 1+\alpha/2)} + |\nabla q|_{D_{\Sigma T}}^{(\alpha, \alpha/2)} + |q|_{D_{\Sigma T}}^{(\gamma, 1+\alpha)}$$
$$\leqslant c\Big(T, |\boldsymbol{v}_0|_{\cup\Omega_0^\pm}^{(2+\alpha)}, |\mathcal{D}_x \boldsymbol{f}|_{Q_T}^{(\alpha, \frac{1+\alpha-\gamma}{2})}\Big)\Big\{|\boldsymbol{f}|_{Q_T}^{(\alpha, \frac{1+\alpha-\gamma}{2})} + |\boldsymbol{v}_0|_{\cup\Omega_0^\pm}^{(2+\alpha)}\Big\} \tag{6.4.2}$$

holds.

We observe that $\partial\Omega = \Sigma$, and the Korn inequality

$$\|\boldsymbol{v}\|_{W_2^1(\Omega)} \leqslant c_0 \|\mathbb{S}(\boldsymbol{v})\|_\Omega \tag{6.4.3}$$

is satisfied for $\boldsymbol{v}$ in Ω because $\boldsymbol{v}\big|_\Sigma = 0$ (see [80]); moreover, $\|\boldsymbol{v}\|_{W_2^1(\Omega)}$ coincides with $\|\boldsymbol{v}\|_{W_2^1(\Omega_t^- \cup \Omega_t^+)}$ due to the fact that $[\boldsymbol{v}]\big|_{\Gamma_t} = 0$.

Proposition 6.4.1 *We assume that a classical solution to problem (1.1.1), (1.0.3) is defined on $[0,T]$ and v_0 satisfies compatibility conditions (6.1.2). In addition, let $\boldsymbol{f}(\cdot, \tau) \in L_2(\Omega)$, and let $\int_0^\infty \|\mathrm{e}^{b\tau} \boldsymbol{f}(\cdot, \tau)\|_\Omega \, \mathrm{d}\tau < \infty$.*

Then

$$\|\boldsymbol{v}(\cdot, t)\|_\Omega \leqslant \mathrm{e}^{-bt}\Big\{\|\boldsymbol{v}_0\|_\Omega + c_1 \int_0^t \|\mathrm{e}^{b\tau} \boldsymbol{f}(\cdot, \tau)\|_\Omega \, \mathrm{d}\tau\Big\}, t \in (0, T], \tag{6.4.4}$$

$$\int_0^T \|\boldsymbol{v}(\cdot,\tau)\|_\Omega \, \mathrm{d}\tau \leqslant \frac{1}{b}\Big\{\|\boldsymbol{v}_0\|_\Omega + c_1 \int_0^\infty \|\boldsymbol{f}(\cdot,\tau)\|_\Omega \, \mathrm{d}\tau\Big\}, \tag{6.4.5}$$

where the constants b and c_1 are independent of t.

Proof We multiply the first equation in (1.1.1) by $\boldsymbol{v}$ and integrate by parts over $\Omega_t^- \cup \Omega_t^+$. Then we have

$$\frac{1}{2}\frac{d}{dt}\|\sqrt{\rho^\pm}\boldsymbol{v}\|_\Omega^2 + \frac{\mu^+}{2}\|\mathbb{S}(\boldsymbol{v})\|_{\Omega_t^+}^2 + \frac{\mu^-}{2}\|\mathbb{S}(\boldsymbol{v})\|_{\Omega_t^-}^2 = \int_\Omega \rho^\pm \boldsymbol{f}\cdot\boldsymbol{v}\, \mathrm{d}x.$$

If we apply (6.4.3) to the left-hand side of this equality and the Hölder inequality to its right-hand side and divide by $\|\boldsymbol{v}\|_\Omega$, we arrive at the estimate

$$\frac{d}{dt}\|\boldsymbol{v}\|_\Omega + b\|\boldsymbol{v}\|_\Omega \leqslant c_1\|\boldsymbol{f}\|_\Omega,$$

where $b = \min\{\nu^+,\nu^-\}/(2c_0^2)$ and $c_1 = \dfrac{\max(\rho^\pm)}{\min(\rho^\pm)}$. Next, we multiply both sides of the inequality by $\mathrm{e}^{b\tau}$ and integrate with respect to τ from 0 to t:

$$\|\boldsymbol{v}\|_\Omega \leqslant \mathrm{e}^{-bt}\Big\{\|\boldsymbol{v}_0\|_\Omega + c_1\int_0^t \mathrm{e}^{b\tau}\|\boldsymbol{f}(\cdot,\tau)\|_\Omega \, \mathrm{d}\tau\Big\}.$$

Finally, we integrate over $t \in (0,T)$ and apply the Fubini theorem:

$$\begin{aligned}\int_0^T \|\boldsymbol{v}(\cdot,t)\|_\Omega \mathrm{d}t &\leqslant \frac{1}{b}\|\boldsymbol{v}_0\|_\Omega + c_1\int_0^\infty\int_0^t \mathrm{e}^{-b(t-\tau)}\|\boldsymbol{f}(\cdot,\tau)\|_\Omega \, \mathrm{d}\tau \mathrm{d}t\\ &\leqslant \frac{1}{b}\|\boldsymbol{v}_0\|_\Omega + c_1\int_0^\infty\int_\tau^\infty \mathrm{e}^{-b(t-\tau)}\|\boldsymbol{f}(\cdot,\tau)\|_\Omega \, \mathrm{d}t\, \mathrm{d}\tau\\ &\leqslant \frac{1}{b}\Big\{\|\boldsymbol{v}_0\|_\Omega + c_1\int_0^\infty \|\boldsymbol{f}(\cdot,\tau)\|_\Omega \, \mathrm{d}\tau\Big\}.\end{aligned}$$

Thus, (6.4.5) is also proved. □

Lemma 6.4.1 *Let $u \in C^{0,\frac{1+\alpha}{2}}(D_{\Sigma T_0})$, $T_0 > 0$, $0 < \theta_1 < T_0^{1/2}$. Then u is subjected to the inequality*

$$\langle u\rangle_{t,D_{\Sigma T_0}}^{(\frac{1+\alpha-\gamma}{2})} \leqslant 2{\theta_1}^\gamma \langle u\rangle_{t,D_{\Sigma T_0}}^{(\frac{1+\alpha}{2})} + c{\theta_1}^{\gamma-\alpha-\frac{9}{2}}\int_0^{T_0}\|u(\cdot,\tau)\|_\Omega \, \mathrm{d}\tau. \tag{6.4.6}$$

Proof Let us consider a smooth function $\zeta(x,t)$ with the support in $M_1 = B_1(0)\times(0,1)$ such that the integral of it over M_1 equals the unite and

$M_{\theta_1} = B_{\theta_1}(0) \times (0, {\theta_1}^2)$, $B_{\theta_1}(0) = \{|x| < \theta_1\}$. The function $\zeta_{\theta_1}(x,t) = \zeta(x/r, t/{\theta_1}^2){\theta_1}^{-5}$ has the following property: $\int_{M_{\theta_1}} \zeta_{\theta_1}(x,t)\, \mathrm{d}x\, \mathrm{d}t = 1$.

We can write

$$u(x,t) = \int_{M_{\theta_1}} \zeta_{\theta_1}(y,\tau)\big\{u(x,t) - u(x+y,t+\tau)\big\}\, \mathrm{d}y\mathrm{d}\tau$$
$$+ \int_{M_{\theta_1}} \zeta_{\theta_1}(y,\tau) u(x+y,t+\tau)\, \mathrm{d}y\mathrm{d}\tau \equiv u_1(x,t) + u_2(x,t).$$

Let $t, t+h \in (0, T_0)$. First, we assume that $0 < h < {\theta_1}^2$. Then the difference

$$u_1(x,t+h) - u_1(x,t) = \int_{M_{\theta_1}} \zeta_{\theta_1}(y,\tau)\big\{u(x,t+h) - u(x,t)$$
$$- u(x+y,t+\tau+h) + u(x+y,t+\tau)\big\}\, \mathrm{d}y\mathrm{d}\tau$$

can be estimated as follows:

$$|u_1(x,t+h) - u_1(x,t)| \leqslant 2\langle u\rangle_{t,M_{2\theta_1}}^{(\frac{1+\alpha}{2})} h^{\frac{1+\alpha}{2}} \leqslant 2\langle u\rangle_{t,M_{2\theta_1}}^{(\frac{1+\alpha}{2})} h^{\frac{1+\alpha-\gamma}{2}} {\theta_1}^{\gamma}. \qquad (6.4.7)$$

In addition, for the difference

$$u_2(x,t+h) - u_2(x,t) = \int_{M_{\theta_1}} \zeta_{\theta_1}(y,\tau)\big\{u(x+y,t+\tau+h) - u(x+y,t+\tau)\big\}\mathrm{d}y\mathrm{d}\tau$$
$$= \int_{M_{\theta_1}} \zeta_{\theta_1}(y,\tau) \int_0^1 \mathcal{D}_s u(x+y,t+\tau+sh)\, \mathrm{d}s\mathrm{d}y\mathrm{d}\tau$$
$$= \int_{M_{\theta_1}} \zeta_{\theta_1}(y,\tau) h \int_0^1 \mathcal{D}_\tau u(x+y,t+\tau+sh)\, \mathrm{d}s\mathrm{d}y\mathrm{d}\tau$$
$$= -h \int_{M_{\theta_1}} \mathcal{D}_\tau \zeta_{\theta_1}(y,\tau) \int_0^1 u(x+y,t+\tau+sh)\, \mathrm{d}s\mathrm{d}y\mathrm{d}\tau$$

we obtain

$$|u_2(x,t+h) - u_2(x,t)| \leqslant ch{\theta_1}^{-\frac{11}{2}} \int_0^{2{\theta_1}^2} \|u(\cdot,\tau)\|_\Omega \mathrm{d}\tau$$
$$\leqslant ch^{\frac{1+\alpha-\gamma}{2}} {\theta_1}^{\gamma-\alpha-\frac{9}{2}} \int_0^{T_0} \|u(\cdot,\tau)\|_\Omega \mathrm{d}\tau, \quad (6.4.8)$$

since $\max \|\mathcal{D}_\tau \zeta_{\theta_1}\|_\Omega \leqslant c{\theta_1}^{-\frac{11}{2}}$.

If $h \geqslant {\theta_1}^2$, we have

$$|u(x,t+h)-u(x,t)| \leqslant |u(x,t+h)|+|u(x,t)| \leqslant c{\theta_1}^{-\frac{7}{2}} \int_0^{T_0} \|u(\cdot,\tau)\|_\Omega \mathrm{d}\tau$$

$$\leqslant ch^{\frac{1+\alpha-\gamma}{2}} {\theta_1}^{\gamma-\alpha-\frac{9}{2}} \int_0^{T_0} \|u(\cdot,\tau)\|_\Omega \mathrm{d}\tau. \tag{6.4.9}$$

Here we have extended u across Γ and across Σ with preservation of class. We divide inequalities (6.4.7), (6.4.8), and (6.4.9) by $h^{\frac{1+\alpha-\gamma}{2}}$ and take supremum first with respect to h and then second with respect to t and x over $D_{\Sigma T_0}$. As a result, we arrive at inequality (6.4.6). □

In a similar way, by using an interpolation inequality with respect to x from [88], one can prove the following lemma.

Lemma 6.4.2 *For arbitrary* $u \in C^{2+\alpha,1+\frac{\alpha}{2}}(D_{\Sigma T_0})$, $0 < \theta_2$, $\varkappa_1 < \min\{\mathrm{diam}\{\Omega\}, T_0^{\frac{1}{2}}\}$, *the inequalities*

$$\langle u\rangle^{(\alpha,\frac{\alpha}{2})}_{D_{\Sigma T_0}} \leqslant 2\theta_2^2 \langle u\rangle^{(2+\alpha,1+\frac{\alpha}{2})}_{D_{\Sigma T_0}} + c{\theta_2}^{-\alpha-\frac{7}{2}} \int_0^{T_0} \|u(\cdot,\tau)\|_\Omega \mathrm{d}\tau, \tag{6.4.10}$$

$$|u|_{D_{\Sigma T_0}} \leqslant c\left\{ \varkappa_1^{1+\alpha} \langle u\rangle^{(\frac{1+\alpha}{2})}_{t,D_{\Sigma T_0}} + \varkappa_1^{-\frac{7}{2}} \int_0^{T_0} \|u(\cdot,\tau)\|_\Omega \,\mathrm{d}\tau \right\} \tag{6.4.11}$$

hold.

In addition, we cite Lemma 5.3.3 on the estimates of the Newtonian potential gradient in the ordinary Hölder classes of functions in a bounded domain.

Lemma 6.4.3 *If* $F \in C^{(0,\frac{1+\alpha-\gamma}{2})}(D_{\Sigma T})$ *vanishes at infinity, then the Newtonian potential gradient*

$$\nabla_x V(x,t) = \nabla_x \int_{\mathbb{R}^3} \mathcal{E}(x,y)F(y,t)\,\mathrm{d}y$$

satisfies the estimates

$$|\nabla V|^{(\gamma,0)}_{D_{\Sigma T}} \leqslant c|F|_{D_{\Sigma T}},$$

$$|\nabla V|^{(\gamma,1+\alpha)}_{D_{\Sigma T}} \leqslant c|F|^{(\frac{1+\alpha-\gamma}{2})}_{t,D_{\Sigma T}}.$$

Proposition 6.4.2 *Let there exist a solution* $(\boldsymbol{v},p)$ *to problem* (1.1.1), (1.0.3) *in the interval* $(0,T]$, *and let the estimate*

$$N_{(0,T)}[\boldsymbol{v}, p] \equiv |\boldsymbol{u}|_{D_{\Sigma T}}^{(2+\alpha,1+\alpha/2)} + |\nabla q|_{D_{\Sigma T}}^{(\alpha,\alpha/2)} + |q|_{D_{\Sigma T}}^{(\gamma,1+\alpha)} \leqslant \mu$$

hold; here $(\boldsymbol{u}, q)$ *is the solution to problem* (1.1.1), (1.0.3) *in the Lagrangian coordinates.*

Then for any $t_0 \in (0, T]$, *there is* $\tau_0 \in (0, t_0/2)$ *such that*

$$N_{(t_0-\tau_0,t_0)}[\boldsymbol{v}, p] \leqslant c\Big(\delta, |\nabla \boldsymbol{f}|_{\cup Q_0'}^{(\alpha,\frac{1+\alpha-\gamma}{2})}\Big)\Big\{|\boldsymbol{f}|_{\cup Q_0'}^{(\alpha,\frac{1+\alpha-\gamma}{2})} + \int\limits_{t_0-2\tau_0}^{t_0} \|\boldsymbol{v}(\cdot,\tau)\|_\Omega \mathrm{d}\tau\Big\}, \tag{6.4.12}$$

where $c(\delta)$ *is a nondecreasing function,* $Q_\beta' = Q_{(t_0-2\tau_0+\beta,t_0)}$*;* τ_0 *depends on* μ, *and* δ *is a constant from* (6.4.18).

Proof We fix arbitrary $t_0 \in (0, T]$. Let $\tau_0 \in (0, t_0/2)$, and let $\eta_\lambda(t)$ be a smooth monotone function of t such that

$$\eta_\lambda(t) = \begin{cases} 0, & \text{if } t \leqslant t_0 - 2\tau_0 + \lambda/2, \\ 1, & \text{if } t \geqslant t_0 - 2\tau_0 + \lambda, \end{cases} \tag{6.4.13}$$

$\lambda \in (0, \tau_0]$, and for $\dot{\eta}_\lambda \equiv \mathrm{d}\eta_\lambda(t)/\mathrm{d}t$, the inequalities

$$\big|\dot{\eta}_\lambda(t)\big|_{\mathbb{R}} \leqslant c\lambda^{-1}, \quad \big\langle \dot{\eta}_\lambda(t)\big\rangle_{\mathbb{R}}^{(\alpha/2)} \leqslant c\lambda^{-1-\alpha/2} \tag{6.4.14}$$

hold.

We introduce the Lagrangian coordinates by the formula

$$\boldsymbol{x} = \boldsymbol{\xi}' + \int_{t_0-2\tau_0}^{t} \boldsymbol{u}(\xi', \tau)\mathrm{d}\tau \equiv \boldsymbol{X}(\xi', t), \quad \xi' \in \cup\Omega^{\pm}_{t_0-2\tau_0}, \quad t > t_0 - 2\tau_0, \tag{6.4.15}$$

where $\boldsymbol{u}(\xi', t) = \boldsymbol{v}(X(\xi', t), t)$.

We consider function couple $\boldsymbol{w} = \boldsymbol{v}\eta_\lambda$, $s = p\eta_\lambda$. It solves the system

$$\begin{aligned} &\mathcal{D}_t\boldsymbol{w} + (\boldsymbol{v}\cdot\nabla)\boldsymbol{w} - \nu^{\pm}\nabla^2\boldsymbol{w} + \frac{1}{\rho^{\pm}}\nabla s = \boldsymbol{f}\eta_\lambda + \boldsymbol{v}\dot{\eta}_\lambda, \\ &\nabla\cdot\boldsymbol{w} = 0 \quad \text{in} \quad \Omega_t^- \cup \Omega_t^+, \ \ t > t_0 - 2\tau_0, \\ &\boldsymbol{w}|_{t=t_0-2\tau_0} = 0 \quad \text{in} \quad \Omega^-_{t_0-2\tau_0} \cup \Omega^+_{t_0-2\tau_0}, \\ &[\boldsymbol{w}]|_{\Gamma_t} = 0, \quad [\mathbb{T}(\boldsymbol{w}, s)\boldsymbol{n}]|_{\Gamma_t} = 0, \quad \boldsymbol{w}|_\Sigma = 0. \end{aligned} \tag{6.4.16}$$

Then we rewrite (6.4.16) in the form

$$\mathcal{D}_t\boldsymbol{w} - \nu^{\pm}\nabla^2_{\boldsymbol{u}}\boldsymbol{w} + \frac{1}{\rho^{\pm}}\nabla_{\boldsymbol{u}} s = \boldsymbol{f}(X, t)\eta_\lambda + \boldsymbol{u}\dot{\eta}_\lambda,$$

$$
\begin{aligned}
&\nabla_{\boldsymbol{u}} \cdot \boldsymbol{w} = 0 \quad \text{ in } \cup\Omega', \quad t > t_0 - 2\tau_0,\\
&\boldsymbol{w}|_{t=t_0-2\tau_0} = 0 \quad \text{ in } \cup\Omega',\\
&[\boldsymbol{w}]|_{\Gamma'} = 0, \quad \boldsymbol{w}|_{\Sigma} = 0,\\
&[\mu^{\pm}\Pi_0\Pi\mathbb{S}_{\boldsymbol{u}}(\boldsymbol{w})\boldsymbol{n}]|_{\Gamma'} = 0, \qquad [\boldsymbol{n}_0 \cdot \mathbb{T}_{\boldsymbol{u}}(\boldsymbol{w}, s)\boldsymbol{n}]|_{\Gamma'} = 0,
\end{aligned} \tag{6.4.17}
$$

Here $\cup\Omega' = \Omega^-_{t_0-2\tau_0} \cup \Omega^+_{t_0-2\tau_0}$, $\Gamma' = \Gamma_{t_0-2\tau_0}$, $\boldsymbol{n}_0$ is the outward normal to Γ', and Π_0 and Π are the projections onto the tangent planes to Γ' and to Γ_t, respectively. Other notation, for example, $\nabla_{\boldsymbol{u}}$, also corresponds to transformation (6.4.15). The functions $\boldsymbol{w}$, s, and $\boldsymbol{f}$ in the Lagrangian coordinates are denoted by the same letters.

In order to apply Theorem 6.3.1 to problem (6.4.17), we need to verify the hypotheses of it. To this end, we choose τ_0 so small that inequality (6.3.2) holds. It is enough to take τ_0 such that

$$
(2\tau_0 + (2\tau_0)^{\gamma/2})\mu \leqslant \delta. \tag{6.4.18}
$$

Since $\nabla_{\boldsymbol{u}} \cdot \boldsymbol{w} = 0$, the third hypothesis

$$
-\nabla_{\boldsymbol{u}} \cdot (\boldsymbol{f}(X, t)\eta_\lambda + \boldsymbol{u}\dot{\eta}_\lambda) = \nabla \cdot \boldsymbol{g}
$$

is satisfied with $\boldsymbol{g} = -\mathbb{A}^*(\boldsymbol{f}(X, t)\eta_\lambda + \boldsymbol{u}\dot{\eta}_\lambda)$, it is obvious that

$$
[(\boldsymbol{g} + \mathbb{A}^*(\boldsymbol{f}(X, t)\eta_\lambda + \boldsymbol{u}\dot{\eta}_\lambda)) \cdot \boldsymbol{n}_0]|_{\Gamma'} = 0.
$$

As $\mathbb{G} = \{G_{ik}\}^3_{i,k=1}$, we take the potential

$$
\mathbb{G}(\xi', t) = \nabla \int_{\mathbb{R}^3} \mathcal{E}(\xi', y)\mathbb{A}^*(\boldsymbol{f}(X(y, t), t)\eta_\lambda + \boldsymbol{u}\dot{\eta}_\lambda)\mathrm{d}y,
$$

where $\mathcal{E}(x, y) = \frac{-1}{4\pi|x-y|}$ and $\boldsymbol{f}$ and $\boldsymbol{u}$ are extended with preservation of class in the whole space $\mathbb{R}^3$ and vanish at infinity.

Assumption 1 is fulfilled because $\boldsymbol{u} \in \boldsymbol{C}^{\alpha,\alpha/2}(D_{\Sigma T})$. As to hypotheses 2 and 4, all the functions in (6.2.5), (6.3.3), and (6.2.4) with $\Omega^{\pm} = \Omega^{\pm}_{t_0-2\tau_0}$ vanish at $t = t_0 - 2\tau_0$. Hence, by (6.3.4)

$$
\begin{aligned}
N_{(t_0-2\tau_0+\lambda, t_0)}[\boldsymbol{u}, q] &\leqslant N_{(t_0-2\tau_0, t_0)}[\boldsymbol{w}, s]\\
&\leqslant c_1(2\tau_0)\Big\{|\boldsymbol{f}(X, t)\eta_\lambda|^{(\alpha,\alpha/2)}_{Q'_0} + |\boldsymbol{u}\dot{\eta}_\lambda|^{(\alpha,\frac{\alpha}{2})}_{D'_0}\\
&\qquad + |\boldsymbol{g}|^{(\alpha,\alpha/2)}_{D'_0} + |\mathbb{G}|^{(\gamma,1+\alpha)}_{D'_0} + |\mathbb{G}|^{(\gamma,0)}_{D'_0}\Big\}.
\end{aligned}
$$

We can estimate the Hölder norm of the composite function $\boldsymbol{f}(X(\xi,t),t)$ as follows:

$$|\boldsymbol{f}(X,t)|^{(\alpha,\alpha/2)}_{Q'_0} \leqslant |\boldsymbol{f}|^{(\alpha,\alpha/2)}_{Q'_0} + |\nabla \boldsymbol{f}|_{Q'_0}\big(2\tau_0|\boldsymbol{u}|^{(\alpha)}_{x,D'_0} + (2\tau_0)^{1-\alpha/2}|\boldsymbol{u}|_{D'_0}\big).$$

Next, we assume that $\lambda < 1$. In view of Lemma 6.4.3 on the estimate of Newtonian potential gradient from [29], we can deduce

$$\begin{aligned}
|\mathbb{G}|^{(\gamma,1+\alpha)}_{D'_0} &\leqslant c\Big\{\frac{1}{\lambda^{\frac{1+\alpha-\gamma}{2}}}|\boldsymbol{f}(X,t)|^{(\frac{1+\alpha-\gamma}{2})}_{t,Q'_0} + \frac{1}{\lambda}\langle \boldsymbol{u}\rangle^{(\frac{1+\alpha-\gamma}{2})}_{t,D'_{\lambda/2}} + \frac{1}{\lambda^{\frac{3+\alpha-\gamma}{2}}}|\boldsymbol{u}|_{D'_{\lambda/2}}\Big\}\\
&\leqslant c\Big\{\frac{1}{\lambda^{\frac{1+\alpha-\gamma}{2}}}\big(|\boldsymbol{f}|^{(\frac{1+\alpha-\gamma}{2})}_{t,Q'_0} + |\nabla \boldsymbol{f}|_{Q'_0}(2\tau_0)^{\gamma/2}|\boldsymbol{u}|_{D'_0}\big)\\
&\qquad + \frac{1}{\lambda}\langle \boldsymbol{u}\rangle^{(\frac{1+\alpha-\gamma}{2})}_{t,D'_{\lambda/2}} + \frac{1}{\lambda^{\frac{3+\alpha-\gamma}{2}}}|\boldsymbol{u}|_{D'_{\lambda/2}}\Big\}.
\end{aligned}$$

The norm $|\mathbb{G}|^{(\gamma,0)}_{D'_0}$ is estimated in a similar way.

Finally, we conclude that

$$\begin{aligned}
N_{(t_0-2\tau_0+\lambda,t_0)}[\boldsymbol{u},q] &\leqslant c_2(1+\delta)\Big\{\frac{1}{\lambda^{\frac{\alpha}{2}}}|\boldsymbol{f}|^{(\alpha)}_{x,Q'_0} + \frac{1}{\lambda^{\frac{1+\alpha-\gamma}{2}}}\big(|\boldsymbol{f}|^{(\frac{1+\alpha-\gamma}{2})}_{t,Q'_0} + (2\tau_0)^{\gamma/2}|\boldsymbol{u}|_{D'_0}\big)\\
&+ \frac{1}{\lambda}\big(\langle \boldsymbol{u}\rangle^{(\alpha,\frac{\alpha}{2})}_{D'_{\lambda/2}} + \langle \boldsymbol{u}\rangle^{(\frac{1+\alpha-\gamma}{2})}_{t,D'_{\lambda/2}}\big) + \big(\frac{1}{\lambda^{1+\frac{\alpha}{2}}} + \frac{1}{\lambda^{\frac{3+\alpha-\gamma}{2}}}\big)|\boldsymbol{u}|_{D'_{\lambda/2}}\Big\}.
\end{aligned} \tag{6.4.19}$$

We take now $\theta_1 = (\varepsilon\lambda)^{1/\gamma}$ and $\theta_2 = (\varepsilon\lambda)^{1/2}$ in estimates (6.4.6) and (6.4.10), respectively. We estimate $|\boldsymbol{u}|_{D'_{\lambda/2}}$ by inequality (6.4.9). As a result, we obtain

$$\begin{aligned}
N_{(t_0-2\tau_0+\lambda,t_0)}[\boldsymbol{u},q] &\leqslant c_3(\delta)\Big\{\varepsilon N_{(t_0-2\tau_0+\frac{\lambda}{2},t_0)}[\boldsymbol{u},q] + \frac{1}{\lambda^{\frac{1+\alpha-\gamma}{2}}}|\boldsymbol{f}|^{(\frac{1+\alpha-\gamma}{2})}_{t,Q'_0}\\
&+ \frac{1}{\lambda^{\frac{\alpha}{2}}}|\boldsymbol{f}|^{(\alpha)}_{x,Q'_0} + c(\varepsilon)\lambda^{-\varkappa}\int\limits_{t_0-2\tau_0}^{t_0}\|\boldsymbol{u}(\cdot,\tau)\|_\Omega \mathrm{d}\tau\Big\},
\end{aligned} \tag{6.4.20}$$

here $\varkappa = \max\{\frac{9}{2\gamma}+\frac{\alpha}{\gamma}, \frac{11}{4}+\frac{\alpha}{2}, 1+\frac{\alpha}{2}, \frac{3+\alpha-\gamma}{2}\} = \frac{9}{2\gamma}+\frac{\alpha}{\gamma}$.

We introduce the function $\Phi(\lambda) = \lambda^{\varkappa}N_{(t_0-2\tau_0+\lambda,t_0)}[\boldsymbol{u},q]$. Then we can rewrite (6.4.20) as follows:

$$\Phi(\lambda) \leqslant c_4\varepsilon\Phi(\lambda/2) + K, \tag{6.4.21}$$

where $c_4 = c_3(\delta)2^{\varkappa}$ and

$$K = c_3(\delta)\Big\{|\boldsymbol{f}|^{(\alpha,\frac{1+\alpha-\gamma}{2})}_{Q'_0} + c(\varepsilon)\int\limits_{t_0-2\tau_0}^{t_0} \|\boldsymbol{u}(\cdot,\tau)\|_\Omega \mathrm{d}\tau\Big\}.$$

We set $\varepsilon = \frac{1}{2c_4}$ in (6.4.21). By the iterations with $\lambda/2,\ldots,\lambda/2^k$, we deduce from inequality (6.4.21) as $k \to \infty$ that

$$\Phi(\lambda) \leqslant 2K.$$

This estimate with $\lambda = \tau_0$ implies (6.4.12). □

Now we can prove Theorem 6.1.1.

Proof By Theorem 6.4.1, there exists a solution $(\boldsymbol{v}, p)$ in the interval $(0, T_0]$. We can choose ε in condition (6.4.1) so small that $T_0 > 1$, for example. Moreover, the norm of the solution will satisfy the inequality

$$N_{(0,T_0)}[\boldsymbol{v}, p] \leqslant \mu$$

with some $\mu > 0$. Then due to Proposition 6.4.2, there exists $\tau_0 < T_0/2$ such that (6.4.18) is satisfied and estimate (6.4.12) holds for $(\boldsymbol{v}, p)$. Together with (6.4.5), it implies that

$$\begin{aligned} N_{(T_0-\tau_0,T_0)}[\boldsymbol{v},p] &\leqslant c(\delta,\tau_0)\Big\{|\boldsymbol{f}|^{(\alpha,\frac{1+\alpha-\gamma}{2})}_{Q(T_0-2\tau_1,T_0)} + \int\limits_{T_0-2\tau_0}^{T_0} \|\boldsymbol{u}(\cdot,\tau)\|_\Omega \mathrm{d}\tau\Big\} \\ &\leqslant c_5(\delta,\tau_0)\mathrm{e}^{-b(T_0-2\tau_0)}\Big\{|\mathrm{e}^{b\tau}\boldsymbol{f}|^{(\alpha,\frac{1+\alpha-\gamma}{2})}_{Q_\infty} + \|\boldsymbol{v}_0\|_\Omega + c_1\int\limits_0^\infty \mathrm{e}^{b\tau}\|\boldsymbol{f}(\cdot,\tau)\|_\Omega \mathrm{d}\tau\Big\} \\ &\leqslant c_6(\delta,\tau_0)\mathrm{e}^{-bT_0}\big(1 + c_1 + |\Omega|^{1/2}\big)\varepsilon, \end{aligned} \tag{6.4.22}$$

where $|\Omega|$ is the measure of Ω.

We take into account the inequality

$$|\boldsymbol{v}(\cdot,T_0)|^{(2+\alpha)}_{\cup\Omega^\pm_{T_0}} \leqslant \mu,$$

as well as the estimate

$$\begin{aligned} \|\boldsymbol{u}(\cdot,T_0)\|_\Omega = \|\boldsymbol{v}(\cdot,T_0)\|_\Omega &\leqslant \mathrm{e}^{-bT_0}\Big(\|\boldsymbol{v}_0\|_\Omega + c_1\int\limits_0^{T_0} \mathrm{e}^{b\tau}\|\boldsymbol{f}(\cdot,\tau)\|_\Omega \mathrm{d}\tau\Big) \\ &\leqslant \big(|\Omega|^{1/2} + c_1\big)\varepsilon \end{aligned} \tag{6.4.23}$$

which follows from (6.4.4). We apply again Theorem 6.4.1, and we get a solution on the interval $(T_0, T_0 + T_1]$, corresponding to the initial velocity $\boldsymbol{v}(\cdot, T_0)$, such that

$$N_{(T_0,T_0+T_1)}[\boldsymbol{v}, p] \leqslant \mu_1.$$

Then by Proposition 6.4.2, there exists $\tau_1 < T_1/2$ such that (6.4.18)holds with τ_1, μ_1. Consequently, similarly to (6.4.22), in view of (6.4.23), we deduce

$$\begin{aligned} N_{(T_0+T_1-\tau_1,T_0+T_1)}[\boldsymbol{v}, p] &\leqslant c_5(\delta, \tau_1)\mathrm{e}^{-b(T_1-2\tau_1)} \\ &\times \Big\{ \mathrm{e}^{-bT_0} |\mathrm{e}^{b\tau} \boldsymbol{f}|^{(\alpha, \frac{1+\alpha-\gamma}{2})}_{Q(T_0+T_1-2\tau_1,T_0+T_1)} + \|\boldsymbol{v}(\cdot, T_0)\|_\Omega + c_1 \int\limits_{T_0}^{\infty} \mathrm{e}^{b\tau} \|\boldsymbol{f}(\cdot, \tau)\|_\Omega \mathrm{d}\tau \Big\} \\ &\leqslant c_6(\delta, \tau_1)\mathrm{e}^{-bT_1} \big(1 + 2c_1 + |\Omega|^{1/2}\big)\varepsilon. \end{aligned} \tag{6.4.24}$$

We take in (6.1.4) ε so small that $c_6(\delta, \tau_1)\big(1 + 2c_1 + |\Omega|^{1/2}\big)\varepsilon < \mu$. Hence, we obtain again that

$$|\boldsymbol{u}(\cdot, T_0 + T_1)|^{(2+\alpha)}_{\cup\Omega^\pm_{T_0+T_1}} \leqslant \mu \mathrm{e}^{-bT_1} \leqslant \mu$$

and

$$\begin{aligned} \|\boldsymbol{v}(\cdot, T_0 + T_1)\|_\Omega &\leqslant \mathrm{e}^{-b(T_0+T_1)} \Big(\|\boldsymbol{v}_0\|_\Omega + c_1 \int\limits_0^{T_0+T_1} \|\mathrm{e}^{b\tau} \boldsymbol{f}(\cdot, \tau)\|_\Omega \mathrm{d}\tau \Big) \\ &\leqslant \big(|\Omega|^{1/2} + c_1\big)\varepsilon. \end{aligned}$$

Since the norms of the initial data do not increase, a solution to problem (1.1.1), (1.0.3) exists on $(T_0 + T_1, T_0 + 2T_1]$

$$N_{(T_0+T_1,T_0+2T_1)}[\boldsymbol{v}, p] \leqslant \mu_1,$$

and inequality (6.4.24) holds for $N_{(T_0+2T_1-\tau_1,T_0+2T_1)}$. Consequently, $|\boldsymbol{u}(\cdot, T_0 + 2T_1)|^{(2+\alpha)}_{\cup\Omega^\pm_{T_0+2T_1}} \leqslant \mu \mathrm{e}^{-bT_1} \leqslant \mu$. In this way, the solution of (1.1.1), (1.0.3) can be extended with respect to t as far as one likes.

We prove the uniqueness of a solution in the same way as it has been done in the proof of local existence theorem in Chap. 5. □

Remark 6.4.1 *In order to estimate the distance between the solid boundary and the interface for* $\sigma = 0$*, we need to evaluate the integral* $I \equiv \int_0^\infty |\boldsymbol{u}(\cdot, t)|_{\Omega_0^+} \,\mathrm{d}t$. *By virtue of* (6.1.5)

$$|\boldsymbol{u}(\cdot, t_0)|_{\Omega_0^+} \leqslant c(\delta, \tau_0)\mathrm{e}^{-bt_0}\varepsilon. \tag{6.4.25}$$

Taking inequality (6.4.2) *into account and integrating* (6.4.25) *with respect to* $t_0 > T_0/2$, *we arrive at estimate*

$$I \leqslant \frac{T_0}{2}|\boldsymbol{u}|_{Q^+_{T_0/2}} + \int_{T_0/2}^{\infty} |\boldsymbol{u}(\cdot, t)|_{\Omega_0^+}\, \mathrm{d}t \leqslant c(\varepsilon, T_0)\varepsilon\frac{T_0}{2} + c(\delta, \tau_0)\frac{1}{b}\varepsilon \leqslant c_7\varepsilon.$$

Thus, if the distance between the surfaces is great than $c_7\varepsilon$, Γ_t *will never intersect the surface* Σ.

Chapter 7
Global Solvability of the Problem Including Capillary Forces: Case of the Hölder Spaces

Abstract Under the assumption of a sufficiently small initial velocity vector field and a small deviation of the initial surface of the drop from a sphere, the unique solvability of the nonlinear problem is proved in the anisotropic Hölder spaces for all $t > 0$. The arguments are based on the local-in-time existence theorem and the Hölder estimates of the solution. We follow the scheme proposed by V. A. Solonnikov for proving the existence of a global solution to the problem governing the motion of a single fluid bounded by a free surface [88].

7.1 Setting of the Problem, Statement of the Main Result

Global classical solvability was obtained of the problem with the homogeneous equations in [30] and in the case of inhomogeneous equations with exponentially decreasing mass forces with respect to time in [24].

Now we state the problem.

At the initial moment, let both fluids fill bounded volumes as in Chap. 6.

For every $t > 0$, we need to find $\Gamma_t = \partial\Omega_t^+$, velocity vector field $\boldsymbol{v}(x,t) = (v_1, v_2, v_3)$, and the function p, which is the deviation from the hydrostatic pressure P_0, which satisfy the following initial–boundary value problem:

$$\begin{aligned}
&\mathcal{D}_t\boldsymbol{v} + (\boldsymbol{v}\cdot\nabla)\boldsymbol{v} - \nu^\pm\nabla^2\boldsymbol{v} + \frac{1}{\rho^\pm}\nabla p = \boldsymbol{f}, \quad \nabla\cdot\boldsymbol{v} = 0 \quad \text{in } \Omega_t^-\cup\Omega_t^+,\ t>0,\\
&\boldsymbol{v}|_{t=0} = \boldsymbol{v}_0 \quad \text{in} \quad \Omega_0^-\cup\Omega_0^+, \quad \boldsymbol{v}|_\Sigma = 0, \qquad (7.1.1)\\
&[\boldsymbol{v}]|_{\Gamma_t} = 0, \quad [\mathbb{T}\boldsymbol{n}]|_{\Gamma_t} = \sigma H\boldsymbol{n}.
\end{aligned}$$

Here we conserve the previous notation, moreover, surface tension coefficient $\sigma > 0$, $\Omega \equiv \Omega_0^+ \cup \overline{\Omega_0^-}$, $Q_T \equiv \Omega\times(0,T)$.

© The Author(s), under exclusive license to Springer Nature Switzerland AG 2021

I. V. Denisova, V. A. Solonnikov, *Motion of a Drop in an Incompressible Fluid*, Advances in Mathematical Fluid Mechanics,
https://doi.org/10.1007/978-3-030-70053-9_7

We exclude mass transportation across the interface Γ_t, so the radius-vector $\boldsymbol{x}(\xi, t)$, $x \in \Gamma_t$, is a solution to the Cauchy problem (1.0.3). This condition is equivalent to the equality

$$V_{\boldsymbol{n}} = \boldsymbol{v} \cdot \boldsymbol{n}|_{\Gamma_t}, \tag{7.1.2}$$

where $V_{\boldsymbol{n}}$ is the speed of the interface Γ_t in the direction of its outward normal.

By using a uniform exponential estimate of the solution, we are going to show that the solution tends to an equilibrium state: the fluid velocity tends to zero, the pressure does to a step function, and drop shape goes to a ball of the certain radius.

Thus, we assume that the drop Ω_0^+ is close to the ball B_{R_0} whose volume is equal to drop volume: $|B_{R_0}| = |\Omega_0^+|$. Without loss of generality, we place the center of this ball into the origin, which coincides with the gravity center of the drop at the initial moment. Moreover, the interface between the liquids is defined as the normal disturbance of the sphere $S_{R_0}(0)$ (Fig. 7.1).

For the convenience of estimating the solution, we introduce a new pressure function: $p_1 = p$ in Ω_t^+ and $p_1 = p + \sigma\frac{2}{R_0}$ in Ω_t^-. Then only the last-named boundary condition in (7.1.1) changes:

$$\begin{aligned} &\mathcal{D}_t \boldsymbol{v} + (\boldsymbol{v} \cdot \nabla)\boldsymbol{v} - \nu^\pm \nabla^2 \boldsymbol{v} + \frac{1}{\rho^\pm}\nabla p_1 = \boldsymbol{f}, \ \nabla \cdot \boldsymbol{v} = 0 \ \text{ in } \ \Omega_t^- \cup \Omega_t^+, \ t > 0, \\ &\boldsymbol{v}|_{t=0} = \boldsymbol{v}_0 \quad \text{in} \quad \Omega_0^- \cup \Omega_0^+, \quad \boldsymbol{v}|_\Sigma = 0, \\ &[\boldsymbol{v}]|_{\Gamma_t} = 0, \quad [\mathbb{T}(\boldsymbol{v}, p_1)\boldsymbol{n}]|_{\Gamma_t} = \sigma\Big(H + \frac{2}{R_0}\Big)\boldsymbol{n}. \end{aligned} \tag{7.1.3}$$

We pass in (7.1.3) from Eulerian to Lagrangian coordinates by formula (1.1.2).

As a result of this transformation and of projecting the last-named boundary condition in (7.1.3) onto the tangent planes first to Γ_t, then to Γ,

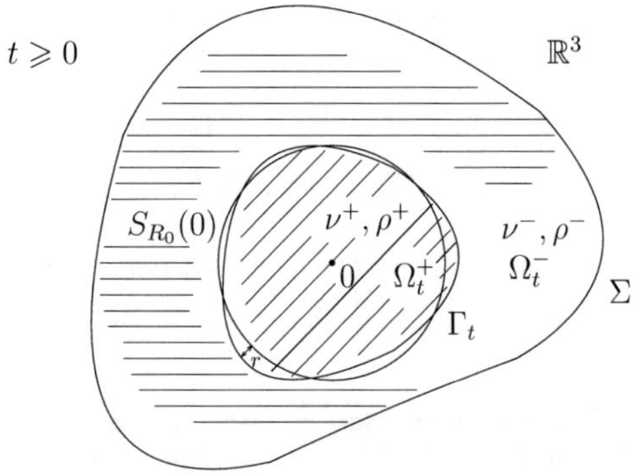

Fig. 7.1 Motion of a drop close to a ball

we arrive at a problem for $\boldsymbol{u}$, $q = p_1(X_{\boldsymbol{u}}, t)$ with the given interface $\Gamma \equiv \Gamma_0$. If the angle between $\boldsymbol{n}$ and the outward normal $\boldsymbol{n}_0$ to Γ is acute, the system obtained is equivalent to the following one:

$$\begin{aligned}
&\mathcal{D}_t \boldsymbol{u} - \nu^{\pm}\nabla_{\boldsymbol{u}}^2 \boldsymbol{u} + \frac{1}{\rho^{\pm}}\nabla_{\boldsymbol{u}} q = \boldsymbol{f}(X_{\boldsymbol{u}}, t), \quad \nabla_{\boldsymbol{u}} \cdot \boldsymbol{u} = 0 \text{ in } Q_T^{\pm} = \Omega_0^{\pm} \times (0,T),\\
&\boldsymbol{u}|_{t=0} = \boldsymbol{v}_0 \text{ in } \Omega_0^- \cup \Omega_0^+, \quad \boldsymbol{u}|_{\Upsilon_T} = 0 \quad (\Upsilon_T = \Sigma \times (0,T)), \qquad (7.1.4)\\
&[\boldsymbol{u}]|_{G_T} = 0, \quad [\mu^{\pm}\Pi_0\Pi\mathbb{S}_{\boldsymbol{u}}(\boldsymbol{u})\boldsymbol{n}]|_{G_T} = 0 \quad (G_T \equiv \Gamma \times (0,T)),\\
&[\boldsymbol{n}_0 \cdot \mathbb{T}_{\boldsymbol{u}}(\boldsymbol{u}, q)\boldsymbol{n}]|_{G_T} = \sigma\Big(H(X_{\boldsymbol{u}}) + \frac{2}{R_0}\Big)\boldsymbol{n}_0 \cdot \boldsymbol{n}.
\end{aligned}$$

In the case where the surface Σ and the term $\sigma\frac{2}{R_0}\boldsymbol{n}_0\cdot\boldsymbol{n}$ in the last boundary condition in problem (7.1.4) are absent, an existence and uniqueness theorem for the system has been proved on a finite time interval whose size is determined by the norms of $\boldsymbol{v}_0$ and $\boldsymbol{f}$, as well as the curvature of Γ (Theorem 5.1.1).

This result was obtained in Hölder spaces with a power weight at infinity. Since the spaces mentioned are equivalent to the ordinary Hölder ones in bounded domains, the solvability of Problem (7.1.4) can be proved in the same way as the solvability of the problem in an infinite domain. To obtain the estimates near the outer boundary, one needs to apply the existence theorem to the Dirichlet problem for the Stokes system in a half-space [76]. The term $\sigma\frac{2}{R_0}\boldsymbol{n}_0 \cdot \boldsymbol{n}$ in the last boundary condition is weak with respect to the left-hand side of the equality, it just changes the curvature of the initial interface on the value of ball curvature with the radius R_0. As we will see below, this term will be included in estimates. Moreover, the norm of the sum of it and the curvature of the initial interface will be small.

Now we state the theorem obtained.

Theorem 7.1.1 (Local classical solvability in a bounded domain) *We assume that* $\Gamma \in C^{3+\alpha}$, $\Sigma \in \boldsymbol{C}^{2+\alpha}$, $\boldsymbol{f}, D_x\boldsymbol{f} \in \boldsymbol{C}^{\alpha,\frac{1+\alpha-\gamma}{2}}(Q_T)$ $\boldsymbol{v}_0 \in \boldsymbol{C}^{2+\alpha}(\Omega_0^- \cup \Omega_0^+)$, $\sigma \in C^{3+\alpha}(\mathbb{R}_+)$, $\sigma > 0$, *with some* $\alpha, \gamma \in (0,1)$, $\gamma < \alpha$. *In addition, let the compatibility conditions be satisfied*:

$$\begin{aligned}
&\nabla \cdot \boldsymbol{v}_0 = 0, \quad \boldsymbol{v}_0|_{\Sigma} = 0, \quad [\boldsymbol{v}_0]|_{\Gamma} = 0,\\
&[\mu^{\pm}\Pi_0\mathbb{S}(\boldsymbol{v}_0)\boldsymbol{n}_0]\big|_{\Gamma} = 0, \qquad [\Pi_0(\nu^{\pm}\nabla^2\boldsymbol{v}_0 - \frac{1}{\rho^{\pm}}\nabla q_0)]\big|_{\Gamma} = 0, \qquad (7.1.5)\\
&\Pi_{\Sigma}\big(\boldsymbol{f}(\xi,0) - \frac{1}{\rho^-}\nabla q_0(\xi) + \nu^-\nabla^2\boldsymbol{v}_0(\xi)\big)\Big|_{\xi\in\Sigma} = 0,
\end{aligned}$$

where $q_0(\xi) \equiv q(\xi, 0)$ *is a solution to diffraction problem*

$$\frac{1}{\rho^{\pm}}\nabla^2 q_0(\xi) = \nabla\cdot(\boldsymbol{f}(\xi,0) - \mathcal{D}_t\mathbb{B}^*(\boldsymbol{v}_0)\boldsymbol{v}_0(\xi)), \qquad \xi\in\Omega_0^-\cup\Omega_0^+,$$

$$[q_0]|_\Gamma = \Big[2\mu^{\pm}\frac{\partial \boldsymbol{v}_0}{\partial \boldsymbol{n}_0}\cdot\boldsymbol{n}_0\Big]\Big|_\Gamma - \sigma\Big(H_0(\xi)+\frac{2}{R_0}\Big), \qquad \xi\in\Gamma,$$

$$\Big[\frac{1}{\rho^{\pm}}\frac{\partial q_0}{\partial \boldsymbol{n}_0}\Big]\Big|_\Gamma = [\nu^{\pm}\boldsymbol{n}_0\cdot\nabla^2\boldsymbol{v}_0]|_\Gamma, \quad \frac{1}{\rho^-}\frac{\partial q_0}{\partial \boldsymbol{n}_\Sigma}\Big|_\Sigma = \boldsymbol{n}_\Sigma\cdot\big(\nu^-\nabla^2\boldsymbol{v}_0+\boldsymbol{f}|_{t=0}\big)|_\Sigma.$$

Here $H_0(\xi) = \boldsymbol{n}_0\cdot\Delta(0)\boldsymbol{\xi}|_\Gamma$ *is twice the mean curvature of* Γ*;* $\mathbb{B}=\mathbb{A}-\mathbb{I}$*, the matrix* $\mathbb{B}^*$ *is the transpose of* $\mathbb{B}$*,* $\boldsymbol{n}_\Sigma$ *is the outward normal to* Σ*,* $\Pi_\Sigma\boldsymbol{\omega}\equiv \boldsymbol{\omega}-\boldsymbol{n}_\Sigma(\boldsymbol{n}_\Sigma\cdot\boldsymbol{\omega})$.

Then for any $T\in(0,\infty)$ *there exists a positive number* $\varepsilon(T)$ *such that problem (7.1.4) has a unique solution* $(\boldsymbol{u},q)$ *with the differential properties:* $\boldsymbol{u}\in \boldsymbol{C}^{2+\alpha,1+\alpha/2}(D_{\Sigma T})$, $q\in C^{(\gamma,1+\alpha)}(D_{\Sigma T})$, $\nabla q\in \boldsymbol{C}^{\alpha,\alpha/2}(D_{\Sigma T})$, *provided that*

$$|\boldsymbol{f}|_{Q_T}^{(\alpha,\frac{1+\alpha-\gamma}{2})} + |\boldsymbol{v}_0|_{\cup\Omega_0^{\pm}}^{(2+\alpha)} + \sigma\Big|H_0+\frac{2}{R_0}\Big|_\Gamma^{(1+\alpha)} \leqslant \varepsilon(T). \tag{7.1.6}$$

In addition, the inequality

$$|\boldsymbol{u}|_{D_{\Sigma T}}^{(2+\alpha,1+\alpha/2)} + |\nabla q|_{D_{\Sigma T}}^{(\alpha,\alpha/2)} + |q|_{D_{\Sigma T}}^{(\gamma,1+\alpha)}$$
$$\leqslant c\big(|\boldsymbol{v}_0|_{\cup\Omega_0^{\pm}}^{(2+\alpha)}, |\nabla\boldsymbol{f}|_{Q_\infty}^{(\alpha,\frac{1+\alpha-\gamma}{2})}\big)\Big\{|\boldsymbol{f}|_{Q_T}^{(\alpha,\frac{1+\alpha-\gamma}{2})} + |\boldsymbol{v}_0|_{\cup\Omega_0^{\pm}}^{(2+\alpha)} + \sigma\Big|H_0+\frac{2}{R_0}\Big|_\Gamma^{(1+\alpha)}\Big\} \tag{7.1.7}$$

holds. Moreover, pressure function q *is unique up to a bounded time function. The interface* Γ_t *belongs to the class* $C^{3+\alpha}$.

A result similar to Theorem 7.1.1 holds also in the Sobolev–Slobodetskiĭ spaces of functions. It is not difficult to obtain it by analogy with the proof of the local solvability of the problem in an unbounded domain (see [13, 27] or [17]), as in the case of the Hölder spaces. We treat this material in Chap. 10. Global-in-time solvability of the problem in the Sobolev–Slobodetskiĭ spaces was first studied by N. Tanaka [96], who considered nonhomogeneous fluids. However, it should be noted that the exposition of the proof in [96] is not convincing. The reference to the local solvability of the problem is given as a particular case of the thermocapillary convection problem [97], where the proofs are also sketchy; necessary additional weak pressure norms being not considered. One needs to estimate the limit position of the barycenter of the inner fluid in order to exclude the possibility that the interface and the solid boundary intersect. In the case of the Hölder spaces, it will be done via an exponential estimate of a solution in the terms of initial data. We obtain a similar estimate in the Sobolev classes of functions in Chap. 12.

The following theorem establishes the existence of a global solution of system (7.1.3), (1.0.3) [30].

Theorem 7.1.2 *Let the assumptions of Theorem* 7.1.1 *hold with* $q_0 = p_1(x,0)$ *and let, at* $t=0$, $\Gamma \in C^{3+\alpha}$ *be given by the equation*

$$|x| = R\Big(\frac{x}{|x|}, 0\Big)$$

on the unit sphere S_1. *We suppose additionally that the initial data are small enough, i. e.,*

$$|e^{b_1 t}\boldsymbol{f}|_{Q_\infty}^{(\alpha,\frac{1+\alpha-\gamma}{2})} + \|e^{b_1 t}\boldsymbol{f}\|_{Q_\infty} + |\mathbf{v}_0|_{\cup\Omega_0^\pm}^{(2+\alpha)} + |r_0|_{S_1}^{(3+\alpha)} \leqslant \varepsilon \ll 1, \qquad (7.1.8)$$

where $b_1 > 0$, $r_0(x/|x|) = R(x/|x|,0) - R_0$, R_0 *being the radius of the ball* $B_{R_0} : |\Omega_0^+| = 4\pi R_0^3/3$.

Then problem (7.1.3), (1.0.3) *is uniquely solvable on the whole positive half-axis* $t > 0$, *and the solution* $(\boldsymbol{v}, p_1)$ *possesses the following properties:* $\boldsymbol{v} \in \boldsymbol{C}^{2+\alpha,1+\alpha/2}$, $p_1 \in C^{(\gamma,1+\alpha)}$, $\nabla p_1 \in \boldsymbol{C}^{\alpha,\alpha/2}$, *the boundary* Γ_t *is given for every* t *by a function* $R(\cdot,t)$ *of the class* $C^{3+\alpha}$:

$$|x - h(t)| = R\Big(\frac{x-h}{|x-h|}, t\Big)$$

(*where* $h(t)$ *is a position of the barycenter of* Ω_t^+ *at the moment* t), Γ_t *tends to the sphere of radius* R_0 *centered at a certain point* h_∞,, *and the pressure is defined up to a bounded function of time. It means that for any* $t_0 \in (0,\infty)$, *the solution* $(\boldsymbol{u}, q)$ *and its derivatives with respect to the Lagrangian coordinates belong to the respective Hölder spaces over* $D_{(t_0,t_0+\tau)} \equiv \cup Q^\pm_{(t_0,t_0+\tau)}$ *for a sufficiently small time interval* $(t_0, t_0+\tau)$. *Moreover, the estimate*

$$|\boldsymbol{u}|_{D_{(t_0,t_0+\tau)}}^{(2+\alpha,1+\alpha/2)} + |\nabla q|_{D_{(t_0,t_0+\tau)}}^{(\alpha,\alpha/2)} + |q|_{D_{(t_0,t_0+\tau)}}^{(\gamma,1+\alpha)} + \sup_{t\in(t_0,t_0+\tau)} |r(\cdot,t)|_{S_1}^{(3+\alpha)}$$
$$\leqslant c\Big(|\boldsymbol{v}_0|_{\cup\Omega_0^\pm}^{(2+\alpha)}, |\nabla \boldsymbol{f}|_{Q_\infty}^{(\alpha,\frac{1+\alpha-\gamma}{2})}\Big) e^{-bt_0}\Big\{|e^{bt}\boldsymbol{f}|_{Q_\infty}^{(\alpha,\frac{1+\alpha-\gamma}{2})} + \|e^{bt}\boldsymbol{f}\|_{Q_\infty}$$
$$+ |\boldsymbol{v}_0|_{\cup\Omega_0^\pm}^{(2+\alpha)} + |r_0|_{S_1}^{(3+\alpha)}\Big\} \qquad (7.1.9)$$

holds, where $0 < b \leqslant b_1$, $r(\omega,t) = R(\omega,t) - R_0$, $\omega \in S_1$.

One can conclude from this theorem that the trivial solution is unique if the initial velocity vanishes and the initial interface coincides with a sphere. The stability of this solution takes place in the sense that for a small deviation of initial data from zero, the solution is close to zero. However, the center of the limiting sphere $S_{R_0}(h_\infty)$ may be displaced with respect to the initial barycenter of Ω_0^+ no matter how small an initial velocity $\boldsymbol{v}_0$ and mass forces $\boldsymbol{f}$ are. This displacement is estimated in inequality (7.4.23) at the end of the paper. There we also give a necessary estimate from below of the initial distance between the outer boundary and fluid interface.

7.2 Energy Estimate of the Solution

The aim of this section is to prove an exponential estimate for the solution of nonlinear problem (7.1.3), (1.0.3) in L_2 by using the notion of generalized energy $\mathcal{E}$ introduced in [58, 87].

Let us assume that we have found the solution in the interval $[0, T]$. (Its existence is guaranteed by Theorem 7.1.1.) Then we know also the trajectory of the barycenter of the drop Ω_t^+. Its coordinates are given by the formula:

$$h_i(t) = \frac{1}{|\Omega_t^+|}\int_{\Omega_t^+} x_i\,dx = \frac{1}{|\Omega_t^+|}\int_0^t\int_{\Omega_t^+} v_i(x,\tau)\,dx\,d\tau, \quad i = 1, 2, 3. \tag{7.2.1}$$

The incompressibility of the fluids implies that the domains $\Omega_t^\pm$ conserve their volumes for $t > 0$:

$$\int_{\Omega_t^+} dx = \int_{B_{R_0}} dx.$$

Hence, passing to the spherical coordinates and integrating with respect to radius, we deduce

$$\frac{1}{3}\int_{S_1} (R^3 - R_0^3)\,d\omega = 0,$$

whence it follows the equality

$$\int_{S_1} r\,d\omega = -\frac{1}{R_0}\int_{S_1} r^2\,d\omega - \frac{1}{3R_0^2}\int_{S_1} r^3\,d\omega \tag{7.2.2}$$

for $r(\omega, t) = R(\omega, t) - R_0$, deviation function of the surface Γ_t from the sphere $S_{R_0}(t) \equiv S_{R_0}(h(t)) = \{|x - h(t)| = R_0\}$. We are going to use this equality later.

We assume (without loss of generality) that the barycenter of the inner liquid coincides with the origin at the initial moment of time: $\boldsymbol{h}(0) = 0$. Moreover, let Γ be defined by the equation

$$x = y + \boldsymbol{N}(y) r_0(\frac{y}{|y|}), \quad y \in S_{R_0}. \tag{7.2.3}$$

(By $\boldsymbol{N}(y)$ we mean the outward normal to $S_{R_0} = \{|y| = R_0\}$, i. e., $\boldsymbol{N}(y) = \boldsymbol{y}/|y|$.) We also assume that for $0 < t \leqslant T$ the surface Γ_t can be defined as

$$x = y + \boldsymbol{N}(y) r(\frac{y}{|y|}, t) + \boldsymbol{h}(t), \quad y \in S_{R_0}. \tag{7.2.4}$$

Equations (7.2.3) and (7.2.4) are equivalent to the relations

$$|x| = R_0 + r_0\Big(\frac{x}{|x|}\Big) \quad \text{and} \quad |x - h(t)| = R_0 + r\Big(\frac{x-h}{|x-h|}, t\Big),$$

respectively, $r(\omega, 0) = r_0(\omega)$, $\omega \in S_1$.

By the kinematic condition (7.1.2), we have

$$\mathcal{D}_t x \cdot \boldsymbol{n} = \boldsymbol{v} \cdot \boldsymbol{n}|_{\Gamma_t}.$$

Consequently,

$$n_{0N}(y)\mathcal{D}_t r|_{t=0} = \boldsymbol{v}_0 \cdot \boldsymbol{n}_0 - \dot{\boldsymbol{h}}(0) \cdot \boldsymbol{n}_0, \quad n_N(y)\mathcal{D}_t r = \boldsymbol{v} \cdot \boldsymbol{n} - \dot{\boldsymbol{h}} \cdot \boldsymbol{n}, \quad y \in S_{R_0}, \tag{7.2.5}$$

where $n_{0N} = \boldsymbol{n}_0 \cdot \boldsymbol{N}$ is the radial component of $\boldsymbol{n}_0$ in the coordinate system with the origin at 0, and $n_N = \boldsymbol{n} \cdot \boldsymbol{N}$ is the radial component of $\boldsymbol{n}$ in the coordinate system with the origin at h. We recall that

$$\dot{\boldsymbol{h}}(t) = |\Omega_t^+|^{-1} \int_{\Omega_t^+} \boldsymbol{v}(x,t)\,\mathrm{d}x. \tag{7.2.6}$$

We note that in the coordinates $\{y\}$, the barycenter of Ω_t^+ always coincides with the origin, the center of the ball B_{R_0}, so that (7.2.1) implies the equality

$$\frac{3}{16\pi R_0^3} \int_{S_1} (R^4 - R_0^4)\omega_i\,\mathrm{d}\omega = 0, \tag{7.2.7}$$

where $\omega_i = y_i/|y|$.

Let again $\Omega = \Omega_0^- \cup \overline{\Omega_0^+} \equiv \Omega_t^- \cup \overline{\Omega_t^+}$, $Q_T = \Omega \times (0,T)$, $T \in (0,\infty)$, and let the norm $\|\cdot\|_{L_2(\Omega)}$ be denoted by $\|\cdot\|_\Omega$.

Proposition 7.2.1 *We assume that the classical solution of problem* (7.1.3), (1.0.3) *is defined in* $[0,T]$ *and* $\boldsymbol{v}_0$ *satisfies compatibility conditions* (7.1.5). *In addition, let* r *be such that*

$$|r(\omega,t)|_{S_1\times(0,T)} + |\nabla_{S_1} r(\omega,t)|_{S_1\times(0,T)} \leqslant \delta_1 R_0 \ll 1. \tag{7.2.8}$$

Then for arbitrary $t \in (0, T]$

$$\|\boldsymbol{v}(\cdot,t)\|_{\Omega}^2 + \|r(\cdot,t)\|^2_{W_2^1(S_1)} \leqslant c\mathrm{e}^{-2bt}\Big\{\|\mathrm{e}^{b\tau}\boldsymbol{f}\|^2_{Q_t} + \|\boldsymbol{v}_0\|^2_{\Omega} + \|r_0\|^2_{W_2^1(S_1)}\Big\}, \tag{7.2.9}$$

$$\int_0^T \big(\|\boldsymbol{v}(\cdot,\tau)\|_{\Omega} + \|r(\cdot,t)\|_{W_2^1(S_1)}\big)\,\mathrm{d}\tau \leqslant c\Big\{\|\mathrm{e}^{b\tau}\boldsymbol{f}\|^2_{Q_\infty} + \|\boldsymbol{v}_0\|_{\Omega} + \|r_0\|_{W_2^1(S_1)}\Big\}, \tag{7.2.10}$$

where b *and* c *are positive constants independent of* T.

Proof We multiply the first equation in (7.1.3) by $\rho^{\pm}\boldsymbol{v}$ and integrate by parts over $\Omega_t^- \cup \Omega_t^+$. Then

$$\frac{1}{2}\frac{\mathrm{d}}{\mathrm{d}t}\|\sqrt{\rho^{\pm}}\boldsymbol{v}\|^2_{\Omega} + \Big\|\sqrt{\frac{\mu^{\pm}}{2}}\mathbb{S}(\boldsymbol{v})\Big\|^2_{\cup\Omega_t^{\pm}} = \int_{\cup\Omega_t^{\pm}} \rho^{\pm}\boldsymbol{f}\cdot\boldsymbol{v}\,\mathrm{d}x + \sigma\int_{\Gamma_t}\Big(H + \frac{2}{R_0}\Big)\boldsymbol{n}\cdot\boldsymbol{v}\,\mathrm{d}\Gamma.$$

In view of (1.1.3) and since $\int_{\Gamma_t}\boldsymbol{v}\cdot\boldsymbol{n}\,\mathrm{d}\Gamma = 0$, surface integral in the right-hand side has the form: $\sigma\int_{\Gamma_t}\boldsymbol{v}\cdot\Delta(t)\boldsymbol{x}\,\mathrm{d}\Gamma$. In [78, p. 148] it is proved that this integral equals $-\sigma\frac{\mathrm{d}}{\mathrm{d}t}|\Gamma_t|$. Therefore we can write

$$\frac{\mathrm{d}}{\mathrm{d}t}\Big\{\frac{1}{2}\|\sqrt{\rho^{\pm}}\boldsymbol{v}\|^2_{\Omega} + \sigma\big(|\Gamma_t| - 4\pi R_0^2\big)\Big\} + \Big\|\sqrt{\frac{\mu^{\pm}}{2}}\mathbb{S}(\boldsymbol{v})\Big\|^2_{\cup\Omega_t^{\pm}} = \int_{\cup\Omega_t^{\pm}}\rho^{\pm}\boldsymbol{f}\cdot\boldsymbol{v}\,\mathrm{d}x.$$

Since $\boldsymbol{v}\big|_{\Sigma} = 0$, the vector field $\boldsymbol{v}$ satisfies the Korn inequality (6.4.3) in Ω. Consequently,

$$\frac{\mathrm{d}}{\mathrm{d}t}\Big\{\frac{1}{2}\|\sqrt{\rho^{\pm}}\boldsymbol{v}\|^2_{\Omega} + \sigma\big(|\Gamma_t| - 4\pi R_0^2\big)\Big\} + c_1\|\boldsymbol{v}\|^2_{W_2^1(\Omega)} \leqslant c\|\boldsymbol{f}\|^2_{\Omega}. \tag{7.2.11}$$

In order to obtain an exponential estimate for the generalized energy mentioned above, we should add to the left-hand side a term of the type $|\Gamma_t| - 4\pi R_0^2$. To this end, we construct an auxiliary solenoidal vector-valued function $\boldsymbol{W}(x,t)$, $x \in \Omega$.

Let $f_0(z)$ be a function defined on the sphere S_{R_0} and $\int_{S_{R_0}} f_0 dS_{R_0} = 0$. We define $\boldsymbol{W}_0(z)$ as a solenoidal vector field in the whole space $\mathbb{R}^3$ such that $\boldsymbol{W}_0\big|_{S_{R_0}} = \boldsymbol{N} f_0(y)$. We also assume that $\mathrm{supp}\boldsymbol{W}_0$ is contained in a ball $B_{R_0+a} = \{|y| \leqslant R_0 + a\}$, where $a > 0$.

In addition,

$$\|\boldsymbol{W}_0\|_{W_2^1(\mathbb{R}^3)} \leqslant c\|f_0\|_{W_2^{1/2}(S_{R_0})},$$

$$\|\boldsymbol{W}_0\|_{\mathbb{R}^3} \leqslant c\|f_0\|_{S_{R_0}}, \tag{7.2.12}$$

and if $f_0 = f_0(y, t)$, then

$$\|\mathcal{D}_t \boldsymbol{W}_0\|_{\mathbb{R}^3} \leqslant c\|\mathcal{D}_t f_0\|_{S_{R_0}}. \tag{7.2.13}$$

In particular, such $\boldsymbol{W}_0$ was constructed in [43, Chap. I].

Further, we set

$$f_0(y, t) = \widetilde{r}(y, t) \equiv r(\omega, t) - \overline{r}(t).$$

Here $\overline{r}(t) = \frac{1}{4\pi}\int_{S_1} r(\omega, t)\, d\omega$. For a and $\boldsymbol{h}$ sufficiently small, the vector field $\boldsymbol{W}_0$ vanishes near Σ for all $t \leqslant T$.

Now we make the following coordinate transformation in the domain Ω:

$$x = y + \boldsymbol{N}^*(y)\widetilde{r}^*(y, t) + \chi(y)\boldsymbol{h}(t) = e_r(y, t), \quad y \in \Omega, \tag{7.2.14}$$

where $\boldsymbol{N}^*$ is the extension of $\boldsymbol{N}$ into Ω, $\widetilde{r}^*(y, t)$ is the extension of $\widetilde{r}(y)$ with the support in the a-neighborhood of S_{R_0}; $\chi(y)$ is a smooth cutoff function equal to 1 in the ball B_{R_0+a} and to 0 near the surface Σ. The number $a > 0$ is chosen so that $\Omega_t^+ \Subset B_{R_0+a} \Subset \Omega$. An estimate for the value a will be given in the end of the chapter. It is easily seen that S_{R_0} goes over into Γ_t and $\widetilde{r}^* = 0$ near Σ.

We note that for small $r^*(y, t)$ and $\mathbf{h}$, transformation (7.2.14) is invertible, and the vector field $\boldsymbol{W}$ can be defined as

$$\boldsymbol{W}(x, t) = \frac{\mathcal{L}(y, t)}{L(y, t)}\boldsymbol{W}_0(y, t)\big|_{y=e_r^{-1}(x)},$$

where $\mathcal{L}$ is the Jacobian matrix of the transformation e_r:

$$\mathcal{L}(y, t) = \Big\{\frac{\partial e_r(y, t)}{\partial y}\Big\} = \Big\{\delta^i_j + \frac{\partial(N_i^*\widetilde{r}^*)}{\partial y_j}\Big\}_{i,j=1}^3,$$

and $L = det\mathcal{L}$. Let $\widehat{\mathcal{L}}$ be cofactor matrix of $\mathcal{L}$, i. e., $\widehat{\mathcal{L}}(y, t) = L\mathcal{L}^{-1}(y, t)$.

The vector field $\boldsymbol{W}$ is divergence-free as a function of x. Indeed,

$$\nabla_x \cdot \boldsymbol{W} = \mathcal{L}^{-1T}\nabla_y \cdot W = L^{-1}\nabla_y \cdot (\widehat{\mathcal{L}}\boldsymbol{W}) = L^{-1}\nabla_y \cdot \Big(\widehat{\mathcal{L}}\frac{\mathcal{L}}{L}\boldsymbol{W}_0\Big) = L^{-1}\nabla_y \cdot \boldsymbol{W}_0 = 0$$

in view of the identity $\nabla_y \cdot \widehat{\mathcal{L}} = 0$ valid for co-factor matrix of the Jacobian matrix of an arbitrary coordinate transformation. (Here the superscript ^{T} means transposition.) In addition, $\boldsymbol{W}(x, t) = 0$ for $x \in \Sigma$ and

$$\boldsymbol{W} \cdot \boldsymbol{n}\big|_{\Gamma_t} = \boldsymbol{W} \cdot \frac{\widehat{\mathcal{L}}^T\boldsymbol{N}}{|\widehat{\mathcal{L}}^T\boldsymbol{N}|} = \frac{\boldsymbol{W}_0 \cdot \boldsymbol{N}}{|\widehat{\mathcal{L}}^T\boldsymbol{N}|} = \frac{\widetilde{r}}{|\widehat{\mathcal{L}}^T\boldsymbol{N}|}\Big|_{y=e_r^{-1}(x)}, \qquad \boldsymbol{n} = \frac{\widehat{\mathcal{L}}^T\boldsymbol{N}}{|\widehat{\mathcal{L}}^T\boldsymbol{N}|}. \tag{7.2.15}$$

From (7.2.12), (7.2.13), it follows that

$$\|\boldsymbol{W}(\cdot,t)\|_{W_2^1(\Omega)} \leqslant c\|\boldsymbol{W}_0(\cdot,t)\|_{W_2^1(B_{R_0+a})} \leqslant c\|r(\cdot,t)\|_{W_2^{1/2}(S_1)}, \quad (7.2.16)$$
$$\|\boldsymbol{W}(\cdot,t)\|_{\Omega} \leqslant c\|r(\cdot,t)\|_{S_1},$$

and since

$$\frac{\mathrm{d}}{\mathrm{d}t}\boldsymbol{W}(e_r(y,t),t) = \mathcal{D}_t\boldsymbol{W}(x,t)\big|_{x=e_r(y,t)} + \nabla_x\boldsymbol{W}(x,t)\big|_{x=e_r(y,t)}\big(\boldsymbol{N}^*\mathcal{D}_t\widetilde{r}^* + \chi(y)\dot{\boldsymbol{h}}(t)\big),$$

$$\mathcal{D}_t\boldsymbol{W}(e_r(y,t),t) = L^{-1}\mathcal{L}\mathcal{D}_t\boldsymbol{W}_0(y,t) + \mathcal{D}_t\big(L^{-1}\mathcal{L}\big)\boldsymbol{W}_0(y,t),$$

we have

$$\begin{aligned}\Big\|\frac{\mathrm{d}}{\mathrm{d}t}\boldsymbol{W}(\cdot,t)\Big\|_{\Omega} &\leqslant c\Big(\sup_{\mathbb{R}^3}|L^{-1}\mathcal{L}|\|\mathcal{D}_t\boldsymbol{W}_0(\cdot,t)\|_{\mathbb{R}^3} + \sup_{\mathbb{R}^3}|\mathcal{D}_t(L^{-1}\mathcal{L})|\|\boldsymbol{W}_0(\cdot,t)\|_{\mathbb{R}^3} \\ &\qquad + \sup_{\mathbb{R}^3}\big|\boldsymbol{N}^*\mathcal{D}_t\widetilde{r}^* + \chi(y)\dot{\boldsymbol{h}}(t)\big|\|\nabla\boldsymbol{W}_0\|_{\mathbb{R}^3}\Big) \\ &\leqslant c\Big(\|\mathcal{D}_t r(\cdot,t)\|_{S_1} + \|r(\cdot,t)\|_{W_2^{1/2}(S_1)}\Big). \end{aligned} \quad (7.2.17)$$

The matrix $\mathcal{L}$ depends on $\nabla_{S_1}r$, hence the constants c in the inequalities (7.2.16) depend on $|\nabla_{S_1}r|_{S_1}$, and in (7.2.17) also on $|\mathcal{D}_t r|_{S_1}$, $|\mathcal{D}_t\nabla_{S_1}r|_{S_1}$. The time derivative of r can be estimated from the condition (7.2.5) on the moving interface Γ_t:

$$\mathcal{D}_t r = \frac{1}{n_N}(\boldsymbol{v}\cdot\boldsymbol{n} - \dot{\boldsymbol{h}}(t)\cdot\boldsymbol{n}),$$

whence

$$|\mathcal{D}_t r|_{S_1} \leqslant c|\boldsymbol{v}|_{\Omega}, \qquad |\mathcal{D}_t\nabla_{S_1}r|_{S_1} \leqslant c|\boldsymbol{v}|^{(1)}_{\cup\Omega_t^{\pm}}.$$

In addition,

$$\|\mathcal{D}_t r(\cdot,t)\|_{S_1} \leqslant c\big\{\|\boldsymbol{v}\|_{\Gamma_t} + \|\boldsymbol{v}\|_{\Omega_t^+}\big\}.$$

Hence, we can continue (7.2.17) as follows:

$$\|\mathcal{D}_t\boldsymbol{W}(\cdot,t)\|_{\Omega} \leqslant c(|\boldsymbol{v}|^{(1)}_{\cup\Omega_t^{\pm}})\big\{\|\boldsymbol{v}\|_{W_2^1(\Omega)} + \|r\|_{W_2^{1/2}(S_1)}\big\}. \quad (7.2.18)$$

Now we multiply the first equation in (7.1.3) by $\rho^{\pm}\boldsymbol{W}$ and integrate by parts over $\Omega_t^- \cup \Omega_t^+$:

$$\frac{\mathrm{d}}{\mathrm{d}t}\int_\Omega \rho^{\pm}\boldsymbol{v}\cdot\boldsymbol{W}\,\mathrm{d}x - \int_\Omega \rho^{\pm}\boldsymbol{v}\cdot(\mathcal{D}_t\boldsymbol{W} + (\boldsymbol{v}\cdot\nabla)\boldsymbol{W})dx + \int_\Omega \frac{\mu^{\pm}}{2}\mathbb{S}(\boldsymbol{v}) : \mathbb{S}(\boldsymbol{W})\,\mathrm{d}x$$
$$-\sigma\int_{\Gamma_t}(H + \frac{2}{R_0})\boldsymbol{n}\cdot\boldsymbol{W}\,\mathrm{d}\Gamma = \int_\Omega \rho^{\pm}\boldsymbol{f}\cdot\boldsymbol{W}\,\mathrm{d}x, \quad (7.2.19)$$

where $\mathbb{S}(\boldsymbol{v}) : \mathbb{S}(\boldsymbol{W}) = S_{ij}(\boldsymbol{v})S_{ij}(\boldsymbol{W})$. We add equality (7.2.19) multiplied by a small ϑ to (7.2.11). Making use of (7.2.15) and of the Korn inequality (6.4.3), we obtain

$$\frac{\mathrm{d}}{dt}\Big\{\frac{1}{2}\|\sqrt{\rho^{\pm}}\boldsymbol{v}\|_\Omega^2 + \vartheta\int_\Omega \rho^{\pm}\boldsymbol{v}\cdot\boldsymbol{W}\,\mathrm{d}x + \sigma\big(|\Gamma_t| - 4\pi R_0^2\big)\Big\} + c_1\|\boldsymbol{v}\|^2_{W_2^1(\Omega)}$$
$$-\vartheta\int_\Omega \rho^{\pm}\boldsymbol{v}\cdot(\mathcal{D}_t\boldsymbol{W} + (\boldsymbol{v}\cdot\nabla)\boldsymbol{W})\,\mathrm{d}x + \vartheta\int_\Omega \frac{\mu^{\pm}}{2}\mathbb{S}(\boldsymbol{v}) : \mathbb{S}(\boldsymbol{W})\,\mathrm{d}x$$
$$-\vartheta\sigma\int_{S_{R_0}}(H + \frac{2}{R_0})\widetilde{r}\,\mathrm{d}S_{R_0} \leqslant c(\varepsilon_1)\|\boldsymbol{f}\|_\Omega^2 + \varepsilon_1\vartheta\|\boldsymbol{W}\|_\Omega^2. \quad (7.2.20)$$

We have used here the equality $\mathrm{d}\Gamma(x) = |\widehat{\mathcal{L}}^T\boldsymbol{N}|\,\mathrm{d}S_{R_0}(y)$ [87, p. 227].

We are going to show that for sufficiently small ϑ, the expression under the derivative sign is positive, it may be called generalized energy

$$\mathcal{E}(t) = \frac{1}{2}\|\sqrt{\rho^{\pm}}\boldsymbol{v}\|_\Omega^2 + \vartheta\int_\Omega \rho^{\pm}\boldsymbol{v}\cdot\boldsymbol{W}\,\mathrm{d}x + \sigma\big(|\Gamma_t| - 4\pi R_0^2\big). \quad (7.2.21)$$

At first, by Theorem 3 from [78] under conditions (7.2.7), (7.2.8), the inequality

$$E_1(R, R_0) \equiv |\Gamma_t| - 4\pi R_0^2 \geqslant c_2\|r\|^2_{W_2^1(S_1)} \quad (7.2.22)$$

holds with a constant c_2 independent of δ_1 and R_0.

Let us make a small digression into differential geometry.

If a surface A is given by the radius-vector $\boldsymbol{r}$ in local coordinates (u, v), then a vector-valued element of the surface

$$\mathrm{d}\boldsymbol{A} = (\boldsymbol{r}'_u \times \boldsymbol{r}'_v)\,\mathrm{d}u\,\mathrm{d}v = \boldsymbol{n}|\,\mathrm{d}\boldsymbol{A}| = \boldsymbol{n}\sqrt{a(u, v)}\,\mathrm{d}u\,\mathrm{d}v,$$

where $\boldsymbol{n}$ is the normal to the surface, $a(u, v) = |\boldsymbol{r}'_u \times \boldsymbol{r}'_v|^2 = EG - F^2$ is the determinant of the first quadratic form for the surface A, and E, F, G are the elements of this quadratic form. The sign $\times$ means the vector product.

We use spherical coordinates θ, φ onto the surface $\Gamma_t = R(\theta, \varphi)$, therefore $\boldsymbol{r}'_\theta = (R'_\theta, 1, 0)$, $\boldsymbol{r}'_\varphi = (R'_\varphi, 0, 1)$, and the metric tensor is $\{g_{ij}\}_{i,j=1}^3 \equiv \{g_{ii}\}_{i=1}^3$, $g_{ij} = 0$ if $i \neq j$, $g_{11} \equiv g_{rr} = 1$, $g_{22} \equiv g_{\theta\theta} = R^2$, $g_{33} \equiv g_{\varphi\varphi} = R^2 \sin^2\theta$, $g \equiv |g_{ii}|_{i=1}^3 = R^4 \sin^2\theta$. Thus, we obtain

$$E = \boldsymbol{r}'_\theta \cdot \boldsymbol{r}'_\theta = g_{ii}(\boldsymbol{r}'_\theta)^i(\boldsymbol{r}'_\theta)^i = {R'_\theta}^2 + R^2,$$

$$F = \boldsymbol{r}'_\theta \cdot \boldsymbol{r}'_\varphi = g_{ii}(\boldsymbol{r}'_\theta)^i(\boldsymbol{r}'_\varphi)^i = R'_\theta R'_\varphi,$$

$$G = \boldsymbol{r}'_\varphi \cdot \boldsymbol{r}'_\varphi = {R'_\varphi}^2 + R^2 \sin^2\theta,$$

$$\boldsymbol{n} = \frac{\boldsymbol{r}'_\theta \times \boldsymbol{r}'_\varphi}{\sqrt{a(\theta,\varphi)}}, \qquad \{n^i\}_{i=1}^3 = \frac{1}{\sqrt{d}}\Big(1, \frac{-R'_\theta}{R^2}, \frac{-R'_\varphi}{R^2\sin^2\theta}\Big),$$

$$a(\theta,\varphi) \equiv EG - F^2 = R^4 \sin^2\theta\Big(1 + \frac{{R'_\theta}^2}{R^2} + \frac{{R'_\varphi}^2}{R^2\sin^2\theta}\Big) \equiv g d(\theta,\varphi).$$

Now we can calculate the area of the surface Γ_t:

$$\int_{\Gamma_t} \mathrm{d}\Gamma = \int_0^{2\pi}\int_0^\pi \sqrt{a(\theta,\varphi)}\,\mathrm{d}\theta\,\mathrm{d}\varphi = \int_0^{2\pi}\int_0^\pi R^2 \sin\theta\sqrt{d(\theta,\varphi)}\,\mathrm{d}\theta\,\mathrm{d}\varphi$$
$$= \int_{S_1} R^2\sqrt{d(\theta,\varphi)}\,\mathrm{d}\omega. \tag{7.2.23}$$

In addition, we will need also the second quadratic form for our surface

$$L\,\mathrm{d}\theta\,\mathrm{d}\theta + 2M\,\mathrm{d}\theta\,\mathrm{d}\varphi + N\,\mathrm{d}\varphi\,\mathrm{d}\varphi,$$

where, as we calculate below,

$$L = -\mathcal{D}'_\theta\boldsymbol{n}\cdot\boldsymbol{r}'_\theta = -g_{ii}(\mathcal{D}'_\theta\boldsymbol{n})^i(\boldsymbol{r}'_\theta)^i = \frac{1}{\sqrt{d}}\{R''_{\theta\theta} - \frac{2{R'_\theta}^2}{R} - R\},$$

$$M = -\mathcal{D}'_\varphi\boldsymbol{n}\cdot\boldsymbol{r}'_\theta = -g_{ii}(\mathcal{D}'_\varphi\boldsymbol{n})^i(\boldsymbol{r}'_\theta)^i = \frac{1}{\sqrt{d}}\{R''_{\theta\varphi} - \frac{2R'_\theta R'_\varphi}{R} - R'_\varphi\cot\theta\}, \tag{7.2.24}$$

$$N = -\mathcal{D}'_\varphi\boldsymbol{n}\cdot\boldsymbol{r}'_\varphi = \frac{1}{\sqrt{d}}\{R''_{\varphi\varphi} - \frac{2{R'_\varphi}^2}{R} + R'_\theta\sin\theta\cos\theta - R\sin^2\theta\}.$$

Here $\mathcal{D}'_\alpha$ means the covariant derivative with respect to the variable with the number $\alpha = 2, 3$. For example, if $\alpha = 3$ we have

$$\big(\mathcal{D}'_\varphi\boldsymbol{n}\big)^i = \frac{1}{\sqrt{g_{ii}}}\frac{\partial\big(\sqrt{g_{ii}}n^i\big)}{\partial\varphi} + \sum_{k=1}^3 n^k\Gamma^i_{k\varphi}, \quad i = 1, 2, 3,$$

where Γ^i_{kj} are the Christoffel symbols; for the spherical coordinates, they are $\Gamma^r_{\theta\theta} = -R$, $\Gamma^r_{\varphi\varphi} = -R\sin^2\theta$, $\Gamma^\theta_{\varphi\varphi} = -\sin\theta\cos\theta$, $\Gamma^\theta_{r\theta} = \frac{1}{R}$, $\Gamma^\varphi_{r\varphi} = \frac{1}{R}$, $\Gamma^\varphi_{\theta\varphi} = \cot\theta$ (and there is symmetry with respect to the lower indices); other coefficients are equal to zero. Thus, in the contravariant coordinates,

$$\mathcal{D}'_\theta \boldsymbol{n} = \left(\frac{\partial}{\partial\theta}\left(\frac{1}{\sqrt{d}}\right) + \frac{R'_\theta}{R\sqrt{d}}, \frac{1}{R}\frac{\partial}{\partial\theta}\left(\frac{-R'_\theta}{R\sqrt{d}}\right) + \frac{1}{R\sqrt{d}}, \frac{1}{R\sin\theta}\frac{\partial}{\partial\theta}\left(\frac{-R'_\theta}{R\sin\theta\sqrt{d}}\right) + \frac{R'_\varphi\cot\theta}{R^2\sqrt{d}}\right),$$

$$\mathcal{D}'_\varphi \boldsymbol{n} = \left(\frac{\partial}{\partial\varphi}\left(\frac{1}{\sqrt{d}}\right) + \frac{R'_\varphi}{R\sqrt{d}}, \frac{1}{R}\frac{\partial}{\partial\varphi}\left(\frac{-R'_\theta}{R\sqrt{d}}\right) + \frac{R'_\varphi\cot\theta}{R^2\sqrt{d}}, \frac{1}{R\sin\theta}\frac{\partial}{\partial\varphi}\left(\frac{-R'_\varphi}{R\sin\theta\sqrt{d}}\right) - \frac{R'_\theta\cot\theta}{R^2\sqrt{d}} + \frac{1}{R\sqrt{d}}\right),$$

which implies formulas (7.2.24).

The doubled mean curvature of the surface is known to be calculated by the formula

$$H = \frac{EN + GL - 2FM}{a} \tag{7.2.25}$$

(see, for example, [41]). We note that all calculations can be verified by expressing the Cartesian coordinates in terms of the spherical ones.

We return again to the expression for the deviation of the area of the surface Γ_t from the area of the sphere (see (7.2.22)). By (7.2.23), we have

$$E_1(R, R_0) = \int_{S_1} (R^2\sqrt{d} - R_0^2)\, d\omega,$$

$d = 1 + \frac{|\nabla_{S_1} R|^2}{R^2}$, the same as before, $\nabla_{S_1} R = R'_\theta \boldsymbol{e}_\theta + \frac{1}{\sin\theta} R'_\varphi \boldsymbol{e}_\varphi$. Using the Taylor expansion of $E_1(R, R_0)$ at the point R_0 and the equality for $r = R - R_0$

$$\int_{S_1} r\omega_i\, d\omega = -\frac{3}{2R_0}\int_{S_1} r^2\omega_i\, d\omega - \frac{1}{R_0^2}\int_{S_1} r^3\omega_i\, d\omega - \frac{1}{4R_0^3}\int_{S_1} r^4\omega_i\, d\omega,\ i = 1, 2, 3, \tag{7.2.26}$$

which follows from (7.2.7) in view of special factoring formulas, we can show that

$$E_1(R, R_0) = \int_{S_1} r^2\, d\omega + \frac{1}{2}\int_{S_1} |\nabla_{S_1} r|^2\, d\omega - \frac{2}{3R_0}\int_{S_1} r^3\, d\omega + I,$$

where I is the integral of the remainder in the Taylor series. It also depends only on r and ∇r and is small in comparison with the other terms (see [78, p. 142]). Therefore, we write $E_1(r) \equiv E_1(R, R_0)$. Under condition (7.2.8), we deduce that

$$E_1(r) \leqslant c_3 \|r\|^2_{W^1_2(S_1)}. \tag{7.2.27}$$

We show now that the surface integral

$$E_2(r) \equiv - \int_{S_{R_0}} (H + \frac{2}{R_0}) \widetilde{r} \, \mathrm{d}S_{R_0}$$

is also positive definite. To this end, we use formula (7.2.25) written in divergence form

$$H[R] = \frac{1}{R} \nabla_{S_1} \cdot \frac{\nabla_{S_1} R}{R\sqrt{d}} - \frac{2}{R\sqrt{d}}, \tag{7.2.28}$$

where $\nabla_{S_1} \cdot \boldsymbol{u} = \frac{1}{\sin\theta} \frac{\partial}{\partial\theta} (\sin\theta u_\theta) + \frac{1}{\sin\theta} \frac{\partial u_\varphi}{\partial\varphi}$ in the spherical coordinates. One easily obtains formula (7.2.28) if one uses the well-known fact that

$$H[R] = -(\mathcal{D}'_\theta \boldsymbol{n})^\theta - (\mathcal{D}'_\varphi \boldsymbol{n})^\varphi.$$

To estimate $E_2(r)$, we take into account that $\nabla_{S_1} R = \nabla_{S_1} r$ and

$$\int_{S_1} \nabla_{S_1} \cdot \frac{\nabla_{S_1} R}{R\sqrt{d}} \, \mathrm{d}\omega = 0.$$

Therefore,

$$\begin{aligned} -R_0^2 \int_{S_1} \frac{\widetilde{r}}{R} \nabla_{S_1} \cdot \frac{\nabla_{S_1} R}{R\sqrt{d}} \, \mathrm{d}\omega &= -R_0^2 (R_0 + \overline{r}) \int_{S_1} \nabla_{S_1} \big(\frac{1}{R}\big) \cdot \frac{\nabla_{S_1} R}{R\sqrt{d}} \, \mathrm{d}\omega \\ &= R_0^2 (R_0 + \overline{r}) \int_{S_1} \frac{|\nabla_{S_1} R|^2}{R^3 \sqrt{d}} \, \mathrm{d}\omega \\ &\geqslant \int_{S_1} |\nabla_{S_1} r|^2 \, \mathrm{d}\omega - \delta_1 c \int_{S_1} (|\nabla_{S_1} r|^2 + r^2) \, \mathrm{d}\omega, \end{aligned}$$

and since

$$\frac{2}{R_0} - \frac{2}{R\sqrt{d}} = \frac{2\big(r(R + R_0) + |\nabla_{S_1} r|^2\big)}{R_0 \big(R^2 + |\nabla_{S_1} r|^2 + R_0 R\sqrt{d}\big)}, \tag{7.2.29}$$

it is not difficult to show, using (7.2.2), that under condition (7.2.8), it holds

$$2R_0^2 \Big| \int_{S_1} \Big(\frac{1}{R_0} - \frac{1}{R\sqrt{d}}\Big) \widetilde{r} \, \mathrm{d}\omega \Big| \leqslant 2 \int_{S_1} r^2 \, \mathrm{d}\omega + \delta_1 c \int_{S_1} (|\nabla_{S_1} r|^2 + r^2) \, \mathrm{d}\omega,$$

and, as a consequence,

$$E_2(r) \geqslant E_0(r) - \delta_1 c \|r\|^2_{W^1_2(S_1)}, \tag{7.2.30}$$

where $E_0(r) = \int_{S_1} (|\nabla_{S_1} r|^2 - 2r^2)\,\mathrm{d}\omega$.

Lemma 2.5 in [81] yields the estimate

$$E_0(r) \geqslant \frac{4}{7} \int_{S_1} (|\nabla_{S_1} r|^2 + r^2)\,\mathrm{d}\omega - \frac{9}{14\pi} \Big(\int_{S_1} r\,\mathrm{d}\omega \Big)^2 - \frac{9}{7\pi} \Big(\int_{S_1} r\omega_i\,\mathrm{d}\omega \Big)^2. \tag{7.2.31}$$

We estimate the second term on the right-hand side taking (7.2.2) into account and using Eq. (7.2.8):

$$\begin{aligned} \frac{9}{14\pi} \Big(\int_{S_1} r\,\mathrm{d}\omega \Big)^2 &\leqslant \frac{c\delta_1^2 R_0^2}{R_0^2} \int_{S_1} r^2\,\mathrm{d}\omega + \frac{c\delta_1^4 R_0^4}{R_0^4} \int_{S_1} r^2\,\mathrm{d}\omega \\ &\leqslant c\delta_1 \int_{S_1} r^2\,\mathrm{d}\omega. \end{aligned}$$

The third term on the right-hand side of (7.2.31) can be evaluated in a similar way by virtue of equality (7.2.26). It is also small for small δ_1. Hence,

$$E_0(r) \geqslant c \int_{S_1} (|\nabla_{S_1} r|^2 + r^2)\,\mathrm{d}\omega.$$

Finally, from this inequality and (7.2.30), we get an analogous estimate for $E_2(r)$:

$$E_2(r) \geqslant c\|r\|^2_{W^1_2(S_1)}. \tag{7.2.32}$$

Thus, for $\mathcal{E}(t)$ defined by (7.2.21) on the basis of (7.2.16), (7.2.22), (7.2.27), we conclude that, for sufficiently small ϑ,

$$c_4(\|\boldsymbol{v}\|^2_\Omega + \|r\|^2_{W^1_2(S_1)}) \leqslant \mathcal{E}(t) \leqslant c_5(\|\boldsymbol{v}\|^2_\Omega + \|r\|^2_{W^1_2(S_1)}). \tag{7.2.33}$$

We denote by $\mathcal{E}_1(t)$ the terms outside the sign of derivative in the left-hand side of (7.2.20). It is easily seen in view of (7.2.16), (7.2.18), (7.2.32), and (7.2.33), that for small ϑ,

$$\mathcal{E}_1(t) \geqslant b\mathcal{E}(t)$$

with a positive constant b. Hence, (7.2.20) implies the estimate

$$\frac{\mathrm{d}}{\mathrm{d}t}\mathcal{E}(t) + b\mathcal{E}(t) \leqslant c\|\boldsymbol{f}\|^2_\Omega.$$

We multiply both sides of this inequality by $e^{b\tau}$ and integrate with respect to τ from 0 to t, which yields

$$\mathcal{E}(t) \leqslant e^{-2bt}\mathcal{E}(0) + \int_0^t e^{-2b(t-\tau)}\|\boldsymbol{f}(\cdot,\tau)\|_\Omega^2 d\tau$$
$$\leqslant c\, e^{-2bt}\{\|\boldsymbol{v}_0\|_\Omega^2 + \|r_0\|^2_{W_2^1(S_1)} + \|e^{b\tau}\boldsymbol{f}\|^2_{Q_t}\},$$

and inequality (7.2.9), in view of (7.2.33).

Inequality (7.2.10) is an obvious consequence of (7.2.9). □

Corollary 7.2.1 *The coordinates of the center of gravity of* Ω_t^+ *satisfy the inequality*

$$|h(t)| \leqslant c\Big\{\|e^{b\tau}\boldsymbol{f}\|_{Q_\infty} + \|\boldsymbol{v}_0\|_\Omega + \|r_0\|_{W_2^1(S_1)}\Big\},\ \forall t\in[0,T]. \tag{7.2.34}$$

Proof From formula (7.2.6), it follows that

$$|h(t)| \leqslant \frac{1}{|\Omega_t^+|^{1/2}}\int_0^t \|\boldsymbol{v}(\cdot,\tau)\|_{\Omega_t^+}\, d\tau,$$

which, together with (7.2.10), implies inequality (7.2.34). □

7.3 A Linearized Problem

Now we consider the linearized problem

$$\begin{gathered}
\mathcal{D}_t\boldsymbol{w} - \nu^\pm\nabla_{\boldsymbol{u}}^2\boldsymbol{w} + \frac{1}{\rho^\pm}\nabla_{\boldsymbol{u}} s = \boldsymbol{f}, \qquad \nabla_{\boldsymbol{u}}\cdot\boldsymbol{w} = r \quad \text{in } D_{\Sigma T},\\
\boldsymbol{w}|_{t=0} = \boldsymbol{w}_0 \qquad \text{in } \Omega_0^+\cup\Omega_0^-,\\
[\boldsymbol{w}]|_{G_T} = 0,\quad \boldsymbol{w}|_{\Upsilon_T} = 0, \qquad [\mu^\pm\Pi_0\Pi\mathbb{S}_{\boldsymbol{u}}(\boldsymbol{w})\boldsymbol{n}]|_{G_T} = \Pi_0\boldsymbol{b},\\
[\boldsymbol{n}_0\cdot\mathbb{T}_{\boldsymbol{u}}(\boldsymbol{w},s)\boldsymbol{n}]|_{G_T} - \sigma\boldsymbol{n}_0\cdot\Delta(t)\int_0^t \boldsymbol{w}\big|_\Gamma d\tau = b + \int_0^t B d\tau \quad \text{on } G_T.
\end{gathered} \tag{7.3.1}$$

Here the functions in the right-hand sides and the vector $\boldsymbol{u}$ are given; moreover, $[\boldsymbol{u}]|_{G_T} = 0$.

Problem (7.3.1) has been studied in Chap. 5, where the unique solvability of the problem has been proved on an arbitrary finite time interval when the surface Σ is absent and the domain $\Omega_0^-\cup\overline{\Omega_0^+}$ is the whole space $\mathbb{R}^3$

(Theorem 5.3.1). This result has been obtained in the Hölder spaces with a power weight at infinity; however, it is valid also in our case. First, the weighted spaces are equivalent to the ordinary ones in a bounded domain. Second, there are respective estimates for the model problem with adhesion condition on the outer boundary Σ [76].

We state the existence theorem for problem (7.3.1).

Theorem 7.3.1 *We assume that for some* $\alpha, \gamma \in (0,1)$, $\gamma < \alpha$, $0 < T < \infty$, *the surfaces* $\Gamma, \Sigma \in C^{2+\alpha}$, *the function* $\sigma \in C^{1+\alpha}(\Gamma)$, $\sigma \geqslant \sigma_0 > 0$, *and a vector* $\boldsymbol{u} \in \boldsymbol{C}^{2+\alpha,1+\alpha/2}(D_{\Sigma T})$ *is such that* $[\boldsymbol{u}]|_{G_T} = 0$ *and the inequality*

$$(T + T^{\gamma/2})|\boldsymbol{u}|^{(2+\alpha,1+\alpha/2)}_{D_{\Sigma T}} \leqslant \delta \tag{7.3.2}$$

holds for sufficiently small $\delta > 0$.

In addition, let the following four sets of conditions be satisfied:

(1) $\boldsymbol{f} \in \boldsymbol{C}^{\alpha,\alpha/2}(D_{\Sigma T})$, $r \in C^{1+\alpha,\frac{1+\alpha}{2}}(D_{\Sigma T})$, $\boldsymbol{w}_0 \in \boldsymbol{C}^{2+\alpha}(\Omega_0^+ \cup \Omega_0^-)$, $\boldsymbol{b} \in \boldsymbol{C}^{1+\alpha,\,\frac{1+\alpha}{2}}(G_T)$, $b \in C^{(\gamma,1+\alpha)}(G_T)$, $B \in C^{\alpha,\alpha/2}(G_T)$;
(2) *compatibility conditions* (6.2.5) *are satisfied with* $\boldsymbol{v}_0 = \boldsymbol{w}_0$, $\boldsymbol{d} = \boldsymbol{b}$ *and* $p_0 = s_0 \equiv s(\xi, 0)$;
(3) *there exist a vector field* $\boldsymbol{g} \in \boldsymbol{C}^{\alpha,\alpha/2}(D_{\Sigma T})$ *and a tensor* $\mathbb{G} = \{G_{ik}\}^3_{i,k=1}$, $G_{ik} \in C^{(\gamma,1+\alpha)}(D_{\Sigma T}) \cap C^{\gamma,0}(D_{\Sigma T})$ *such that representation formulas*

$$\mathcal{D}_t r - \nabla_{\boldsymbol{u}} \cdot \boldsymbol{f} = \nabla \cdot \boldsymbol{g}, \quad \boldsymbol{g} = \nabla \cdot \mathbb{G} \quad (g_i = \partial G_{ik}/\partial \xi_k, \quad i = 1, 2, 3),$$

hold (in a generalized sense) and, moreover,

$$[(\boldsymbol{g} + \mathbb{A}^* \boldsymbol{f}) \cdot \boldsymbol{n}_0]|_{G_T} = 0;$$

(4) $s_0(\xi)$ *is a solution to the equation*

$$\frac{1}{\rho^\pm} \nabla^2 s_0(\xi) = \nabla \cdot \big(\mathcal{D}_t \mathbb{B}^*|_{t=0} \boldsymbol{w}_0(\xi) - \boldsymbol{g}(\xi, 0)\big) \equiv \nabla \cdot \boldsymbol{d} \quad in \quad \Omega_0^+ \cup \Omega_0^- \tag{7.3.3}$$

with boundary conditions (6.2.4), *where* $\boldsymbol{v}_0 = \boldsymbol{w}_0$ *and* $p_0 = s_0$.

Under these assumptions, problem (7.3.1) *has a unique solution* $(\boldsymbol{w}, s)$, $\boldsymbol{w} \in \boldsymbol{C}^{2+\alpha,1+\alpha/2}(D_{\Sigma T})$, $s \in C^{(\gamma,1+\alpha)}(D_{\Sigma T})$, $\nabla s \in \boldsymbol{C}^{\alpha,\alpha/2}(D_{\Sigma T})$ *(the pressure is defined up to a bounded function of time), and for arbitrary* $t' \in (0, T]$, *the inequality*

$$\begin{aligned} N_{t'}[\boldsymbol{w}, s] &\equiv |\boldsymbol{w}|^{(2+\alpha,1+\alpha/2)}_{D_{\Sigma t'}} + |\nabla s|^{(\alpha,\alpha/2)}_{D_{\Sigma t'}} + |s|^{(\gamma,1+\alpha)}_{D_{\Sigma t'}} \\ &\leqslant c_1(t') \Big\{ |\boldsymbol{f}|^{(\alpha,\alpha/2)}_{D_{\Sigma t'}} + |r|^{(1+\alpha,\frac{1+\alpha}{2})}_{D_{\Sigma t'}} + |\boldsymbol{w}_0|^{(2+\alpha)}_{\cup\Omega_0^\pm} + |\boldsymbol{g}|^{(\alpha,\alpha/2)}_{D_{\Sigma t'}} \end{aligned}$$

$$+ |\mathbb{G}|^{(\gamma,1+\alpha)}_{D_{\Sigma t'}} + |\mathbb{G}|^{(\gamma,0)}_{D_{\Sigma t'}} + |\boldsymbol{b}|^{(1+\alpha,\frac{1+\alpha}{2})}_{G_{t'}} + |b|_{G_{t'}} + |b|^{(\gamma,1+\alpha)}_{G_{t'}}$$

$$+ |\nabla_\Gamma b|^{(\alpha,\alpha/2)}_{G_{t'}} + |B|^{(\alpha,\alpha/2)}_{G_{t'}} + P_{t'}[\boldsymbol{u}]|\boldsymbol{w}_0|^{(1)}_{\cup\Omega_0^\pm}\Big\} \tag{7.3.4}$$

$$\equiv c_1(t')\{F(t') + P_{t'}[\boldsymbol{u}]|\boldsymbol{w}_0|^{(1)}_{\cup\Omega_0^\pm}\}$$

holds, where $c_1(t')$ *is a nondecreasing function of* $t' \leqslant T$, $\nabla_\Gamma = \Pi_0\nabla$, *and* $P_t[\boldsymbol{u}] = t^{\frac{1-\alpha}{2}}|\nabla\boldsymbol{u}|_{D_{\Sigma t}} + |\nabla\boldsymbol{u}|^{(\alpha,\alpha/2)}_{D_{\Sigma t}}$.

In what follows, we prove an important estimate (7.4.4) for the solution of problem (7.1.3), (1.0.3) using inequality (7.3.4). As above, the presence of the term $\sigma\frac{2}{R_0}\boldsymbol{n}_0\cdot\boldsymbol{n}$ brings no difficulties in the proof.

Remark 7.3.1 *An estimate of the pressure* s_0 *at the initial moment* $t = 0$ *for a bounded domain will be given in Sect.* 9.2.

7.4 Global Classical Solvability of Problem (7.1.3), (1.0.3)

The aim of this section is to prove the solvability of problem (7.1.3), (1.0.3) in the whole time interval $t > 0$. In the proof, we use Lemmas 6.4.1, 6.4.2.

Making use of the interpolation inequalities, one can prove the following statement.

Lemma 7.4.1 *An arbitrary function* $u \in C^{2+\alpha,1+\frac{\alpha}{2}}(D_{\Sigma T_0})$, $0 < \theta_3, \theta_4, \theta_5 < \min\{\operatorname{diam}\Omega, T_0^{\frac{1}{2}}\}$, *satisfies the inequalities*

$$\langle\nabla u\rangle^{(\alpha,\frac{\alpha}{2})}_{D_{\Sigma T_0}} \leqslant c\Big\{\theta_3\langle u\rangle^{(2+\alpha,1+\frac{\alpha}{2})}_{D_{\Sigma T_0}} + \theta_3^{-\alpha-\frac{9}{2}}\int_0^{T_0}\|u(\cdot,\tau)\|_\Omega d\tau\Big\}, \tag{7.4.1}$$

$$|u|^{(2)}_{x,D_{\Sigma T_0}} \leqslant c\Big\{\theta_4^\alpha\langle u\rangle^{(2+\alpha,1+\frac{\alpha}{2})}_{D_{\Sigma T_0}} + \theta_4^{-\frac{11}{2}}\int_0^{T_0}\|u(\cdot,\tau)\|_\Omega d\tau\Big\}, \tag{7.4.2}$$

$$|\nabla u|_{D_{\Sigma T_0}} \leqslant c\Big\{\theta_5^{1+\alpha}\langle u\rangle^{(2+\alpha,1+\frac{\alpha}{2})}_{D_{\Sigma T_0}} + \theta_5^{-\frac{9}{2}}\int_0^{T_0}\|u(\cdot,\tau)\|_\Omega d\tau\Big\}. \tag{7.4.3}$$

Proposition 7.4.1 *Let a solution* $(\boldsymbol{v}, p_1)$ *of Problem* (7.1.3), (1.0.3) *be defined in the interval* $(0, T]$ *and let the estimate*

$$N_{(0,T)}[\boldsymbol{v}, p_1] \equiv |\boldsymbol{u}|^{(2+\alpha,1+\alpha/2)}_{D_{\Sigma T}} + |\nabla q|^{(\alpha,\alpha/2)}_{D_{\Sigma T}} + |q|^{(\gamma,1+\alpha)}_{D_{\Sigma T}} \leqslant \mu$$

hold, where $(\boldsymbol{u}, q)$ *is the solution* $(\boldsymbol{v}, p_1)$ *written in the Lagrangian coordinates.*

Then for any $t_0 \in (0, T]$ *there is* $\tau_0 \in (0, t_0/2)$ *such that*

$$N_{(t_0-\tau_0,t_0)}[\boldsymbol{v},p_1,r] \equiv N_{(t_0-\tau_0,t_0)}[\boldsymbol{v},p_1]+\sup_{t_0-\tau_0<\tau<t_0}|r(\cdot,\tau)|_{S_1}^{(3+\alpha)}$$

$$\leqslant c\big(\delta,\tau_0,|\nabla \boldsymbol{f}|_{D_{\Sigma T}}^{(\alpha,\frac{1+\alpha-\gamma}{2})}\big)\Big\{|\boldsymbol{f}|_{D_{\Sigma(t_0-2\tau_0,t_0)}}^{(\alpha,\frac{1+\alpha-\gamma}{2})}+\int\limits_{t_0-2\tau_0}^{t_0}\big(\|\boldsymbol{v}(\cdot,\tau)\|_{\Omega}+\|r(\cdot,\tau)\|_{W_2^1(S_1)}\big)\,\mathrm{d}\tau\Big\},$$

(7.4.4)

τ_0 depends on μ and on the constant δ in (6.4.18).

Proof We fix an arbitrary $t_0 \in (0,T]$. Let $\tau_0 \in (0,t_0/2)$, and let $\eta_\lambda(t)$ be a smooth monotone cutoff function (6.4.13) whose derivatives satisfy estimates (6.4.14).

We consider the couple $\boldsymbol{w}=\boldsymbol{v}\eta_\lambda$, $s=p\eta_\lambda$. It is a solution to the system

$$\mathcal{D}_t\boldsymbol{w}+(\boldsymbol{v}\cdot\nabla)\boldsymbol{w}-\nu^\pm\nabla^2\boldsymbol{w}+\frac{1}{\rho^\pm}\nabla s=\boldsymbol{f}\eta_\lambda+\boldsymbol{v}\dot\eta_\lambda,$$

$$\nabla\cdot\boldsymbol{w}=0 \quad\text{in}\quad \Omega_t^-\cup\Omega_t^+,\ \ t>t_0-2\tau_0,$$

$$\boldsymbol{w}\big|_{t=t_0-2\tau_0}=0 \quad\text{in}\quad \cup\,\Omega'\equiv\Omega^-_{t_0-2\tau_0}\cup\Omega^+_{t_0-2\tau_0},$$

$$[\boldsymbol{w}]|_{\Gamma_t}=0,\quad [\mathbb{T}(\boldsymbol{w},s)\boldsymbol{n}]\big|_{\Gamma_t}=\sigma\big(H+\frac{2}{R_0}\big)\boldsymbol{n}\eta_\lambda\big|_{\Gamma_t},\quad \boldsymbol{w}\big|_\Sigma=0,$$

where $\dot\eta_\lambda\equiv \mathrm{d}\eta_\lambda(t)/\mathrm{d}t$.

We introduce the Lagrangian coordinates by formula (6.4.15). The functions $\boldsymbol{w}$ and s written in the Lagrangian coordinates will be denoted by the same symbols. They satisfy the system

$$\begin{aligned}
&\mathcal{D}_t\boldsymbol{w}-\nu^\pm\nabla^2_{\boldsymbol{u}}\boldsymbol{w}+\frac{1}{\rho^\pm}\nabla_{\boldsymbol{u}}s=\boldsymbol{f}(X,t)\eta_\lambda+\boldsymbol{u}\dot\eta_\lambda,\\
&\nabla_{\boldsymbol{u}}\cdot\boldsymbol{w}=0 \qquad\text{in}\ \cup\,\Omega',\quad t>t_0-2\tau_0,\\
&\boldsymbol{w}|_{t=t_0-2\tau_0}=0 \qquad \text{in}\ \cup\,\Omega',\\
&[\boldsymbol{w}]|_{\Gamma'}=0,\quad [\mu^\pm\Pi_0'\Pi\mathbb{S}_{\boldsymbol{u}}(\boldsymbol{w})\boldsymbol{n}]\big|_{\Gamma'}=0,\qquad \boldsymbol{w}|_\Sigma=0,\\
&[\boldsymbol{n}_0'\cdot\mathbb{T}_{\boldsymbol{u}}(\boldsymbol{w},s)\boldsymbol{n}]\big|_{\Gamma'}=\sigma\eta_\lambda\big(H+\frac{2}{R_0}\big)\boldsymbol{n}_0'\cdot\boldsymbol{n}\big|_{\Gamma'}.
\end{aligned}\tag{7.4.5}$$

Here $\Gamma'=\Gamma_{t_0-2\tau_0}$, $\boldsymbol{n}_0'$ is the outward normal to Γ', Π_0' and Π are the projections on the tangent planes to Γ' and to Γ_t, respectively. The other notation, for instance, $\nabla_{\boldsymbol{u}}$, also corresponds to transformation (6.4.15).

We write the last boundary condition in (7.4.5) in the form:

$$[\boldsymbol{n}_0'\cdot\mathbb{T}_{\boldsymbol{u}}(\boldsymbol{w},s)\boldsymbol{n}]\big|_{\Gamma'}-\sigma\eta_\lambda\big(H+\frac{2}{R_0}\big)\boldsymbol{n}_0'\cdot\boldsymbol{n}\big|_{\Gamma'}$$

$$-\int_{t_0-2\tau_0}^{t}\dot{\eta}_\lambda\Big\{[\boldsymbol{n}_0'\cdot\mathbb{T}_{\boldsymbol{u}}(\boldsymbol{w},s)\boldsymbol{n}]\big|_{\Gamma'}-\sigma\big(H+\frac{2}{R_0}\big)\boldsymbol{n}_0'\cdot\boldsymbol{n}\big|_{\Gamma'}\Big\}\,\mathrm{d}\tau=0,$$

then we use relation (1.1.3) and integrate by parts. We obtain

$$[\boldsymbol{n}_0'\cdot\mathbb{T}_{\boldsymbol{u}}(\boldsymbol{w},s)\boldsymbol{n}]\big|_{\Gamma'}-\sigma\boldsymbol{n}_0'\cdot\int_{t_0-2\tau_0}^{t}\Delta(\tau)\boldsymbol{w}\big|_{\Gamma'}\,\mathrm{d}\tau=\int_{t_0-2\tau_0}^{t}\Big\{\dot{\eta}_\lambda[\boldsymbol{n}_0'\cdot\mathbb{T}_{\boldsymbol{u}}(\boldsymbol{u},q)\boldsymbol{n}]\big|_{\Gamma'}$$
$$+\frac{2\sigma}{R_0}\eta_\lambda\boldsymbol{n}_0'\cdot\frac{\mathrm{d}}{\mathrm{d}t}\boldsymbol{n}\big|_{\Gamma'}+\sigma\eta_\lambda(\tau)\boldsymbol{n}_0'\cdot\dot{\Delta}(\tau)\boldsymbol{X}(\xi',\tau)\Big\}\,\mathrm{d}\tau,$$

where $\dot{\Delta}(t)$ is the operator obtained from (1.1.4) by time differentiating its coefficients (see Sect. 5.3). Replacing $\boldsymbol{X}$ according to formula (6.4.15), we obtain

$$[\boldsymbol{n}_0'\cdot\mathbb{T}_{\boldsymbol{u}}(\boldsymbol{w},s)\boldsymbol{n}]\big|_{\Gamma'}-\sigma\boldsymbol{n}_0'\cdot\Delta(t)\int_{t_0-2\tau_0}^{t}\boldsymbol{w}\big|_{\Gamma'}\,\mathrm{d}\tau=\int_{t_0-2\tau_0}^{t}B(\xi',\tau)\,\mathrm{d}\tau,\qquad \xi'\in\Gamma',\tag{7.4.6}$$

where $B(\xi',\tau)=\frac{2\sigma}{R_0}\eta_\lambda\boldsymbol{n}_0'\cdot\frac{\mathrm{d}}{\mathrm{d}t}\boldsymbol{n}\big|_{\Gamma'}+\dot{\eta}_\lambda[\boldsymbol{n}_0'\cdot\mathbb{T}_{\boldsymbol{u}}(\boldsymbol{u},q)\boldsymbol{n}]|_{\Gamma'}+\sigma\eta_\lambda(\tau)\boldsymbol{n}_0'\cdot\dot{\Delta}(\tau)\boldsymbol{\xi}'+\sigma\eta_\lambda(\tau)\boldsymbol{n}_0'\cdot\dot{\Delta}(\tau)\int_{t_0-2\tau_0}^{\tau}\boldsymbol{u}(\xi',\tau')\mathrm{d}\tau'-\sigma\boldsymbol{n}_0'\cdot\int_{\tau}^{t}\dot{\Delta}(\tau')\,\mathrm{d}\tau'\boldsymbol{w}\big|_{\Gamma'}$. We have used here the equality $\Delta(\tau)-\Delta(t)=-\int_\tau^t\dot{\Delta}(\tau')\,\mathrm{d}\tau'$.

In order to apply Theorem 7.3.1 to Problem (7.4.5), (7.4.6), we need to verify its assumptions. We can do it similarly to verifying the application of Theorem 6.3.1 in Sect. 6.4. We again choose τ_0 satisfying inequality (6.4.18). The vector field $\boldsymbol{u}$ is extended with preservation of class in the whole space $\mathbb{R}^3$ so it vanishes at infinity.

Hence, by (7.3.4)

$$N_{(t_0-2\tau_0+\lambda,t_0)}[\boldsymbol{v},p]\leqslant N_{(t_0-2\tau_0,t_0)}[\boldsymbol{w},s]$$
$$\leqslant c_1(2\tau_0)\Big\{|\boldsymbol{f}(X,t)\eta_\lambda|_{Q_0'}^{(\alpha,\alpha/2)}+|\boldsymbol{u}\dot{\eta}_\lambda|_{D_0'}^{(\alpha,\frac{\alpha}{2})}+|\boldsymbol{g}|_{D_0'}^{(\alpha,\alpha/2)}+|\mathbb{G}|_{D_0'}^{(\gamma,0)}$$
$$+|\mathbb{G}|_{D_0'}^{(\gamma,1+\alpha)}+|B|_{G_0'}^{(\alpha,\alpha/2)}\Big\}.\tag{7.4.7}$$

Here we have used the notation: $D_\beta'=\cup Q^{\pm}_{(t_0-2\tau_0+\beta,t_0)}$, $G'_\beta=\Gamma'\times(t_0-2\tau_0+\beta,t_0)$ with $\beta\in[0,2\tau_0)$.

We set $\lambda < 1$. Now we estimate the norms of the functions in the right-hand side of inequality (7.4.7) in a similar way to that in Sect. 6.4. The only difference is the presence of the norm $|B|^{(\alpha,\alpha/2)}_{G'_0}$. Let us dwell on its estimate. The term containing $\frac{\mathrm{d}}{\mathrm{d}t}\boldsymbol{n}$, by formula (1.1.6), is estimated in an obvious way:

$$|\eta_\lambda \boldsymbol{n}'_0 \cdot \frac{\mathrm{d}}{\mathrm{d}t}\boldsymbol{n}|^{(\alpha,\alpha/2)}_{G'_0} \leqslant c(\delta)\Big\{|\nabla \boldsymbol{u}|^{(\alpha,\alpha/2)}_{D'_{\lambda/2}} + \frac{1}{\lambda^{\frac{\alpha}{2}}}|\nabla \boldsymbol{u}|_{D'_{\lambda/2}}\Big\}.$$

Applying Lemma 5.4.3 with $\boldsymbol{u}' = 0$ and using the estimates of the other terms in B (see [54, Lemma 5.2]), one can write

$$\begin{aligned}|B|^{(\alpha,\alpha/2)}_{G'_0} \leqslant c\Big\{&\frac{1}{\lambda}\Big(\langle\nabla \boldsymbol{u}\rangle^{(\alpha,\alpha/2)}_{D'_{\lambda/2}} + \langle[q]|_\Gamma\rangle^{(\alpha,\alpha/2)}_{G'_{\lambda/2}}\Big) + \frac{1}{\lambda^{1+\frac{\alpha}{2}}}\big(|\nabla \boldsymbol{u}|_{D'_{\lambda/2}} + |[q]|_\Gamma|_{G'_{\lambda/2}}\big)\\ &+\Big(1+\frac{1}{\lambda^{1+\frac{\alpha}{2}}}\Big)|\nabla \boldsymbol{u}|^{(\alpha,\alpha/2)}_{D'_{\lambda/2}} + \Big(\tau_0^{(1-\alpha/2)} + \frac{1}{\lambda^{\frac{\alpha}{2}}}\Big)|\boldsymbol{u}|^{(2)}_{\xi',D'_{\lambda/2}} + \\ &+\tau_0|\boldsymbol{u}|^{(2+\alpha)}_{\xi',D'_{\lambda/2}} + \big(\tau_0^{(1-\alpha/2)} + \tau_0\big)|\boldsymbol{w}|^{(2+\alpha,\alpha/2)}_{\cup Q'_0}\Big\}.\end{aligned}$$

As (6.4.19), inequality (7.4.7) can be continued for small τ_0 as follows:

$$\begin{aligned}N_{(t_0-2\tau_0+\lambda,t_0)}[\boldsymbol{v},p_1] \leqslant c_2\big(\delta, |\nabla \boldsymbol{f}|^{(\alpha,\frac{1+\alpha-\gamma}{2})}_{Q'_0}\big)\Big\{&\frac{1}{\lambda^{\frac{\alpha}{2}}}|\boldsymbol{f}|^{(\alpha)}_{x,Q'_0} + \frac{1}{\lambda^{\frac{1+\alpha-\gamma}{2}}}|\boldsymbol{f}|^{(\frac{1+\alpha-\gamma}{2})}_{t,Q'_0}\\ &+ \tau_0|\boldsymbol{u}|^{(2+\alpha)}_{\xi',D'_{\lambda/2}} + \frac{1}{\lambda}\Big(\langle\nabla \boldsymbol{u}\rangle^{(\alpha,\frac{\alpha}{2})}_{D'_{\lambda/2}} + \langle[q]|_\Gamma\rangle^{(\alpha,\frac{\alpha}{2})}_{G'_{\lambda/2}}\Big)\\ &+ \frac{1}{\lambda}\Big(\langle\boldsymbol{u}\rangle^{(\alpha,\frac{\alpha}{2})}_{D'_{\lambda/2}} + \langle\boldsymbol{u}\rangle^{(\frac{1+\alpha-\gamma}{2})}_{t,D'_{\lambda/2}}\Big) + \frac{1}{\lambda^{\frac{3+\alpha-\gamma}{2}}}|\boldsymbol{u}|_{D'_{\lambda/2}}\\ &+ \frac{1}{\lambda^{\frac{\alpha}{2}}}|\boldsymbol{u}|^{(2)}_{\xi',D'_{\lambda/2}} + \frac{1}{\lambda^{1+\frac{\alpha}{2}}}\Big(|\nabla \boldsymbol{u}|_{D'_{\lambda/2}} + |[q]|_\Gamma|_{G'_{\lambda/2}}\Big)\Big\}.\end{aligned} \qquad (7.4.8)$$

We estimate the jump of the pressure using the last boundary condition in system (7.1.4) and inequality (7.4.1):

$$\begin{aligned}\langle[q]|_\Gamma\rangle^{(\alpha)}_{\xi',G'_{\lambda/2}} &\leqslant c(\delta)\{|\nabla \boldsymbol{u}|^{(\alpha)}_{\xi',G'_{\lambda/2}} + |H + \frac{2}{R_0}|^{(\alpha)}_{\xi',G'_{\lambda/2}}\}\\ &\leqslant c(\delta)\{|\nabla \boldsymbol{u}|^{(\alpha)}_{\xi',D'_{\lambda/2}} + |r|^{(2+\alpha)}_{y,S_1\times(t_0-2\tau_0+\lambda/2,t_0)}\}\\ &\leqslant c(\delta)\Big\{\theta_3\big(|\boldsymbol{u}|^{(2+\alpha,1+\alpha/2)}_{D'_{\lambda/2}} + |r|^{(3+\alpha)}_{y,S_1\times(t_0-2\tau_0+\lambda/2,t_0)}\big)\\ &\quad+c\,\theta_3^{-9/2-\alpha}\int_{t_0-2\tau_0}^{t_0}\Big(\|\boldsymbol{u}(\cdot,\tau)\|_\Omega + \|r(\cdot,\tau)\|_{W_2^1(S_1)}\Big)\,\mathrm{d}\tau\Big\},\end{aligned} \qquad (7.4.9)$$

$|r|^{(2+\alpha)}_{y,S_1\times(t_0-2\tau_0+\lambda/2,t_0)} \equiv \sup_{t_0-2\tau_0+\lambda/2<\tau<t_0} |r(\cdot,\tau)|^{(2+\alpha)}_{S_1}$.

The norm of r in $C^{3+\alpha}(S_1)$ can be estimated by substituting relation (7.2.28) into the last boundary condition in (7.1.4). It is obvious that the function r is a solution of a quasilinear elliptic equation, hence its norm satisfies the inequality

$$|r(\cdot,t)|^{(3+\alpha)}_{S_1} \leqslant c\{|\nabla_{\Gamma'}[q(\cdot,t)]|_{\Gamma'}|^{(\alpha)}_{\Gamma'} + |\boldsymbol{u}(\cdot,t)|^{(2+\alpha)}_{\cup\Omega'} + \|r(\cdot,t)\|_{S_1}\} \qquad (7.4.10)$$

(see [46]). The estimate of the semi-norm $\langle[q]|_{\Gamma}\rangle^{(\alpha/2)}_{t,G'_{\lambda/2}}$ follows from the same boundary condition but in a modified form:

$$[q]|_{\Gamma}\boldsymbol{n}'_0\cdot\boldsymbol{n} = 2[\mu^{\pm}\boldsymbol{n}'_0\cdot\frac{\partial\boldsymbol{u}}{\partial\boldsymbol{n}}]|_{\Gamma'} - \sigma\boldsymbol{n}'_0\cdot\Delta(t)\int\limits_{t_0-2\tau_0+\frac{\lambda}{2}}^{t}\boldsymbol{u}\,\mathrm{d}\tau - \sigma\boldsymbol{n}'_0\cdot\int\limits_{t_0-2\tau_0+\frac{\lambda}{2}}^{t}\dot{\Delta}(\tau)\boldsymbol{\xi}'\,\mathrm{d}\tau$$

$$-\sigma\Big(H_1+\frac{2}{R_0}\Big) - \sigma\frac{2}{R_0}(\boldsymbol{n}-\boldsymbol{n}'_0)\cdot\boldsymbol{n}'_0,$$

where H_1 is twice the mean curvature of the surface $\Gamma_{t_0-2\tau_0+\lambda/2}$.

Since $|\boldsymbol{n}'_0\cdot\boldsymbol{n}| \geqslant c$ for sufficiently small τ_0, we obtain, in view of Lemmas 7.4.1, 5.4.3 and formula (1.1.6),

$$\langle[q]|_{\Gamma'}\rangle^{(\frac{\alpha}{2})}_{t,G'_{\lambda/2}} \leqslant c(\delta)\Big\{\tau_0^{1-\frac{\alpha}{2}}|\boldsymbol{u}|^{(2)}_{\xi',D'_{\lambda/2}} + \theta_3|\boldsymbol{u}|^{(2+\alpha,1+\frac{\alpha}{2})}_{D'_{\lambda/2}} + c\,\theta_3^{-\frac{9}{2}-\alpha}\int\limits_{t_0-2\tau_0}^{t_0}\|\boldsymbol{u}(\cdot,\tau)\|_{\Omega}\,\mathrm{d}\tau\Big\}. \qquad (7.4.11)$$

In addition, we have

$$|[q]|_{\Gamma'}|_{G'_{\lambda/2}} \leqslant c(\delta)\Big\{|\nabla\boldsymbol{u}|_{D'_{\lambda/2}} + \tau_0|\boldsymbol{u}|^{(2)}_{\xi',D'_{\lambda/2}} + \tau_0|\nabla\boldsymbol{u}|^{(\alpha,\alpha/2)}_{D'_{\lambda/2}} + |r(\cdot,t_0-2\tau_0+\frac{\lambda}{2})|^{(2)}_{S_1}\Big\}. \qquad (7.4.12)$$

We estimate the lower order norms of $\boldsymbol{u}$ in (7.4.8) and (7.4.12) by Lemmas 6.4.1, 6.4.2, 7.4.1, setting $\theta_1 = (\varepsilon\lambda)^{1/\gamma}$, $\theta_2 = (\varepsilon\lambda)^{1/2}$, $\theta_3 = \varepsilon\lambda$, $\theta_4 = (\varepsilon\lambda^{\frac{\alpha}{2}})^{\frac{1}{\alpha}}$, $\theta_5 = (\varepsilon\lambda^{1+\frac{\alpha}{2}})^{\frac{1}{1+\alpha}}$ in inequalities (6.4.6), (6.4.10), (7.4.1)–(7.4.3), respectively. Next, we evaluate the last term in (7.4.12) using an interpolation inequality similar to (7.4.2); finally, $|\boldsymbol{u}|_{D'_{\lambda/2}}$ can be estimated by the inequality

$$|\boldsymbol{u}|_{D'_{\lambda/2}} \leqslant c\Big\{\theta_6^{1+\alpha}\langle\boldsymbol{u}\rangle^{(\frac{1+\alpha}{2})}_{t,D'_{\lambda/2}} + \theta_6^{-\frac{7}{2}}\int_{t_0-2\tau_0}^{t_0}\|\boldsymbol{u}(\cdot,\tau)\|_{\Omega}\mathrm{d}\tau\Big\} \qquad (7.4.13)$$

with $\theta_6 = (\varepsilon\lambda^{\frac{3+\alpha-\gamma}{2}})^{\frac{1}{1+\alpha}}$.

As a result, we deduce from the estimates (7.4.8)–(7.4.13) that

$$N_{(t_0-2\tau_0+\lambda,t_0)}[\boldsymbol{v},p_1,r]\leqslant c_3\big(\delta,|\nabla\boldsymbol{f}|_{Q_0'}^{(\alpha,\frac{1+\alpha-\gamma}{2})}\big)\Big\{(\varepsilon+\tau_0)N_{(t_0-2\tau_0+\frac{\lambda}{2},t_0)}[\boldsymbol{v},p_1,r]$$

$$+\lambda^{-\frac{\alpha}{2}}|\boldsymbol{f}|_{x,Q_0'}^{(\alpha)}+\lambda^{-\frac{1+\alpha-\gamma}{2}}|\boldsymbol{f}|_{t,Q_0'}^{(\frac{1+\alpha-\gamma}{2})}$$

$$+c(\varepsilon)\lambda^{-\varkappa}\int_{t_0-2\tau_0}^{t_0}\Big(\|\boldsymbol{u}(\cdot,\tau)\|_{\Omega}+\|r(\cdot,\tau)\|_{W_2^1(S_1)}\Big)\,\mathrm{d}\tau\Big\},\qquad(7.4.14)$$

here $N_{(t_0-2\tau_0+\lambda,t_0)}[\boldsymbol{v},p_1,r]=N_{(t_0-2\tau_0+\lambda,t_0)}[\boldsymbol{v},p_1]+|r|_{\xi',S_1\times(t_0-2\tau_0+\lambda,t_0)}^{(3+\alpha)}$, $\varkappa=\max\{\frac{9}{2\gamma}+\frac{\alpha}{\gamma},\frac{11}{4}+\frac{\alpha}{2},(\frac{11}{2}+\alpha)(1+\frac{\alpha}{2}),\frac{2+\alpha}{2}\big(1+\frac{9}{2(1+\alpha)}\big),\frac{3+\alpha-\gamma}{2}\big(1+\frac{7}{2(1+\alpha)}\big)\}$.

We introduce the function $\Phi(\lambda)=\lambda^{\varkappa}N_{(t_0-2\tau_0+\lambda,t_0)}[\boldsymbol{v},p,r]$ and write (7.4.14) in the following way:

$$\Phi(\lambda)\leqslant c_4(\varepsilon+\tau_0)\Phi(\lambda/2)+K,\qquad(7.4.15)$$

where $c_4=c_3(\delta)2^{\varkappa}$,

$$K=c_3\big(\delta,|\nabla\boldsymbol{f}|_{Q_0'}^{(\alpha,\frac{1+\alpha-\gamma}{2})}\big)\Big\{|\boldsymbol{f}|_{Q_0'}^{(\alpha,\frac{1+\alpha-\gamma}{2})}+c(\varepsilon)\int_{t_0-2\tau_0}^{t_0}(\|\boldsymbol{u}(\cdot,\tau)\|_{\Omega}+\|r(\cdot,\tau)\|_{W_2^1(S_1)})\mathrm{d}\tau\Big\}.$$

Setting $\varepsilon+\tau_0=\frac{1}{2c_4}$, we deduce from (7.4.15), by iterating it with $\lambda/2,\ldots,\lambda/2^k$ and taking the limit as $k\to\infty$, the inequality

$$\Phi(\lambda)\leqslant 2K.$$

This estimate with $\lambda=\tau_0$ implies (7.4.4). □

Lemma 7.4.2 *Let $r_0\in C^{1+\alpha}(S_1)$ and $\boldsymbol{u}\in C^{1+\alpha,0}(D_{\Sigma T_0})$, $\alpha\in(0,1)$. Then $r(\cdot,t)\in C^{1+\alpha}(S_1)$ for every $t\in(0,T_0)$ and the inequality*

$$|r(\cdot,t)|_{S_1}^{(1+\alpha)}\leqslant c_5\big(|r_0|_{S_1}^{(1+\alpha)}+t|\boldsymbol{u}|_{\xi,D_{\Sigma t}}^{(1+\alpha)}\big)\qquad(7.4.16)$$

holds provided that the norms of r_0 and $\boldsymbol{u}$ are small.

Similar to Proposition 4.5 in [89], this lemma is proved by passing to Lagrangian coordinates (1.1.2).

Proof We introduce distance function to the sphere S_{R_0} $R_1(x)=\pm dist\{x,S_{R_0}\}$, where $R_1(x)<0$ if x lies inside S_{R_0}, and $R_1(x)>0$ if x lies outside S_{R_0}. The equation $R_1(x)=r_0\big(\frac{x}{|x|}\big)$, $x\in\Gamma$, defines the surface Γ, and the function $r_1(\xi,t)\equiv R_1(X(\xi,t)-h(t))=|\boldsymbol{X}(\xi,t)-\boldsymbol{h}(t)|-R_0$, $\xi\in\Gamma$, defines the surface Γ_t. We note that $\xi=x$ for $t=0$, and in view of (7.2.3), it is clear that $r_1(\xi,0)\equiv|\boldsymbol{\xi}|-R_0=r_0(\frac{x}{|x|})$, $\xi\in\Gamma$. By (1.1.2), we have

$$\boldsymbol{X}(\xi,t)-\boldsymbol{h}(t)=\boldsymbol{\xi}+\int_0^t(\boldsymbol{u}(\xi,\tau)-\dot{\boldsymbol{h}}(\tau))\,\mathrm{d}\tau$$

from which (in view of (7.2.6)) it follows that

$$r_1(\xi,t)=R_1(\xi)+R_1(X(\xi,t)-h(t))-R_1(\xi)=R_1(\xi)+\int_0^1\frac{\mathrm{d}}{\mathrm{d}s}R_1(X_s)\,\mathrm{d}s$$
$$=R_1(\xi)+\int_0^1\nabla_{X_s}R_1(X_s)\,\mathrm{d}s\cdot\int_0^t\Big(\boldsymbol{u}(\xi,\tau)-\frac{1}{|\Omega_0^+|}\int_{\Omega_0^+}\boldsymbol{u}(\xi',\tau)\,\mathrm{d}\xi'\Big)\mathrm{d}\tau,$$

where $\boldsymbol{X}_s(\xi,t)=\boldsymbol{\xi}+s\int_0^t\Big(\boldsymbol{u}(\xi,\tau)-\frac{1}{|\Omega_0^+|}\int_{\Omega_0^+}\boldsymbol{u}(\xi',\tau)\,\mathrm{d}\xi'\Big)\mathrm{d}\tau.$

It is obvious that

$$|r_1(\cdot,t)|_\Gamma^{(1+\alpha)}\leqslant c\Big\{|r_0|_{S_1}^{(1+\alpha)}+t\max_{s\in[0,1]}|\nabla_{X_s}R_1(X_s)|_\Gamma^{(1+\alpha)}\times$$
$$\times\Big(|\boldsymbol{u}|_{\xi,G_t}^{(1+\alpha)}+\frac{1}{|\Omega_0^+|}\max_{\tau\in(0,t)}\Big|\int_{\Omega_0^+}\boldsymbol{u}(\xi',\tau)\,\mathrm{d}\xi'\Big|\Big)\Big\}.$$

Since $R_1(x)$ is a smooth function of x in the neighborhood of Γ_t, and the norm of the argument of X_s is bounded, the inequality

$$|\nabla_{X_s}R_1(X_s)|_\Gamma^{(1+\alpha)}\leqslant c|X_s(\cdot,t)|_\Gamma^{(1+\alpha)}\leqslant c\Big(1+t\Big(|\boldsymbol{u}|_{\xi,G_t}^{(1+\alpha)}+\max_{\tau\in(0,t)}\Big|\int_{\Omega_0^+}\boldsymbol{u}(\xi',\tau)\,\mathrm{d}\xi'\Big|\Big)\Big)\leqslant c$$

holds. Thus,

$$|r_1(\cdot,t)|_\Gamma^{(1+\alpha)}\leqslant c\big(|r_0|_{S_1}^{(1+\alpha)}+t|\boldsymbol{u}|_{\xi,D_{\Sigma t}}^{(1+\alpha)}\big). \tag{7.4.17}$$

We consider the transformation

$$y=\frac{X(\xi,t)-h(t)}{|X(\xi,t)-h(t)|}R_0\equiv\mathcal{T}(\xi,t),\quad \xi\in\Gamma.$$

For sufficiently small $\boldsymbol{u}$, this is an invertible mapping of Γ on S_{R_0} of class $C^{1+\alpha}$ in ξ, $r_1(\mathcal{T}^{-1}(y,t),t)$ being equal to $r(\frac{y}{R_0},t)$. Therefore

$$|r(\cdot,t)|_{S_1}^{(1+\alpha)}\leqslant c|r_1|_\Gamma^{(1+\alpha)},$$

which, together with (7.4.17), implies (7.4.16). □

Now we can prove Theorem 7.1.2.

Proof By Theorem 7.1.1 there exists a solution $(\boldsymbol{u}, q)$ to Problem (7.1.4) on an interval $(0, T_0]$. Let ε in condition (7.1.6) be so small that the value of $T_0 > 1$. The norm of the solution $(\boldsymbol{u}, q)$ satisfies estimate (7.1.7). After returning to the Eulerian coordinates, we have

$$N_{(0,T_0)}[\boldsymbol{v}, p_1] \leqslant c\big(|\mathcal{D}_x \boldsymbol{f}|_{Q_{T_0}}^{(\alpha, \frac{1+\alpha-\gamma}{2})}\big)\Big(|\boldsymbol{f}|_{Q_{T_0}}^{(\alpha, \frac{1+\alpha-\gamma}{2})} + |\boldsymbol{v}_0|_{\cup\Omega_0^\pm}^{(2+\alpha)} + |H_0 + \frac{2}{R_0}|_\Gamma^{(1+\alpha)}\Big)$$
$$\leqslant c\Big(|\boldsymbol{f}|_{Q_{T_0}}^{(\alpha, \frac{1+\alpha-\gamma}{2})} + |\boldsymbol{v}_0|_{\cup\Omega_0^\pm}^{(2+\alpha)} + |r_0|_{S_1}^{(3+\alpha)}\Big) \leqslant c_3\varepsilon \equiv \mu. \tag{7.4.18}$$

In the last inequality we have used the estimate

$$\Big|H_0 + \frac{2}{R_0}\Big|_\Gamma^{(1+\alpha)} \leqslant c|r_0|_{S_1}^{(3+\alpha)}$$

obtained from formulas (7.2.28), (7.2.29) for $t = 0$.

According to Proposition 7.4.1, there exists $\tau_0 < T_0/2$ such that (6.4.18) is satisfied and estimate (7.4.4) holds for $(\boldsymbol{v}, p_1)$ and T_0. Lemma 7.4.2 guarantees the inequality

$$|r|_{y,S_1\times(0,T_0)}^{(1+\alpha)} \leqslant c_5\big(|r_0|_{S_1}^{(1+\alpha)} + \mu T_0\big) \leqslant \delta_1 R_0$$

($|r|_{y,S_1\times(0,T_0)}^{(1+\alpha)} \equiv \sup_{0<\tau<T_0} |r(\cdot,\tau)|_{S_1}^{(1+\alpha)}$) if ε in condition (7.1.8) is sufficiently small. This allows us to apply Proposition 7.2.1, and (7.4.4) combined with (7.2.9) implies that

$$N_{(t_0-\tau_0,t_0)}[\boldsymbol{v}, p_1, r] \leqslant c_6\Big(|\nabla \boldsymbol{f}|_{Q_0'}^{(\alpha, \frac{1+\alpha-\gamma}{2})}\Big)\, \mathrm{e}^{-b(t_0-2\tau_0)}\Big\{|\mathrm{e}^{bt}\boldsymbol{f}|_{Q_0'}^{(\alpha, \frac{1+\alpha-\gamma}{2})} + \|\mathrm{e}^{b\tau}\boldsymbol{f}\|_{Q_0'}^2$$
$$+ \|\boldsymbol{v}_0\|_\Omega + \|r_0\|_{W_2^1(S_1)}\Big\}$$
$$\leqslant c_7(\tau_0)\mathrm{e}^{-bt_0}\big(1 + |\Omega|^{\frac{1}{2}} + 2\pi^{\frac{1}{2}}\big)\varepsilon,$$

where $|\Omega|$ is the measure of Ω, $t_0 \in (2\tau_0, T_0]$.

For $t_0 = T_0$, it follows from (7.4.18) that

$$|\boldsymbol{v}(\cdot, T_0)|_{\cup\Omega_{T_0}^\pm}^{(2+\alpha)} + |r(\cdot, T_0)|_{S_1}^{(3+\alpha)} \leqslant \mu. \tag{7.4.19}$$

Therefore we can apply Theorem 7.1.1 again and get a solution in the interval $(T_0, T_0 + T_1]$ for the initial data $\boldsymbol{v}(\cdot, T_0)$, $r(\cdot, T_0)$. The norm of the solution satisfies the inequality

$$N_{(T_0,T_0+T_1)}[\boldsymbol{v}, p_1] \leqslant \mu_1.$$

By Proposition 7.4.1, we can find $0 < \tau_1 < T_1/2$ such that estimate (6.4.18) holds and

$$N_{(T_0+T_1-\tau_1,T_0+T_1)}[\boldsymbol{v},p_1,r] \leqslant c(\delta,\tau_1)\Big\{|\boldsymbol{f}|^{(\alpha,\frac{1+\alpha-\gamma}{2})}_{Q_{(T_0+T_1-2\tau_1,T_0+T_1)}} + \int\limits_{T_0+T_1-2\tau_1}^{T_0+T_1} \big(\|\boldsymbol{v}(\cdot,\tau)\|_{\Omega}+\|r(\cdot,0)\|_{W_2^1(S_1)}\big)\,\mathrm{d}\tau\Big\}. \tag{7.4.20}$$

Since in view of (7.4.16) and (7.4.19)

$$|r|^{(1+\alpha,0)}_{S_1\times(T_0,T_0+T_1)} \leqslant c_5\big(|r(\cdot,0)|^{(1+\alpha)}_{S_1}+T_1|\boldsymbol{u}|^{(1+\alpha)}_{\xi,D_{(T_0,T_0+T_1)}}\big) \leqslant c_2(c_1\varepsilon+T_1\mu_1) \leqslant \delta_1 R_0,$$

then similar to (6.4.24), we can continue (7.4.20) because of Proposition 7.2.1 as follows:

$$N_{(T_0+T_1-\tau_1,T_0+T_1)}[\boldsymbol{v},p_1,r] \leqslant c_7(\tau_1)\mathrm{e}^{-b_1(T_0+T_1)}\big(1+|\Omega|^{1/2}+2\pi^{1/2}\big)\varepsilon.$$

We choose ε such small that $c_7(\tau_1)\big(1+|\Omega|^{1/2}+2\pi^{1/2}\big)\varepsilon \leqslant \mu$.

Hence,

$$|\boldsymbol{v}(\cdot,T_0+T_1)|^{(2+\alpha)}_{\cup\Omega^{\pm}_{T_0+T_1}} + |r(\cdot,T_0+T_1)|^{(3+\alpha)}_{S_1} \leqslant \mu\mathrm{e}^{-b_1(T_0+T_1)}.$$

Thus, we have proved that the norms of the initial data do not increase. Therefore we can extend the solution on the interval $(T_0+T_1,T_0+2T_1]$.

When repeating our arguments for obtaining exponential estimates, we need to pass to the Lagrangian coordinates according to the formula

$$\boldsymbol{X}(\xi^{(1)},t) = \boldsymbol{\xi}^{(1)} + \int\limits_{T_0}^{t} \widetilde{\boldsymbol{u}}(\xi^{(1)},\tau)\,\mathrm{d}\tau, \qquad \xi^{(1)}\in\cup\Omega^{\pm}_{T_0}, \quad t\in(T_0,T_0+T_1). \tag{7.4.21}$$

In fact, it follows from the additivity of the integral, that formula (7.4.21) coincides with (1.1.2):

$$\boldsymbol{X}(\xi,t) = \boldsymbol{\xi} + \int\limits_{0}^{T_0} \boldsymbol{u}(\xi,\tau)\,\mathrm{d}\tau + \int\limits_{T_0}^{t} \boldsymbol{u}(\xi,\tau)\,\mathrm{d}\tau, \qquad \xi\in\cup\Omega^{\pm}_{0}, \quad t\in(T_0,T_0+T_1),$$

because $\widetilde{\boldsymbol{u}}(\xi^{(1)},\tau) = \boldsymbol{u}(\xi,\tau) \equiv \boldsymbol{v}(\boldsymbol{X}(\xi,t))$.

The same remark applies to the barycenter of the inner fluid, since the volume of the fluid is conserved:

$$\begin{aligned}
\boldsymbol{h}(t) &= \boldsymbol{h}(T_0) + \int\limits_{T_0}^{t} \frac{1}{|\Omega_\tau^+|} \int\limits_{\Omega_\tau^+} \boldsymbol{v}(x,\tau)\,\mathrm{d}x\,\mathrm{d}\tau \\
&= \int\limits_{0}^{T_0} \frac{1}{|\Omega_\tau^+|} \int\limits_{\Omega_\tau^+} \boldsymbol{v}(x,\tau)\,\mathrm{d}x\,\mathrm{d}\tau + \int\limits_{T_0}^{t} \frac{1}{|\Omega_\tau^+|} \int\limits_{\Omega_\tau^+} \boldsymbol{v}(x,\tau)\,\mathrm{d}x\,\mathrm{d}\tau \\
&= \frac{3}{4\pi R_0^3} \int\limits_{0}^{t} \int\limits_{\Omega_\tau^+} \boldsymbol{v}(x,\tau)\,\mathrm{d}x\,\mathrm{d}\tau, \qquad t > T_0. \qquad (7.4.22)
\end{aligned}$$

The solution of system (7.1.3), (1.0.3) can thus be extended to any time interval, moreover, inequality (7.1.9) holds for it.

The limiting position of gravity center is estimated on the basis of Corollary 7.2.1 and (7.4.22):

$$\begin{aligned}
|h_\infty| &\leqslant \frac{c_8}{|\Omega_0^+|^{1/2} b} \Big\{ \|\mathrm{e}^{bt}\boldsymbol{f}\|_{Q_\infty} + \|\boldsymbol{v}_0\|_\Omega + \|r_0\|_{W_2^1(S_1)} \Big\} \\
&\leqslant c_9 \Big\{ \|\mathrm{e}^{bt}\boldsymbol{f}\|_{Q_\infty} + |\boldsymbol{v}_0|^{(2+\alpha)}_{\cup\Omega_0^\pm} + |r_0|^{(3+\alpha)}_{S_1} \Big\} \leqslant c_9\varepsilon. \qquad (7.4.23)
\end{aligned}$$

Consequently, the initial distance between the surface Γ and Σ should be strictly larger than

$$c_9 \Big(\|\mathrm{e}^{bt}\boldsymbol{f}\|_{Q_\infty} + |\boldsymbol{v}_0|^{(2+\alpha)}_{\cup\Omega_0^\pm} + |r_0|^{(3+\alpha)}_{S_1} \Big) + \delta_1 R_0$$

with δ_1, R_0 from (7.2.8), to exclude the intersection of these surfaces in the future.

The uniqueness of the solution is proved in the same way as it has been done for the local case in Chap. 5. □

Chapter 8
Thermocapillary Convection Problem

Abstract This chapter deals with the unsteady motion of two viscous incompressible fluids separated by a closed unknown interface Γ_t. Both liquids have a finite volume; they are bounded by a given surface Σ where the adhesion condition holds. On interface Γ_t, we take into account surface tension depending on the temperature. The boundaries Γ_t and Σ are assumed to have no intersection.

8.1 Setting of the Problem and Statement of Results

The motion under the influence of the tangential gradient of surface tension is called thermocapillary convection, or the Marangoni effect, since K. Marangoni was one of the first scientists who studied and explained this phenomenon. It is known that the Marangoni effect plays a principal role in thermodynamics under low gravity.

The main result of our study is local-in-time unique solvability of the problem described in the Hölder spaces of functions, it is firstly published in [19]. The proof of this fact applies the technique developed for the investigation of thermocapillary convection problem for a drop in vacuum [48], and it is based on the existence theorem for the case of constant temperature, i.e., we rely on the results of Chaps. 2, 4, and 5. We observe that the problem of thermocapillary convection in the Sobolev spaces was considered by N. Tanaka [97]. He also took advantage of the ideas from [48] and even formally transferred the form of boundary conditions for one fluid to the two-phase case, which corresponds to the physics of the process only in the zero approximation.

We give a more precise mathematical setting of the problem governing the thermocapillary convection for two fluids in a reservoir [63, 64]. We retain the notation of the previous chapters.

© The Author(s), under exclusive license to Springer Nature Switzerland AG 2021
I. V. Denisova, V. A. Solonnikov, *Motion of a Drop in an Incompressible Fluid*, Advances in Mathematical Fluid Mechanics,
https://doi.org/10.1007/978-3-030-70053-9_8

Let, as in the previous chapter, the fluids have bounded volume, and, at the initial instant $t = 0$, Γ_0 and $\Sigma \equiv \partial(\Omega_0^+ \cup \Gamma_0 \cup \Omega_0^-)$ be given closed surfaces, $\Sigma \cap \Gamma_0 = \emptyset$.

For every $t > 0$, it is necessary to find the interface Γ_t between domains Ω_t^+ and Ω_t^-, as well as the velocity vector field $\boldsymbol{v}(x,t) = (v_1, v_2, v_3)$, the pressure function p, and the temperature function θ of both fluids satisfying the following initial-boundary value problem

$$\begin{aligned}
&\mathcal{D}_t\boldsymbol{v} + (\boldsymbol{v}\cdot\nabla)\boldsymbol{v} - \nu^\pm\nabla^2\boldsymbol{v} + \frac{1}{\rho^\pm}\nabla p = \boldsymbol{f}, \quad \nabla\cdot\boldsymbol{v} = 0,\\
&\mathcal{D}_t\theta + (\boldsymbol{v}\cdot\nabla)\theta - k^\pm\nabla^2\theta = 0 \quad \text{in} \quad \Omega_t^-\cup\Omega_t^+, \quad t>0,\\
&\boldsymbol{v}|_{t=0} = \boldsymbol{v}_0, \quad \theta|_{t=0} = \theta_0 \quad \text{in} \quad \Omega_0^-\cup\Omega_0^+, \qquad\qquad (8.1.1)\\
&[\boldsymbol{v}]|_{\Gamma_t} = 0, \quad [\theta]|_{\Gamma_t} = 0, \quad \boldsymbol{v}|_\Sigma = 0, \quad \theta|_\Sigma = a,\\
&[\mathbb{T}\boldsymbol{n}]|_{\Gamma_t} = \sigma(\theta)H\boldsymbol{n} + \nabla_{\Gamma_t}\sigma(\theta), \qquad \left[k^\pm\frac{\partial\theta}{\partial\boldsymbol{n}}\right]\Big|_{\Gamma_t} + \varkappa\theta\nabla_{\Gamma_t}\cdot\boldsymbol{v} = 0 \ \text{on}\ \Gamma_t.
\end{aligned}$$

Here $\boldsymbol{f}$ is the given vector field of mass forces, $\boldsymbol{v}_0$, θ_0 are the initial distributions of viscosity and temperature, respectively, a is the given temperature on the surface Σ, $\mu^\pm = \nu^\pm\rho^\pm$, $\sigma(\theta) = \sigma_1 - \varkappa(\theta - \theta_1) > 0$ is the coefficient of the surface tension, $\sigma_1, \varkappa, \theta_1$ are the positive constants, $k^\pm$ is the step functions of thermal conductivity. The temperature coefficient of the surface tension $\varkappa$ is a sufficiently small constant compared to σ_1, so that for a limited change of θ, from crystallization temperature to the boiling point, and for a correct choice of the parameters σ_1 and θ_1, $\sigma(\theta)$ always remains a positive function.

As before, we assume the absence of mass transfer across Γ_t, i.e., the validity of (1.0.3). Hence, $\Gamma_t = \{x(\xi,t)|\,\xi\in\Gamma_0\}$, $\Omega_t^\pm = \{x(\xi,t)|\,\xi\in\Omega_0^\pm\}$.

We pass from the Eulerian to Lagrangian coordinates by formula (1.1.2), apply formula for twice the mean curvature (1.1.3) and divide the boundary condition for the jump of normal stresses into the normal and tangential parts. As a result, we arrive at a problem for the velocity $\boldsymbol{u}$ and the pressure $q = p(X_{\boldsymbol{u}}, t)$ in the Lagrangian coordinates with the known interface $\Gamma \equiv \Gamma_0$. If the angle between $\boldsymbol{n}$ and the outer normal $\boldsymbol{n}_0$ to Γ is acute, the original problem is equivalent to the following system:

$$\begin{aligned}
&\mathcal{D}_t\boldsymbol{u} - \nu^\pm\nabla_{\boldsymbol{u}}^2\boldsymbol{u} + \frac{1}{\rho^\pm}\nabla_{\boldsymbol{u}}q = \boldsymbol{f}(X_{\boldsymbol{u}}, t),\\
&\nabla_{\boldsymbol{u}}\cdot\boldsymbol{u} = 0, \qquad \mathcal{D}_t\hat{\theta} - k^\pm\nabla_{\boldsymbol{u}}^2\hat{\theta} = 0 \qquad \text{in} \quad D_{\Sigma T},\\
&\boldsymbol{u}|_{t=0} = \boldsymbol{v}_0, \qquad \hat{\theta}|_{t=0} = \theta_0 \qquad \text{in} \quad \Omega_0^-\cup\Omega_0^+, \qquad\qquad (8.1.2)\\
&[\boldsymbol{u}]|_{G_T} = 0, \quad [\hat{\theta}]|_{G_T} = 0, \quad \boldsymbol{u}|_{\Upsilon_T} = 0, \quad \hat{\theta}|_{\Upsilon_T} = \hat{a} \quad (\Upsilon_T = \Sigma\times(0,T)),
\end{aligned}$$

$$[\mu^{\pm}\Pi_0\Pi\mathbb{S}_{\boldsymbol{u}}(\boldsymbol{u})\boldsymbol{n}]|_{G_T} = \Pi_0\Pi\nabla_{\boldsymbol{u}}\sigma(\hat{\theta}),$$

$$[\boldsymbol{n}_0\cdot\mathbb{T}_{\boldsymbol{u}}(\boldsymbol{u},q)\boldsymbol{n}]|_{G_T} - \sigma(\hat{\theta})\boldsymbol{n}_0\cdot\Delta(t)\int_0^t \boldsymbol{u}\big|_{\Gamma}\mathrm{d}\tau = \sigma(\hat{\theta})H_0(\xi) +$$

$$+\sigma(\hat{\theta})\boldsymbol{n}_0\cdot\int_0^t \dot{\Delta}(\tau)\boldsymbol{\xi}\big|_{\Gamma}\mathrm{d}\tau + \boldsymbol{n}_0\cdot\Pi\nabla_{\boldsymbol{u}}\sigma(\hat{\theta}),$$

$$[k^{\pm}\boldsymbol{n}\cdot\nabla_{\boldsymbol{u}}\hat{\theta}]|_{G_T} + \varkappa\hat{\theta}\Pi\nabla_{\boldsymbol{u}}\cdot\boldsymbol{u} = 0 \qquad \text{on} \quad G_T = \Gamma\times(0,T).$$

Here we use the same notation as in system (1.1.5), $D_{\Sigma T} \equiv (\Omega_0^+\cup\Omega_0^-)\times(0,T)$ is a bounded cylinder for $T>0$.

The main result of this chapter is the theorem as follows:

Theorem 8.1.1 *We assume that* $\Gamma \in C^{3+\alpha}$, $\Sigma \in C^{2+\alpha}$, $\boldsymbol{f}, D_x\boldsymbol{f} \in \boldsymbol{C}^{\alpha,(1+\alpha-\gamma)/2}(\mathbb{R}^3\times(0,T))$, $\boldsymbol{v}_0 \in \boldsymbol{C}^{2+\alpha}(\Omega_0^-\cup\Omega_0^+)$, $\theta_0 \in C^{2+\alpha}(\Omega_0^-\cup\Omega_0^+)$, $\sigma \in C^{3+\alpha}(\mathbb{R}_+)$, $\sigma \geqslant \sigma_0 > 0$, $a \in C^{2+\alpha,1+\alpha/2}(\Upsilon_T)$, $a>0$, *with some* $\alpha \in (0,1)$, $\gamma \in (0,\alpha)$, $T<\infty$. *In addition, let compatibility conditions hold:*

$$\nabla\cdot\boldsymbol{v}_0 = 0,\ \ \boldsymbol{v}_0|_{\Sigma} = 0,\ \ [\boldsymbol{v}_0]|_{\Gamma} = 0,\ \ [\theta_0]|_{\Gamma} = 0,\ \ \theta_0|_{\Sigma} = a|_{t=0},$$

$$[\mu^{\pm}\Pi_0\mathbb{S}(\boldsymbol{v}_0)\boldsymbol{n}_0]\big|_{\Gamma} = \Pi_0\nabla\sigma(\theta_0), \qquad [\Pi_0(\nu^{\pm}\nabla^2\boldsymbol{v}_0 - \frac{1}{\rho^{\pm}}\nabla q_0)]\big|_{\Gamma} = 0,$$

$$\big(\Pi_{\Sigma}(\nu^-\nabla^2\boldsymbol{v}_0 - \frac{1}{\rho^-}\nabla q_0 + \boldsymbol{f}\big|_{t=0})\big)\Big|_{\Sigma} = 0, \quad k^-\nabla^2\theta_0|_{\Sigma} = \mathcal{D}_t a\big|_{t=0}, \tag{8.1.3}$$

$$[k^{\pm}\nabla^2\theta_0]|_{\Gamma} = 0, \quad \left[k^{\pm}\frac{\partial\theta_0}{\partial\boldsymbol{n}_0}\right]\Big|_{\Gamma} + \varkappa\theta_0(\Pi_0\nabla)\cdot\boldsymbol{v}_0 = 0,$$

where $q_0(\xi) \equiv q(\xi,0)$ *is a solution to diffraction problem*

$$\frac{1}{\rho^{\pm}}\nabla^2 q_0(\xi) = \nabla\cdot(\boldsymbol{f}(\xi,0) - \mathcal{D}_t\mathbb{B}^*\big|_{t=0}\boldsymbol{v}_0(\xi)), \qquad \xi \in \Omega_0^-\cup\Omega_0^+,$$

$$[q_0]|_{\Gamma} = \left[2\mu^{\pm}\frac{\partial\boldsymbol{v}_0}{\partial\boldsymbol{n}_0}\cdot\boldsymbol{n}_0\right]\Big|_{\Gamma} - \sigma(\theta_0)H_0(\xi), \qquad \xi\in\Gamma,$$

$$\left[\frac{1}{\rho^{\pm}}\frac{\partial q_0}{\partial\boldsymbol{n}_0}\right]\Big|_{\Gamma} = [\nu^{\pm}\boldsymbol{n}_0\cdot\nabla^2\boldsymbol{v}_0]\big|_{\Gamma}, \tag{8.1.4}$$

$$\frac{1}{\rho^-}\frac{\partial q_0}{\partial\boldsymbol{n}_{\Sigma}}\Big|_{\Sigma} = \nu^-\boldsymbol{n}_{\Sigma}\cdot\nabla^2\boldsymbol{v}_0\big|_{\Sigma} + \boldsymbol{n}_{\Sigma}\cdot\boldsymbol{f}\big|_{\Sigma,t=0}.$$

Here $H_0(\xi) = \boldsymbol{n}_0\cdot\Delta(0)\xi\big|_{\Gamma}$, $\mathbb{B} = \mathbb{A}-\mathbb{I}$, $\mathbb{B}^*$ *is the transpose to* $\mathbb{B}$, $\boldsymbol{n}_{\Sigma}$ *is the outward normal to* Σ, $\Pi_{\Sigma}\boldsymbol{\omega} \equiv \boldsymbol{\omega} - \boldsymbol{n}_{\Sigma}(\boldsymbol{n}_{\Sigma}\cdot\boldsymbol{\omega})$.

Then there exists a positive constant $T_0 \leqslant T$ *such that problem* (8.1.2) *has a unique solution* $(\boldsymbol{u}, q, \hat{\theta})$ *with the differential properties:* $\boldsymbol{u} \in \boldsymbol{C}^{2+\alpha,1+\alpha/2}(D_{\Sigma T_0})$, $q \in C^{(\gamma,1+\alpha)}(D_{\Sigma T_0})$, $\nabla q \in \boldsymbol{C}^{\alpha,\alpha/2}(D_{\Sigma T_0})$, $\hat{\theta} \in$

$C^{2+\alpha,1+\alpha/2}(D_{\Sigma T_0})$. *The solution satisfies the inequality*

$$|\boldsymbol{u}|^{(2+\alpha,1+\alpha/2)}_{D_{\Sigma T_0}} + |\nabla q|^{(\alpha,\alpha/2)}_{D_{\Sigma T_0}} + |q|^{(\gamma,1+\alpha)}_{D_{\Sigma T_0}} + |\hat{\theta}|^{(2+\alpha,1+\alpha/2)}_{D_{\Sigma T_0}}$$
$$\leqslant c(T_0)\Big\{|\boldsymbol{f}|^{(\alpha,\frac{1+\alpha-\gamma}{2})}_{\mathbb{R}^3_{T_0}} + |\boldsymbol{v}_0|^{(2+\alpha)}_{\cup\Omega^\pm} + |\theta_0|^{(2+\alpha)}_{\cup\Omega^\pm}$$
$$+ |a|^{(2+\alpha,1+\alpha/2)}_{\Upsilon_{T_0}} + |H_0|^{(1+\alpha)}_{\Gamma}\Big\}, \qquad (8.1.5)$$

where $c(T_0)$ *depends on the norms of the given functions mentioned in Theorem assumptions. The value of* T_0 *also depends on data norms, on the curvature of* Γ *and on the distance between the surfaces* Γ *and* Σ.

Remark 8.1.1 *We note that the function* σ *should be close to a linear one, given in system* (8.1.1)*, otherwise the setting of the problem becomes more complicated.*

The proof of Theorem 8.1.1 is based on the solvability of auxiliary linearized problems.

8.2 Linearized Problems

First, we discuss system (7.3.1). This is a linearized problem under the assumption that the temperature is constant. Theorem 7.3.1 is valid for it.

Second, we also need the solvability of the problem for temperature function ψ:

$$\begin{aligned}
&\mathcal{D}_t\psi - k^\pm\nabla^2_{\boldsymbol{u}}\psi = f \quad \text{in } D_{\Sigma T},\\
&\psi|_{t=0} = \psi_0 \quad \text{in } \Omega^-\cup\Omega^+,\\
&[\psi]|_{G_T} = 0, \qquad \psi|_{\Upsilon_T} = \varphi, \\
&[k^\pm\boldsymbol{n}\cdot\nabla_{\boldsymbol{u}}\psi]|_{G_T} + \varkappa\psi(\Pi\nabla_{\boldsymbol{u}})\cdot\boldsymbol{w} = d \quad \text{on } G_T,
\end{aligned} \qquad (8.2.1)$$

where $\boldsymbol{u}$, $\boldsymbol{w}$ are given vectors.

Remark 8.2.1 *We observe that the differential operator* $\Pi\nabla_{\boldsymbol{u}} = A_{ij}\frac{\partial}{\partial\xi_j} - n_in_kA_{kj}\frac{\partial}{\partial\xi_j}$ *does not contain the derivative* $\frac{\partial}{\partial\boldsymbol{n}_0}$. *Indeed, let us consider the coefficient by* $\frac{\partial}{\partial\boldsymbol{n}_0}$ *in this expression. It is equal to* $A_{ij}n_{0j} - n_in_kA_{kj}n_{0j} = |\mathbb{A}\boldsymbol{n}_0|(n_i - n_in_kn_k) = 0$. *Hence, the quantity* $(\Pi\nabla_{\boldsymbol{u}})\cdot\boldsymbol{w}$ *is well defined on the boundary* Γ*, the vector field* $\boldsymbol{w}$ *being continuous when passing across* Γ.

Theorem 8.2.1 *Let surfaces* $\Gamma, \Sigma \in C^{2+\alpha}$*, and vectors* $\boldsymbol{u}, \boldsymbol{w} \in \boldsymbol{C}^{2+\alpha,1+\frac{\alpha}{2}}(D_{\Sigma T})$ *for* $T > 0$, $[\boldsymbol{u}]|_\Gamma = [\boldsymbol{w}]|_\Gamma = 0$*, satisfy inequality* (7.3.2) *with some* $\delta > 0$. *Then for arbitrary* $f \in C^{\alpha,\alpha/2}(D_{\Sigma T}), \varphi \in C^{\alpha,\alpha/2}(\Upsilon_T), \psi_0 \in C^{2+\alpha}(\Omega^- \cup \Omega^+), d \in C^{1+\alpha,(1+\alpha)/2}(G_T)$ *which satisfy compatibility conditions*

$$\begin{aligned} &[\psi_0]|_\Gamma = 0, \quad \psi_0|_\Sigma = \varphi|_{t=0}, \quad -[k^\pm \nabla^2 \psi_0]|_\Gamma = [f|_{t=0}]|_\Gamma, \\ &[k^\pm \frac{\partial \psi_0}{\partial \boldsymbol{n}_0}]|_\Gamma + \varkappa \psi_0 \nabla_\Gamma \cdot \boldsymbol{w}(\xi, 0) = d(\xi, 0), \quad \xi \in \Gamma, \\ &\mathcal{D}_t \varphi|_{t=0} = k^- \nabla^2 \psi_0|_\Sigma + f|_{\Sigma,t=0}, \end{aligned} \tag{8.2.2}$$

problem (8.2.1) *has a unique solution* $\psi \in C^{2+\alpha,1+\alpha/2}(D_{\Sigma T})$ *and the estimate*

$$\begin{aligned} |\psi|^{(2+\alpha,1+\alpha/2)}_{D_{\Sigma T}} \leqslant c_2(T)\{&|f|^{(\alpha,\alpha/2)}_{D_{\Sigma T}} + |\psi_0|^{(2+\alpha)}_{\cup\Omega^\pm} + |\varphi|^{(2+\alpha,1+\alpha/2)}_{\Upsilon_T} + |d|^{(1+\alpha,\frac{1+\alpha}{2})}_{G_T} \\ &+ T^{\frac{1-\alpha}{2}} |\nabla \boldsymbol{u}|_{D_{\Sigma T}} \left(|\psi_0|^{(1)}_{\cup\Omega^\pm} + |\nabla \boldsymbol{w}|_{D_{\Sigma T}} |\psi_0|_{\cup\Omega^\pm}\right)\} \end{aligned} \tag{8.2.3}$$

holds. Here c_2 *is a nondecreasing function of* T.

Proof We write system (8.2.1) in the form:

$$\begin{aligned} &\mathcal{D}_t \psi - k^\pm \nabla^2 \psi = f + h_1(\psi) \quad \text{in } D_{\Sigma T}, \\ &\psi|_{t=0} = \psi_0 \quad \text{in } \Omega^- \cup \Omega^+, \\ &[\psi]|_{G_T} = 0, \quad \psi|_{\Upsilon_T} = \varphi, \\ &[k^\pm \boldsymbol{n}_0 \cdot \nabla \psi]|_{G_T} + \varkappa \psi (\Pi_0 \nabla) \cdot \boldsymbol{w}|_{G_T} = d + h_2(\psi) + h_3(\psi, \boldsymbol{w}), \end{aligned} \tag{8.2.4}$$

where

$$\begin{aligned} h_1(\psi) &= k^\pm (\nabla^2_{\boldsymbol{u}} \psi - \nabla^2 \psi), \\ h_2(\psi) &= [k^\pm (\boldsymbol{n}_0 \cdot \nabla \psi - \boldsymbol{n} \cdot \nabla_{\boldsymbol{u}} \psi)]\big|_\Gamma, \\ h_3(\psi, \boldsymbol{w}) &= \varkappa \psi (\Pi_0 \nabla - \Pi \nabla_{\boldsymbol{u}}) \cdot \boldsymbol{w}\big|_\Gamma. \end{aligned} \tag{8.2.5}$$

We note that the multiplier $\varkappa(\Pi_0 \nabla) \cdot \boldsymbol{w}|_{G_T}$ of ψ in the third boundary condition belongs to $C^{1+\alpha,\frac{1+\alpha}{2}}(G_T)$. Therefore, under hypotheses (8.2.2), problem (8.2.4) with $h_1(\psi) = h_2(\psi) = h_3(\psi) = 0$ is solvable in $C^{2+\alpha,1+\frac{\alpha}{2}}(D_{\Sigma T})$ [43].

Now, we write problem (8.2.4) in operator form:

$$\begin{aligned} \psi &= \mathcal{L}[f + h_1(\psi), \psi_0, \varphi, d + h_2(\psi) + h_3(\psi, \boldsymbol{w})] \\ &\equiv \mathcal{L}[f, \psi_0, \varphi, d] + \mathcal{K}_{\boldsymbol{u},\boldsymbol{w}}(\psi). \end{aligned} \tag{8.2.6}$$

Here $\mathcal{L}$ is a linear continuous operator from the subspace of $C^{\alpha,\frac{\alpha}{2}}(D_{\Sigma T}) \times C^{2+\alpha}(\Omega^- \cup \Omega^+) \times C^{2+\alpha,1+\frac{\alpha}{2}}(\Upsilon_T) \times C^{1+\alpha,\frac{1+\alpha}{2}}(\Gamma)$, whose elements (f, ψ_0, φ, d) satisfy (8.2.2), into the Hölder space $C^{2+\alpha,1+\frac{\alpha}{2}}(D_{\Sigma T})$; $\mathcal{K}_{\boldsymbol{u},\boldsymbol{w}}(\psi) =$

$\mathcal{L}[h_1(\psi), 0, 0, h_2(\psi)+h_3(\psi, \boldsymbol{w})]$. To every quadruple (f, ψ_0, φ, d), the operator $\mathcal{L}$ assigns a solution of system (8.2.4) with $h_1(\psi) = h_2(\psi) = h_3(\psi) = 0$.

We take into account that

$$\begin{aligned} h_1(\psi) &= k^{\pm}(\mathbb{A}\nabla \cdot \mathbb{B}\nabla)\psi + k^{\pm}\mathbb{B}\nabla \cdot \nabla\psi, \\ h_2(\psi) &= [k^{\pm}((\boldsymbol{n}_0 - \boldsymbol{n}) \cdot \nabla\psi - \boldsymbol{n} \cdot \mathbb{B}\nabla\psi)]\big|_{\Gamma}, \\ h_3(\psi, \boldsymbol{w}) &= \varkappa\psi\{\boldsymbol{n}(\boldsymbol{n} \cdot \nabla) - \boldsymbol{n}_0(\boldsymbol{n}_0 \cdot \nabla) - (\Pi\mathbb{B}\nabla)\} \cdot \boldsymbol{w}\big|_{\Gamma}. \end{aligned}$$

Then Lemma 5.3.1 implies the inequality

$$\begin{aligned} |h_1(\psi)|_{D_{\Sigma T}}^{(\alpha,\alpha/2)} + |h_2(\psi)|_{G_T}^{(1+\alpha,\frac{1+\alpha}{2})} &\leqslant c\{\delta|\psi|_{D_{\Sigma T}}^{(2+\alpha,1+\alpha/2)} + \\ &\quad + T^{\frac{1-\alpha}{2}}|\nabla \boldsymbol{u}|_{D_{\Sigma T}}|\psi_0|_{\cup\Omega^{\pm}}^{(1)}\}, \\ |h_3(\psi, \boldsymbol{w})|_{G_T}^{(1+\alpha,\frac{1+\alpha}{2})} &\leqslant c\{\delta|\psi|_{D_{\Sigma T}}^{(2+\alpha,1+\alpha/2)}|\boldsymbol{w}|_{D_{\Sigma T}}^{(2+\alpha,1+\frac{\alpha}{2})} + \\ &\quad + T^{\frac{1-\alpha}{2}}|\nabla \boldsymbol{u}|_{D_{\Sigma T}}|\psi_0|_{\cup\Omega^{\pm}}|\nabla \boldsymbol{w}|_{D_{\Sigma T}}\}. \end{aligned} \tag{8.2.7}$$

Hence, the operator $\mathcal{K}_{\boldsymbol{u},\boldsymbol{w}}(\psi)$ is a contraction for small δ, i.e., for arbitrary $\psi,\ \psi' \in C^{2+\alpha,1+\frac{\alpha}{2}}(D_{\Sigma T})$ such that $\psi - \psi'|_{t=0} = 0, \quad \psi - \psi'|_{\Upsilon_T} = 0$, the estimate

$$|\mathcal{K}_{\boldsymbol{u},\boldsymbol{w}}(\psi) - \mathcal{K}_{\boldsymbol{u},\boldsymbol{w}}(\psi')|_{D_{\Sigma T}}^{(2+\alpha,1+\alpha/2)} \leqslant \varepsilon|\psi - \psi'|_{D_{\Sigma T}}^{(2+\alpha,1+\alpha/2)}, \quad \varepsilon < 1,$$

holds. Consequently, Eq. (8.2.6) and, hence, problem (8.2.4) are uniquely solvable. Inequality (8.2.3) follows from (8.2.7) and the boundedness of the operator $\mathcal{L}$. □

Now we linearize the boundary conditions in (8.1.2):

$$[\mu^{\pm}\Pi_0\Pi\mathbb{S}_{\boldsymbol{u}}(\boldsymbol{w})\boldsymbol{n}]|_{G_T} = \Pi_0\Pi\nabla_{\boldsymbol{u}}\sigma(\hat{\theta}) \equiv \boldsymbol{K}_1(\hat{\theta}),$$

$$[\boldsymbol{n}_0 \cdot \mathbb{T}_{\boldsymbol{u}}(\boldsymbol{w}, s)\boldsymbol{n}]|_{G_T} - \sigma(\theta_0)\boldsymbol{n}_0 \cdot \Delta(t)\int_0^t \boldsymbol{w}\big|_{\Gamma}\mathrm{d}\tau = K_2(\hat{\theta}) + \int_0^t K_3(\boldsymbol{w}, \hat{\theta})\mathrm{d}\tau \quad \text{on} \quad G_T.$$

Here we have used the notation:

$$K_2(\hat{\theta}) = \sigma(\hat{\theta})H_0(\xi) + (\boldsymbol{n}_0 - \boldsymbol{n}) \cdot \Pi\nabla_{\boldsymbol{u}}\sigma(\hat{\theta}),$$

$$\begin{aligned} K_3(\boldsymbol{w}, \hat{\theta}) &= \sigma(\hat{\theta})\boldsymbol{n}_0 \cdot \dot{\Delta}(t)\boldsymbol{\xi} + (\sigma(\hat{\theta}) - \sigma(\theta_0))\boldsymbol{n}_0 \cdot \Big\{\Delta(t)\boldsymbol{w}\big|_{\Gamma} + \dot{\Delta}(t)\int_0^t \boldsymbol{w}\big|_{\Gamma}\mathrm{d}\tau\Big\} \\ &\quad + \mathcal{D}_t\sigma(\hat{\theta})\boldsymbol{n}_0 \cdot \Big\{\int_0^t \dot{\Delta}(\tau)\boldsymbol{\xi}\mathrm{d}\tau + \Delta(t)\int_0^t \boldsymbol{w}\big|_{\Gamma}\mathrm{d}\tau\Big\}. \end{aligned}$$

Since the operators K_i coincide with ones corresponding to the case of a single liquid, we cite two lemmas concerning them in [48] (see also [54]).

Lemma 8.2.1 *Let a continuous vector* $\boldsymbol{u}$ *across* G_T *be subjected to* (7.3.2). *Then increments of the operators* K_i *with respect to the temperature* $\theta \in C^{2+\alpha,1+\alpha/2}(D_{\Sigma T})$ *and the velocity* $\boldsymbol{w} \in \boldsymbol{C}^{2+\alpha,1+\alpha/2}(D_{\Sigma T})$, *which are also continuous across* G_T,

$$\begin{aligned}\boldsymbol{K}_1(\theta) - \boldsymbol{K}_1(\theta') =& \Pi_0 \Pi \nabla_{\boldsymbol{u}} \big(\sigma(\theta) - \sigma(\theta')\big),\\ K_2(\theta) - K_2(\theta') =& \big(\sigma(\theta) - \sigma(\theta')\big) H_0(\xi) + (\boldsymbol{n}_0 - \boldsymbol{n}) \cdot \Pi \nabla_{\boldsymbol{u}} \big(\sigma(\theta) - \sigma(\theta')\big),\end{aligned}$$

$$\begin{aligned}&K_3(\boldsymbol{w}, \theta) - K_3(\boldsymbol{w}', \theta') =\\ &\quad = (\sigma(\theta) - \sigma(\theta'))\boldsymbol{n}_0 \cdot \left\{ \dot{\Delta}(t)\boldsymbol{\xi} + \Delta(t)\boldsymbol{w}\big|_\Gamma + \dot{\Delta}(t) \int_0^t \boldsymbol{w}\big|_\Gamma \mathrm{d}\tau \right\}\\ &\qquad + (\sigma(\theta') - \sigma(\theta_0))\boldsymbol{n}_0 \cdot \left\{ \Delta(t)(\boldsymbol{w} - \boldsymbol{w}')\big|_\Gamma + \dot{\Delta}(t) \int_0^t (\boldsymbol{w} - \boldsymbol{w}')\big|_\Gamma \mathrm{d}\tau \right\}\\ &\qquad + \big(\mathcal{D}_t\sigma(\theta) - \mathcal{D}_t\sigma(\theta')\big)\boldsymbol{n}_0 \cdot \left(\int_0^t \dot{\Delta}(\tau)\boldsymbol{\xi}\mathrm{d}\tau + \Delta(t) \int_0^t \boldsymbol{w}\big|_\Gamma \mathrm{d}\tau \right) +\\ &\qquad + \mathcal{D}_t\sigma(\theta')\boldsymbol{n}_0 \cdot \Delta(t) \int_0^t (\boldsymbol{w} - \boldsymbol{w}')\big|_\Gamma \mathrm{d}\tau,\end{aligned}$$

where $\theta|_{t=0} = \theta'|_{t=0} = \theta_0$, *satisfy the inequalities*

$$\begin{aligned}&|\boldsymbol{K}_1(\theta) - \boldsymbol{K}_1(\theta')|_{G_T}^{(1+\alpha, \frac{1+\alpha}{2})} + |K_2(\theta) - K_2(\theta')|_{G_T}^{(1+\alpha, \frac{1+\alpha}{2})} \leqslant\\ &\qquad \leqslant c\Big(1 + |\theta|_{D_{\Sigma T}}^{(2+\alpha, 1+\frac{\alpha}{2})} + |\theta'|_{D_{\Sigma T}}^{(2+\alpha, 1+\frac{\alpha}{2})}\Big)^{2+\alpha} \times\\ &\qquad\qquad \times \big(1 + |H_0|_\Gamma^{(1+\alpha)}\big) |\theta - \theta'|_{D_{\Sigma T}}^{(2+\alpha, 1+\frac{\alpha}{2})},\end{aligned}$$

$$\begin{aligned}&|K_3(\boldsymbol{w}, \theta) - K_3(\boldsymbol{w}', \theta')|_{G_T}^{(\alpha, \frac{\alpha}{2})} \leqslant\\ &\quad \leqslant c(T + T^{1/2}) \Big\{ \Big(1 + |\theta|_{D_{\Sigma T}}^{(2+\alpha, 1+\frac{\alpha}{2})} + |\theta'|_{D_{\Sigma T}}^{(2+\alpha, 1+\frac{\alpha}{2})}\Big)^2 \times\\ &\qquad \times \Big(|\nabla \boldsymbol{u}|_{D_{\Sigma T}} + |\boldsymbol{w}|_{D_{\Sigma T}}^{(2+\alpha, 1+\frac{\alpha}{2})} \Big) |\theta - \theta'|_{D_{\Sigma T}}^{(2+\alpha, 1+\frac{\alpha}{2})}\\ &\qquad + \big|\mathcal{D}_t\theta'\big|_{D_{\Sigma T}}^{(\alpha, \frac{\alpha}{2})} (1 + \langle\theta\rangle_{D_{\Sigma T}}^{(\alpha, \frac{\alpha}{2})}) |\boldsymbol{w} - \boldsymbol{w}'|_{D_{\Sigma T}}^{(2+\alpha, 1+\frac{\alpha}{2})} \Big\}\\ &\quad + c|\theta'|_{D_{\Sigma T}}^{(2+\alpha, 1+\frac{\alpha}{2})} \Big(1 + |\theta'|_{D_{\Sigma T}}^{(2+\alpha, 1+\frac{\alpha}{2})}\Big) M_T[\boldsymbol{w} - \boldsymbol{w}'].\end{aligned}$$

Here

$$M_T(\boldsymbol{v}) = \int_0^T |\boldsymbol{v}|^{(2+\alpha)}_{\cup\Omega^\pm}\mathrm{d}t + \sup_{0<\tau<t<T} \tau^{-\alpha/2}\int_{t-\tau}^t (|\nabla \boldsymbol{v}|_{\cup\Omega^\pm} + |\nabla\nabla \boldsymbol{v}|_{\cup\Omega^\pm})\mathrm{d}\tau'. \tag{8.2.8}$$

Lemma 8.2.2 *For operator differences $K_i - K_i'$, where K_i, K_i' correspond to the vectors $\boldsymbol{u}$ and $\boldsymbol{u}'$ satisfying* (7.3.2), *respectively,*

$$\begin{aligned} \boldsymbol{K}_1(\theta) - \boldsymbol{K}_1'(\theta) &= \Pi_0(\Pi\Pi')\mathbb{A}\nabla\sigma(\theta) - \Pi_0\Pi'(\mathbb{B}-\mathbb{B}')\nabla\sigma(\theta), \\ K_2(\theta) - K_2'(\theta) &= \boldsymbol{n}_0\cdot(\Pi-\Pi')\mathbb{A}\nabla\sigma + (\boldsymbol{n}_0-\boldsymbol{n}')\cdot\Pi'(\mathbb{A}-\mathbb{A}')\nabla\sigma - \\ &\quad -(\boldsymbol{n}_0-\boldsymbol{n}')\cdot\Pi\mathbb{A}\nabla\sigma - \boldsymbol{n}'\cdot(\Pi-\Pi')\mathbb{A}\nabla\sigma, \end{aligned}$$

$$\begin{aligned} K_3(\boldsymbol{w},\theta) - K_3'(\boldsymbol{w},\theta) &= \sigma(\theta)\boldsymbol{n}_0\cdot(\dot\Delta(t) - \dot\Delta'(t))\boldsymbol{\xi} + \\ &+(\sigma(\theta)-\sigma(\theta_0))\boldsymbol{n}_0\cdot\Big\{(\Delta(t)-\Delta'(t))\boldsymbol{w} + (\dot\Delta(t)-\dot\Delta'(t))\int_0^t \boldsymbol{w}\mathrm{d}\tau\Big\} \\ &+\mathcal{D}_t\sigma(\theta)\boldsymbol{n}_0\cdot\Big\{\int_0^t(\dot\Delta(\tau)-\dot\Delta'(\tau))\boldsymbol{\xi}\mathrm{d}\tau + (\Delta(t)-\Delta'(t))\int_0^t \boldsymbol{w}\mathrm{d}\tau\Big\}, \end{aligned}$$

the estimates

$$\begin{aligned} &|\boldsymbol{K}_1(\theta) - \boldsymbol{K}_1'(\theta)|^{(1+\alpha,\frac{1+\alpha}{2})}_{G_T} + |K_2(\theta) - K_2'(\theta)|^{(1+\alpha,\frac{1+\alpha}{2})}_{G_T} \\ &\qquad \leqslant c|\nabla\sigma(\theta)|^{(1+\alpha,\frac{1+\alpha}{2})}_{G_T} M_T[\boldsymbol{u}-\boldsymbol{u}'] + cT^{\frac{1-\alpha}{2}}|\nabla\sigma|_{G_T}|\nabla(\boldsymbol{u}-\boldsymbol{u}')|_{D_{\Sigma T}}, \end{aligned}$$

$$\begin{aligned} |K_3(\boldsymbol{w},\theta) &- K_3'(\boldsymbol{w},\theta)|^{(\alpha,\frac{\alpha}{2})}_{G_T} \leqslant c|\sigma(\theta)|^{(\alpha,\frac{\alpha}{2})}_{G_T}|\nabla(\boldsymbol{u}-\boldsymbol{u}')|_{D_{\Sigma T}} \\ &+ c(T+T^{\frac12})|\mathcal{D}_t\theta|^{(\alpha,\frac{\alpha}{2})}_{D_{\Sigma T}}(1+\langle\theta\rangle^{(\alpha,\frac{\alpha}{2})}_{D_{\Sigma T}})|\boldsymbol{w}|^{(2+\alpha,1+\frac{\alpha}{2})}_{D_{\Sigma T}}|\boldsymbol{u}-\boldsymbol{u}'|^{(2+\alpha,1+\frac{\alpha}{2})}_{D_{\Sigma T}} \\ &+ c|\theta|^{(2+\alpha,1+\frac{\alpha}{2})}_{D_{\Sigma T}}(1+|\theta|^{(2+\alpha,1+\frac{\alpha}{2})}_{D_{\Sigma T}})\times \\ &\qquad\times\Big\{M_T[\boldsymbol{u}-\boldsymbol{u}'] + (T+T^{\frac12})|\boldsymbol{w}|^{(2+\alpha,1+\frac{\alpha}{2})}_{D_{\Sigma T}}|\boldsymbol{u}-\boldsymbol{u}'|^{(2+\alpha,1+\frac{\alpha}{2})}_{D_{\Sigma T}}\Big\} \end{aligned}$$

hold.

One also can prove the following proposition.

Lemma 8.2.3 *Let $\psi\in C^{2+\alpha,1+\alpha/2}(D_{\Sigma T})$, $[\psi]|_\Gamma = 0$. Operator differences*

$$\begin{aligned} h_1(\psi) - h_1'(\psi) &= k^\pm(\nabla^2_{\boldsymbol{u}}\psi - \nabla^2_{\boldsymbol{u}'}\psi) = \\ &= k^\pm\Big\{\mathbb{A}\nabla\cdot(\mathbb{B}-\mathbb{B}')\nabla\psi + (\mathbb{B}-\mathbb{B}')\nabla\cdot\mathbb{A}'\psi\Big\}, \end{aligned}$$

$$h_2(\psi) - h_2'(\psi) = \left[k^{\pm}(\boldsymbol{n}' \cdot \nabla_{\boldsymbol{u}'}\psi - \boldsymbol{n} \cdot \nabla_{\boldsymbol{u}}\psi)\right]\Big|_{\Gamma}$$

$$= \left[k^{\pm}\{(\boldsymbol{n}' - \boldsymbol{n}) \cdot \mathbb{A}\nabla\psi + \boldsymbol{n} \cdot (\mathbb{B} - \mathbb{B}')\nabla\psi\}\right]\Big|_{\Gamma},$$

$$h_3(\psi, \boldsymbol{w}) - h_3'(\psi, \boldsymbol{w}) = \varkappa\psi\Big(\Pi'\nabla_{\boldsymbol{u}'} - \Pi\nabla_{\boldsymbol{u}}\Big) \cdot \boldsymbol{w}\big|_{\Gamma}$$

$$= \varkappa\psi\Big\{(\Pi' - \Pi)\mathbb{A}'\nabla \cdot \boldsymbol{w}\big|_{\Gamma} + \Pi(\mathbb{B}' - \mathbb{B})\nabla \cdot \boldsymbol{w}\big|_{\Gamma}\Big\}$$

satisfy the inequalities

$$|h_1(\psi) - h_1'(\psi)|_{D_{\Sigma T}}^{(\alpha,\frac{\alpha}{2})} \leqslant c_6|\psi|_{D_{\Sigma T}}^{(2+\alpha,1+\frac{\alpha}{2})} M_T[\boldsymbol{u} - \boldsymbol{u}'],$$

$$|h_2(\psi) - h_2'(\psi)|_{G_T}^{(1+\alpha,\frac{1+\alpha}{2})} \leqslant$$

$$\leqslant c_7|\psi|_{D_{\Sigma T}}^{(2+\alpha,1+\frac{\alpha}{2})} M_T[\boldsymbol{u} - \boldsymbol{u}'] + c_8 T^{\frac{1-\alpha}{2}} |\nabla\psi|_{D_{\Sigma T}} |\nabla(\boldsymbol{u} - \boldsymbol{u}')|_{D_{\Sigma T}},$$

$$|h_3(\psi, \boldsymbol{w}) - h_3'(\psi, \boldsymbol{w})|_{G_T}^{(1+\alpha,\frac{1+\alpha}{2})} \leqslant$$

$$\leqslant c_9|\psi|_{D_{\Sigma T}}^{(1+\alpha,\frac{1+\alpha}{2})} \Big\{|\nabla\boldsymbol{w}|_{D_{\Sigma T}}^{(1+\alpha,\frac{1+\alpha}{2})} M_T[\boldsymbol{u} - \boldsymbol{u}'] +$$

$$+ T^{\frac{1-\alpha}{2}} |\nabla\boldsymbol{w}|_{D_{\Sigma T}} |\nabla(\boldsymbol{u} - \boldsymbol{u}')|_{D_{\Sigma T}}\Big\},$$

where $M_T[\boldsymbol{v}]$ *is calculated by formula* (8.2.8).

8.3 The Solvability of Problem (8.1.2)

In this section we prove Theorem 8.1.1.

Proof We set $(\boldsymbol{u}^{(0)}, q^{(0)}, \hat{\theta}^{(0)}) = (\boldsymbol{v}_0, q_0, \theta_0)$. Next, we define $(\boldsymbol{u}^{(m+1)}, q^{(m+1)})$, $m = 0, 1, \ldots$, as a solution of the following problem:

$$\mathcal{D}_t\boldsymbol{u}^{(m+1)} - \nu^{\pm}\nabla_m^2\boldsymbol{u}^{(m+1)} + \frac{1}{\rho^{\pm}}\nabla_m q^{(m+1)} = \boldsymbol{f}(X_m, t), \quad \nabla_m \cdot \boldsymbol{u}^{(m+1)} = 0 \quad \text{in } D_{\Sigma T},$$

$$\boldsymbol{u}^{(m+1)}|_{t=0} = \boldsymbol{v}_0 \qquad \text{in } \Omega^- \cup \Omega^+,$$

$$[\boldsymbol{u}^{(m+1)}]|_{G_T} = 0, \quad \boldsymbol{u}^{(m+1)}|_{\Upsilon_T} = 0, \tag{8.3.1}$$

$$[\mu^{\pm}\Pi_0\Pi_m \mathbb{S}_m(\boldsymbol{u}^{(m+1)})\boldsymbol{n}_m]|_{G_T} = \boldsymbol{K}_1^{(m)}(\hat{\theta}^{(m)}),$$

$$[\boldsymbol{n}_0 \cdot \mathbb{T}_m(\boldsymbol{u}^{(m+1)}, q^{(m+1)})\boldsymbol{n}_m]|_{G_T} - \sigma(\theta_0)\boldsymbol{n}_0 \cdot \Delta_m(t) \int_0^t \boldsymbol{u}^{(m+1)}\big|_\Gamma \,\mathrm{d}\tau =$$

$$= K_2^{(m)}(\hat{\theta}^{(m)}) + \int_0^t K_3^{(m)}(\boldsymbol{u}^{(m)}, \hat{\theta}^{(m)})\,\mathrm{d}\tau \qquad \text{on} \quad G_T,$$

where $\nabla_m = \nabla_{\boldsymbol{u}^{(m)}}$, $\Pi_m \boldsymbol{\omega} = \boldsymbol{\omega} - \boldsymbol{n}_m(\boldsymbol{n}_m \cdot \boldsymbol{\omega})$, $\boldsymbol{n}_m$ is the outward normal to $\Gamma_m = \{\boldsymbol{x} = \boldsymbol{X}_m(\xi, t),\ \xi \in \Gamma\}$, $\boldsymbol{X}_m = \boldsymbol{X}_{\boldsymbol{u}^{(m)}}$; $\mathbb{S}_m = \mathbb{S}_{\boldsymbol{u}^{(m)}}$, $\mathbb{T}_m = \mathbb{T}_{\boldsymbol{u}^{(m)}}$; Δ_m is the Laplace–Beltrami operator on Γ_m, $\boldsymbol{K}_1^{(m)} = \Pi_0\Pi_m\nabla_m\sigma$ etc.

Finally, we determine $\hat{\theta}^{(m+1)}$, $m = 0, 1, \ldots$, as a solution to the problem

$$\begin{aligned}
&\mathcal{D}_t\hat{\theta}^{(m+1)} - k^{\pm}\nabla_m^2\hat{\theta}^{(m+1)} = 0 \quad \text{in } D_{\Sigma T},\\
&\hat{\theta}^{(m+1)}|_{t=0} = \theta_0 \quad \text{in } \Omega^- \cup \Omega^+,\\
&[\hat{\theta}^{(m+1)}]|_{G_T} = 0, \qquad \hat{\theta}^{(m+1)}|_{\Upsilon_T} = \hat{a},\\
&[k^{\pm}\boldsymbol{n}_m \cdot \nabla_m\hat{\theta}^{(m+1)}]|_{G_T} + \varkappa\hat{\theta}^{(m+1)}(\Pi_m\nabla_m) \cdot \boldsymbol{u}^{(m+1)} = 0 \ \text{ on } G_T.
\end{aligned} \tag{8.3.2}$$

We suppose that the solution $\boldsymbol{u}^{(m)}$, $q^{(m)}$ exists in the interval $(0, T_m]$, and $\boldsymbol{u}^{(m)}$ satisfies inequality (7.3.2) on it. We will successively apply Theorem 7.3.1 to problems (8.3.1). The first and fourth assumptions of this theorem follow from hypotheses (8.1.3), (8.1.4) (see Sect. 5.4). As to the third one, we can put

$$\boldsymbol{g} \equiv \boldsymbol{g}_m = -\mathbb{A}^{(m)*}\boldsymbol{f}, \qquad \mathbb{G} \equiv \mathbb{G}_m(\xi, t) = \nabla \int_{\cup\Omega^{\pm}} \mathcal{E}(\xi, \eta)\mathbb{A}^{(m)*}\boldsymbol{f}(\eta, t)\,\mathrm{d}\eta, \tag{8.3.3}$$

where $\mathbb{A}^{(m)} = \mathbb{A}(\boldsymbol{u}^{(m)})$. The first equality in (8.3.3) follows from the identity $\frac{\partial A_{ij}}{\partial \xi_j} = 0$, which is valid for the cofactor matrix of the Jacobian matrix of an arbitrary transformation and which implies that $\mathbb{A}\nabla \cdot \boldsymbol{w} = \nabla \cdot \mathbb{A}^*\boldsymbol{w}$. Moreover, it is obvious that $\big[\boldsymbol{n}_0 \cdot \big(\boldsymbol{g}_m + \mathbb{A}^{(m)*}\boldsymbol{f}\big)\big]\,\big|_\Gamma = 0$.

Next, estimates (5.4.5) hold for weightless norms [54]:

$$\begin{aligned}
|\boldsymbol{g}_m|_{D_{\Sigma T_m}}^{(\alpha,\alpha/2)} &+ |\mathbb{G}_m|_{D_{\Sigma T_m}}^{(\gamma,1+\alpha)} + |\mathbb{G}_m|_{D_{\Sigma T_m}}^{(\gamma,0)} \leqslant \\
&\leqslant c(T)\Big(1 + (1+T)|\boldsymbol{u}^{(m)}|_{D_{\Sigma T_m}}^{(2+\alpha,1+\alpha/2)}\Big)|\boldsymbol{f}|_{t,D_{\Sigma T_m}}^{(\frac{1+\alpha-\gamma}{2})}\\
&\leqslant c_1(T)|\boldsymbol{f}|_{D_{\Sigma T_m}}^{(\alpha,\frac{1+\alpha-\gamma}{2})},
\end{aligned} \tag{8.3.4}$$

$$\begin{aligned}
|\sigma(\hat{\theta}^{(m)})H_0(\xi)|_{G_{T_m}}^{(1+\alpha,\frac{1+\alpha}{2})} &\leqslant c|\sigma(\hat{\theta}^{(m)})|_{G_{T_m}}^{(1+\alpha,\frac{1+\alpha}{2})}|H_0|_\Gamma^{(1+\alpha)} \leqslant\\
&\leqslant c_2\Big(1 + |\hat{\theta}^{(m)}|_{D_{\Sigma T_m}}^{(1+\alpha,\frac{1+\alpha}{2})}\Big)|H_0|^{(1+\alpha)},
\end{aligned} \tag{8.3.5}$$

$$
\begin{aligned}
|\sigma(\hat{\theta}^{(m)})\boldsymbol{n}_0\cdot\dot{\Delta}_m(\tau)\boldsymbol{\xi}|_{G_{T_m}}^{(\alpha,\alpha/2)} &\leqslant c|\sigma(\hat{\theta}^{(m)})|_{G_{T_m}}^{(\alpha,\alpha/2)}|\nabla\boldsymbol{u}^{(m)}|_{G_{T_m}}^{(\alpha,\alpha/2)}\\
&\leqslant c_3\Big(1+|\hat{\theta}^{(m)}|_{D_{\Sigma T_m}}^{(\alpha,\alpha/2)}\Big)|\boldsymbol{u}^{(m)}|_{D_{\Sigma T_m}}^{(2+\alpha,1+\alpha/2)}.
\end{aligned}
$$

Thus, we have verified that $(\boldsymbol{u}^{(m+1)},q^{(m+1)})$ exist, and inequalities (7.3.4), (8.3.4), (8.3.5) and Lemma 8.2.2 with $\boldsymbol{u}'=0$ imply the estimate as follows:

$$
\begin{aligned}
N_T[\boldsymbol{u}^{(m+1)},q^{(m+1)}]\leqslant c_4(T)\Big\{&\big(1+|\boldsymbol{u}^{(m)}|_{D_{\Sigma T_m}}^{(2+\alpha,1+\frac{\alpha}{2})}\big)|\boldsymbol{f}|_{Q_{T_m}}^{(\alpha,\frac{1+\alpha-\gamma}{2})}+|\boldsymbol{v}_0|_{\cup\Omega^\pm}^{(2+\alpha)}\Big\}\\
&+c_5\big(1+|\hat{\theta}^{(m)}|_{D_{\Sigma T_m}}^{(1+\alpha,\frac{1+\alpha}{2})}\big)\Big(|\boldsymbol{u}^{(m)}|_{D_{\Sigma T_m}}^{(2+\alpha,1+\frac{\alpha}{2})}+|H_0|_\Gamma^{(1+\alpha)}\Big)\\
&+c_6T^{\frac{1-\alpha}{2}}|\nabla\sigma(\hat{\theta}^{(m)})|_{G_T}|\nabla\boldsymbol{u}^{(m)}|_{D_{\Sigma T_m}}+\\
&+c_7|\hat{\theta}^{(m)}|_{D_{\Sigma T_m}}^{(2+\alpha,1+\frac{\alpha}{2})}\Big(1+|\hat{\theta}^{(m)}|_{D_{\Sigma T_m}}^{(2+\alpha,1+\frac{\alpha}{2})}\Big)\times\\
&\times\Big\{M_T[\boldsymbol{u}^{(m)}]+(T+T^{1/2})\Big(|\boldsymbol{u}^{(m)}|_{D_{\Sigma T_m}}^{(2+\alpha,1+\frac{\alpha}{2})}\Big)^2\Big\}\\
&+c_8\Big(T^{\frac{1-\alpha}{2}}|\nabla\boldsymbol{u}^{(m)}|_{D_{\Sigma T_m}}+|\nabla\boldsymbol{u}^{(m)}|_{D_{\Sigma T_m}}^{(\alpha,\frac{\alpha}{2})}\Big)|\boldsymbol{v}_0|_{\cup\Omega^\pm}^{(2+\alpha)}.
\end{aligned}
$$

The boundedness of the norm of velocity vector $\boldsymbol{u}^{(m+1)}$ implies the existence of such $T_{m+1}\leqslant T_m$ that $\boldsymbol{u}^{(m+1)}$ satisfies inequality (7.3.2) on $(0,T_{m+1}]$ with fixed δ.

For $m=0$ we obtain that

$$
\begin{aligned}
N_{T_0}[\boldsymbol{u}^{(1)},q^{(1)}]\leqslant c_4(T)\Big\{&\big(1+|\boldsymbol{v}_0|_{\cup\Omega^\pm}^{(2+\alpha)}\big)|\boldsymbol{f}|_{Q_{T_0}}^{(\alpha,\frac{1+\alpha-\gamma}{2})}+|\boldsymbol{v}_0|_{\cup\Omega^\pm}^{(2+\alpha)}\Big\}+\\
&+c_9\Big\{\Big(1+|\theta_0|_{\cup\Omega^\pm}^{(2+\alpha)}\Big)\Big(|\boldsymbol{v}_0|_{\cup\Omega^\pm}^{(2+\alpha)}+|H_0|_\Gamma^{(1+\alpha)}\Big)+\\
&+T_0^{\frac{1-\alpha}{2}}|\boldsymbol{v}_0|_{\cup\Omega^\pm}^{(1)}\Big(|\theta_0|_{\cup\Omega^\pm}^{(1)}+|\boldsymbol{v}_0|_{\cup\Omega^\pm}^{(1+\alpha)}\Big)+\\
&+(T_0+T_0^{1/2})|\theta_0|_{\cup\Omega^\pm}^{(2+\alpha)}\Big(1+|\theta_0|_{\cup\Omega^\pm}^{(2+\alpha)}\Big)\times\\
&\times|\boldsymbol{v}_0|_{\cup\Omega^\pm}^{(2+\alpha)}\Big(1+|\boldsymbol{v}_0|_{\cup\Omega^\pm}^{(2+\alpha)}\Big)\Big\}.
\end{aligned}
\tag{8.3.6}
$$

Compatibility conditions (8.2.2) for problem (8.3.2) follow from (8.1.3). Therefore Theorem 8.2.1 for every $m=0,1,2,\dots$ guarantees us the existence of $\hat{\theta}^{(m+1)}$ and the estimate

$$|\hat{\theta}^{(m+1)}|^{(2+\alpha,1+\frac{\alpha}{2})}_{D_{\Sigma T_{m+1}}} \leqslant c_{10}(T_{m+1})\Big\{|\hat{a}|^{(2+\alpha,1+\frac{\alpha}{2})}_{\Sigma_{T_{m+1}}} + |\theta_0|^{(2+\alpha)}_{\cup\Omega^\pm} + \tag{8.3.7}$$
$$+ \big(\delta + T_{m+1}^{\frac{1-\alpha}{2}}|\nabla \boldsymbol{u}^{(m)}|_{\cup\Omega^\pm}\big)\Big(|\theta_0|^{(1)}_{\cup\Omega^\pm} + |\nabla \boldsymbol{u}^{(m+1)}|_{D_{\Sigma T_{m+1}}}|\theta_0|_{\cup\Omega^\pm}\Big)\Big\}.$$

In this way, we deduce by induction that, for every $m \geqslant 0$, system (8.3.1), (8.3.2) has a solution $(\boldsymbol{u}^{(m+1)}, q^{(m+1)}, \hat{\theta}^{(m+1)})$ in the interval $(0, T_{m+1}] \subset (0, T]$ such that the preceding approximation $(\boldsymbol{u}^{(m)}, q^{(m)}), \hat{\theta}^{(m)}$ is also defined on this interval, while $\boldsymbol{u}^{(m+1)}$ satisfies condition (7.3.2) on it with small δ.

Now it is necessary to show that there exists such $T' > 0$ that, for all $m \in \mathbb{N}$ $T_m \geqslant T'$, the norms $N_T[\boldsymbol{u}^{(m)}, q^{(m)}, \hat{\theta}^{(m)}] \equiv N_{T'}[\boldsymbol{u}^{(m)}, q^{(m)}] + |\hat{\theta}^{(m)}|^{(2+\alpha,1+\frac{\alpha}{2})}_{D_{\Sigma T'}}$ are uniformly bounded and the sequence $\{\boldsymbol{u}^{(m)}, q^{(m)}, \hat{\theta}^{(m)}\}, m > 0$, converges to a solution of problem (8.1.2).

To this end, we compose the difference between systems (8.3.1) corresponding to $j+1$ and to j. We consider the functions $\boldsymbol{w}^{(j+1)} = \boldsymbol{u}^{(j+1)} - \boldsymbol{u}^{(j)}$, $s^{(j+1)} = q^{(j+1)} - q^{(j)}$, $j = 1, 2 \ldots$, which solve the problem

$$\mathcal{D}_t \boldsymbol{w}^{(j+1)} - \nu^\pm \nabla_j^2 \boldsymbol{w}^{(j+1)} + \frac{1}{\rho^\pm}\nabla_j s^{(j+1)} =$$
$$= \boldsymbol{l}_1^{(j)}(\boldsymbol{u}^{(j)}, q^{(j)}) - \boldsymbol{l}_1^{(j-1)}(\boldsymbol{u}^{(j)}, q^{(j)}) + \boldsymbol{f}(X_j, t) - \boldsymbol{f}(X_{j-1}, t) \equiv \boldsymbol{f}^{(j)},$$
$$\nabla_j \cdot \boldsymbol{w}^{(j+1)} = l_2^{(j)}(\boldsymbol{u}^{(j)}) - l_2^{(j-1)}(\boldsymbol{u}^{(j)}) \equiv r^{(j)} \quad \text{in} \quad D_{\Sigma T_{m+1}},$$
$$\boldsymbol{w}^{(j+1)}|_{t=0} = 0 \qquad \text{in } \Omega^- \cup \Omega^+,$$
$$[\boldsymbol{w}^{(j+1)}]|_{G_{T_{m+1}}} = 0, \quad \boldsymbol{w}^{(j+1)}|_{\Upsilon_{T_{m+1}}} = 0, \tag{8.3.8}$$
$$[\mu^\pm \Pi_0 \Pi_j \mathbb{S}_j(\boldsymbol{w}^{(j+1)})\boldsymbol{n}_j]|_{G_{T_{m+1}}} = \boldsymbol{l}_3^{(j)}(\boldsymbol{u}^{(j)}) - \boldsymbol{l}_3^{(j-1)}(\boldsymbol{u}^{(j)}) +$$
$$+ \boldsymbol{K}_1^{(j)}(\hat{\theta}^{(j)}) - \boldsymbol{K}_1^{(j-1)}(\hat{\theta}^{(j-1)}),$$
$$[\boldsymbol{n}_0 \cdot \mathbb{T}_j(\boldsymbol{w}^{(j+1)}, s^{(j+1)})\boldsymbol{n}_j]|_{G_{Tm+1}} - \sigma(\theta_0)\boldsymbol{n}_0 \cdot \Delta_j(t)\int_0^t \boldsymbol{w}^{(j+1)}|_\Gamma \mathrm{d}\tau =$$
$$= l_4^{(j)}(\boldsymbol{u}^{(j)}, q^{(j)}) - l_4^{(j-1)}(\boldsymbol{u}^{(j)}, q^{(j)}) + K_2^{(j)}(\hat{\theta}^{(j)}) -$$
$$- K_2^{(j-1)}(\hat{\theta}^{(j-1)}) + \sigma(\theta_0)\int_0^t (l_5^{(j)}(\boldsymbol{u}^{(j)}) - l_5^{(j-1)}(\boldsymbol{u}^{(j)}))\mathrm{d}\tau +$$
$$+ \int_0^t \Big\{K_3^{(j)}(\boldsymbol{u}^{(j)}, \hat{\theta}^{(j)}) - K_3^{(j-1)}(\boldsymbol{u}^{(j-1)}, \hat{\theta}^{(j-1)})\Big\}\,\mathrm{d}\tau \quad \text{on } G_{T_{m+1}}.$$

Here

$$
\begin{aligned}
\boldsymbol{l}_1^{(j)}(\boldsymbol{w},s) &= \nu^{\pm}(\nabla_j^2-\nabla^2)\boldsymbol{w}+\frac{1}{\rho^{\pm}}(\nabla-\nabla_j)s \equiv \frac{\partial L_{1k}^{(j)}(\boldsymbol{w},s)}{\partial \xi_k},\\
l_2^{(j)}(\boldsymbol{w}) &= (\nabla-\nabla_j)\cdot\boldsymbol{w} = -\mathbb{B}^{(j)}\nabla\cdot\boldsymbol{w}\equiv\nabla\cdot L_2^{(j)}(\boldsymbol{w}),\\
\boldsymbol{l}_3^{(j)}(\boldsymbol{w}) &= \big[\mu^{\pm}\Pi_0\big(\mathbb{S}(\boldsymbol{w})\boldsymbol{n}_0-\Pi_j\mathbb{S}_j(\boldsymbol{w})\boldsymbol{n}_j\big)\big]\big|_{\Gamma},\\
l_4^{(j)}(\boldsymbol{w},s) &= \big[\boldsymbol{n}_0\cdot\big(\mathbb{T}(\boldsymbol{w},s)\boldsymbol{n}_0-\mathbb{T}_j(\boldsymbol{w},s)\boldsymbol{n}_j\big)\big]\big|_{\Gamma} \qquad (8.3.9)\\
&= \big[\boldsymbol{n}_0\cdot(\boldsymbol{n}_j-\boldsymbol{n}_0)s+\mu^{\pm}\boldsymbol{n}_0\cdot\big(\mathbb{S}(\boldsymbol{w})\boldsymbol{n}_0-\mathbb{S}_j(\boldsymbol{w})\boldsymbol{n}_j\big)\big]\big|_{\Gamma},\\
l_5^{(j)}(\boldsymbol{w}) &= \boldsymbol{n}_0\cdot\mathcal{D}_t\Big\{\big(\Delta_j(t)-\Delta(0)\big)\int_0^t \boldsymbol{w}|_{\Gamma}\mathrm{d}t'\Big\}\\
&= \boldsymbol{n}_0\cdot\Big\{\big(\Delta_j(t)-\Delta(0)\big)\boldsymbol{w}+\dot{\Delta}_j(t)\int_0^t \boldsymbol{w}|_{\Gamma}\,\mathrm{d}t'\Big\},
\end{aligned}
$$

where

$$
\begin{aligned}
L_{1k}^{(j)}(\boldsymbol{w},s) &= \nu^{\pm}\big(B_{ik}^{(j)}A_{im}^{(j)}+B_{km}^{(j)}\big)\frac{\partial\boldsymbol{w}}{\partial\xi_m}-\mathbb{B}^{(j)}\boldsymbol{e}_k\frac{s}{\rho^{\pm}},\\
L_2^{(j)}(\boldsymbol{w}) &= -\mathbb{B}^{(j)*}\boldsymbol{w},
\end{aligned}
$$

$\boldsymbol{e}_k$ is the basis vector in the direction of ξ_k.

In addition, we observe that the function $\psi^{(j+1)}=\hat{\theta}^{(j+1)}-\hat{\theta}^{(j)}$, $j=1,\ldots,$ is a solution to the problem which is the difference of systems (8.3.2) corresponding to the neighboring indices $j+1$ and j:

$$
\begin{aligned}
\mathcal{D}_t\psi^{(j+1)}-k^{\pm}\nabla_j^2\psi^{(j+1)} &= k^{\pm}\nabla_j^2\hat{\theta}^{(j)}-k^{\pm}\nabla_{j-1}^2\hat{\theta}^{(j)}\\
&\equiv h_1^{(j)}(\hat{\theta}^{(j)})-h_1^{(j-1)}(\hat{\theta}^{(j)}) \qquad \text{in} \quad D_{\Sigma T_{m+1}},\\
\psi^{(j+1)}|_{t=0} &= 0 \qquad \text{in} \quad \Omega^-\cup\Omega^+,\\
[\psi^{(j+1)}]|_{G_{T_{m+1}}} &= 0, \qquad \psi^{(j+1)}|_{\Upsilon_{T_{m+1}}}=0,\\
[k^{\pm}\boldsymbol{n}_j\cdot\nabla_j\psi^{(j+1)}]|_{\Gamma}+\varkappa\psi^{(j+1)}(\Pi_j\nabla_j)\cdot\boldsymbol{u}^{(j+1)} &= \qquad (8.3.10)\\
&= -[k^{+}\boldsymbol{n}_j\cdot\nabla_j\hat{\theta}^{(j)}]|_{\Gamma}-\varkappa\hat{\theta}^{(j)}(\Pi_j\nabla_j)\cdot\boldsymbol{u}^{(j+1)}+\\
&\quad+[k^{\pm}\boldsymbol{n}_{j-1}\cdot\nabla_{j-1}\hat{\theta}^{(j)}]|_{\Gamma}+\varkappa\hat{\theta}^{(j)}(\Pi_{j-1}\nabla_{j-1})\cdot\boldsymbol{u}^{(j)}\equiv\\
&\equiv h_2^{(j)}(\hat{\theta}^{(j)})-h_2^{(j-1)}(\hat{\theta}^{(j)})+h_3^{(j)}(\hat{\theta}^{(j)},\boldsymbol{w}^{(j+1)})+\\
&\quad+h_3^{(j)}(\hat{\theta}^{(j)},\boldsymbol{u}^{(j)})-h_3^{(j-1)}(\hat{\theta}^{(j)},\boldsymbol{u}^{(j)}) \qquad \text{on} \quad G_{T_{m+1}}.
\end{aligned}
$$

Here $h_i^{(j)}$ are calculated by (8.2.5), where $\boldsymbol{u}$ is replaced with $\boldsymbol{u}^{(j)}$.

In order to apply Theorem 7.3.1 to problem (8.3.8), we need to verify the third assumption of this theorem. Following § 5.4, we compute the potentials $\boldsymbol{g}^{(j)}$ and $G_k^{(j)}$ by formulas (5.4.7) and (5.4.8), respectively.

We assume that $(\boldsymbol{u}^{(j)}, q^{(j)}, \hat{\theta}^{(j)})$, $j = 1, \dots, m$, satisfy the inequality

$$(T + T^{\gamma/2}) N_T[\boldsymbol{u}^{(j)}, q^{(j)}, \hat{\theta}^{(j)}] < \delta, \quad T \leqslant T_{m+1}, \tag{8.3.11}$$

which is stronger than (7.3.2).

The norms of the right-hand sides of the system (8.3.8) has been estimated in Sect. 5.4 (Lemmas 5.4.1–5.4.4). For a bounded domain, inequality (5.4.10) takes the form:

$$\begin{aligned}
&|\boldsymbol{f}^{(j)}|_{D_{\Sigma T}}^{(\alpha,\alpha/2)} + |r^{(j)}|_{D_{\Sigma T}}^{(1+\alpha,\frac{1+\alpha}{2})} + |\boldsymbol{g}^{(j)}|_{D_{\Sigma T}}^{(\alpha,\alpha/2)} + |\mathbb{G}^{(j)}|_{D_{\Sigma T}}^{(\gamma,1+\alpha)} \\
&\quad + |\mathbb{G}^{(j)}|_{D_{\Sigma T}}^{(\gamma,0)} + |\boldsymbol{l}_3^{(j)} - \boldsymbol{l}_3^{(j-1)}|_{G_T}^{(1+\alpha,\frac{1+\alpha}{2})} + |l_4^{(j)} - l_4^{(j-1)}|_{G_T} \\
&\quad + |\nabla_\Gamma (l_4^{(j)} - l_4^{(j-1)})|_{G_T}^{(\alpha,\alpha/2)} + ||l_4^{(j)} - l_4^{(j-1)}|_{G_T}^{(\gamma,1+\alpha)} + |l_5^{(j)} - l_5^{(j-1)}|_{G_T}^{(\alpha,\alpha/2)} \\
&\leqslant c_{11}\Big\{(T + T^{\gamma/2})|\boldsymbol{w}^{(j)}|_{D_{\Sigma T}}^{(2+\alpha,1+\alpha/2)} N_T[\boldsymbol{u}^{(j)}, q^{(j)}] + |\nabla \boldsymbol{w}^{(j)}|_{D_{\Sigma T}}^{(\alpha,\alpha/2)} \\
&\quad + (T + T^{1/2})|\boldsymbol{w}^{(j)}|_{D_{\Sigma T}}^{(\alpha,\alpha/2)} + P_T[\boldsymbol{w}^{(j)}]|\boldsymbol{u}^{(j)}(\cdot, 0)|_{\Omega^- \cup \Omega^+}^{(1)}\Big\}.
\end{aligned} \tag{8.3.12}$$

The norms of the remaining terms K_i in (8.3.8) can be estimated by Lemmas 8.2.1 and 8.2.2:

$$\begin{aligned}
&|\boldsymbol{K}_1^{(j)}(\hat{\theta}^{(j)}) - \boldsymbol{K}_1^{(j-1)}(\hat{\theta}^{(j-1)})|_{G_T}^{(1+\alpha,\frac{1+\alpha}{2})} + |K_2^{(j)}(\hat{\theta}^{(j)}) - K_2^{(j-1)}(\hat{\theta}^{(j-1)})|_{G_T}^{(1+\alpha,\frac{1+\alpha}{2})} \\
&\qquad \leqslant c_{12}\Big\{\Big(|\hat{\theta}^{(j)}|_{D_{\Sigma T}}^{(2+\alpha,1+\frac{\alpha}{2})} + |\hat{\theta}^{(j-1)}|_{D_{\Sigma T}}^{(2+\alpha,1+\frac{\alpha}{2})}\Big)^{2+\alpha}\Big(1 + |H_0|_\Gamma^{(1+\alpha)}\Big) \times \\
&\qquad\qquad \times |\psi^{(j)}|_{D_{\Sigma T}}^{(2+\alpha,1+\frac{\alpha}{2})} + |\nabla\sigma(\hat{\theta}^{(j)})|_{G_T}^{(1+\alpha,\frac{1+\alpha}{2})} M_T[\boldsymbol{w}^{(j)}] + \\
&\qquad\qquad + T^{\frac{1-\alpha}{2}} |\nabla\sigma(\hat{\theta}^{(j)})|_{G_T} |\nabla \boldsymbol{w}^{(j)}|_{D_{\Sigma T}}\Big\},
\end{aligned}$$

$$\begin{aligned}
&|K_3^{(j)}(\boldsymbol{u}^{(j)}, \hat{\theta}^{(j)}) - K_3^{(j-1)}(\boldsymbol{u}^{(j-1)}, \hat{\theta}^{(j-1)})|_{G_T}^{(\alpha,\frac{\alpha}{2})} \leqslant \\
&\qquad \leqslant |K_3^{(j)}(\boldsymbol{u}^{(j)}, \hat{\theta}^{(j)}) - K_3^{(j-1)}(\boldsymbol{u}^{(j)}, \hat{\theta}^{(j)})|_{G_T}^{(\alpha,\frac{\alpha}{2})} + \\
&\qquad\quad + |K_3^{(j-1)}(\boldsymbol{u}^{(j)}, \hat{\theta}^{(j)}) - K_3^{(j-1)}(\boldsymbol{u}^{(j-1)}, \hat{\theta}^{(j-1)})|_{G_T}^{(\alpha,\frac{\alpha}{2})} \\
&\qquad \leqslant c_{13}(T + T^{1/2})\Big\{\Big(1 + |\hat{\theta}^{(j)}|_{D_{\Sigma T}}^{(2+\alpha,1+\frac{\alpha}{2})} + |\hat{\theta}^{(j-1)}|_{D_{\Sigma T}}^{(2+\alpha,1+\frac{\alpha}{2})}\Big)^2 \times \\
&\qquad\qquad \times \Big(|\nabla \boldsymbol{u}^{(j-1)}|_{D_{\Sigma T}} + |\boldsymbol{u}^{(j)}|_{D_{\Sigma T}}^{(2+\alpha,1+\frac{\alpha}{2})}\Big)|\psi^{(j)}|_{D_{\Sigma T}}^{(2+\alpha,1+\frac{\alpha}{2})}
\end{aligned}$$

$$+ |\hat{\theta}^{(j-1)}|_{D_{\Sigma T}}^{(2+\alpha,1+\frac{\alpha}{2})}\Big(1+|\hat{\theta}^{(j)}|_{D_{\Sigma T}}^{(2+\alpha,1+\frac{\alpha}{2})}\Big)|\boldsymbol{w}^{(j)}|_{D_{\Sigma T}}^{(2+\alpha,1+\frac{\alpha}{2})}\Big\}$$

$$+ c_{14}\Big\{|\sigma(\hat{\theta}^{(j)})|_{G_T}^{(\alpha,\frac{\alpha}{2})}|\nabla\boldsymbol{w}^{(j)}|_{D_{\Sigma T}}+$$

$$+ |\hat{\theta}^{(j-1)}|_{D_{\Sigma T}}^{(2+\alpha,1+\frac{\alpha}{2})}\Big(1+|\hat{\theta}^{(j-1)}|_{D_{\Sigma T}}^{(2+\alpha,1+\frac{\alpha}{2})}\Big)$$

$$\times\Big\{M_T[\boldsymbol{w}^{(j)}]+(T+T^{1/2})|\boldsymbol{u}^{(j)}|_{D_{\Sigma T}}^{(2+\alpha,1+\frac{\alpha}{2})}|\boldsymbol{w}^{(j)}|_{D_{\Sigma T}}^{(2+\alpha,1+\frac{\alpha}{2})}\Big\}\Big\}.$$

We estimate the solution of problem (8.3.10) using Theorem 8.2.1, Lemma 8.2.3 and inequality (8.2.7):

$$\begin{aligned}
&|\psi^{(j+1)}|_{D_{\Sigma T}}^{(2+\alpha,1+\frac{\alpha}{2})}\\
&\quad\leqslant |h_1^{(j)}(\hat{\theta}^{(j)})-h_1^{(j-1)}(\hat{\theta}^{(j)})|_{D_{\Sigma T}}^{(\alpha,\frac{\alpha}{2})}\\
&\qquad+|h_2^{(j)}(\hat{\theta}^{(j)})-h_2^{(j-1)}(\hat{\theta}^{(j)})|_{G_T}^{(1+\alpha,\frac{1+\alpha}{2})}+|h_3^{(j)}(\hat{\theta}^{(j)},\boldsymbol{w}^{(j+1)})|_{G_T}^{(1+\alpha,\frac{1+\alpha}{2})}\\
&\qquad+|h_3^{(j)}(\hat{\theta}^{(j)},\boldsymbol{u}^{(j)})-h_3^{(j-1)}(\hat{\theta}^{(j)},\boldsymbol{u}^{(j)})|_{G_T}^{(1+\alpha,\frac{1+\alpha}{2})}\\
&\quad\leqslant c_{15}\Big\{\Big(|\hat{\theta}^{(j)}|_{D_{\Sigma T}}^{(2+\alpha,1+\frac{\alpha}{2})}+|\hat{\theta}^{(j)}|_{D_{\Sigma T}}^{(1+\alpha,\frac{1+\alpha}{2})}|\nabla\boldsymbol{u}^{(j)}|_{D_{\Sigma T}}^{(1+\alpha,\frac{1+\alpha}{2})}\Big)M_T[\boldsymbol{w}^{(j)}]\\
&\qquad+T^{\frac{1-\alpha}{2}}|\hat{\theta}^{(j)}|_{D_{\Sigma T}}^{(1+\alpha,\frac{1+\alpha}{2})}(1+|\nabla\boldsymbol{u}^{(j)}|_{D_{\Sigma T}})|\nabla\boldsymbol{w}^{(j)}|_{D_{\Sigma T}}+\\
&\qquad+\delta|\hat{\theta}^{(j)}|_{D_{\Sigma T}}^{(2+\alpha,1+\frac{\alpha}{2})}|\boldsymbol{w}^{(j+1)}|_{D_{\Sigma T}}^{(2+\alpha,1+\frac{\alpha}{2})}\\
&\qquad+T^{\frac{1-\alpha}{2}}|\nabla\boldsymbol{u}^{(j)}|_{D_{\Sigma T}}|\theta_0|_{\cup\Omega^\pm}^{(1)}|\nabla\boldsymbol{w}^{(j+1)}|_{D_{\Sigma T}}\Big\},\ j=1,2,\ldots
\end{aligned}\tag{8.3.13}$$

For $j=0$, we have

$$\begin{aligned}
|\psi^{(1)}|_{D_{\Sigma T}}^{(2+\alpha,1+\frac{\alpha}{2})}&\leqslant|\hat{\theta}^{(1)}|_{D_{\Sigma T}}^{(2+\alpha,1+\frac{\alpha}{2})}+|\theta_0|_{\cup\Omega^\pm}^{(2+\alpha)}\leqslant\\
&\leqslant c(T)\Big\{|\hat{a}|_{\Upsilon_T}^{(2+\alpha,1+\frac{\alpha}{2})}+\big(\delta+T^{\frac{1-\alpha}{2}}|\boldsymbol{v}_0|_{\cup\Omega^\pm}^{(1)}\big)|\theta_0|_{\cup\Omega^\pm}^{(1)}|\boldsymbol{u}^{(1)}|_{D_{\Sigma T}}^{(1)}\\
&\qquad+|\theta_0|_{\cup\Omega^\pm}^{(2+\alpha)}\Big\}\equiv\Psi[T].
\end{aligned}\tag{8.3.14}$$

Since $\boldsymbol{w}^{(j)}|_{t=0}=0$, $\nabla\boldsymbol{w}^{(j)}|_{t=0}=0$, we obtain

$$\begin{aligned}
P_T[\boldsymbol{w}^{(j)}]|\boldsymbol{v}_0|_{\cup\Omega^\pm}^{(1)}&\leqslant c_{13}(T+T^{1/2})|\boldsymbol{w}^{(j)}|_{D_{\Sigma T}}^{(2+\alpha,1+\frac{\alpha}{2})}|\boldsymbol{v}_0|_{\cup\Omega^\pm}^{(1)}\leqslant c_{13}\delta|\boldsymbol{w}^{(j)}|_{D_{\Sigma T}}^{(2+\alpha,1+\frac{\alpha}{2})},\\
M_T[\boldsymbol{w}^{(j)}]&\leqslant 2(T+T^{1/2})|\boldsymbol{w}^{(j)}|_{D_{\Sigma T}}^{(2+\alpha,1+\frac{\alpha}{2})},
\end{aligned}$$

$$|\boldsymbol{w}^{(j)}|_{D_{\Sigma T}} \leqslant \int_0^T |\mathcal{D}_t \boldsymbol{w}^{(j)}(\cdot,t)|_{\cup\Omega^\pm}\,\mathrm{d}t \leqslant \int_0^T N_t[\boldsymbol{w}^{(j)}, s^{(j)}]\,\mathrm{d}t. \qquad (8.3.15)$$

Next, we use the interpolation inequality

$$|\boldsymbol{w}^{(j)}|^{(\alpha,\alpha/2)}_{D_{\Sigma T}} + |\nabla \boldsymbol{w}^{(j)}|^{(\alpha,\alpha/2)}_{D_{\Sigma T}} \leqslant \varepsilon |\boldsymbol{w}^{(j)}|^{(2+\alpha,1+\frac{\alpha}{2})}_{D_{\Sigma T}} + c_{16}(\varepsilon)|\boldsymbol{w}^{(j)}|_{D_{\Sigma T}}.$$

For $j = 0, 1, \ldots$, estimates (8.3.7), (8.3.11) yield

$$|\hat{\theta}^{(j)}|^{(2+\alpha,1+\frac{\alpha}{2})}_{D_{\Sigma T}} \leqslant c_{10}(T)\Big\{|\hat{a}|^{(2+\alpha,1+\frac{\alpha}{2})}_{\Upsilon_T} + |\theta_0|^{(2+\alpha)}_{\cup\Omega^\pm} + \\ +\big(\delta^2 + \delta|\boldsymbol{v}_0|^{(1)}_{\cup\Omega^\pm} + T^{\frac{1-\alpha}{2}}|\boldsymbol{v}_0|^{(1)}_{\cup\Omega^\pm}|\boldsymbol{v}_0|_{\cup\Omega^\pm}\big)|\theta_0|^{(1)}_{\cup\Omega^\pm}\Big\} \equiv \Theta[T].$$

Collecting all the inequalities beginning with (8.3.12), we arrive at the estimate:

$$N_T^{(j+1)} \equiv N_T[\boldsymbol{w}^{(j+1)}, s^{(j+1)}, \psi^{(j+1)}] \leqslant \\ \leqslant \Big\{c_{11}\{(1+c_{13})\delta + (1+T+T^{1/2})\varepsilon\} + c_{14}(c_{17}+\Theta[T])\varepsilon + 2\delta(c_{12}+c_{14}) \\ + \delta(1+\Theta[T])\{c_{13} + c_{14}\Theta[T] + 3c_{15}\}\Big\}|\boldsymbol{w}^{(j)}|^{(2+\alpha,1+\frac{\alpha}{2})}_{D_{\Sigma T}} \\ + c_{18}(\delta,\Theta,H_0)|\psi^{(j)}|^{(2+\alpha,1+\frac{\alpha}{2})}_{D_{\Sigma T}} \qquad (8.3.16) \\ + 2c_{15}\delta\Theta[T]|\boldsymbol{w}^{(j+1)}|^{(2+\alpha,1+\frac{\alpha}{2})}_{D_{\Sigma T}} + (c_{11}+c_{14})c_{16}(\varepsilon)|\boldsymbol{w}^{(j)}|_{D_{\Sigma T}},$$

where $c_{18}(\delta,\Theta,H_0) = c_{12}(2\Theta[T])^{2+\alpha}\big(1 + |H_0|^{(+1+\alpha)}_{\Gamma}\big) + c_{13}\delta(1+\Theta[T])^2$.

We choose δ so small that

$$2c_{15}\delta\Theta[T_{m+1}] < \frac{1}{2},$$

and we take (8.3.13) and (8.3.15) for $j-1$ instead of j into account. As a result, we have

$$N_T^{(j+1)} \leqslant \varkappa_1 N_T^{(j)} + \varkappa_2 N_T^{(j-1)} + c_{19}\int_0^T N_t^{(j)}\,\mathrm{d}t, \qquad (8.3.17)$$

where $\varkappa_1(T) = 2\Big\{c_{11}\{(1+c_{13})\delta + (1+T+T^{1/2})\varepsilon\} + 2c_{12}\delta + c_{13}\delta(1+\Theta) + c_{14}\{(c_{17}+\Theta)\varepsilon + 2\delta(c_{12}+c_{14}) + \delta(1+\Theta[T])\{c_{13}+c_{14}\Theta[T]+3c_{15}\}\Big\} + 4c_{15}\delta\Theta c_{18}(\delta,\Theta,H_0)$, $\varkappa_2(T) = 6c_{15}\delta(1+\Theta)c_{18}(\delta,\Theta,H_0)$ for $j = 2, 3\ldots$.

For $j = 1$, we deduce from (8.3.14), (8.3.16) that

$$N_T^{(2)} \leqslant \hat{\varkappa}_1 N_T^{(1)} + 2c_{18}(\delta, \Theta, H_0)\Psi[T] + c_{19} \int_0^T N_t^{(1)}\,\mathrm{d}t, \quad \hat{\varkappa}_1 < \varkappa_1. \tag{8.3.18}$$

Finally, we choose δ, ε such that $\varkappa \equiv \varkappa_1(T_{m+1}) + \varkappa_2(T_{m+1}) < 1$. Summing up inequalities (8.3.17) from $j = 2$ to $j = m$ and adding (8.3.18), for the expression

$$\Sigma_{m+1}(T) = \sum_{j=1}^{m+1} N_T^{(j)},$$

we get the estimate

$$\begin{aligned}\Sigma_{m+1}(T) &\leqslant \varkappa_1 \Sigma_m(T) + \varkappa_2 \Sigma_{m-1}(T) + c_{19} \int_0^T \Sigma_m(t)\,\mathrm{d}t + F_1(T) \\ &\leqslant \varkappa \Sigma_{m+1}(T) + c_{19} \int_0^T \Sigma_{m+1}(t)\,\mathrm{d}t + F_1(T),\end{aligned}$$

where $F_1(T) = N_T^{(1)} + 2c_{18}(\delta, \Theta, H_0)\Psi[T]$.

Finally, the Gronwall lemma applied to the inequality

$$\Sigma_{m+1}(T) \leqslant c_{20} \int_0^T \Sigma_{m+1}(t)\,\mathrm{d}t + \frac{1}{1-\varkappa} F_1(T),$$

implies that

$$\begin{aligned}N_T[\boldsymbol{u}^{(m+1)}, q^{(m+1)}, \hat{\theta}^{(m+1)}] &\leqslant \Sigma_{m+1}(T) + N_T[\boldsymbol{u}^{(0)}, q^{(0)}, \hat{\theta}^{(0)}] \\ &\leqslant \frac{F_1(T)}{1-\varkappa} e^{c_{20}T} + |\boldsymbol{v}_0|_{\cup\Omega^\pm}^{(2+\alpha)} + |\nabla q_0|_{\cup\Omega^\pm}^{(\alpha)} + \langle q_0 \rangle_{\cup\Omega^\pm}^{(\gamma)} + |\theta_0|_{\cup\Omega^\pm}^{(2+\alpha)}.\end{aligned} \tag{8.3.19}$$

Since $(\boldsymbol{w}^{(1)}, s^{(1)}, \psi^{(1)}) = (\boldsymbol{u}^{(1)} - \boldsymbol{v}_0, q^{(1)} - q_0, \hat{\theta}^{(1)} - \theta_0)$, the norm $N_T^{(1)}$ can be evaluated by (8.3.6), (8.3.14); in addition, the initial pressure, as a solution to problem (8.2.1), will be estimated in the next chapter (see paragraph 9.2, estimate (9.2.18)). Hence, $F_1(T)$ is bounded by the norms of the given functions. If

$$(T + T^{\gamma/2})\Big\{F_1(T)(1-\varkappa)^{-1} e^{c_{20}T} + |\boldsymbol{v}_0|_{\cup\Omega^\pm}^{(2+\alpha)} + |\nabla q_0|_{\cup\Omega^\pm}^{(\alpha)} + \langle q_0 \rangle_{\cup\Omega^\pm}^{(\gamma)} + |\theta_0|_{\cup\Omega^\pm}^{(2+\alpha)}\Big\} \leqslant \delta, \tag{8.3.20}$$

condition (8.3.11) holds for $\boldsymbol{u}^{(m+1)}, q^{(m+1)}, \hat{\theta}^{(m+1)}$ too. It is clear that there exists a number $T = T_0$ satisfying (8.3.20).

As the right-hand side of (8.3.19) does not depend on m, for every $m \in \mathbb{N}$ the functions $\boldsymbol{u}^{(m)}, q^{(m)}, \hat{\theta}^{(m)}$ are defined in the interval $(0, T_0]$ and satisfy the uniform estimate (8.3.19). Consequently, the series $\sum_{j=1}^{\infty} N_{T_0}^{(j)}$ is convergent, whence it follows that the sequence $\{\boldsymbol{u}^{(m)}, q^{(m)}, \hat{\theta}^{(m)}\}$ is also convergent in the norm N_{T_0}. Passing to the limit in (8.3.1), (8.3.2) as $m \to \infty$, we make sure that $(\boldsymbol{u}, q, \hat{\theta}) = \lim_{m\to\infty}(\boldsymbol{u}^{(m)}, q^{(m)}, \hat{\theta}^{(m)})$ is a solution of Eq. (8.1.2). Estimate (8.1.5) follows from (8.3.19) and (9.2.18) as a limit.

Now we prove the uniqueness of the solution obtained. We suppose that $(\boldsymbol{u}, q, \hat{\theta})$ and $(\boldsymbol{u}', q', \hat{\theta}')$ are two solutions of (8.1.2) and consider the difference $\boldsymbol{w} = \boldsymbol{u} - \boldsymbol{u}'$, $s = q - q'$, $\psi = \hat{\theta} - \hat{\theta}'$. The triple $(\boldsymbol{w}, s, \psi)$ satisfies a problem of type (8.3.8), (8.3.10):

$$\mathcal{D}_t \boldsymbol{w} - \nu^{\pm}\nabla_{\boldsymbol{u}}^2 \boldsymbol{w} + \frac{1}{\rho^{\pm}}\nabla_{\boldsymbol{u}} s = \boldsymbol{l}_1(\boldsymbol{u}', q') - \boldsymbol{l}_1'(\boldsymbol{u}', q') + \boldsymbol{f}(X_{\boldsymbol{u}}, t) - \boldsymbol{f}(X_{\boldsymbol{u}'}, t),$$

$$\nabla_{\boldsymbol{u}} \cdot \boldsymbol{w} = l_2(\boldsymbol{u}') - l_2'(\boldsymbol{u}'), \qquad \mathcal{D}_t\psi - k^{\pm}\nabla_{\boldsymbol{u}}^2\psi = h_1(\hat{\theta}') - h_1'(\hat{\theta}') \quad \text{in } D_{\Sigma T_0},$$

$$\boldsymbol{w}|_{t=0} = 0, \quad \psi|_{t=0} = 0 \quad \text{in } \Omega^- \cup \Omega^+,$$

$$[\boldsymbol{w}]|_{G_{T_0}} = 0, \qquad [\psi]|_{G_{T_0}} = 0, \qquad \boldsymbol{w}|_{\Upsilon_{T_0}} = 0, \qquad \psi|_{\Upsilon_{T_0}} = 0,$$

$$[\mu^{\pm}\Pi_0\Pi\mathbb{S}_{\boldsymbol{u}}(\boldsymbol{w})\boldsymbol{n}]|_{\Gamma} = \boldsymbol{l}_3(\boldsymbol{u}') - \boldsymbol{l}_3'(\boldsymbol{u}') + \boldsymbol{K}_1(\hat{\theta}) - \boldsymbol{K}_1'(\hat{\theta}'),$$

$$[\boldsymbol{n}_0 \cdot \mathbb{T}_{\boldsymbol{u}}(\boldsymbol{w}, s)\boldsymbol{n}]|_{\Gamma} - \sigma \boldsymbol{n}_0 \cdot \Delta(t)\int_0^t \boldsymbol{w}\,\mathrm{d}\tau|_{\Gamma} = l_4(\boldsymbol{u}', q') - l_4'(\boldsymbol{u}', q') +$$

$$+ K_2(\hat{\theta}) - K_2'(\hat{\theta}') + \int_0^t \sigma\big(l_5(\boldsymbol{u}') - l_5'(\boldsymbol{u}')\big) + K_3(\boldsymbol{u}, \hat{\theta}) - K_3'(\boldsymbol{u}', \hat{\theta}')\,\mathrm{d}\tau,$$

$$[k^{\pm}\boldsymbol{n}\cdot\nabla_{\boldsymbol{u}}\psi]|_{\Gamma} + \varkappa\psi(\Pi\nabla_{\boldsymbol{u}})\cdot\boldsymbol{u} = h_2(\hat{\theta}') - h_2'(\hat{\theta}') + h_3(\hat{\theta}', \boldsymbol{w}) +$$
$$+ h_3(\hat{\theta}', \boldsymbol{u}') - h_3'(\hat{\theta}', \boldsymbol{u}') \qquad \text{on } G_{T_0}.$$

Repeating the above arguments, we arrive at the inequality similar to (8.3.17):

$$N_T[\boldsymbol{w}, s, \psi] \leqslant \varkappa N_T[\boldsymbol{w}, s, \psi] + c_{21}\int_0^t N_t[\boldsymbol{w}, s, \psi]\,\mathrm{d}t, \quad \varkappa < 1,$$

which, by the Gronwall lemma, implies that $\boldsymbol{w} = 0$, $s = 0$, $\psi = 0$.

Theorem 8.1.1 is completely proved. □

8.4 The Problem in $\mathbb{R}^3$ with a Constant Temperature Value at Infinity

In the conclusion of this chapter devoted to the Marangoni effect, we would like to discuss also the problem in the whole space $\mathbb{R}^3$. In doing so, we will assume that the temperature function θ tends to a constant as $|x| \to \infty$. A similar problem with a constant temperature gradient at infinity was considered in [5] in a linear approximation.

In the complete setting, the problem of thermocapillary convection in the whole space differs from (8.1.1) by the fact that, instead of boundary conditions on the surface Σ, the behavior of the unknown functions at infinity is set:

$$\boldsymbol{v} \xrightarrow[|x|\to\infty]{} 0, \qquad \theta \xrightarrow[|x|\to\infty]{} \theta_\infty,$$

where θ_∞ is some nonnegative constant.

We consider the new temperature function $\vartheta = \theta - \theta_\infty$. We denote ϑ at $t = 0$ by ϑ_0 and in the Lagrangian coordinates by $\hat{\vartheta}$.

We assume that initial data vanish at infinity:

$$\boldsymbol{v}_0 \xrightarrow[|x|\to\infty]{} 0, \qquad \vartheta_0 \xrightarrow[|x|\to\infty]{} 0. \tag{8.4.1}$$

We pass to the Lagrangian coordinates. For the decreasing given functions, the problem

$$\begin{aligned}
&\mathcal{D}_t \boldsymbol{u} - \nu^\pm \nabla_{\boldsymbol{u}}^2 \boldsymbol{u} + \frac{1}{\rho^\pm} \nabla_{\boldsymbol{u}} q = \boldsymbol{f}(X_{\boldsymbol{u}}, t), \quad \nabla_{\boldsymbol{u}} \cdot \boldsymbol{u} = 0,\\
&\qquad \mathcal{D}_t \hat{\vartheta} - k^\pm \nabla_{\boldsymbol{u}}^2 \hat{\vartheta} = 0 \qquad \text{in} \quad D_T,\\
&\boldsymbol{u}|_{t=0} = \boldsymbol{v}_0, \qquad \hat{\vartheta}|_{t=0} = \vartheta_0 + \theta_\infty \qquad \text{in} \quad \Omega_0^- \cup \Omega_0^+,\\
&[\boldsymbol{u}]|_{G_T} = 0, \quad [\hat{\vartheta}]|_{G_T} = 0, \quad \boldsymbol{u} \xrightarrow[|\xi|\to\infty]{} 0, \quad \hat{\vartheta} \xrightarrow[|\xi|\to\infty]{} 0,\\
&[\mu^\pm \Pi_0 \Pi \mathbb{S}_{\boldsymbol{u}}(\boldsymbol{u})\boldsymbol{n}]|_{G_T} = \Pi_0 \Pi \nabla_{\boldsymbol{u}} \widehat{\sigma}(\hat{\vartheta}),\\
&[\boldsymbol{n}_0 \cdot \mathbb{T}_{\boldsymbol{u}}(\boldsymbol{u}, q)\boldsymbol{n}]|_{G_T} - \widehat{\sigma}(\hat{\vartheta})\boldsymbol{n}_0 \cdot \Delta(t) \int_0^t \boldsymbol{u}\big|_\Gamma \mathrm{d}\tau = \widehat{\sigma}(\hat{\vartheta}) H_0(\xi) +\\
&\qquad\qquad + \widehat{\sigma}(\hat{\vartheta})\boldsymbol{n}_0 \cdot \int_0^t \dot{\Delta}(\tau)\boldsymbol{\xi}\big|_\Gamma \mathrm{d}\tau + \boldsymbol{n}_0 \cdot \Pi \nabla_{\boldsymbol{u}} \widehat{\sigma}(\hat{\vartheta}),\\
&[k^\pm \boldsymbol{n} \cdot \nabla_{\boldsymbol{u}} \hat{\vartheta}]|_{G_T} + \varkappa(\hat{\vartheta} + \theta_\infty) \Pi \nabla_{\boldsymbol{u}} \cdot \boldsymbol{u} = 0 \quad \text{on } G_T,
\end{aligned} \tag{8.4.2}$$

where $\widehat{\sigma}(\hat{\vartheta}) = \sigma(\hat{\vartheta} + \theta_\infty)$, has a solution $(\boldsymbol{u}, q, \hat{\vartheta})$ also vanishing at infinity. We state this in the form of a theorem.

Theorem 8.4.1 *We suppose that* $\Gamma \in C^{3+\alpha}$, $\boldsymbol{f}, D_x\boldsymbol{f} \in \boldsymbol{C}_2^{\alpha,(1+\alpha-\gamma)/2}(\mathbb{R}^3 \times (0,T))$, $\boldsymbol{v}_0 \in \boldsymbol{C}^{2+\alpha}_{1+\gamma}(\Omega_0^- \cup \Omega_0^+)$, $\vartheta_0 \in C^{2+\alpha}_{1+\gamma}(\Omega_0^- \cup \Omega_0^+)$, $\sigma \in C^{3+\alpha}(\mathbb{R}_+)$, $\sigma \geqslant \sigma_0 > 0$, *with some* $\alpha \in (0,1)$, $\gamma \in (0,\alpha)$, $T < \infty$, *the constant* $\theta_\infty > 0$. *In addition, let compatibility conditions* (8.4.1) *and*

$$\nabla \cdot \boldsymbol{v}_0 = 0, \quad [\boldsymbol{v}_0]|_\Gamma = 0, \quad [\vartheta_0]|_\Gamma = 0, \quad \big[k^\pm \nabla^2 \vartheta_0\big]\big|_\Gamma = 0,$$

$$\big[\mu^\pm \Pi_0 \mathbb{S}(\boldsymbol{v}_0)\boldsymbol{n}_0\big]\big|_\Gamma = \Pi_0 \nabla \widehat{\sigma}(\vartheta_0 + \theta_\infty), \quad \big[\Pi_0(\nu^\pm \nabla^2 \boldsymbol{v}_0 - \frac{1}{\rho^\pm}\nabla q_0)\big]\big|_\Gamma = 0,$$

$$\left[k^\pm \frac{\partial \vartheta_0}{\partial \boldsymbol{n}_0}\right]\bigg|_\Gamma + \varkappa(\vartheta_0 + \theta_\infty)(\Pi_0 \nabla) \cdot \boldsymbol{v}_0 = 0 \quad \text{on} \quad \Gamma$$

hold, where $q_0(\xi) \equiv q(\xi, 0)$ *is a solution to diffraction problem* (8.1.4) *without the last condition.*

Then there exists a positive value $T_0 \leqslant T$ *such that problem* (8.4.2) *has a unique solution* $(\boldsymbol{u}, q, \hat{\vartheta})$ *in the interval* $(0, T_0]$ *with following differentiability properties:* $\boldsymbol{u} \in \boldsymbol{C}^{2+\alpha,1+\alpha/2}_{1+\gamma}(D_{T_0})$, $q \in C^{(\gamma,1+\alpha)}_{1,\gamma}(D_{T_0})$, $\nabla q \in \boldsymbol{C}^{\alpha,\alpha/2}_{1+\gamma}(D_{T_0})$, $\hat{\vartheta} \in C^{2+\alpha,1+\alpha/2}_{1+\gamma}(D_{T_0})$, *and this solution satisfies the inequality*

$$|\boldsymbol{u}|^{(2+\alpha,1+\alpha/2)}_{1+\gamma,D_{T_0}} + |\nabla q|^{(\alpha,\alpha/2)}_{1+\gamma,D_{T_0}} + |q|^{(\gamma,1+\alpha)}_{1,\gamma,D_{T_0}} + |\hat{\vartheta}|^{(2+\alpha,1+\alpha/2)}_{1+\gamma,D_{T_0}}$$
$$\leqslant c(T_0)\Big\{|\boldsymbol{f}|^{(\alpha,\frac{1+\alpha-\gamma}{2})}_{\mathbb{R}^3_{T_0}} + |\boldsymbol{v}_0|^{(2+\alpha)}_{1+\gamma,\cup\Omega^\pm}$$
$$+ |\vartheta_0|^{(2+\alpha)}_{1+\gamma,\cup\Omega^\pm} + \theta_\infty + |H_0|^{(1+\alpha)}_\Gamma\Big\}.$$

The value T_0 *depends on the norms of the given functions and on the curvature of* Γ. *The function* $c(T_0)$ *contains the norms of the given functions too.*

Theorem 8.4.1 is proved similarly to Theorem 8.1.1, on the basis of the solvability of a linearized problem independent of the temperature (Theorem 5.3.1). In addition, instead of problem (8.2.1), we need to consider the problem in the whole space:

$$\mathcal{D}_t\psi - k^\pm \nabla^2_{\boldsymbol{u}}\psi = f \quad \text{in} \ D_T, \tag{8.4.3}$$
$$\psi|_{t=0} = \psi_0 \ \text{in} \ \Omega^- \cup \Omega^+, \quad [\psi]|_{G_T} = 0, \quad \psi \xrightarrow[|\xi|\to\infty]{} 0,$$
$$[k^\pm \boldsymbol{n} \cdot \nabla_{\boldsymbol{u}}\psi]|_{G_T} + \varkappa\psi(\Pi\nabla_{\boldsymbol{u}}) \cdot \boldsymbol{w} = d \quad \text{on} \ G_T,$$

which is uniquely solvable in $C_\beta^{2+\alpha,1+\frac{\alpha}{2}}(D_T)$ for any $\beta > 0$ provided that the given functions have a respective decay at infinity. In our case, one should set $\beta = 1 + \gamma$.

Theorem 8.4.2 *Let a surface* $\Gamma \in C^{2+\alpha}$, *and let vectors* $\boldsymbol{u}, \boldsymbol{w} \in \boldsymbol{C}_\beta^{2+\alpha,1+\alpha/2}(D_T)$ *with some* $\beta > 0$, $[\boldsymbol{u}]|_\Gamma = [\boldsymbol{w}]|_\Gamma = 0$, *and satisfy the inequalities*

$$(T + T^{\beta/2})|\boldsymbol{u}|_{\beta,D_T}^{(2+\alpha,1+\alpha/2)} \leqslant \delta, \qquad (T + T^{\beta/2})|\boldsymbol{w}|_{\beta,D_T}^{(2+\alpha,1+\alpha/2)} \leqslant \delta.$$

Then for functions $f \in C_\beta^{\alpha,\frac{\alpha}{2}}(D_T)$, $\psi_0 \in C_\beta^{2+\alpha}(\Omega^- \cup \Omega^+)$, $d \in C^{1+\alpha,\frac{1+\alpha}{2}}(G_T)$ *which satisfy compatibility conditions*

$$[\psi_0]|_\Gamma = 0, \;\; \psi_0 \xrightarrow[|\xi|\to\infty]{} 0, \quad [k^\pm \nabla^2 \psi_0]|_\Gamma = [f|_{t=0}]|_\Gamma,$$

$$\left[k^\pm \frac{\partial \psi_0}{\partial \boldsymbol{n}_0}\right]\Big|_\Gamma + \varkappa \psi_0(\xi) \nabla_\Gamma \cdot \boldsymbol{w}(\xi, 0) = d(\xi, 0), \;\; \xi \in \Gamma,$$

problem (8.4.3) *has a unique solution* $\psi \in C_\beta^{2+\alpha,1+\alpha/2}(D_T)$ *and the estimate*

$$|\psi|_{\beta,D_T}^{(2+\alpha,1+\alpha/2)} \leqslant c_2(T)\big\{|f|_{\beta,D_T}^{(\alpha,\alpha/2)} + |\psi_0|_{\beta,\cup\Omega^\pm}^{(2+\alpha)} + |d|_{G_T}^{(1+\alpha,\frac{1+\alpha}{2})} + \\ + T^{\frac{1-\alpha}{2}} |\nabla \boldsymbol{u}|_{\beta,D_T} \big(|\psi_0|_{\beta,\cup\Omega^\pm}^{(1)} + |\nabla \boldsymbol{w}|_{\beta,D_T} |\psi_0|_{\beta,\cup\Omega^\pm}\big)\big\}$$

holds. Here c_2 *is a nondecreasing function of* T.

This theorem can be proved in the same way as Theorem 8.2.1. Estimates of a decreasing solution can be obtained on the basis of the solvability of the heat equation in the class of the bounded functions, by analogy with the procedure developed in Sect. 5.2.

Chapter 9
Motion of Two Fluids in the Oberbeck-Boussinesq Approximation

Abstract The chapter concerns the unsteady motion of two fluids separated by a closed unknown interface Γ_t in the Oberbeck-Boussinesq approximation (see, for example, [49]). It means that the right-hand side of the problem depends on the temperature in a specific way. On the interface between the liquids, the surface tension is taken into account. This problem is investigated in the Hölder classes of functions, where local existence theorem for the problem is proved at first. As the solvability of the temperature-independent problem has been already obtained and the diffraction problem for the heat equation is studied by well-known methods, the existence of a solution to the complete problem is proved by successive approximations.

Next, the global solvability of the problem in the Oberbeck-Boussinesq approximation is established by applying the technique of Chap. 7, where the isothermal case has been considered.

9.1 Setting of the Problem and the Statement of Its Local Solvability

We use here the material of the articles [25, 26], where the solvability of the problem was first proved.

In this chapter, the liquids occupy a finite volume with rigid boundary Σ where the nonslip condition holds. Surface tension force can act on the interface Γ_t.

For every $t > 0$, it is necessary to find the interface Γ_t between the domains Ω_t^+ and Ω_t^-, as well as the velocity vector field $\boldsymbol{v}(x,t) = (v_1, v_2, v_3)$, the function p, that represents the deviation from the hydrostatic pressure P_0, and the function θ' that is the deviation from the average temperature value for both fluids; these functions satisfy the following initial–boundary value problem:

© The Author(s), under exclusive license to Springer Nature Switzerland AG 2021
I. V. Denisova, V. A. Solonnikov, *Motion of a Drop in an Incompressible Fluid*, Advances in Mathematical Fluid Mechanics,
https://doi.org/10.1007/978-3-030-70053-9_9

$$\begin{aligned}
&\mathcal{D}_t \boldsymbol{v} + (\boldsymbol{v}\cdot\nabla)\boldsymbol{v} - \nu^{\pm}\nabla^2\boldsymbol{v} + \frac{1}{\rho^{\pm}}\nabla p = \boldsymbol{f}(x,t) - \beta^{\pm}\boldsymbol{g}\theta, \quad \nabla\cdot\boldsymbol{v} = 0,\\
&\mathcal{D}_t\theta + (\boldsymbol{v}\cdot\nabla)\theta - k^{\pm}\nabla^2\theta = 0 \quad \text{in} \quad \Omega_t^- \cup \Omega_t^+, \ t>0,\\
&\boldsymbol{v}|_{t=0} = \boldsymbol{v}_0, \ \theta|_{t=0} = \theta_0 \quad \text{in} \quad \Omega_0^- \cup \Omega_0^+, \qquad (9.1.1)
\end{aligned}$$

$$\begin{aligned}
&[\boldsymbol{v}]|_{\Gamma_t} = 0, \quad \boldsymbol{v}|_{\Sigma} = 0, \quad [\theta]|_{\Gamma_t} = 0, \qquad \left[k^{\pm}\frac{\partial\theta}{\partial\boldsymbol{n}}\right]\Big|_{\Gamma_t} = 0, \quad \theta|_{\Sigma} = a,\\
&[\mathbb{T}\boldsymbol{n}]|_{\Gamma_t} = \sigma H\boldsymbol{n} \quad \text{on}\ \Gamma_t. \qquad (9.1.2)
\end{aligned}$$

Here we have used the notation of §§ 2 and 8, in addition, $\beta^{\pm} > 0$ is a step function of the temperature expansion coefficient, $\boldsymbol{g} = g(0,0,1)$, g is the acceleration of gravity; $\boldsymbol{v}_0$ is the initial distribution of the velocity and θ_0 is the initial distribution of the temperature deviation, $k^{\pm}$ is a step function of thermal conductivity. Condition (1.0.3) completes problem (9.1.1) and (9.1.2).

We pass from the Eulerian to Lagrangian coordinates by formula (1.1.2) and apply the well-known relation for the doubled mean curvature (1.1.3). As a result, we arrive at a problem for $\boldsymbol{u}$, $q = p(X_{\boldsymbol{u}}, t)$ and $\hat{\theta} = \theta(X_{\boldsymbol{u}}, t)$ with the given interface $\Gamma \equiv \Gamma_0$. If the angle between $\boldsymbol{n}$ and the outward normal $\boldsymbol{n}_0$ to Γ is acute, this system is equivalent to the following one:

$$\begin{aligned}
&\mathcal{D}_t\boldsymbol{u} - \nu^{\pm}\nabla^2_{\boldsymbol{u}}\boldsymbol{u} + \frac{1}{\rho^{\pm}}\nabla_{\boldsymbol{u}} q = \boldsymbol{f}(X_{\boldsymbol{u}},t) - \beta^{\pm}\hat{\theta}\boldsymbol{g}(X_{\boldsymbol{u}}), \quad \nabla_{\boldsymbol{u}}\cdot\boldsymbol{u} = 0,\\
&\mathcal{D}_t\hat{\theta} - k^{\pm}\nabla^2_{\boldsymbol{u}}\hat{\theta} = 0 \quad \text{in} \quad D_{\Sigma T} = Q_T^- \cup Q_T^+, \quad Q_T^{\pm} = \Omega^{\pm}\times(0,T),\\
&\boldsymbol{u}|_{t=0} = \boldsymbol{v}_0, \qquad \hat{\theta}|_{t=0} = \theta_0 \quad \text{in} \quad \Omega^-\cup\Omega^+, \quad \Omega^{\pm}\equiv\Omega_0^{\pm}, \qquad (9.1.3)\\
&[\boldsymbol{u}]|_{G_T} = 0, \quad [\hat{\theta}]|_{G_T} = 0, \quad \boldsymbol{u}|_{\Upsilon_T} = 0, \quad \hat{\theta}|_{\Upsilon_T} = a, \quad \Upsilon_T \equiv \Sigma\times(0,T),\\
&[\Pi_0\Pi\mathbb{S}_{\boldsymbol{u}}(\boldsymbol{u})\boldsymbol{n}]|_{G_T} = 0, \qquad [k^{\pm}\boldsymbol{n}\cdot\nabla_{\boldsymbol{u}}\hat{\theta}]|_{G_T} = 0 \qquad \text{on} \quad G_T = \Gamma\times(0,T).\\
&[\boldsymbol{n}_0\cdot\mathbb{T}_{\boldsymbol{u}}(\boldsymbol{u},q)\boldsymbol{n}]|_{G_T} - \sigma\boldsymbol{n}_0\cdot\Delta(t)\int_0^t \boldsymbol{u}\big|_{\Gamma}\mathrm{d}\tau = \sigma H_0(\xi) + \sigma\boldsymbol{n}_0\cdot\int_0^t \dot{\Delta}(\tau)\boldsymbol{\xi}\big|_{\Gamma}\mathrm{d}\tau. \qquad (9.1.4)
\end{aligned}$$

Here we have used again the notation of Chap. 2.

The result of the study is the local-in-time unique solvability of the problem described in the Hölder spaces of functions. The proof is based on the existence Theorem 7.3.1 for problem (7.3.1) in the case of a constant temperature (Chap. 5).

Theorem 9.1.1 (Local Existence Theorem) *We assume that* $\Gamma \in C^{3+\alpha}$, $\Sigma \in C^{2+\alpha}$, $\boldsymbol{f}, D_x\boldsymbol{f} \in \boldsymbol{C}^{\alpha,\frac{1+\alpha-\gamma}{2}}(\mathbb{R}^3\times(0,T))$, $\boldsymbol{v}_0 \in \boldsymbol{C}^{2+\alpha}(\Omega^+\cup\Omega^-)$, $a \in C^{2+\alpha,1+\alpha/2}(\Upsilon_T)$, $\theta_0 \in C^{2+\alpha}(\Omega^+\cup\Omega^-)$ *for some* $\alpha\in(0,1)$, $\gamma\in(0,\alpha)$, $0 < T < \infty$. *In addition, let compatibility conditions*

$$\nabla\cdot\boldsymbol{v}_0=0,\;\; [\boldsymbol{v}_0]|_\Gamma=0,\;\; [\theta_0]|_\Gamma=0,\;\; \boldsymbol{v}_0|_\Sigma=0,\quad \theta_0|_\Sigma=a|_{t=0},$$
$$[\mu^\pm\Pi_0\mathbb{S}(\boldsymbol{v}_0)\boldsymbol{n}_0]|_\Gamma=0,$$
$$[\Pi_0(\nu^\pm\nabla^2\boldsymbol{v}_0-\frac{1}{\rho^\pm}\nabla q_0)]|_\Gamma=[\beta^\pm\theta_0\Pi_0\boldsymbol{g}]|_\Gamma,\qquad k^-\nabla^2\theta_0|_\Sigma=\mathcal{D}_t a\big|_{t=0},$$
$$\Pi_\Sigma(\nu^-\frac{\partial^2\boldsymbol{v}_0}{\partial\boldsymbol{n}_\Sigma^2}-\frac{1}{\rho^-}\nabla q_0)\big|_\Sigma=\Pi_\Sigma(\beta^-a\boldsymbol{g}-\boldsymbol{f})\big|_{\Sigma,t=0}\quad\Big(\frac{\partial}{\partial\boldsymbol{n}_\Sigma}=\boldsymbol{n}_\Sigma\cdot\nabla\Big),$$
$$[k^\pm\nabla^2\theta_0]\big|_\Gamma=0,\;\;\Big[k^\pm\frac{\partial\theta_0}{\partial\boldsymbol{n}_0}\Big]\Big|_\Gamma=0\quad\Big(\frac{\partial}{\partial\boldsymbol{n}_0}=\boldsymbol{n}_0\cdot\nabla\Big)\tag{9.1.5}$$

be satisfied, where $q_0(\xi)\equiv q(\xi,0)$ *is a solution to diffraction problem*

$$\frac{1}{\rho^\pm}\nabla^2 q_0(\xi)=\nabla\cdot\big(\boldsymbol{f}(\xi,0)-\beta^\pm\theta_0\boldsymbol{g}-\mathcal{D}_t\mathbb{B}^*\big|_{t=0}\boldsymbol{v}_0(\xi)\big),\;\;\xi\in\Omega^\pm,$$
$$[q_0]|_\Gamma=\Big[2\mu^\pm\frac{\partial\boldsymbol{v}_0}{\partial\boldsymbol{n}_0}\cdot\boldsymbol{n}_0\Big]\Big|_\Gamma-\sigma H_0,\tag{9.1.6}$$
$$\Big[\frac{1}{\rho^\pm}\frac{\partial q_0}{\partial\boldsymbol{n}_0}\Big]\Big|_\Gamma=[\nu^\pm\boldsymbol{n}_0\cdot\nabla^2\boldsymbol{v}_0]\big|_\Gamma-[\boldsymbol{n}_0\cdot\beta^\pm\theta_0\boldsymbol{g}]\big|_\Gamma,$$
$$\frac{1}{\rho^-}\frac{\partial q_0}{\partial\boldsymbol{n}_\Sigma}\Big|_\Sigma=\nu^-\boldsymbol{n}_\Sigma\cdot\nabla^2\boldsymbol{v}_0\big|_\Sigma+\boldsymbol{n}_\Sigma\cdot(\boldsymbol{f}-\beta^-a\boldsymbol{g})\big|_{\Sigma,\,t=0}.$$

Here $\mathbb{B}=\mathbb{A}-\mathbb{I}$, $\mathbb{I}$ *is the identity matrix,* $\mathbb{B}^*$ *is the transpose to* $\mathbb{B}$, $\boldsymbol{n}_\Sigma$ *is the outward normal vector to* Σ, $\Pi_\Sigma\boldsymbol{\omega}\equiv\boldsymbol{\omega}-\boldsymbol{n}_\Sigma(\boldsymbol{n}_\Sigma\cdot\boldsymbol{\omega})$.

Then there exists a positive constant $T_*\leqslant T$ *such that problem* (9.1.3) *and* (9.1.4) *has a unique solution* $(\boldsymbol{u},q,\hat{\theta})$ *with the following properties:* $\boldsymbol{u}\in\boldsymbol{C}^{2+\alpha,1+\frac{\alpha}{2}}(D_{\Sigma T_*})$, $q\in C^{(\gamma,1+\alpha)}(D_{\Sigma T_*})$, $\nabla q\in\boldsymbol{C}^{\alpha,\alpha/2}(D_{\Sigma T_*})$, $\hat{\theta}\in C^{2+\alpha,1+\alpha/2}(D_{\Sigma T_*})$, *moreover,* $\boldsymbol{u}$ *and* $\hat{\theta}$ *are determined in a unique way, whereas* q *is done up to a bounded time dependent function. The value of* T^* *depends on the data norms, on the curvature of* Γ *and on the distance between* Γ *and* Σ. *The solution* $(\boldsymbol{u},q,\hat{\theta})$ *satisfies the inequality*

$$|\boldsymbol{u}|^{(2+\alpha,1+\alpha/2)}_{\mathrm{D}_{\Sigma T_*}}+|\nabla q|^{(\alpha,\alpha/2)}_{\mathrm{D}_{\Sigma T_*}}+|q|^{(\frac{1+\alpha-\gamma}{2})}_{t,\mathrm{D}_{\Sigma T_*}}+\langle q\rangle^{(1+\alpha,\gamma)}_{\mathrm{D}_{\Sigma T_*}}+|\hat{\theta}|^{(2+\alpha,1+\frac{\alpha}{2})}_{\mathrm{D}_{\Sigma T_*}}$$
$$\leqslant c_4\Big(T_*,|\nabla\boldsymbol{f}|^{(\alpha,\frac{1+\alpha-\gamma}{2})}_{\mathrm{D}_{\Sigma T_*}}\Big)\Big\{|\boldsymbol{f}|^{(\alpha,\frac{\alpha}{2})}_{\mathrm{D}_{\Sigma T_*}}+|\boldsymbol{f}|^{(\frac{1+\alpha-\gamma}{2})}_{t,\mathrm{D}_{\Sigma T_*}}+|H_0|^{(1+\alpha)}_\Gamma+|a|^{(2+\alpha,1+\frac{\alpha}{2})}_{\Upsilon_{T_*}}$$
$$+c\big(T_*^{\frac{1-\gamma}{2}},|\boldsymbol{v}_0|^{(1+\alpha)}_{\cup\Omega_0^\pm}\big)\Big(|\boldsymbol{v}_0|^{(2+\alpha)}_{\cup\Omega_0^\pm}+|\theta_0|^{(2+\alpha)}_{\cup\Omega_0^\pm}\Big)\Big\},\tag{9.1.7}$$

where $c_4(T)$ *is a nondecreasing function of* T.

As in the case of thermocapillary convection, we prove existence theorem relying on the solvability of the auxiliary linearized problems.

9.2 A Linearized Problem and Estimates of the Initial Pressure

In the proof of Theorem 9.1.1, we will use the solvability of linear problem (7.3.1) (Theorem 7.3.1) and that of the initial–boundary value problem for the heat equation:

$$\begin{aligned}
&\mathcal{D}_t\psi - k^{\pm}\nabla_{\boldsymbol{u}}^2\psi = f \quad \text{in} \quad D_{\Sigma T},\\
&\psi|_{t=0} = \psi_0 \quad \text{in} \quad \Omega^+\cup\Omega^-,\\
&[\psi]|_{G_T} = 0, \qquad \psi|_{\Upsilon_T} = \varphi,\\
&[k^{\pm}\boldsymbol{n}\cdot\nabla_{\boldsymbol{u}}\psi]|_{G_T} = d \quad \text{on} \quad G_T.
\end{aligned} \tag{9.2.1}$$

Theorem 9.2.1 *We assume that surfaces* $\Gamma,\Sigma \in C^{2+\alpha}$, *a vector-valued function* $\boldsymbol{u} \in \boldsymbol{C}^{2+\alpha,1+\alpha/2}(D_{\Sigma T})$ *satisfies the condition* $[\boldsymbol{u}]|_\Gamma = 0$ *and inequality* (7.3.2)*. Then for any* $f \in C^{\alpha,\alpha/2}(D_{\Sigma T}), \psi_0 \in C^{2+\alpha}(\cup\Omega^{\pm})$, $\varphi \in C^{2+\alpha,1+\alpha/2}(\Upsilon_T)$, $d \in C^{1+\alpha,(1+\alpha)/2}(G_T)$, *that are subject to compatibility conditions*

$$\begin{aligned}
&[\psi_0]|_\Gamma = 0, \quad \psi_0|_\Sigma = \varphi|_{t=0}, \quad -[k^{\pm}\nabla^2\psi_0]|_\Gamma = [f|_{t=0}]|_\Gamma,\\
&[k^{\pm}\frac{\partial\psi_0}{\partial\boldsymbol{n}_0}]|_\Gamma = d(\xi,0), \quad \xi\in\Gamma, \quad \mathcal{D}_t\varphi|_{t=0} - k^-\nabla^2\psi_0|_\Sigma = f|_{\Sigma,\,t=0},
\end{aligned} \tag{9.2.2}$$

problem (9.2.1) *has a unique solution* $\psi \in C^{2+\alpha,1+\alpha/2}(D_{\Sigma T})$ *and the estimate*

$$\begin{aligned}
|\psi|_{D_{\Sigma T}}^{(2+\alpha,1+\alpha/2)} \leqslant c_2(T)\big\{|f|_{D_{\Sigma T}}^{(\alpha,\alpha/2)} + |\psi_0|_{\cup\Omega^\pm}^{(2+\alpha)} + |\varphi|_{\Upsilon_T}^{(2+\alpha,1+\alpha/2)}&\\
+ |d|_{G_T}^{(1+\alpha,\frac{1+\alpha}{2})} + T^{\frac{1-\alpha}{2}}|\nabla\boldsymbol{u}|_{D_{\Sigma T}}|\psi_0|_{\cup\Omega^\pm}^{(1)}\big\}&
\end{aligned} \tag{9.2.3}$$

holds. Here c_2 *is a nondecreasing function of* T.

The proof of this theorem repeats the proof of Theorem 8.2.1 for problem (8.2.1) when $\varkappa = 0$.

Now we estimate pressure function at the initial moment.

We demonstrate the existence of a solution $s_0 = p_0$ of problem (7.3.3) and (6.2.4) with $\boldsymbol{w}_0 = \boldsymbol{v}_0$ and estimate this solution in C^α and $C^{1+\alpha}$. We represent s_0 in the form $s_0 = s_{00} + \rho^{\pm}s_1 + s_2$, where s_{00} is a solution of the Poisson equation

$$\frac{1}{\rho^+}\nabla^2 s_{00} = \nabla\cdot\boldsymbol{d} \quad \text{in } \Omega^+, \tag{9.2.4}$$
$$\frac{1}{\rho^-}\nabla^2 s_{00} = \nabla\cdot\boldsymbol{d} \quad \text{in } \mathbb{R}^3\setminus\overline{\Omega^+},$$

moreover, the vector $\boldsymbol{d}$ is extended into the entire space $\mathbb{R}^3$ with preservation of class and it has the form: $\boldsymbol{d} = \nabla\cdot\mathbb{D}$, while s_1, s_2 are solutions to diffraction problems with homogeneous equations. The function s_1 is a solution of (4.2.14) with $q_0 = p_{00}$, and s_2 solves the following system:

$$\begin{aligned}
&\frac{1}{\rho^\pm}\nabla^2 s_2 = 0 \quad \text{in } \Omega^\pm,\\
&[s_2]|_\Gamma = 0,\\
&\left[\frac{1}{\rho^\pm}\frac{\partial s_2}{\partial \boldsymbol{n}_0}\right]\bigg|_\Gamma = p_{01} - \left[\frac{1}{\rho^\pm}\frac{\partial s_{00}}{\partial \boldsymbol{n}_0}\right]\bigg|_\Gamma \equiv q_1,\\
&\frac{1}{\rho^-}\frac{\partial s_2}{\partial \boldsymbol{n}_\Sigma}\bigg|_\Sigma = p_{02} - \frac{1}{\rho^-}\frac{\partial s_1}{\partial \boldsymbol{n}_\Sigma}\bigg|_\Sigma - \frac{1}{\rho^-}\frac{\partial s_{00}}{\partial \boldsymbol{n}_\Sigma}\bigg|_\Sigma \equiv q_2.
\end{aligned} \tag{9.2.5}$$

The functions s_{00}, s_1 have been constructed and estimated in § 4.2 where the inequalities

$$\begin{aligned}
|s_{00}|^{(\gamma)}_{\cup\Omega^\pm} + |\nabla s_{00}|^{(\alpha)}_{\cup\Omega^\pm} &\leqslant c\{\langle\mathbb{D}\rangle^{(\gamma)}_{\cup\Omega^\pm} + |\boldsymbol{d}|^{(\alpha)}_{\cup\Omega^\pm}\}, \quad \gamma\in(0,1),\\
|s_1|^{(\gamma)}_{\cup\Omega^\pm} &\leqslant c|p_{00}|^{(\gamma)}_\Gamma,\\
|\nabla s_1|^{(\alpha)}_{\cup\Omega^\pm} &\leqslant c|p_{00}|^{(1+\alpha)}_\Gamma
\end{aligned} \tag{9.2.6}$$

have been obtained.

We investigate problem (9.2.5) following N. Günther [37] who studied the solvability of the interior Neumann problem and obtained estimates of its solution.

We look for a solution of this problem in the form of the sum of single-layer potentials

$$s_2(x) = \int_\Gamma \mathcal{E}(x,y)\omega_1(y)\mathrm{d}\Gamma(y) + \int_\Sigma \mathcal{E}(x,y)\omega_2(y)\mathrm{d}\Sigma(y),$$

where $\mathcal{E}(x,y) = -\frac{1}{4\pi|x-y|}$ is the fundamental solution of the Laplace equation, and the densities ω_1, ω_2 satisfy the system of the equations

$$\begin{aligned}
&-\frac{\rho^+-\rho^-}{\rho^+\rho^-}\Big\{\int_\Gamma \frac{\partial\mathcal{E}(x,y)}{\partial\boldsymbol{n}_0(x)}\omega_1(y)\mathrm{d}\Gamma(y) + \int_\Sigma \frac{\partial\mathcal{E}(x,y)}{\partial\boldsymbol{n}_0(x)}\omega_2(y)\mathrm{d}\Sigma(y)\Big\}\\
&\quad + \frac{\rho^++\rho^-}{2\rho^+\rho^-}\omega_1(x) = q_1(x), \quad x\in\Gamma,
\end{aligned} \tag{9.2.7}$$

$$\frac{1}{\rho^-}\Big\{\int_\Gamma \frac{\partial\mathcal{E}(x,y)}{\partial \boldsymbol{n}_\Sigma(x)}\omega_1(y)\mathrm{d}\Gamma(y)+\int_\Sigma \frac{\partial\mathcal{E}(x,y)}{\partial \boldsymbol{n}_\Sigma(x)}\omega_2(y)\mathrm{d}\Sigma(y)\Big\}$$
$$+\frac{1}{2\rho^-}\omega_2(x)=q_2(x),\quad x\in\Sigma.$$

Since the kernels in the integral operators in (9.2.7) have only weak singularities, the matrix-valued operator of the system is a Fredholm one of index zero and the solvability condition for system (9.2.7) is the orthogonality of the right-hand side (q_1,q_2) to the kernel of the conjugate operator. We find this kernel.

We write system (9.2.7) in operator form

$$K_{11}\omega_1+K_{12}\omega_2+\frac{\rho^++\rho^-}{2\rho^+\rho^-}\omega_1=q_1\qquad\text{on }\Gamma,$$
$$K_{21}\omega_1+K_{22}\omega_2+\frac{1}{2\rho^-}\omega_2=q_2\qquad\text{on }\Sigma.\tag{9.2.8}$$

We need to find all the solutions Ψ_1, Ψ_2 of the homogeneous problem conjugate to (9.2.8), i. e.,

$$\mathbb{K}^*\begin{pmatrix}\Psi_1\\ \Psi_2\end{pmatrix}=\begin{pmatrix}-\frac{\rho^++\rho^-}{2\rho^+\rho^-}\Psi_1\\ -\frac{1}{2\rho^-}\Psi_2\end{pmatrix},\tag{9.2.9}$$

where

$$\mathbb{K}^*=\begin{pmatrix}K_{11}^T & K_{21}^T\\ K_{12}^T & K_{22}^T\end{pmatrix},$$

and

$$(K_{11}^T\Psi_1)(x)=\frac{\rho^+-\rho^-}{\rho^+\rho^-}\int_\Gamma\frac{\partial\mathcal{E}(x,y)}{\partial\boldsymbol{n}_0(y)}\Psi_1(y)\mathrm{d}\Gamma(y),\quad x\in\Gamma,$$

$$(K_{21}^T\Psi_2)(x)=-\frac{1}{\rho^-}\int_\Sigma\frac{\partial\mathcal{E}(x,y)}{\partial\boldsymbol{n}_\Sigma(y)}\Psi_2(y)\mathrm{d}\Sigma(y),\quad x\in\Gamma,$$

$$(K_{12}^T\Psi_1)(x)=\frac{\rho^+-\rho^-}{\rho^+\rho^-}\int_\Gamma\frac{\partial\mathcal{E}(x,y)}{\partial\boldsymbol{n}_0(y)}\Psi_1(y)\mathrm{d}\Gamma(y),\quad x\in\Sigma,$$

$$(K_{22}^T\Psi_2)(x)=-\frac{1}{\rho^-}\int_\Sigma\frac{\partial\mathcal{E}(x,y)}{\partial\boldsymbol{n}_\Sigma(y)}\Psi_2(y)\mathrm{d}\Sigma(y),\quad x\in\Sigma.$$

We seek a solution of system (9.2.9) in the form of two constants $(\Psi_1, \Psi_2) = (c_1, c_2)$. If we substitute this couple in (9.2.9) and take the Gauss formulas into account, we obtain the system

$$\frac{\rho^+ - \rho^-}{\rho^+\rho^-}\frac{c_1}{2} - \frac{c_2}{\rho^-} = -\frac{\rho^+ + \rho^-}{2\rho^+\rho^-}c_1,$$

$$0 - \frac{1}{\rho^-}\frac{c_2}{2} = -\frac{1}{2\rho^-}c_2,$$

whence we find its nontrivial solution $c_1 = c_2 = 1$.

Now we show that there exists no other nontrivial solution of system (9.2.9). This is equivalent, by the Fredholm theorem, to the fact that the kernel of the operator $\mathbb{K}$ is one-dimensional, i. e., any two solutions of the homogeneous system (9.2.8) are linearly dependent.

We consider problem (9.2.5), which is equivalent to (9.2.8), for $q_1 = 0$ and $q_2 = 0$. We multiply the equation in (9.2.5) by s_2 and integrate it by parts over $\Omega^+ \cup \Omega^-$:

$$0 = -\int_{\Omega^+\cup\Omega^-} \frac{1}{\rho^\pm}|\nabla s_2|^2 dx + \int_\Gamma \Big[\frac{1}{\rho^\pm}\frac{\partial s_2}{\partial \boldsymbol{n}_0}\Big]\Big|_\Gamma s_2 \mathrm{d}\Gamma + \frac{1}{\rho^-}\int_\Sigma \frac{\partial s_2}{\partial \boldsymbol{n}_\Sigma} s_2 \mathrm{d}\Sigma.$$

This implies that $|\nabla s_2| = 0$ in $\Omega^+ \cup \Omega^-$, and, by continuity of s_2, it equals a constant in $\overline{\Omega^+ \cup \Omega^-}$. We denote it by b. Let the densities ω_1, ω_2 correspond to it and $\boldsymbol{\omega} \equiv (\omega_1, \omega_2)$. Now, assume that $\overline{s_2} = \overline{b}$ is another solution of homogeneous problem (9.2.5) and $\overline{\boldsymbol{\omega}} = (\overline{\omega_1}, \overline{\omega_2})$ correspond to it. We show that $\widetilde{\boldsymbol{\omega}} \equiv \overline{b}\boldsymbol{\omega} - b\overline{\boldsymbol{\omega}} = 0$. To this end, we consider the function

$$\widetilde{s}_2 = \int_\Gamma \mathcal{E}\widetilde{\omega}_1 \mathrm{d}\Gamma + \int_\Sigma \mathcal{E}\widetilde{\omega}_2 \mathrm{d}\Sigma.$$

It is obvious that, by linearity, $\widetilde{s}_2 = \overline{b}s_2 - b\overline{s}_2 = 0$ inside $\overline{\Omega^+} \cup \overline{\Omega^-}$. Consequently,

$$\widetilde{\omega}_1 = \Big[\frac{\partial s_2}{\partial \boldsymbol{n}_0}\Big]\Big|_\Gamma = 0.$$

Outside Σ, $s_2 \to 0$ as a sum of simple layer potentials; therefore, by continuity, $s_2 = 0$ inside Σ. Hence,

$$\widetilde{\omega}_2 = \Big[\frac{\partial s_2}{\partial \boldsymbol{n}_\Sigma}\Big]\Big|_\Sigma = 0.$$

Therefore, $\widetilde{\boldsymbol{\omega}} = 0$, which implies that $\boldsymbol{\omega}$ and $\overline{\boldsymbol{\omega}}$ are linearly dependent.

Thus we have derived that the right-hand side of system (9.2.8) (q_1, q_2) must be orthogonal to $(1,1)$, i.e.,

$$\int_\Gamma q_1 \mathrm{d}\Gamma = -\int_\Sigma q_2 \mathrm{d}\Sigma. \tag{9.2.10}$$

This is a necessary and sufficient condition for the solvability of (9.2.8), and two solutions of this system may differ only by a constant.

We fix the function s_2 by the condition that the densities (ω_1, ω_2) corresponding to it are orthogonal to the kernel of the conjugate operator, i. e., they also satisfy relation (9.2.10):

$$\int_\Gamma \omega_1 \mathrm{d}\Gamma = -\int_\Sigma \omega_2 \mathrm{d}\Sigma.$$

For this solution of Problem (9.2.5), by analogy with the interior Neumann problem, the estimates

$$|s_2|^{(\gamma)}_{\cup\Omega^\pm} \leqslant c(|q_1|_\Gamma + |q_2|_\Sigma) \tag{9.2.11}$$

(see [37]) and

$$|\nabla s_2|^{(\alpha)}_{\cup\Omega^\pm} \leqslant c(|q_1|^{(\alpha)}_\Gamma + |q_2|^{(\alpha)}_\Sigma) \tag{9.2.12}$$

(see [71])) hold.

Summing the corresponding inequalities in (9.2.6), (9.2.11), and (9.2.12), we obtain

$$|s_0|^{(\gamma)}_{\cup\Omega^\pm} \leqslant c\Big\{\langle\mathbb{D}\rangle^{(\gamma)}_{\cup\Omega^\pm} + |\boldsymbol{d}|^{(\alpha)}_{\cup\Omega^\pm} + |p_{00}|^{(\gamma)}_\Gamma + |p_{01}|_\Gamma + |p_{02}|_\Sigma\Big\},$$

$$|\nabla s_0|^{(\alpha)}_{\cup\Omega^\pm} \leqslant c\Big\{|d|^{(\alpha)}_{\cup\Omega^\pm} + |p_{00}|^{(1+\alpha)}_\Gamma + |p_{01}|^{(\alpha)}_\Gamma + |p_{02}|^{(\alpha)}_\Sigma\Big\}. \tag{9.2.13}$$

From (9.2.10), the solvability condition for problem (7.3.3), (6.2.4) with $\boldsymbol{v}_0 = \boldsymbol{w}_0$ and $p_0 = s_0$, the equation

$$\int_\Gamma p_{01} \mathrm{d}\Gamma - \int_{\Omega^+\cup\Omega^-} \nabla\cdot\boldsymbol{d}\,\mathrm{d}x = -\int_\Sigma p_{02}\,\mathrm{d}\Sigma \tag{9.2.14}$$

follows. Here we have used integration by parts, and we have taken equations (9.2.4) and (4.2.14) into account.

In view of the relations

$$\mathcal{D}_t r = \mathcal{D}_t(\nabla_{\boldsymbol{u}}\cdot\boldsymbol{w}) = \nabla\cdot(\mathcal{D}_t\mathbb{B}^*\boldsymbol{w} + \mathbb{A}^*\mathcal{D}_t\boldsymbol{w}),$$

$$[\mathcal{D}_t\boldsymbol{w}]|_\Gamma = 0, \quad \mathcal{D}_t\boldsymbol{w}|_\Sigma = 0,$$

it is not difficult to verify that (9.2.14) reduces to the equality

$$\int_\Gamma \boldsymbol{n}_0 \cdot [\nu^\pm \nabla^2 \boldsymbol{w}_0]|_\Gamma \mathrm{d}\Gamma = -\int_\Sigma \boldsymbol{n}_\Sigma \cdot \nu^- \nabla^2 \boldsymbol{w}_0 \mathrm{d}\Sigma, \tag{9.2.15}$$

which is valid because $\nabla \cdot \boldsymbol{w}_0 = 0$ in $\Omega^+ \cup \Omega^-$.

By similar arguments, it can be shown that a relation analogous to (9.2.14) holds for system (9.1.6) as well. Indeed, taking (9.2.15) for $\boldsymbol{v}_0$ into account, we have

$$-\int_\Gamma [\boldsymbol{n}_0 \cdot \beta^\pm \theta_0 \boldsymbol{g}]|_\Gamma \mathrm{d}\Gamma - \int_{\Omega^+\cup\Omega^-} \nabla \cdot \Big(\boldsymbol{f}(\xi,0) - \beta^\pm\theta_0\boldsymbol{g} - \mathcal{D}_t \mathbb{B}^*|_{t=0}\boldsymbol{v}_0\Big)\mathrm{d}\xi$$
$$= -\int_\Sigma \boldsymbol{n}_\Sigma \cdot \big(\boldsymbol{f}(\xi,0) - \beta^-\theta_0\boldsymbol{g}\big)\mathrm{d}\Sigma.$$

We carry out integration by parts in the volume integral, and we cancel equal terms in the surface integrals. Since $\boldsymbol{f}$ is assumed to be a continuous vector-valued function when passing across Γ, we obtain

$$\int_\Gamma [\boldsymbol{n}_0 \cdot \mathcal{D}_t \mathbb{B}^*|_{t=0}\boldsymbol{v}_0]|_\Gamma \mathrm{d}\Gamma = -\int_\Sigma \boldsymbol{n}_\Sigma \cdot \mathcal{D}_t \mathbb{B}^*|_{t=0}\boldsymbol{v}_0 \mathrm{d}\Sigma.$$

The right-hand side of this relation equals zero, because $\boldsymbol{v}_0|_\Sigma = 0$. In the integral on the left-hand side, we use the relation

$$[\boldsymbol{n}_0 \cdot \mathcal{D}_t \mathbb{B}^*|_{t=0}\boldsymbol{v}_0]|_\Gamma = \boldsymbol{v}_0 \cdot [\mathcal{D}_t \mathbb{B}|_{t=0}\boldsymbol{n}_0]|_\Gamma.$$

The latter jump is equal to zero because the vector $\mathbb{B}\boldsymbol{n}_0$ is continuous across Γ for any $t > 0$. This has been shown in Chap. 5 (see Remark 5.3.1).

Thus we conclude that problem (9.1.6) is uniquely solvable up to a constant, and there exists a solution such that the estimates

$$|q_0|^{(\gamma)}_{\cup\Omega^\pm} \leqslant c\Big\{\langle\mathbb{D}\rangle^{(\gamma)}_{\cup\Omega^\pm} + |\boldsymbol{f}|_{t=0}|^{(\alpha)}_{\mathbb{R}^3} + |\theta_0|^{(\alpha)}_{\cup\Omega^\pm} + |\boldsymbol{v}_0|^{(2)}_{\cup\Omega^\pm}$$
$$+ |\mathcal{D}_t\mathbb{B}^*|_{t=0}\boldsymbol{v}_0|^{(\alpha)}_{\cup\Omega^\pm} + |H_0|^{(\gamma)}_\Gamma + |a|_\Sigma\Big\}, \tag{9.2.16}$$

$$|\nabla q_0|^{(\alpha)}_{\cup\Omega^\pm} \leqslant c\Big\{|\boldsymbol{f}|_{t=0}|^{(\alpha)}_{\mathbb{R}^3} + |\theta_0|^{(\alpha)}_{\cup\Omega^\pm} + |\boldsymbol{v}_0|^{(2+\alpha)}_{\cup\Omega^\pm}$$
$$+ |\mathcal{D}_t\mathbb{B}^*|_{t=0}\boldsymbol{v}_0|^{(\alpha)}_{\cup\Omega^\pm} + |H_0|^{(1+\alpha)}_\Gamma + |a|^{(\alpha)}_\Sigma\Big\} \tag{9.2.17}$$

hold, where $\mathbb{D} = \{\boldsymbol{D}_k\}_{k=1}^3$, $\boldsymbol{D}_k = \frac{\partial \boldsymbol{W}}{\partial \xi_k}$,

$$\boldsymbol{W} = \int_{\Omega^+\cup\Omega^-} \mathcal{E}(\xi,\eta)\big(\boldsymbol{f}|_{t=0} - \beta^\pm\theta_0\boldsymbol{g} - \mathcal{D}_t\mathbb{B}^*|_{t=0}\boldsymbol{v}_0\big)(\eta)\mathrm{d}\eta.$$

The estimate of the Newtonian potential gradient for a bounded domain implies that

$$\langle \mathbb{D} \rangle^{(\gamma)}_{\cup\Omega^\pm} \leqslant c\Big\{|\boldsymbol{f}|_{\mathbb{R}^3} + |\theta_0|_{\cup\Omega^\pm} + |\mathcal{D}_t\mathbb{B}^*|_{t=0}\boldsymbol{v}_0|_{\cup\Omega^\pm}\Big\}.$$

Since the inequality in (5.3.13) yields

$$\big|\mathcal{D}_t\mathbb{B}^*\big|_{t=0}\boldsymbol{v}_0\big|^{(\alpha)}_{\cup\Omega^\pm} \leqslant |\mathcal{D}_t\mathbb{B}^*|^{(\alpha)}_{\cup\Omega^\pm}|\boldsymbol{v}_0|^{(\alpha)}_{\cup\Omega^\pm} \leqslant c|\nabla\boldsymbol{v}_0|^{(\alpha)}_{\cup\Omega^\pm}|\boldsymbol{v}_0|^{(\alpha)}_{\cup\Omega^\pm},$$

we finally deduce from (9.2.16) and (9.2.17) that

$$|q_0|^{(\gamma)}_{\cup\Omega^\pm} + |\nabla q_0|^{(\alpha)}_{\cup\Omega^\pm} \leqslant c\Big\{|\boldsymbol{f}(\cdot,0)|^{(\alpha)}_{\mathbb{R}^3} + |\theta_0|^{(\alpha)}_{\cup\Omega^\pm} + |H_0|^{(1+\alpha)}_{\Gamma} + |a|^{(\alpha)}_{\Sigma}$$
$$+ \big(1 + |\nabla\boldsymbol{v}_0|^{(\alpha)}_{\cup\Omega^\pm}\big)|\boldsymbol{v}_0|^{(2+\alpha)}_{\cup\Omega^\pm}\Big\}. \qquad (9.2.18)$$

Remark 9.2.1 *In conclusion, we note, as an explanation to the proof of Theorem 7.3.1, that estimates similar to (9.2.13) hold for the pressure s at every time moment t. Considering pressure function increments with respect to t, one can derive from (9.2.13) an estimate of the semi-norm $|s|^{(\gamma,1+\alpha)}_{D_{\Sigma T}}$ in terms of a weak norm of the velocity vector $\boldsymbol{w}$ (see § 4.2).*

9.3 Local Solvability of Problem (9.1.3), (9.1.4)

In this section, we prove Theorem 9.1.1.

Proof We set $(\boldsymbol{u}^{(0)}, q^{(0)}, \hat{\theta}^{(0)}) = (\boldsymbol{v}_0, q_0, \hat{\theta}_0)$, where q_0 is a solution of problem (9.1.6). We define the next approximation $(\boldsymbol{u}^{(m+1)}, q^{(m+1)})$, $m = 0, 1, \ldots$, as a solution of the problem

$$\mathcal{D}_t\boldsymbol{u}^{(m+1)} - \nu^\pm\nabla^2_m\boldsymbol{u}^{(m+1)} + \frac{1}{\rho^\pm}\nabla_m q^{(m+1)} = \boldsymbol{f}(X_m, t) - \beta^\pm\hat{\theta}^{(m)}\boldsymbol{g},$$

$$\nabla_m \cdot \boldsymbol{u}^{(m+1)} = 0 \quad \text{in } D_{\Sigma T},$$

$$\boldsymbol{u}^{(m+1)}|_{t=0} = \boldsymbol{v}_0 \qquad \text{in } \Omega^- \cup \Omega^+, \qquad (9.3.1)$$

$$[\boldsymbol{u}^{(m+1)}]|_{G_T} = 0, \quad \boldsymbol{u}^{(m+1)}|_{\Upsilon_T} = 0, \quad [\mu^\pm\Pi_0\Pi_m\mathbb{S}_m(\boldsymbol{u}^{(m+1)})\boldsymbol{n}_m]|_{G_T} = 0,$$

$$[\boldsymbol{n}_0 \cdot \mathbb{T}_m(\boldsymbol{u}^{(m+1)}, q^{(m+1)})\boldsymbol{n}_m]|_{G_T} - \sigma\boldsymbol{n}_0 \cdot \Delta_m(t)\int_0^t \boldsymbol{u}^{(m+1)}\big|_\Gamma \,\mathrm{d}\tau =$$

$$= \sigma H_0(\xi) + \sigma\boldsymbol{n}_0 \cdot \int_0^t \dot{\Delta}_m(\tau)\boldsymbol{\xi}\big|_\Gamma \mathrm{d}\tau \quad \text{on } G_T.$$

Here $\nabla_m = \nabla_{\boldsymbol{u}^{(m)}}$, $\Pi_m \boldsymbol{\omega} = \boldsymbol{\omega} - \boldsymbol{n}_m(\boldsymbol{n}_m \cdot \boldsymbol{\omega})$, $\boldsymbol{n}_m$ is the outward normal to $\Gamma_m = \{\boldsymbol{x} = \boldsymbol{X}_m(\xi, t),\ \xi \in \Gamma\}$, $\boldsymbol{X}_m = \boldsymbol{X}_{\boldsymbol{u}^{(m)}}$; $\mathbb{S}_m = \mathbb{S}_{\boldsymbol{u}^{(m)}}$, $\mathbb{T}_m = \mathbb{T}_{\boldsymbol{u}^{(m)}}$; $\Delta_m(t)$ is the Laplace–Beltrami operator on Γ_m.

We define the temperature approximation $\hat{\theta}^{(m+1)}$, $m = 0, 1, \ldots$, as a solution to the problem

$$
\begin{aligned}
&\mathcal{D}_t \hat{\theta}^{(m+1)} - k^{\pm} \nabla_m^2 \hat{\theta}^{(m+1)} = 0 \quad \text{in} \quad D_{\Sigma T},\\
&\hat{\theta}^{(m+1)}|_{t=0} = \theta_0 \quad \text{in} \quad \Omega^- \cup \Omega^+,\\
&[\hat{\theta}^{(m+1)}]|_{G_T} = 0, \qquad \hat{\theta}^{(m+1)}|_{\Upsilon_T} = a,\\
&[k^{\pm} \boldsymbol{n}_m \cdot \nabla_m \hat{\theta}^{(m+1)}]|_{G_T} = 0 \quad \text{on } G_T.
\end{aligned}
\tag{9.3.2}
$$

We assume that $\boldsymbol{u}^{(m)}$ exists and satisfies inequality (7.3.2) in the interval $(0, T_m] \subset (0, T]$. In order to apply Theorem 7.3.1 to problem (9.3.1), we verify its hypotheses. The first, second and fourth groups of hypotheses follow from the assumptions of Theorem 9.1.1 including (9.1.5) and (9.1.6). As to the third hypothesis, we can set

$$
\begin{aligned}
&\boldsymbol{h} \equiv \boldsymbol{h}_m = -\mathbb{A}^{(m)*}\big(\boldsymbol{f} - \beta^{\pm} \boldsymbol{g} \hat{\theta}^{(m)}\big),\\
&\mathbb{G} \equiv \mathbb{G}_m(\xi, t) = -\nabla \int_{\cup\Omega^{\pm}} \mathcal{E}(\xi, \eta) \mathbb{A}^{(m)*}\big(\boldsymbol{f}(\eta, t) - \beta^{\pm} \boldsymbol{g} \hat{\theta}^{(m)}\big)\, \mathrm{d}\eta,
\end{aligned}
\tag{9.3.3}
$$

where $\mathbb{A}^{(m)} = \mathbb{A}(\boldsymbol{u}^{(m)})$. The first equality in (9.3.3) follows from the relation $\mathbb{A}^{(m)} \nabla \cdot \boldsymbol{w} = \nabla \cdot \mathbb{A}^{(m)*} \boldsymbol{w}$, which arises from the identity $\dfrac{\partial A_{ij}^{(m)}}{\partial \xi_j} = 0$, true for the cofactor matrix of the Jacobi matrix of an arbitrary transformation. Moreover, it is obvious that $\big[\boldsymbol{n}_0 \cdot \big(\boldsymbol{h}_m + \mathbb{A}^{(m)*}(\boldsymbol{f} - \beta^{\pm} \hat{\theta}^{(m)} \boldsymbol{g})\big)\big]\big|_{\Gamma} = 0$.

We evaluate the potentials $\boldsymbol{h}_m$ and $\mathbb{G}_m$ on the basis of the results of Chap. 5 (see also [54]):

$$
\begin{aligned}
&|\boldsymbol{h}_m|_{D_{\Sigma T_m}}^{(\alpha, \alpha/2)} + |\mathbb{G}_m|_{D_{\Sigma T_m}}^{(\gamma, 1+\alpha)} + |\mathbb{G}_m|_{D_{\Sigma T_m}}^{(\gamma, 0)} \leqslant\\
&\leqslant c(T_m)\Big(1 + (T_m + T_m^{\gamma/2})|\boldsymbol{u}^{(m)}|_{D_{\Sigma T_m}}^{(2+\alpha, 1+\alpha/2)}\Big)\Big(|\boldsymbol{f}|_{t, D_{\Sigma T_m}}^{(\frac{1+\alpha-\gamma}{2})} + |\hat{\theta}^{(m)}|_{D_{\Sigma T_m}}^{(\gamma, 1+\alpha)}\Big).
\end{aligned}
\tag{9.3.4}
$$

Next, we have

$$
\begin{aligned}
&|\sigma H_0(\xi)|_{G_{T_m}}^{(1+\alpha, \frac{1+\alpha}{2})} \leqslant c|H_0|_{\Gamma}^{(1+\alpha)},\\
&|\sigma \boldsymbol{n}_0 \cdot \dot{\Delta}_m(\tau) \boldsymbol{\xi}|_{G_{T_m}}^{(\alpha, \alpha/2)} \leqslant c|\nabla \boldsymbol{u}^{(m)}|_{G_{T_m}}^{(\alpha, \alpha/2)} \leqslant c_3 |\boldsymbol{u}^{(m)}|_{D_{\Sigma T_m}}^{(1+\alpha, \alpha/2)}.
\end{aligned}
\tag{9.3.5}
$$

Thus, Theorem 7.3.1 and inequalities (9.3.4) and (9.3.5) imply that there exists a unique solution $(\boldsymbol{u}^{(m+1)}, q^{(m+1)})$ to problem (9.3.1) in the interval $(0, T_m]$ and the estimate

$$N_{T_m}[\boldsymbol{u}^{(m+1)},q^{(m+1)}]\leqslant$$

$$\leqslant c_4(T_m)\Big\{(1+\delta)\Big(|\boldsymbol{f}|^{(\alpha,\frac{\alpha}{2})}_{D_{\Sigma T_m}}+|\boldsymbol{f}|^{(\frac{1+\alpha-\gamma}{2})}_{t,D_{\Sigma T_m}}+|\hat{\theta}^{(m)}|^{(\alpha,\frac{\alpha}{2})}_{D_{\Sigma T_m}}+|\hat{\theta}^{(m)}|^{(\gamma,1+\alpha)}_{D_{\Sigma T_m}}\Big)$$

$$+|H_0|^{(1+\alpha)}_{\Gamma}+|\boldsymbol{u}^{(m)}|^{(1+\alpha,\frac{\alpha}{2})}_{D_{\Sigma T_m}}$$

$$+\Big(1+T_m^{\frac{1-\alpha}{2}}|\nabla\boldsymbol{u}^{(m)}|_{D_{\Sigma T_m}}+|\nabla\boldsymbol{u}^{(m)}|^{(\alpha,\frac{\alpha}{2})}_{D_{\Sigma T_m}}\Big)|\boldsymbol{v}_0|^{(2+\alpha)}_{\cup\Omega^{\pm}}\Big\}$$

holds for it.

Let us choose T_{m+1} so that $\boldsymbol{u}^{(m+1)}$ satisfies inequality (7.3.2).

For $m=0$, we have

$$N_{T_0}[\boldsymbol{u}^{(1)},q^{(1)}]\leqslant c_4(T_0)\Big\{(1+\delta)\Big(|\boldsymbol{f}|^{(\alpha,\frac{\alpha}{2})}_{D_{\Sigma T_0}}+|\boldsymbol{f}|^{(\frac{1+\alpha-\gamma}{2})}_{t,D_{\Sigma T_0}}+|\theta_0|^{(2+\alpha)}_{\cup\Omega^{\pm}}\Big)+|H_0|^{(1+\alpha)}_{\Gamma}$$
$$+\big(2+T_0^{\frac{1-\alpha}{2}}|\boldsymbol{v}_0|^{(1)}_{\cup\Omega^{\pm}}+|\boldsymbol{v}_0|^{(1+\alpha)}_{\cup\Omega^{\pm}}\big)|\boldsymbol{v}_0|^{(2+\alpha)}_{\cup\Omega^{\pm}}\Big\}. \quad (9.3.6)$$

Now we estimate temperature approximations.

Compatibility conditions (9.2.2) for problem (9.3.2) follow from relations (9.1.5); therefore, for every $m=0,1,2,\dots$ Theorem 9.2.1 yields the estimate

$$|\hat{\theta}^{(m+1)}|^{(2+\alpha,1+\frac{\alpha}{2})}_{D_{\Sigma T_m}}\leqslant c_5(T_m)\Big\{|a|^{(2+\alpha,1+\frac{\alpha}{2})}_{\Sigma_{T_m}}+\big(1+\delta+T_m^{\frac{1-\alpha}{2}}|\nabla\boldsymbol{u}^{(m)}|_{D_{\Sigma T_m}}\big)|\theta_0|^{(2+\alpha)}_{\cup\Omega^{\pm}}\Big\}$$
$$\equiv\Theta[T_m]. \quad (9.3.7)$$

In this way, by induction, we deduce that for every $m\geqslant 0$, Systems (9.3.1) and (9.3.2) have solutions $(\boldsymbol{u}^{(m+1)},q^{(m+1)})$ and $\hat{\theta}^{(m+1)}$, respectively, on a time interval $(0,T_{m+1}]\subset(0,T]$ such that the preceding approximation $(\boldsymbol{u}^{(m)},q^{(m)})$, $\hat{\theta}^{(m)}$ is also defined in this interval, and velocity vector field $\boldsymbol{u}^{(m+1)}$ satisfies (7.3.2) with some small δ.

Now, it is necessary to show that there exists $T^*>0$ such that for all $m\in\mathbb{N}$ $T_m\geqslant T^*$, the norms $N_{T^*}[\boldsymbol{u}^{(m)},q^{(m)},\hat{\theta}^{(m)}]\equiv N_{T^*}[\boldsymbol{u}^{(m)},q^{(m)}]+|\hat{\theta}^{(m)}|^{(2+\alpha,1+\frac{\alpha}{2})}_{D_{\Sigma T^*}}$ are uniformly bounded and the sequence $\{\boldsymbol{u}^{(m)},q^{(m)},\hat{\theta}^{(m)}\},m>0$, converges to a solution of problem (9.1.3) and (9.1.4).

To this end, we compose the difference between systems (9.3.1), corresponding to $m=j$ and $m=j-1$. The functions $\boldsymbol{w}^{(j+1)}=\boldsymbol{u}^{(j+1)}-\boldsymbol{u}^{(j)}$, $s^{(j+1)}=q^{(j+1)}-q^{(j)}$, $j=1,2\dots$, satisfy the following problem:

$$\mathcal{D}_t\boldsymbol{w}^{(j+1)}-\nu^{\pm}\nabla_j^2\boldsymbol{w}^{(j+1)}+\frac{1}{\rho^{\pm}}\nabla_j s^{(j+1)}=\boldsymbol{f}(X_j,t)-\boldsymbol{f}(X_{j-1},t)-$$

$$-\beta^{\pm}\boldsymbol{g}\psi^{(j)}+\boldsymbol{l}_1^{(j)}(\boldsymbol{u}^{(j)},q^{(j)})-\boldsymbol{l}_1^{(j-1)}(\boldsymbol{u}^{(j)},q^{(j)})\equiv\boldsymbol{f}^{(j)},$$

$$\nabla_j\cdot\boldsymbol{w}^{(j+1)}=l_2^{(j)}(\boldsymbol{u}^{(j)})-l_2^{(j-1)}(\boldsymbol{u}^{(j)})\equiv r^{(j)}\quad\text{in}\quad D_{\Sigma T_{m+1}},$$

$$\boldsymbol{w}^{(j+1)}|_{t=0}=0\quad\text{in }\Omega^-\cup\Omega^+,\qquad\big[\boldsymbol{w}^{(j+1)}\big]\big|_{G_{T_{m+1}}}=0,\quad\boldsymbol{w}^{(j+1)}\big|_{\Upsilon_{T_{m+1}}}=0,$$

$$\big[\mu^{\pm}\Pi_0\Pi_j\mathbb{S}_j(\boldsymbol{w}^{(j+1)})\boldsymbol{n}_j\big]\big|_{G_{T_{m+1}}}=\boldsymbol{l}_3^{(j)}(\boldsymbol{u}^{(j)})-\boldsymbol{l}_3^{(j-1)}(\boldsymbol{u}^{(j)}),\tag{9.3.8}$$

$$\big[\boldsymbol{n}_0\cdot\mathbb{T}_j(\boldsymbol{w}^{(j+1)},s^{(j+1)})\boldsymbol{n}_j\big]\big|_{G_{Tm+1}}-\sigma\boldsymbol{n}_0\cdot\Delta_j(t)\int_0^t\boldsymbol{w}^{(j+1)}\big|_\Gamma\mathrm{d}\tau=$$

$$=l_4^{(j)}(\boldsymbol{u}^{(j)},q^{(j)})-l_4^{(j-1)}(\boldsymbol{u}^{(j)},q^{(j)})+\sigma\int_0^t\big(l_5^{(j)}(\boldsymbol{u}^{(j)})-l_5^{(j-1)}(\boldsymbol{u}^{(j)})\big)\mathrm{d}\tau$$

$$+\sigma\boldsymbol{n}_0\cdot\int_0^t\big(\dot{\Delta}_j(\tau)-\dot{\Delta}_{j-1}(\tau)\big)\boldsymbol{\xi}\big|_\Gamma\mathrm{d}\tau\quad\text{on }G_{T_{m+1}},$$

where thee operators $l_i^{(j)}$ are calculated by formulas (8.3.9).

In addition, the function $\psi^{(j+1)}=\hat{\theta}^{(j+1)}-\hat{\theta}^{(j)}$, $j=1,\ldots$, is a solution to the problem that is the result of subtraction of systems (9.3.2) with indices $j+1$ and j:

$$\mathcal{D}_t\psi^{(j+1)}-k^{\pm}\nabla_j^2\psi^{(j+1)}=k^{\pm}\nabla_j^2\hat{\theta}^{(j)}-k^{\pm}\nabla_{j-1}^2\hat{\theta}^{(j)}$$

$$\equiv\big(h_1^{(j)}-h_1^{(j-1)}\big)(\hat{\theta}^{(j)})\quad\text{in }D_{\Sigma T_{m+1}},$$

$$\psi^{(j+1)}\big|_{t=0}=0\quad\text{in }\Omega^-\cup\Omega^+,\qquad\big[\psi^{(j+1)}\big]\big|_\Gamma=0,\qquad\psi^{(j+1)}\big|_\Sigma=0,\tag{9.3.9}$$

$$\big[k^{\pm}\boldsymbol{n}_j\cdot\nabla_j\psi^{(j+1)}\big]\big|_\Gamma=-\big[k^{\pm}\boldsymbol{n}_j\cdot\nabla_j\hat{\theta}^{(j)}\big]\big|_\Gamma+\big[k^{\pm}\boldsymbol{n}_{j-1}\cdot\nabla_{j-1}\hat{\theta}^{(j)}\big]\big|_\Gamma$$

$$\equiv\big(h_2^{(j)}-h_2^{(j-1)}\big)(\hat{\theta}^{(j)})\quad\text{on }G_{T_{m+1}}.$$

Here $h_i^{(k)}$, $i=1,2$, are calculated by (8.2.5), where $\boldsymbol{u}$ is replaced by $\boldsymbol{u}^{(k)}$.

We evaluate the solution of problem (9.3.9) using Theorem 9.2.1 and Lemma 8.2.3:

$$|\psi^{(j+1)}|_{D_{\Sigma T}}^{(2+\alpha,1+\frac{\alpha}{2})}\leqslant|\big(h_1^{(j)}-h_1^{(j-1)}\big)(\hat{\theta}^{(j)})|_{D_{\Sigma T}}^{(\alpha,\frac{\alpha}{2})}+|\big(h_2^{(j)}-h_2^{(j-1)}\big)(\hat{\theta}^{(j)})|_{G_T}^{(1+\alpha,\frac{1+\alpha}{2})}$$

$$\leqslant c_5\Big\{|\hat{\theta}^{(j)}|_{D_{\Sigma T}}^{(2+\alpha,1+\frac{\alpha}{2})}M_T[\boldsymbol{w}^{(j)}]+T^{\frac{1-\alpha}{2}}|\nabla\hat{\theta}^{(j)}|_{D_{\Sigma T}}|\nabla\boldsymbol{w}^{(j)}|_{D_{\Sigma T}}\Big\},\tag{9.3.10}$$

where $j=1,2,\ldots,$ $T\leqslant T_{m+1}$, and $M_T[\boldsymbol{v}]$ are computed by formula (8.2.8).

For $j=0$, in view of inequality (9.3.7), we have

$$|\psi^{(1)}|^{(2+\alpha,1+\frac{\alpha}{2})}_{D_{\Sigma T_0}} \leqslant |\hat{\theta}^{(1)}|^{(2+\alpha,1+\frac{\alpha}{2})}_{D_{\Sigma T_0}} + |\theta_0|^{(2+\alpha)}_{\cup\Omega^\pm} \leqslant 2\Theta[T_0]. \tag{9.3.11}$$

We apply again Theorem 7.3.1, now to problem (9.3.8). Let us consider the functions on the right-hand side of this system.

The operators $l_i^{(j)}$ (8.3.9) and their differences for different j have been estimated in §§ 5.3, 5.4 (Lemmas 5.3.2–5.3.5, and 5.4.2). Following Chap. 5, we set

$$\boldsymbol{h}^{(j)} = (\mathbb{B}^{(j-1)*} - \mathbb{B}^{(j)*})\mathcal{D}_t\boldsymbol{u}^{(j)} + \mathcal{D}_t(\mathbb{B}^{(j-1)*} - \mathbb{B}^{(j)*})\boldsymbol{u}^{(j)} - \mathbb{A}^{(j)*}\boldsymbol{f}^{(j)}$$

in the identity

$$\mathcal{D}_t r^{(j)} - \nabla_j \cdot \boldsymbol{f}^{(j)} = \nabla \cdot \boldsymbol{h}^{(j)}.$$

It is easy to see that

$$[(\boldsymbol{h}^{(j)} + \mathbb{A}^{(j)*}\boldsymbol{f}^{(j)}) \cdot \boldsymbol{n}_0]|_{G_t} = 0.$$

Since

$$\begin{aligned}\boldsymbol{f}^{(j)} = \partial\big(\boldsymbol{L}_{1k}^{(j)}(\boldsymbol{u}^{(j)}, q^{(j)}) - \boldsymbol{L}_{1k}^{(j-1)}(\boldsymbol{u}^{(j)}, q^{(j)})\big)/\partial\xi_k - \beta^\pm \boldsymbol{g}\psi^{(j)}\\ +\boldsymbol{f}(X_j, t) - \boldsymbol{f}(X_{j-1}, t)\end{aligned}$$

and

$$\mathcal{D}_t\boldsymbol{u}^{(j)} = \partial \boldsymbol{M}_k^{(j)}/\partial\xi_k + \boldsymbol{f}(X_{j-1}, t) - \beta^\pm \boldsymbol{g}_{j-1}\hat{\theta}^{(j-1)},$$

where $\boldsymbol{M}_k^{(j)} = \nu^\pm(\mathbb{A}^{(j-1)*}\boldsymbol{e}_k \cdot \nabla_{j-1})\boldsymbol{u}^{(j)} - \mathbb{A}^{(j-1)*}\boldsymbol{e}_k q^{(j)}/\rho^\pm$, we can write

$$\boldsymbol{h}^{(j)} = \nabla \cdot \mathbb{G}^{(j)} \equiv \partial \boldsymbol{G}_k^{(j)}/\partial\xi_k$$

with the potentials

$$\begin{aligned}\boldsymbol{G}_k^{(j)} &= (\mathbb{B}^{(j-1)*} - \mathbb{B}^{(j)*})\boldsymbol{M}_k^{(j)} - \mathbb{A}^{(j)*}\big(\boldsymbol{L}_{1k}^{(j)}(\boldsymbol{u}^{(j)}, q^{(j)}) - \boldsymbol{L}_{1k}^{(j-1)}(\boldsymbol{u}^{(j)}, q^{(j)})\big)\\ &\quad + \frac{\partial \boldsymbol{W}^{(j)}}{\partial \xi_k},\\ \boldsymbol{W}^{(j)} &= \int\limits_{\Omega^-\cup\Omega^+} \mathcal{E}(\xi,\eta)\Big\{\frac{\partial}{\partial\eta_i}(\mathbb{B}^{(j-1)*} - \mathbb{B}^{(j)*})\boldsymbol{M}_i^{(j)}\\ &\quad - (\mathbb{B}^{(j-1)*} - \mathbb{B}^{(j)*})\Big(\boldsymbol{f}(X_{j-1}, t) - \beta^\pm \boldsymbol{g}\hat{\theta}^{(j-1)}\Big)\end{aligned}$$

$$- \mathcal{D}_t\big(\mathbb{B}^{(j-1)*} - \mathbb{B}^{(j)*}\big)\boldsymbol{u}^{(j)} - \frac{\partial \mathbb{A}^{(j)*}}{\partial \eta_i}\Big(\boldsymbol{L}_{1i}^{(j)}(\boldsymbol{u}^{(j)}, q^{(j)}) - \boldsymbol{L}_{1i}^{(j-1)}(\boldsymbol{u}^{(j)}, q^{(j)})\Big)$$
$$+ \mathbb{A}^{(j)*}\Big(\boldsymbol{f}(X_j, t) - \boldsymbol{f}(X_{j-1}, t) - \beta^{\pm}\boldsymbol{g}\psi^{(j)}\Big)\Big\}\,\mathrm{d}\eta.$$

We assume that $(\boldsymbol{u}^{(j)}, q^{(j)})$, $\hat{\theta}^{(j)}$, $j = 0, \dots, m$, satisfy the inequality

$$(T + T^{\gamma/2})N_T[\boldsymbol{u}^{(j)}, q^{(j)}, \hat{\theta}^{(j)}] < \delta, \quad T \leqslant T_{m+1}, \tag{9.3.12}$$

which is stronger than (7.3.2).

We note that

$$|\beta^{\pm}\boldsymbol{g}\psi^{(j)}|_{D_{\Sigma T}}^{(\alpha, \frac{\alpha}{2})} \leqslant c|\psi^{(j)}|_{D_{\Sigma T}}^{(2+\alpha, 1+\frac{\alpha}{2})}. \tag{9.3.13}$$

To estimate the right-hand sides of system (9.3.8), we apply Lemmas 5.4.1–5.4.4, taking (9.3.13) into account:

$$|\boldsymbol{f}^{(j)}|_{D_{\Sigma T}}^{(\alpha,\alpha/2)} + |r^{(j)}|_{D_{\Sigma T}}^{(1+\alpha, \frac{1+\alpha}{2})} + |\boldsymbol{h}^{(j)}|_{D_{\Sigma T}}^{(\alpha,\alpha/2)} + |\mathbb{G}^{(j)}|_{D_{\Sigma T}}^{(\gamma,1+\alpha)}$$
$$+ |\mathbb{G}^{(j)}|_{D_{\Sigma T}}^{(\gamma,0)} + |\boldsymbol{l}_3^{(j)} - \boldsymbol{l}_3^{(j-1)}|_{G_T}^{(1+\alpha, \frac{1+\alpha}{2})} + |l_4^{(j)} - l_4^{(j-1)}|_{G_T} + |\nabla_\Gamma(l_4^{(j)} - l_4^{(j-1)})|_{G_T}^{(\alpha,\alpha/2)}$$
$$+ |l_4^{(j)} - l_4^{(j-1)}|_{G_T}^{(\gamma,1+\alpha)} + |l_5^{(j)} - l_5^{(j-1)}|_{G_T}^{(\alpha,\alpha/2)} + |\boldsymbol{n}_0 \cdot (\dot{\Delta}_j - \dot{\Delta}_{j-1})\boldsymbol{\xi}|_{G_T}^{(\alpha,\alpha/2)}$$
$$\leqslant c_6\Big\{(T + T^{\gamma/2})N_T[\boldsymbol{u}^{(j)}, q^{(j)}, \hat{\theta}^{(j)}]|\boldsymbol{w}^{(j)}|_{D_{\Sigma T}}^{(2+\alpha,1+\alpha/2)} + (T + T^{\gamma/2})|\boldsymbol{w}^{(j)}|_{D_{\Sigma T}}^{(\alpha,\alpha/2)}$$
$$+ |\nabla \boldsymbol{w}^{(j)}|_{D_{\Sigma T}}^{(\alpha,\alpha/2)} + P_T[\boldsymbol{w}^{(j)}]|\boldsymbol{u}^{(j)}(\cdot, 0)|_{\Omega^- \cup \Omega^+}^{(1)} + |\psi^{(j)}|_{D_{\Sigma T}}^{(2+\alpha,1+\frac{\alpha}{2})}$$
$$+ (T + T^{\gamma/2})|\hat{\theta}^{(j-1)}|_{D_{\Sigma T}}^{(2+\alpha,1+\alpha/2)}|\boldsymbol{w}^{(j)}|_{D_{\Sigma T}}^{(2+\alpha,1+\alpha/2)}\Big\}.$$

Here we have estimated the norms $\nabla \boldsymbol{W}^{(j)}$ by Lemma 6.4.3 applying it to a bounded domain.

Since $\boldsymbol{w}^{(j)}|_{t=0} = 0$, $\nabla \boldsymbol{w}^{(j)}|_{t=0} = 0$, we obtain for $P_T[\boldsymbol{w}] = T^{\frac{1-\alpha}{2}}|\nabla \boldsymbol{w}|_{D_{\Sigma T}} + |\nabla \boldsymbol{w}|_{D_{\Sigma T}}^{(\alpha, \frac{\alpha}{2})}$ the estimate

$$P_T[\boldsymbol{w}^{(j)}]|\boldsymbol{v}_0|_{\cup\Omega^\pm}^{(1)} \leqslant c_7(T + T^{1/2})|\boldsymbol{w}^{(j)}|_{D_{\Sigma T}}^{(2+\alpha,1+\frac{\alpha}{2})}|\boldsymbol{v}_0|_{\cup\Omega^\pm}^{(1)} \leqslant c_7\delta|\boldsymbol{w}^{(j)}|_{D_{\Sigma T}}^{(2+\alpha,1+\frac{\alpha}{2})}.$$

In addition,

$$|\boldsymbol{w}^{(j)}|_{D_{\Sigma T}}^{(\alpha,\alpha/2)} + |\nabla \boldsymbol{w}^{(j)}|_{D_{\Sigma T}}^{(\alpha,\alpha/2)} \leqslant \varepsilon|\boldsymbol{w}^{(j)}|_{D_{\Sigma T}}^{(2+\alpha,1+\frac{\alpha}{2})} + c_1(\varepsilon)|\boldsymbol{w}^{(j)}|_{D_{\Sigma T}}.$$

Hence, the norms of the right-hand sides of system (9.3.8) are bounded, and, by Theorem 7.3.1, we can conclude that its solution $(\boldsymbol{w}^{(j+1)}, s^{(j+1)})$ satisfies the inequality

$$N_T[\boldsymbol{w}^{(j+1)}, s^{(j+1)}] \leqslant c_8(T)\Big\{(\delta+\varepsilon)N_T[\boldsymbol{w}^{(j)}, s^{(j)}] + c_1(\varepsilon)|\boldsymbol{w}^{(j)}|_{D_{\Sigma T}}\Big\} \quad (9.3.14)$$
$$+c_9|\psi^{(j)}|^{(2+\alpha,1+\frac{\alpha}{2})}_{D_{\Sigma T}}, \qquad T \leqslant T_{m+1}$$

with a nondecreasing function c_8 of T.

Taking into account the estimate (9.3.10) and the fact that

$$M_T[\boldsymbol{w}^{(j)}] \leqslant 2(T+T^{1/2})|\boldsymbol{w}^{(j)}|^{(2+\alpha,1+\frac{\alpha}{2})}_{D_{\Sigma T}}$$

and

$$|\nabla \boldsymbol{w}^{(j-1)}|_{D_{\Sigma T}} \leqslant T^{\frac{1+\alpha}{2}} \langle \nabla \boldsymbol{w}^{(j-1)}\rangle^{(\frac{1+\alpha}{2})}_{t,D_{\Sigma T}}, \qquad j \geqslant 2,$$

we conclude

$$|\psi^{(j)}|^{(2+\alpha,1+\frac{\alpha}{2})}_{D_{\Sigma T}} \leqslant 3c_5\delta N_T[\boldsymbol{w}^{(j-1)}, s^{(j-1)}]. \quad (9.3.15)$$

Summing inequalities (9.3.14) and (9.3.10) and keeping (9.3.15) in mind, we derive that for $j \geqslant 2$ and $T \leqslant T_{m+1}$

$$N_T^{(j+1)} \equiv N_T[\boldsymbol{w}^{(j+1)}, s^{(j+1)}, \psi^{(j+1)}] \leqslant c_{10}(T)\big\{(\delta+\varepsilon)N_T^{(j)} + c_1(\varepsilon)|\boldsymbol{w}^{(j)}|_{D_{\Sigma T}}\big\}$$
$$+ 3c_5\delta\big\{N_T^{(j-1)} + N_T^{(j)}\big\}.$$

We use the inequality

$$|\boldsymbol{w}^{(j)}|_{D_{\Sigma T}} \leqslant \int_0^T |\mathcal{D}_t \boldsymbol{w}^{(j)}(\cdot,t)|_{\cup\Omega^\pm}\, dt \leqslant \int_0^T N_t[\boldsymbol{w}^{(j)}, s^{(j)}]\, dt \quad (9.3.16)$$

to obtain the final estimate

$$N_T^{(j+1)} \leqslant \varkappa_1 N_T^{(j)} + \varkappa_2 N_T^{(j-1)} + c_{11}\int_0^T N_t^{(j)}\, dt, \quad j = 2,3\ldots, \quad (9.3.17)$$

where $\varkappa_1(T) = c_{10}(T)(\delta+\varepsilon) + 3c_5\delta$, $\varkappa_2(T) = 3c_5\delta$.

If $j = 1$, from (9.3.7), (9.3.14), and (9.3.16), we deduce the inequality

$$N_T^{(2)} \leqslant \varkappa_1 N_T^{(1)} + c_{12}\Theta[T_0] + c_{11}\int_0^T N_t^{(1)}\, dt. \quad (9.3.18)$$

If we sum inequalities (9.3.17) over j from 2 to m, and add (9.3.18) and $N_T^{(1)}$, estimated by (9.3.6) and (9.3.11), then the sum

$$\Sigma_{m+1}(T) = \sum_{j=1}^{m+1} N_T^{(j)}$$

can be evaluated as follows:

$$\Sigma_{m+1}(T) \leqslant \varkappa_1 \Sigma_m(T) + \varkappa_2 \Sigma_{m-1}(T) + c_{11} \int_0^T \Sigma_m(t)\,dt + F_1(T)$$
$$\leqslant \varkappa \Sigma_{m+1}(T) + c_{11} \int_0^T \Sigma_{m+1}(t)\,dt + F_1(T),$$

where $F_1(T) = N_T^{(1)} + c_{12}\Theta[T_0]$.

Finally, we choose δ and ε in such a way that $\varkappa \equiv \varkappa_1(T_{m+1}) + \varkappa_2(T_{m+1}) < 1$. The Gronwall lemma applied to the inequality

$$\Sigma_{m+1}(T) \leqslant c_{13} \int_0^T \Sigma_{m+1}(t)\,dt + \frac{1}{1-\varkappa} F_1(T),$$

yields

$$N_T[\boldsymbol{u}^{(m+1)}, q^{(m+1)}, \hat{\theta}^{(m+1)}] \leqslant \Sigma_{m+1}(T) + N_T[\boldsymbol{u}^{(0)}, q^{(0)}, \hat{\theta}^{(0)}]$$
$$\leqslant \frac{F_1(T)}{1-\varkappa} e^{c_{13}T} + |\boldsymbol{v}_0|^{(2+\alpha)}_{\cup\Omega^\pm} + |\nabla q_0|^{(\alpha)}_{\cup\Omega^\pm} + |q_0|^{(\gamma)}_{\cup\Omega^\pm} + |\theta_0|^{(2+\alpha)}_{\cup\Omega^\pm}. \qquad (9.3.19)$$

Since $(\boldsymbol{w}^{(1)}, s^{(1)}, \psi^{(1)}) = (\boldsymbol{u}^{(1)} - \boldsymbol{v}_0, q^{(1)} - q_0, \hat{\theta}^{(1)} - \theta_0)$, the norm $N_T^{(1)}$ can be evaluated by means of (9.3.6), (9.3.11), and (9.2.18). Hence, $F_1(T)$ is bounded by the norms of given functions. If we choose T in such a way that

$$(T+T^{\gamma/2})\Big\{F_1(T)(1-\varkappa)^{-1}e^{c_{13}T} + |\boldsymbol{v}_0|^{(2+\alpha)}_{\cup\Omega^\pm} + |\nabla q_0|^{(\alpha)}_{\cup\Omega^\pm} + |q_0|^{(\gamma)}_{\cup\Omega^\pm} + |\theta_0|^{(2+\alpha)}_{\cup\Omega^\pm}\Big\} \leqslant \delta,$$

condition (9.3.12) is fulfilled also for $\boldsymbol{u}^{(m+1)}, q^{(m+1)}, \hat{\theta}^{(m+1)}$. It is clear that such a number $T \leqslant T_{m+1}$ exists. We denote it by T_*.

Since the right-hand side of (9.3.19) is independent of m, we conclude that the functions $\boldsymbol{u}^{(m)}, q^{(m)}, \hat{\theta}^{(m)}$ are defined in the interval $(0, T_*]$ for every $m \in \mathbb{N}$ and satisfy a uniform estimate. Consequently, the series $\sum_{j=1}^{\infty} N_{T_*}^{(j)}$ is convergent, whence it follows that the sequence $\{\boldsymbol{u}^{(m)}, q^{(m)}, \hat{\theta}^{(m)}\}_{m=0}^{\infty}$ is also convergent in the norm N_{T_*}. Passing to the limit as $m \to \infty$ in (9.3.1) and (9.3.2), we make sure that $(\boldsymbol{u}, q, \hat{\theta}) = \lim_{m\to\infty}(\boldsymbol{u}^{(m)}, q^{(m)}, \hat{\theta}^{(m)})$ is a solution of (9.1.3) and (9.1.4). Estimate (9.1.7) is valid as a limit of inequalities (9.3.19) by (9.2.18).

It remains to prove the uniqueness of the solution obtained. We suppose that $(\boldsymbol{u}, q, \hat{\theta})$ and $(\boldsymbol{u}', q', \hat{\theta}')$ are two solutions of problem (9.1.3), (9.1.4) and consider their difference: $\boldsymbol{w} = \boldsymbol{u} - \boldsymbol{u}'$, $s = q - q'$, $\psi = \hat{\theta} - \hat{\theta}'$. The triple $(\boldsymbol{w}, s, \psi)$ satisfies a system of type (9.3.8), (9.3.9):

$$\mathcal{D}_t \boldsymbol{w} - \nu^{\pm} \nabla^2_{\boldsymbol{u}} \boldsymbol{w} + \frac{1}{\rho^{\pm}} \nabla_{\boldsymbol{u}} s = \boldsymbol{l}_1(\boldsymbol{u}', q') - \boldsymbol{l}'_1(\boldsymbol{u}', q') + \boldsymbol{f}(X_{\boldsymbol{u}}, t) - \boldsymbol{f}(X_{\boldsymbol{u}'}, t) - \beta^{\pm} \boldsymbol{g} \psi,$$

$$\nabla_{\boldsymbol{u}} \cdot \boldsymbol{w} = l_2(\boldsymbol{u}') - l'_2(\boldsymbol{u}'), \qquad \mathcal{D}_t \psi - k^{\pm} \nabla^2_{\boldsymbol{u}} \psi = h_1(\hat{\theta}') - h'_1(\hat{\theta}') \quad \text{in} \quad D_{\Sigma T_*},$$

$$\boldsymbol{w}|_{t=0} = 0, \quad \psi|_{t=0} = 0 \quad \text{in} \quad \Omega^- \cup \Omega^+,$$

$$[\boldsymbol{w}]|_{\Gamma} = 0, \qquad [\psi]|_{\Gamma} = 0, \qquad \boldsymbol{w}|_{\Sigma} = 0, \qquad \psi|_{\Sigma} = 0,$$

$$[\mu^{\pm} \Pi_0 \Pi \mathbb{S}_{\boldsymbol{u}}(\boldsymbol{w}) \boldsymbol{n}]|_{\Gamma} = \boldsymbol{l}_3(\boldsymbol{u}') - \boldsymbol{l}'_3(\boldsymbol{u}'),$$

$$[\boldsymbol{n}_0 \cdot \mathbb{T}_{\boldsymbol{u}}(\boldsymbol{w}, s) \boldsymbol{n}]|_{\Gamma} - \sigma \boldsymbol{n}_0 \cdot \Delta(t) \int_0^t \boldsymbol{w} \, \mathrm{d}\tau|_{\Gamma} = l_4(\boldsymbol{u}', q') - l'_4(\boldsymbol{u}', q') +$$

$$+ \sigma \int_0^t (l_5(\boldsymbol{u}') - l'_5(\boldsymbol{u}')) \, \mathrm{d}\tau + \sigma \int_0^t \boldsymbol{n}_0 \cdot (\dot{\Delta}(\tau) - \dot{\Delta}'(\tau)) \boldsymbol{\xi} \, \mathrm{d}\tau|_{\Gamma},$$

$$[k^{\pm} \boldsymbol{n} \cdot \nabla_{\boldsymbol{u}} \psi]|_{\Gamma} = h_2(\hat{\theta}') - h'_2(\hat{\theta}') \qquad \text{on } G_{T_*}.$$

Repeating the above arguments, we arrive at an inequality similar to (9.3.17):

$$N_{T_*}[\boldsymbol{w}, s, \psi] \leqslant \varkappa N_{T_*}[\boldsymbol{w}, s, \psi] + c_{14} \int_0^{T_*} N_t[\boldsymbol{w}, s, \psi] \, \mathrm{d}t, \quad \varkappa < 1,$$

which, by the Gronwall lemma, implies that $\boldsymbol{w} = 0$, $\nabla s = 0$, $\psi = 0$.

Theorem 9.1.1 is completely proved. □

9.4 Global Solvability of the Oberbek-Boussinesq Problem

In this subsection, we show that a small perturbation of the rest state in the Oberbeck-Boussinesq approximation is damped in time. It is worth noting that steady fall (or uprising) of a drop in a liquid medium under gravity force was analyzed in the isothermal case [84]. Fluid densities were considered there to be close to each other.

In order to prove the global solvability of the problem in the Oberbeck-Boussinesq approximation, we use the technique of Chap. 7, where the isothermal case has been considered. We note that condition (1.0.3) is equivalent to equality (7.1.2).

We suppose as in Chap. 7 that the drop Ω_0^+ at an initial moment is close to the ball B_{R_0} whose volume equals the volume of the drop. We recall that the domains $\Omega_t^{\pm}$ conserve their volumes for all $t > 0$ because of the incompressibility of the fluids. In particularly, for the drop, we have $|\Omega_t^+| = |\Omega_0^+| = \frac{4}{3}\pi R_0^3$.

In order to simplify estimates, we introduce the new pressure function: $p_1 = p$ in Ω_t^+ and $p_1 = p + \sigma\frac{2}{R_0}$ in Ω_t^-. Then only boundary condition (9.1.2) changes in interface problem (9.1.1), (9.1.2):

$$[\mathbb{T}(\boldsymbol{v}, p_1)\boldsymbol{n}]\big|_{\Gamma_t} = \sigma\Big(H + \frac{2}{R_0}\Big)\boldsymbol{n}. \tag{9.4.1}$$

To exclude the intersection between the interface Γ_t and the outer boundary Σ, we need to control the position of the barycenter of the inner fluid Ω_t^+ (7.2.1).

We denote deviation function of Γ_t from the sphere $\mathrm{S}_{R_0}(t) \equiv \mathrm{S}_{R_0}(h(t)) = \{|x - h(t)| = R_0\}$ by $r(\omega, t)$. Next, we assume (without restriction of generality) that $\boldsymbol{h}(0) = 0$ and that $\Gamma \equiv \Gamma_0$ is defined by the equation

$$|x| = R_0 + r_0\left(\frac{x}{|x|}\right) \tag{9.4.2}$$

on the unit sphere S_1.

The main step of the proof is a uniform exponential L_2-estimate for fluid velocity and temperature deviation from the mean value. Next, we prove step by step that their Hölder norms decay exponentially to zero, pressure function tending to a step-function and the interface between the liquids doing to a sphere with the radius R_0.

As usual, we pass to the Lagrangian coordinates by formula (1.1.2), where $\boldsymbol{u}(\xi, t)$ is the velocity vector in the Lagrangian coordinates. In addition, we apply relation (1.1.3).

As a result, we arrive at an equivalent system to (9.1.3) for $\boldsymbol{u}$, $q = p_1(X_{\boldsymbol{u}}, t)$, $\hat{\theta} = \theta(X_{\boldsymbol{u}}, t)$ with the given interface $\Gamma \equiv \Gamma_0$ and with the boundary condition:

$$[\boldsymbol{n}_0 \cdot \mathbb{T}_{\boldsymbol{u}}(\boldsymbol{u}, q)\boldsymbol{n}]\big|_{G_T} - \sigma\boldsymbol{n}_0 \cdot \Delta(t)\int_0^t \boldsymbol{u}|_\Gamma \mathrm{d}\tau = \sigma H_0 + \frac{2\sigma}{R_0} + \sigma\boldsymbol{n}_0 \cdot \int_0^t \dot{\Delta}(\tau)\boldsymbol{\xi}\big|_\Gamma \mathrm{d}\tau \text{ on } \mathrm{G}_T.$$

Here $H_0(\xi) = \boldsymbol{n}_0 \cdot \Delta(0)\boldsymbol{\xi}$ is twice the mean curvature of Γ.

Let $T \in (0, \infty]$, $t, \tau > 0$. We introduce the notation:

$$\Omega = \overline{\Omega_0^+} \cup \Omega_0^- \equiv \overline{\Omega_t^+} \cup \Omega_t^-, \quad Q_T = \Omega \times (0, T);$$

$$\mathrm{Q}^{\pm}_{(t,\ t+\tau)} = \Omega_t^{\pm} \times (t, t+\tau),\ \mathrm{D}_{(t,t+\tau)} = \cup \mathrm{Q}^{\pm}_{(t,t+\tau)}.$$

We recall that $\|\cdot\|_{L_2(\Omega)} \equiv \|\cdot\|_\Omega$; $\|a\|_{\mathrm{W}_2^1(\mathrm{S}_1)} \equiv \|a\|_{\mathrm{S}_1} + \|\nabla a\|_{\mathrm{S}_1}$.

The additional term $\sigma\frac{2}{R_0}\boldsymbol{n}_0\cdot\boldsymbol{n}$ in boundary condition (9.1.4) is weak with respect to the left-hand side of the equality. Therefore Theorem 9.1.1 remains valid for problem (9.1.1) and (9.4.1), moreover, the norm of H_0 in (9.1.7) should be changed by the norm of $H_0+\frac{2}{R_0}$.

Theorem 9.4.1 (Global Existence Theorem) *Let the hypotheses of Theorem 9.1.1 hold. Assume, in addition, that for $t=0$ interface Γ is given by (9.4.2) on the unit sphere, and the initial data are small enough, i. e., for some $b>0$*

$$|\mathrm{e}^{bt}\boldsymbol{f}|_{\mathrm{Q}_\infty}^{(\alpha,\frac{1+\alpha-\gamma}{2})}+\|\mathrm{e}^{bt}\boldsymbol{f}\|_{\mathrm{Q}_\infty}+\|\mathrm{e}^{bt}a\|_{\mathrm{W}_2^{3/2,3/4}(\Upsilon_\infty)}$$
$$+|\mathrm{e}^{bt}a|_{\Upsilon_\infty}^{(2+\alpha,1+\alpha/2)}+|\boldsymbol{v}_0|_{\cup\Omega_0^\pm}^{(2+\alpha)}+|\theta_0|_{\cup\Omega_0^\pm}^{(2+\alpha)}+|r_0|_{\mathrm{S}_1}^{(3+\alpha)}\leqslant\varepsilon\ll 1. \qquad (9.4.3)$$

Then problem (9.1.1), (9.4.1), *and* (1.0.3) *is uniquely solvable for all $t>0$, and the solution $(\boldsymbol{v},p_1)$ has the properties: $\boldsymbol{v}\in\boldsymbol{C}^{2+\alpha,1+\alpha/2}$, $p_1\in C^{(\gamma,1+\alpha)}$, $\nabla p_1\in\boldsymbol{C}^{\alpha,\alpha/2}$, $\theta\in C^{2+\alpha,1+\alpha/2}$, the function p_1 being defined up to a bounded time dependent function. The interface Γ_t is given for any $t>0$ by a function $r\in C^{3+\alpha}$:*

$$|x-h(t)|=R_0+r\left(\frac{x-h}{|x-h|},t\right)$$

(where $h(t)$ is a position of the barycenter of Ω_t^+ at the moment t), it tends to a sphere of the radius R_0 with the center in a certain point h_∞; $r_0(\omega)\equiv r(\omega,0)$. It means that for every $t_0\in(0,\infty)$, the solution $(\boldsymbol{u},q,\hat{\theta})$ and its derivatives in local Lagrangian coordinates belong to the corresponding Hölder spaces for a sufficiently small time interval $(t_0,t_0+\tau)$; it is subjected to the estimate

$$|\boldsymbol{u}|_{\mathrm{D}_{(t_0,t_0+\tau)}}^{(2+\alpha,1+\frac{\alpha}{2})}+|\nabla q|_{\mathrm{D}_{(t_0,t_0+\tau)}}^{(\alpha,\frac{\alpha}{2})}+|q|_{\mathrm{D}_{(t_0,t_0+\tau)}}^{(\gamma,1+\alpha)}+|\hat{\theta}|_{\mathrm{D}_{(t_0,t_0+\tau)}}^{(2+\alpha,1+\frac{\alpha}{2})}+\sup_{t\in(t_0,t_0+\tau)}|r(\cdot,t)|_{\mathrm{S}_1}^{(3+\alpha)}$$
$$\leqslant c\mathrm{e}^{-b_1t_0}\Big\{|\mathrm{e}^{bt}\boldsymbol{f}|_{\mathrm{Q}_\infty}^{(\alpha,\frac{1+\alpha-\gamma}{2})}+\|\mathrm{e}^{bt}\boldsymbol{f}\|_{\mathrm{Q}_\infty}+\|\mathrm{e}^{bt}a\|_{\mathrm{W}_2^{\frac{3}{2},\frac{3}{4}}(\Upsilon_\infty)}$$
$$+|\mathrm{e}^{bt}a|_{\Upsilon_\infty}^{(2+\alpha,1+\alpha/2)}+|\boldsymbol{v}_0|_{\cup\Omega_0^\pm}^{(2+\alpha)}+|\theta_0|_{\cup\Omega_0^\pm}^{(2+\alpha)}+|r_0|_{\mathrm{S}_1}^{(3+\alpha)}\Big\}, \qquad (9.4.4)$$

where the values $b_1\in(0,b]$, τ and c are independent of t_0, the constant c containing the norm $|\nabla\boldsymbol{f}|_{\mathrm{Q}_\infty}^{(\alpha,\frac{1+\alpha-\gamma}{2})}$. (The definition of the norm $\|\cdot\|_{\mathrm{W}_2^{\frac{3}{2},\frac{3}{4}}}$ is in § 10.1.)

One can conclude from this theorem that the solution of problem (9.1.1), (9.4.1), and (1.0.3) is stable in the sense that it is close to zero provided that the deviations of the data from zero and an initial interface from a sphere are small. However, the center of the limit sphere $S_{R_0}(h_\infty)$ may be displaced with respect to the initial barycenter of Ω_0^+ no matter how small the initial velocity $\boldsymbol{v}_0$ and the temperature θ_0 are. This displacement is evaluated in inequality (9.4.24) at the end of the paper. We also give there an estimate from below of the distance between the outer boundary and the initial fluid interface which guarantees the absence of the intersection between these surfaces in time.

Remark 9.4.1 *We note that global solvability of a similar problem governing the motion of two fluids in the Oberbeck-Boussinesq approximation without including surface tension may be obtained on the basis of the results in* § 6.4. *In this case, one can reduce the necessary smoothness of the initial interface and take* $\Gamma \in C^{2+\alpha}$.

9.4.1 Energy Estimate of the Solution

In this subsection, we prove an exponential estimate in L_2 for a solution of nonlinear problem (9.1.1), (9.4.1), and (1.0.3).

We will use Proposition 7.2.1 proved for system (9.1.1) and (9.4.1) in the isothermal case.

We suppose that the L_2-norms of mass forces and of the temperature on the outer boundary decrease with respect to time in an exponential way with a certain constant $b > 0$.

Proposition 9.4.1 *We assume that a solution of problem* (9.1.1), (9.4.1), *and* (1.0.3) *is defined in* $[0, T]$. *Let, in addition,* r *satisfies the smallness condition* (7.2.8).

Then for any $t \in (0, T]$

$$\|\boldsymbol{v}(\cdot,t)\|_\Omega + \|r(\cdot,t)\|_{W_2^1(S_1)} \leqslant c_2 \mathrm{e}^{-b_1 t}\Big\{\|\boldsymbol{v}_0\|_\Omega + \|\theta_0\|_\Omega + \|r_0\|_{W_2^1(S_1)} + \big\|\mathrm{e}^{b\tau}\boldsymbol{f}\big\|_{Q_t} + \big\|\mathrm{e}^{b\tau}a\big\|_{W_2^{3/2,3/4}(\Upsilon_t)}\Big\}, \tag{9.4.5}$$

$$\|\theta(\cdot,t)\|_\Omega \leqslant c_3 \mathrm{e}^{-b_2 t}\Big\{\|\theta_0\|_\Omega + \big\|\mathrm{e}^{b_2\tau}a\big\|_{W_2^{3/2,3/4}(\Upsilon_t)}\Big\} \tag{9.4.6}$$

with constants b_1, b_2, c_2 *and* c_3 *independent of* t;*and* $b_1 = \min\{b, b_2\}$; b_2 *is the constant from* (9.4.8).

Proof In order to apply Proposition 7.2.1 to system (9.1.1) and (9.4.1), we should verify that the L_2-norm of θ decays exponentially in a certain way.

We extend the function a with preservation of class from the surface Σ into the domain Ω in such a way that the extension $a^* = 0$ in a neighborhood of Ω_t^+ and inside it. We consider the difference $\widetilde{\theta} = \theta - a^*$. It solves the problem

$$\begin{aligned} &\mathcal{D}_t\widetilde{\theta} + (\boldsymbol{v}\cdot\nabla)\widetilde{\theta} - k^{\pm}\nabla^2\widetilde{\theta} = \mathcal{D}_t a^* - k^-\nabla^2 a^* \quad in \quad \Omega_t^+\cup\Omega_t^-, \ \ t>0,\\ &\widetilde{\theta}|_{t=0} = \widetilde{\theta}_0 - a^*|_{t=0} \quad in \quad \Omega_0^+\cup\Omega_0^-, \\ &[\widetilde{\theta}]|_{\Gamma_t} = 0, \quad \Big[k^{\pm}\frac{\partial\widetilde{\theta}}{\partial \boldsymbol{n}}\Big]\Big|_{\Gamma_t} = 0, \quad \widetilde{\theta}|_{\Sigma} = 0. \end{aligned} \tag{9.4.7}$$

To obtain an exponential estimate for $\widetilde{\theta}$, we multiply heat equation in (9.4.7) by $\widetilde{\theta}$ and integrate by parts over $\Omega_t^+\cup\Omega_t^-$:

$$\frac{1}{2}\frac{\mathrm{d}}{\mathrm{d}t}\|\widetilde{\theta}\|^2_{\cup\Omega_t^{\pm}} + \|\sqrt{k^{\pm}}\nabla\widetilde{\theta}\|^2_{\cup\Omega_t^{\pm}} = \int_{\Omega_t^-}\Big(\frac{\mathrm{d}a^*}{\mathrm{d}t} - k^-\nabla^2 a^*\Big)\widetilde{\theta}\,\mathrm{d}x.$$

Since $[\widetilde{\theta}]|_{\Gamma_t} = 0$, we have: $\|\nabla\widetilde{\theta}\|^2_{\cup\Omega_t^{\pm}} = \|\nabla\widetilde{\theta}\|^2_{\Omega}$, $\Omega = \overline{\Omega_t^+}\cup\Omega_t^-$, and

$$\frac{1}{2}\frac{\mathrm{d}}{\mathrm{d}t}\|\widetilde{\theta}\|^2_{\Omega} + \min\{k^{\pm}\}\|\nabla\widetilde{\theta}\|^2_{\Omega} \leqslant \Big\|\frac{\mathrm{d}a^*}{\mathrm{d}t} - k^-\nabla^2 a^*\Big\|_{\Omega_t^-}\|\widetilde{\theta}\|_{\Omega_t^-}.$$

We can apply the Poincaré inequality to the function $\widetilde{\theta}$ because of $\widetilde{\theta}|_{\Sigma} = 0$. Then

$$\frac{\mathrm{d}}{\mathrm{d}t}\|\widetilde{\theta}\|^2_{\Omega} + 2b_2\|\widetilde{\theta}\|^2_{\Omega} \leqslant c\Big\|\frac{\mathrm{d}a^*}{\mathrm{d}t} - k^-\nabla^2 a^*\Big\|^2_{\Omega_t^-}, \tag{9.4.8}$$

which gives us

$$\begin{aligned} \|\widetilde{\theta}(\cdot,t)\|^2_{\Omega} &\leqslant \mathrm{e}^{-2b_2 t}\|\theta_0\|^2_{\Omega} + c\int_0^t \mathrm{e}^{-2b_2(t-\tau)}\Big\|\frac{\mathrm{d}a^*}{\mathrm{d}t} - k^-\nabla^2 a^*\Big\|^2_{\Omega}\,\mathrm{d}\tau \\ &\leqslant c\mathrm{e}^{-2b_2 t}\Big(\|\theta_0\|^2_{\Omega} + \|\mathrm{e}^{b_2\tau}a^*\|^2_{W_2^{2,1}(Q_t)}\Big) \\ &\leqslant c\mathrm{e}^{-2b_2 t}\Big(\|\theta_0\|^2_{\Omega} + \|\mathrm{e}^{b_2\tau}a\|^2_{W_2^{3/2,3/4}(\Upsilon_t)}\Big). \end{aligned}$$

Finally,

$$\|\theta(\cdot,t)\|_{\Omega} \leqslant \|\widetilde{\theta}(\cdot,t)\|_{\Omega} + \|a^*(\cdot,t)\|_{\Omega} \leqslant \mathrm{e}^{-b_2 t}\Big(\|\theta_0\|_{\Omega} + \|a(\cdot,0)\|_{\Sigma} + \|\mathrm{e}^{b_2\tau}a\|_{W_2^{3/2,3/4}(\Upsilon_t)}\Big).$$

Now it is evident that inequality (9.4.5) follows from (9.4.6) and estimate (7.2.9) for the first equation in (9.1.1). □

Corollary 9.4.1 *The coordinates of the barycenter of Ω_t^+ satisfy the inequality*

$$|\boldsymbol{h}(t)| \leqslant c\Big\{\|e^{b\tau}\boldsymbol{f}\|_{Q_T} + \big\|e^{b\tau}a\big\|_{W_2^{3/2,3/4}(\Upsilon_T)} + \|\boldsymbol{v}_0\|_\Omega + \|\theta_0\|_\Omega + \|r_0\|_{W_2^1(S_1)}\Big\} \tag{9.4.9}$$

for any $t \in [0,T]$.

Inequality (9.4.9) is proved in the same way as the estimate in Corollary 7.2.1.

9.4.2 Global Solvability of Problem (9.1.1), (9.4.1), (1.0.3)

The aim of this subsection is to prove the solvability of problems (9.1.1), (9.4.1), and (1.0.3) in the whole time interval $t > 0$.

We use Proposition 7.4.1 in the proof of the following statement.

Proposition 9.4.2 *We assume that a solution* $(\boldsymbol{v}, p_1, \theta)$ *of problem* (9.1.1), (9.4.1), and (1.0.3) *exist in the interval* $(0,T]$ *and the inequality*

$$N_{(0,T)}[\boldsymbol{v}, p_1, \theta] \equiv |\boldsymbol{u}|^{(2+\alpha,1+\alpha/2)}_{D_{\Sigma T}} + |\nabla q|^{(\alpha,\alpha/2)}_{D_{\Sigma T}} + |q|^{(\gamma,1+\alpha)}_{D_{\Sigma T}} + |\widehat{\theta}|^{(2+\alpha,1+\alpha/2)}_{D_{\Sigma T}} \leqslant \mu_1$$

holds. Here the triple $(\boldsymbol{u}, q, \widehat{\theta})$ *denotes the solution written as a function of the Lagrangian coordinates.*

Then

$$\begin{aligned} N_{(t_0-\tau_0/2,t_0)}[\boldsymbol{v}, p_1, \theta, r] &\equiv N_{(t_0-\tau_0/2,t_0)}[\boldsymbol{v}, p_1, \theta] + \sup_{t_0-\tau_0/2<\tau<t_0} |r(\cdot,\tau)|^{(3+\alpha)}_{S_1} \\ &\leqslant c\Big(\delta, \tau_0, |\nabla \boldsymbol{f}|^{(\alpha,\frac{1+\alpha-\gamma}{2})}_{D'_0}\Big)\Big\{|\boldsymbol{f}|^{(\alpha,\frac{1+\alpha-\gamma}{2})}_{D'_0} + |a|^{(2+\alpha,1+\alpha/2)}_{\Upsilon'_0} \\ &+ \int\limits_{t_0-2\tau_0}^{t_0} \big(\|\boldsymbol{v}(\cdot,\tau)\|_\Omega + \|\theta(\cdot,\tau)\|_\Omega + \|r(\cdot,\tau)\|_{W_2^1(S_1)}\big)\,d\tau\Big\}, \end{aligned} \tag{9.4.10}$$

where $t_0 \in (0,T]$, $\tau_0 \in (0, t_0/2)$, τ_0 *depends on* μ *and on constant* δ *in* (7.3.2). *In addition, we have used the notation:* $D'_\beta = D_{(t_0-2\tau_0+\beta,t_0)}$, $\Upsilon'_\beta = \Sigma \times (t_0 - 2\tau_0 + \beta, t_0)$ *with* $\beta \in [0, 2\tau_0)$.

Proof We fix an arbitrary $t_0 \in (0,T]$. For $\tau_0 \in (0, t_0/2)$, we denote by $\eta_\lambda(t)$ a smooth monotone function of t such that

$$\eta_\lambda(t) = \begin{cases} 0, & \text{if } t \leqslant t_0 - 2\tau_0 + \lambda/2, \\ 1, & \text{if } t \geqslant t_0 - 2\tau_0 + \lambda, \end{cases}$$

$\lambda \in (0, \tau_0]$, and

$$|\dot\eta_\lambda(t)|_{\mathbb{R}} \leqslant c\lambda^{-1}, \quad \langle\dot\eta_\lambda(t)\rangle_{\mathbb{R}}^{(\alpha/2)} \leqslant c\lambda^{-1-\alpha/2},$$

where $\dot\eta_\lambda \equiv \mathrm{d}\eta_\lambda(t)/\mathrm{d}t$.

We consider the triple $\boldsymbol{w} = \boldsymbol{v}\eta_\lambda$, $s = p_1\eta_\lambda$, $\vartheta = \theta\eta_\lambda$. It satisfies the system

$$\begin{aligned}
&\mathcal{D}_t\boldsymbol{w} + (\boldsymbol{v}\cdot\nabla)\boldsymbol{w} - \nu^\pm\nabla^2\boldsymbol{w} + \frac{1}{\rho^\pm}\nabla s = \boldsymbol{f}\eta_\lambda + \boldsymbol{v}\dot\eta_\lambda - \beta^\pm\boldsymbol{g}\vartheta,\\
&\mathcal{D}_t\vartheta + (\boldsymbol{v}\cdot\nabla)\vartheta - k^\pm\nabla^2\vartheta = \theta\dot\eta_\lambda,\\
&\nabla\cdot\boldsymbol{w} = 0 \quad \text{in} \quad \Omega_t^-\cup\Omega_t^+, \quad t > t_0 - 2\tau_0,\\
&\boldsymbol{w}|_{t=t_0-2\tau_0} = 0, \ \ \theta|_{t=t_0-2\tau_0} = 0 \quad \text{in} \quad \cup\Omega'_\pm \equiv \cup\Omega^\pm_{t_0-2\tau_0},\\
&\qquad \boldsymbol{w}|_\Sigma = 0, \quad \vartheta|_\Sigma = a\eta_\lambda,\\
&\qquad [\vartheta]\big|_{\Gamma_t} = 0, \quad \Big[k^\pm\frac{\partial\vartheta}{\partial\boldsymbol{n}}\Big]\Big|_{\Gamma_t} = 0,\\
&[\boldsymbol{w}]\big|_{\Gamma_t} = 0, \quad [\mathbb{T}(\boldsymbol{w}, s)\boldsymbol{n}]\big|_{\Gamma_t} = \sigma\big(H + \frac{2}{R_0}\big)\boldsymbol{n}\eta_\lambda\big|_{\Gamma_t}.
\end{aligned} \tag{9.4.11}$$

We introduce the Lagrangian coordinates by formula (6.4.15), where the vector $\boldsymbol{v}(X(\xi', t), t)$ is denoted by $\boldsymbol{u}(\xi', t)$. We transform problem (9.4.11) by (6.4.15). The functions $\boldsymbol{w}$, s and ϑ written in the Lagrangian coordinates will be denoted by the same symbols.

The function ϑ solves the system

$$\begin{aligned}
&\mathcal{D}_t\vartheta - k^\pm\nabla_{\boldsymbol{u}}^2\vartheta = \widehat{\theta}\dot\eta_\lambda \quad \text{in} \quad \cup\Omega'_\pm, \quad t > t_0 - 2\tau_0,\\
&\vartheta|_{t=t_0-2\tau_0} = 0 \quad \text{in} \quad \cup\Omega'_\pm, \qquad [\vartheta]\big|_{\Gamma'} = 0,\\
&\qquad \Big[k^\pm\boldsymbol{n}'\cdot\nabla_{\boldsymbol{u}}\vartheta\Big]\Big|_{\Gamma'} = 0, \qquad \vartheta|_\Sigma = a\eta_\lambda.
\end{aligned} \tag{9.4.12}$$

Here $\Gamma' = \Gamma_{t_0-2\tau_0}$, all the notation, say $\nabla_{\boldsymbol{u}}$, correspond to transformation (6.4.15).

We verify the assumptions of Theorem 9.2.1 to apply them to problem (9.4.12). First, we choose τ_0 so small that the inequality (7.3.2) holds. It suffices to take it satisfying inequality (6.4.18).

Next, we note that compatibility conditions (9.2.2) are fulfiled by virtue of (9.1.5). Hence, for ϑ, estimate (9.2.3) holds. All the functions in (9.4.12) vanish for $t = t_0 - 2\tau_0$. Thus,

$$|\widehat{\theta}|_{\mathrm{D}'_\lambda}^{(2+\alpha,1+\alpha/2)} \leqslant |\vartheta|_{\mathrm{D}'_{\tau_0}}^{(2+\alpha,1+\alpha/2)} \leqslant c_2(2\tau_0)\big\{|\widehat{\theta}\dot\eta_\lambda|_{\mathrm{D}'_{\tau_0}}^{(\alpha,\alpha/2)} + |a\eta_\lambda|_{\Upsilon'_0}^{(2+\alpha,1+\alpha/2)}\big\}. \tag{9.4.13}$$

Hence, for $\lambda \leqslant \tau_0 < 1$, inequality (9.4.13) can be extended as follows:

$$|\widehat{\theta}|^{(2+\alpha,1+\alpha/2)}_{D'_\lambda} \leqslant c_2(2\tau_0)\Big\{\frac{1}{\lambda}|\widehat{\theta}|^{(\alpha,\alpha/2)}_{D'_{\lambda/2}} + \frac{1}{\lambda^{1+\frac{\alpha}{2}}}|\widehat{\theta}|_{D'_{\lambda/2}} + \frac{1}{\lambda^{1+\frac{\alpha}{2}}}|a|^{(2+\alpha,1+\alpha/2)}_{\Upsilon'_{\lambda/2}}\Big\}. \tag{9.4.14}$$

We estimate the lower order norms of $\widehat{\theta}$ in (9.4.14) by Lemma 6.4.2 setting $\theta_2 = (\epsilon\lambda)^{1/2}$ in (6.4.10). Next, we evaluate $|\widehat{\theta}|_{D'_{\lambda/2}}$ by inequality (6.4.11) with $\varkappa_1 = (\epsilon\lambda^{1+\frac{\alpha}{2}})^{\frac{1}{1+\alpha}}$.

As a result, we deduce the inequality

$$|\widehat{\theta}|^{(2+\alpha,1+\alpha/2)}_{D'_\lambda} \leqslant c_3(\delta)\Big\{\varepsilon|\widehat{\theta}|^{(2+\alpha,1+\alpha/2)}_{D'_{\lambda/2}} + c(\varepsilon)\lambda^{-\kappa}\int\limits_{t_0-2\tau_0}^{t_0} \|\widehat{\theta}(\cdot,\tau)\|_\Omega \,\mathrm{d}\tau + \frac{1}{\lambda^{1+\frac{\alpha}{2}}}|a|^{(2+\alpha,1+\alpha/2)}_{\Upsilon'_{\lambda/2}}\Big\} \tag{9.4.15}$$

with $\kappa = \max\Big\{\frac{11}{4} + \frac{\alpha}{2}, \Big(1+\frac{\alpha}{2}\Big)\Big(1 + \frac{7}{2(1+\alpha)}\Big)\Big\}$.

We introduce the function $\Phi(\lambda) = \lambda^\kappa|\widehat{\theta}|^{(2+\alpha,1+\alpha/2)}_{D'_\lambda}$. Since $\kappa > 1 + \frac{\alpha}{2}$, we can write (9.4.15) as follows:

$$\Phi(\lambda) \leqslant c_4\varepsilon\Phi(\lambda/2) + K, \tag{9.4.16}$$

where $c_4 = c_3(\delta)2^\kappa$,

$$K = c_3(\delta)\Big\{c(\varepsilon)\int\limits_{t_0-2\tau_0}^{t_0} \|\widehat{\theta}(\cdot,\tau)\|_\Omega \,\mathrm{d}\tau + |a|^{(2+\alpha,1+\alpha/2)}_{\Upsilon'_0}\Big\}.$$

We set $\varepsilon = \frac{1}{2c_4}$ in (9.4.16). By iterations with $\lambda/2,\dots,\lambda/2^k$ and by passing to the limit as $k \to \infty$, we deduce that

$$\Phi(\lambda) \leqslant 2K.$$

This estimate with $\lambda = \tau_0$ implies the inequality

$$|\widehat{\theta}|^{(2+\alpha,1+\alpha/2)}_{D'_{\tau_0}} \leqslant c_5(\delta)\Big\{\int\limits_{t_0-2\tau_0}^{t_0} \|\widehat{\theta}\|_\Omega \,\mathrm{d}\tau + |a|^{(2+\alpha,1+\alpha/2)}_{\Upsilon'_0}\Big\}. \tag{9.4.17}$$

We apply Proposition 7.4.1 with $\tau_1 = \tau_0/2$ to the part of problem (9.1.1) and (9.4.1) which contains $\boldsymbol{v}, p_1$ and consider θ as a known function in the right-hand side of the first equation in (9.1.1). We choose τ_0 so small that (6.4.18) is satisfied for τ_1. Then, by (7.4.4), we have

$$N_{(t_0-\tau_1,t_0)}[\boldsymbol{v},p_1,r] \leqslant c_2(\delta,\tau_0)\Big\{|\boldsymbol{f}|^{(\alpha,\frac{1+\alpha-\gamma}{2})}_{\mathrm{D}'_{\tau_0}} + |\widehat{\theta}|^{(\alpha,\frac{1+\alpha-\gamma}{2})}_{\mathrm{D}'_{\tau_0}} + \int_{t_0-2\tau_1}^{t_0} \big(\|\boldsymbol{v}(\cdot,\tau)\|_{\Omega} + \|r(\cdot,\tau)\|_{\mathrm{W}_2^1(S_1)}\big)\,\mathrm{d}\tau\Big\}. \tag{9.4.18}$$

We note that c_2 depends on the norms $|\nabla\boldsymbol{f}|^{(\alpha,\frac{1+\alpha-\gamma}{2})}_{\mathrm{D}'_{\tau_0}}$, $|\nabla\widehat{\theta}|^{(\alpha,\frac{1+\alpha-\gamma}{2})}_{\mathrm{D}'_{\tau_0}}$. In view of (9.4.17), we deduce from (9.4.18) the inequality

$$N_{(t_0-\tau_1,t_0)}[\boldsymbol{v},p_1,r] \leqslant c_2(\delta,\tau_0)\Big\{|\boldsymbol{f}|^{(\alpha,\frac{1+\alpha-\gamma}{2})}_{\mathrm{D}'_{\tau_0}} + |a|^{(2+\alpha,1+\alpha/2)}_{\Upsilon'_0} + \int_{t_0-2\tau_0}^{t_0} \|\widehat{\theta}(\cdot,\tau)\|_{\Omega}\,\mathrm{d}\tau + \int_{t_0-2\tau_1}^{t_0} \big(\|\boldsymbol{v}(\cdot,\tau)\|_{\Omega} + \|r(\cdot,\tau)\|_{\mathrm{W}_2^1(S_1)}\big)\,\mathrm{d}\tau\Big\}$$

and estimate (9.4.10). □

Now we can prove Theorem 9.4.1.

Proof If ε in (9.4.3) is sufficiently small, by Theorem 9.1.1 there exists a local solution $(\boldsymbol{v},p_1,\theta)$ on an interval $(0,T_*]$, $T_*>1$. The norm of the solution $(\boldsymbol{v},p_1,\theta)$ satisfies estimate (9.1.7), therefore,

$$N_{(0,T_*)}[\boldsymbol{v},p_1,\theta] \leqslant c\Big(|\boldsymbol{f}|^{(\alpha,\frac{1+\alpha-\gamma}{2})}_{Q_{T_*}} + |\boldsymbol{v}_0|^{(2+\alpha)}_{\cup\Omega_0^\pm} + |\theta_0|^{(2+\alpha)}_{\cup\Omega_0^\pm} + |a|^{(2+\alpha,1+\frac{\alpha}{2})}_{\Upsilon_{T_*}} + |r_0|^{(3+\alpha)}_{\mathrm{S}_1}\Big) \leqslant c_3\varepsilon \equiv \mu_1 \tag{9.4.19}$$

with small $\mu_1>0$. In inequality (9.4.19), we have used the estimate

$$\Big|H_0 + \frac{2}{R_0}\Big|^{(1+\alpha)}_{\Gamma} \leqslant c|r_0|^{(3+\alpha)}_{\mathrm{S}_1}.$$

By Proposition 9.4.2, there exists $\tau_0 < T_*/2$ such that condition (6.4.18) is satisfied and for $(\boldsymbol{v},p_1,\theta)$ estimate (9.4.10) holds with T replaced by T_*. Lemma 7.4.2 guarantees the inequality

$$|r|^{(1+\alpha,0)}_{\mathrm{S}_1\times(0,T_*)} \leqslant c_2\big(|r_0|^{(1+\alpha)}_{\mathrm{S}_1} + c_3\varepsilon T_*\big) \leqslant \delta_1 R_0$$

($|r|^{(1+\alpha,0)}_{\mathrm{S}_1\times(0,T_*)} \equiv \sup_{0<\tau<T_*}|r(\cdot,\tau)|^{(1+\alpha)}_{\mathrm{S}_1}$), when ε is sufficiently small. This allows us to apply Proposition 9.4.1. Inequality (9.4.10) combined with (9.4.5), (9.4.6) leads to the estimate

$$
\begin{aligned}
N_{(t_0-\tau_0/2,t_0)}[\boldsymbol{v},p_1,\theta,r] &\leqslant c_4\,\mathrm{e}^{-b_1(t_0-2\tau_0)}\Big\{|\mathrm{e}^{b_1t}\boldsymbol{f}|^{(\alpha,\frac{1+\alpha-\gamma}{2})}_{\mathrm{D}'_0}+|\mathrm{e}^{b_1t}a|^{(2+\alpha,1+\frac{\alpha}{2})}_{\Upsilon'_0}\\
&\qquad+\|\boldsymbol{v}_0\|_\Omega+\|\theta_0\|_\Omega+\|r_0\|_{\mathrm{W}^1_2(\mathrm{S}_1)}\Big\}\\
&\leqslant c_5(\tau_0)\mathrm{e}^{-b_1t_0}\big(|\Omega|^{\frac12}+2\sqrt{\pi}+1\big)\varepsilon, \qquad (9.4.20)
\end{aligned}
$$

where $|\Omega|$ is the measure of Ω, and $t_0\in(2\tau_0,T_*]$.

For $t_0=T_*$, estimate (9.4.19) implies that

$$
|\boldsymbol{v}(\cdot,T_*)|^{(2+\alpha)}_{\cup\Omega^\pm_{T_*}}+|\theta(\cdot,T_*)|^{(2+\alpha)}_{\cup\Omega^\pm_{T_*}}+|r(\cdot,T_*)|^{(3+\alpha)}_{\mathrm{S}_1}\leqslant\mu_1. \qquad (9.4.21)
$$

Next, we use Theorem 9.1.1 again to obtain a solution in $(T_*,T_*+T_1]$ for the initial data $\boldsymbol{v}(\cdot,T_*)$, $\theta(\cdot,T_*)$, $r(\cdot,T_*)$. The norm of this solution is bounded:

$$
N_{(T_*,T_*+T_1)}[\boldsymbol{v},p_1,\theta]\leqslant\mu_2. \qquad (9.4.22)
$$

Due to Proposition 9.4.2, we can find $0<\tau_1<T_1/2$ such that satisfies condition (6.4.18) and

$$
\begin{aligned}
&N_{(T_*+T_1-\tau_1/2,T_*+T_1)}[\boldsymbol{v},p_1,\theta,r]\\
&\leqslant c(\delta,\tau_1)\Big\{|\boldsymbol{f}|^{(\alpha,\frac{1+\alpha-\gamma}{2})}_{\mathrm{Q}_{(T_*+T_1-2\tau_1,T_*+T_1)}}+|a|^{(2+\alpha,1+\alpha/2)}_{\Upsilon_{(T_*+T_1-2\tau_1,T_*+T_1)}}\\
&+\int\limits_{T_*+T_1-2\tau_1}^{T_*+T_1}\big(\|\boldsymbol{v}(\cdot,\tau)\|_\Omega+\|\theta(\cdot,\tau)\|_\Omega+\|r(\cdot,\tau)\|_{\mathrm{W}^1_2(\mathrm{S}_1)}\big)\,\mathrm{d}\tau\Big\}. \qquad (9.4.23)
\end{aligned}
$$

We apply again Lemma 7.4.2. Then in view of (7.4.16), (9.4.21), and (9.4.22), we conclude:

$$
|r|^{(1+\alpha,0)}_{\mathrm{S}_1\times(T_*,T_*+T_1)}\leqslant c_2\big(|r(\cdot,T_*)|^{(1+\alpha)}_{\mathrm{S}_1}+T_1|\boldsymbol{u}|^{(1+\alpha)}_{\xi,\mathrm{D}_{(T_*,T_*+T_1)}}\big)\leqslant c_2(\mu_1+T_1\mu_2).
$$

We can make μ_1 and T_1 so small that

$$
c_2(\mu_1+T_1\mu_2)\leqslant\delta_1R_0.
$$

Consequently, similarly to (9.4.20), inequality (9.4.23) may be continued by virtue of Proposition 9.4.1 as follows:

$$
N_{(T_*+T_1-\tau_1,T_*+T_1)}[\boldsymbol{v},p_1,\theta,r]\leqslant c_6(\delta,\tau_1)\mathrm{e}^{-b_1T_1}\big(1+|\Omega|^{1/2}+2\sqrt{\pi}\big)\varepsilon.
$$

Let us choose ε so small that $c_6(\delta,\tau_1)\big(1+|\Omega|^{1/2}+2\sqrt{\pi}\big)\varepsilon \leqslant \mu_1$. Hence,

$$|\boldsymbol{v}(\cdot,T_*+T_1)|^{(2+\alpha)}_{\cup\Omega^{\pm}_{T_*+T_1}}+|\theta(\cdot,T_*+T_1)|^{(2+\alpha)}_{\cup\Omega^{\pm}_{T_*+T_1}}+|r(\cdot,T_*+T_1)|^{(3+\alpha)}_{S_1}\leqslant \mu_1 \mathrm{e}^{-b_1T_1}.$$

Thus, the norms of the initial data do not increase. Therefore we can extend the solution in the interval $(T_*+T_1, T_*+2T_1]$. This procedure may be repeated again and again as long as we like.

As to the Lagrangian coordinates and coordinates of inner fluid barycenter on every step, they are calculated by formulas (7.4.21) and (7.4.22). The explanations are given at the end of § 7.4.

Hence, the solution of system (9.1.1), (9.4.1), and (1.0.3) may be extended in t as far as necessary. After that, inequality (9.4.4) follows for any finite time interval from Propositions 9.4.1 and 9.4.2.

The limiting position of the barycenter is evaluated from Corollary 9.4.1:

$$|h_\infty|\leqslant d\leqslant c_8\varepsilon, \tag{9.4.24}$$

where

$$d=c_7\Big\{\|\mathrm{e}^{b_1t}\boldsymbol{f}\|_{Q_\infty}+\|\mathrm{e}^{b_1t}a\|^2_{W_2^{3/2,3/4}(\Upsilon_\infty)}+\|\boldsymbol{v}_0\|_\Omega+\|\theta_0\|_\Omega+\|r_0\|_{W_2^{\frac{1}{2}}(S_1)}\Big\}.$$

Inequality (9.4.24) implies an estimate concerning the distance between the surfaces Γ and Σ at the initial moment. It is clear that, in order to exclude the intersection of Γ_t and Σ in the future, an initial distance between boundaries should be strictly greater than the sum $d+\delta_1R_0$ with δ_1 and R_0 in (7.2.8).

The uniqueness follows from that of local solutions. □

Chapter 10
Local L_2-Solvability of the Problem with Nonnegative Coefficient of Surface Tension

Abstract The purpose of this chapter is to give a detailed proof of the local existence theorem in the Sobolev-Slobodetskiĭ spaces in the case of the entire space. First, we give the definition of these spaces and then estimate the solution of the model problem with the plane interface between the fluids found in Chap. 2. Next, we construct a solution to the linear problem with the closed interface and obtain estimates of the solution in L_2-setting. Finally, we solve the nonlinear problem in a small enough time interval.

We study problem (1.1.1), (1.1.5), and (1.1.7) in the Sobolev-Slobodetskiĭ spaces, both in weighted and in ordinary ones. In addition, we present the main stages of proving the local-in-time solvability of problem (1.1.1), (1.0.3) with a nonnegative coefficient of surface tension. We observe that an additional smoothness of the initial interface Γ is not required when $\sigma = 0$. It is sufficient to assume the usual regularity: $\Gamma \in W_2^{3/2+l}$, the same as for a solid surface. (The material is presented according to works [15, 28], see also [13, 14, 17, 23].)

We also note that, in the Sobolev spaces, the existence of local and global solutions to the problem governing the motion of a single fluid drop in vacuum was obtained in [78, 83].

10.1 The Definition of the Sobolev-Slobodetskiĭ Spaces and the Introduction of Equivalent Norms

We recall the definition of the ordinary Sobolev spaces.

Let Ω be a domain in $\mathbb{R}^n$, $n \in \mathbb{N}$. In this chapter, $\mathcal{D}_x^{\boldsymbol{r}}$, $\mathcal{D}_t^s$ mean the Sobolev derivatives, where $\boldsymbol{r}$ is a multi-index. For $m > 0$, we define the

© The Author(s), under exclusive license to Springer Nature Switzerland AG 2021

I. V. Denisova, V. A. Solonnikov, *Motion of a Drop in an Incompressible Fluid*, Advances in Mathematical Fluid Mechanics,
https://doi.org/10.1007/978-3-030-70053-9_10

Sobolev–Slobodetskiĭ space $W_2^m(\Omega)$ as a space of functions u with the finite norm

$$\|u\|_{W_2^m(\Omega)} = \Big(\sum_{|\boldsymbol{r}|<m} \|\mathcal{D}^{\boldsymbol{r}} u\|_\Omega^2 + \|u\|^2_{\dot{W}_2^m(\Omega)} \Big)^{1/2},$$

where $\|\cdot\|_\Omega$ is the norm in L_2,

$$\|u\|^2_{\dot{W}_2^m(\Omega)} = \sum_{|\boldsymbol{r}|=m} \|\mathcal{D}^{\boldsymbol{r}} u\|_\Omega^2 \quad \text{for} \quad m \in \mathbb{N}$$

and

$$\|u\|^2_{\dot{W}_2^m(\Omega)} = \sum_{|\boldsymbol{r}|=[m]} \int_\Omega \int_\Omega \frac{|\mathcal{D}^{\boldsymbol{r}} u(x) - \mathcal{D}^{\boldsymbol{r}} u(y)|^2}{|x-y|^{n+2(m-[m])}} \, dx \, dy$$

for non-integral m, $[m]$ denotes here the integral part of m.

The anisotropic space $W_2^{m,m/2}(Q_T)$ consists of functions defined in the cylinder $Q_T = \Omega \times (0,T)$, $0 < T \leqslant \infty$, and having finite norm

$$\|u\|_{W_2^{m,m/2}(Q_T)} = \Big(\int_0^T \|u\|^2_{W_2^m(\Omega)} \, dt + \int_\Omega \|u\|^2_{W_2^{m/2}(0,T)} \, dx \Big)^{1/2}.$$

The Sobolev–Slobodetskiĭ spaces with exponential weight were introduced by M. S. Agranovich and M. I. Vishik in [3]; their properties were studied in [82].

Let $\gamma \geqslant 0$. The weighted space $H_\gamma^{m,m/2}(Q_T)$ is the space of functions admitting zero extension in the domain $t < 0$ without loss of smoothness. It is equipped with the norm

$$\|u\|_{H_\gamma^{m,m/2}(Q_T)} = \Big(\|u\|^2_{H_\gamma^{m,0}(Q_T)} + \|u\|^2_{H_\gamma^{0,m/2}(Q_T)} \Big)^{1/2}.$$

Here

$$\|u\|^2_{H_\gamma^{m,0}(Q_T)} = \int_0^T e^{-2\gamma t} \Big(\|u\|^2_{\dot{W}_2^m(\Omega)} + \gamma^l \|u\|_\Omega^2 \Big) \, dt,$$

$$\|u\|^2_{H_\gamma^{0,m/2}(Q_T)} = \int_0^T e^{-2\gamma t} \left\| \frac{\partial^{m/2} u}{\partial t^{m/2}} \right\|_\Omega^2 dt \quad \text{for} \quad [m/2] = m/2$$

and

$$\|u\|^2_{H^{0,m/2}_\gamma(Q_T)} = \int_0^T e^{-2\gamma t} \int_0^\infty \left\| \frac{\partial^k u_0(\cdot,t)}{\partial t^k} - \frac{\partial^k u_0(\cdot,t-\tau)}{\partial t^k} \right\|^2_\Omega \frac{d\tau}{\tau^{1+m-2k}}\, dt$$

for $k = [m/2] < m/2$, where u_0 is the function u continued by zero in the domain $t < 0$. In addition, for $m > 1$,

$$\left.\frac{\partial^i u}{\partial t^i}\right|_{t=0} = 0, \quad i = 0, \ldots, \left[\frac{m-1}{2}\right];$$

for integral $\frac{m}{2}$, the function u_0 has generalized derivative $\mathcal{D}_t^{m/2}$ in $\Omega \times (-\infty, T)$.

All these norms can be introduced on any smooth manifold by means of local maps and the partition of unity.

In addition, on the surface $G_T = \Gamma \times (0,T)$, we will also use the space $H_\gamma^{m+1/2,1/2,m/2}(G_T)$ with the norm whose square is

$$\|u\|^2_{H_\gamma^{m+1/2,1/2,m/2}(G_T)} = \int_0^T e^{-2\gamma t}\Bigg\{ \|u\|^2_{W_2^{1/2+m}(\Gamma)} + \gamma^l \|u\|^2_{W_2^{1/2}(\Gamma)}$$

$$+ \int_0^\infty \left\| \frac{\partial^k u_0(\cdot,t)}{\partial t^k} - \frac{\partial^k u_0(\cdot,t-\tau)}{\partial t^k} \right\|^2_{W_2^{1/2}(\Gamma)} \frac{d\tau}{\tau^{1+m-2k}} \Bigg\} dt, \quad k = [m/2] < m/2.$$

As usual, we set

$$\|u\|_{W_2^m(\cup\Omega^\pm)} = \|u\|_{W_2^m(\Omega^+)} + \|u\|_{W_2^m(\Omega^-)},$$

and

$$\|u\|_{W_2^{m,m/2}(D_T)} = \|u\|_{W_2^{m,m/2}(Q_T^+)} + \|u\|_{W_2^{m,m/2}(Q_T^-)},$$

$$\|u\|_{H_\gamma^{m,m/2}(D_T)} = \|u\|_{H_\gamma^{m,m/2}(Q_T^+)} + \|u\|_{H_\gamma^{m,m/2}(Q_T^-)}, \qquad D_T \equiv Q_T^+ \cup Q_T^-.$$

We define the norm of a vector as the sum of the norms of its components.

We will need another normalization of $W_2^{l,l/2}(Q_T)$, $l \in (0,1)$, $0 < T < \infty$:

$$\|u\|^{(l,l/2)}_{Q_T} = \left(\|u\|^2_{W_2^{l,l/2}(Q_T)} + T^{-l}\|u\|^2_{Q_T} \right)^{1/2},$$

$$\|u\|^{(0,l/2)}_{Q_T} = \left(\int_\Omega \|u\|^2_{W_2^{l/2}(0,T)} dx + T^{-l}\|u\|^2_{Q_T} \right)^{1/2}.$$

The equivalent norm in $W_2^{2+l,1+l/2}(Q_T)$ is as follows:

$$\left(\|u\|_{Q_T}^{(2+l,1+l/2)}\right)^2 = \|u\|^2_{W_2^{2+l,1+l/2}(Q_T)} + T^{-l}\Big\{\|\mathcal{D}_t u\|^2_{Q_T} + \sum_{|\boldsymbol{r}|=2} \|\mathcal{D}_x^{\boldsymbol{r}} u\|^2_{Q_T}\Big\} + \\ + \sup_{t\in[0,T]} \|u(\cdot,t)\|^2_{W_2^{1+l}(\Omega)}.$$

For $0 < T < \infty$, it is obvious that

$$\|u\|_{Q_T}^{(l,l/2)} \leqslant c\left(1 + T^{\frac{1-l}{2}}\right)\langle\!\langle u\rangle\!\rangle_T^{(l,l/2)},$$

$$\langle\!\langle u\rangle\!\rangle_T^{(l,l/2)} \equiv \|u\|_{W_2^{l,l/2}(Q_T)} + \sup_{t\in[0,T]} \|u(\cdot,t)\|_\Omega.$$

We will also work in two half-spaces $D^\pm_\infty \equiv \mathbb{R}^3_\pm \times (0,\infty)$. We introduce the notation: $D^3_\infty \equiv D^+_\infty \cup D^-_\infty$, $\mathbb{R}^2_\infty \equiv \mathbb{R}^2 \times (0,\infty)$.

Using the Parseval equality for transformation (2.2.2)

$$\int\limits_{-\infty}^{\infty}\int\limits_{\mathbb{R}^2} |\widetilde{f}(\xi, x_3, \gamma + \mathrm{i}\xi_0)|^2 \,\mathrm{d}\xi \,\mathrm{d}\xi_0 = (2\pi)^3 \int\limits_0^\infty e^{-2\gamma t} \int\limits_{\mathbb{R}^2} |f(x,t)|^2 \,\mathrm{d}x' \,\mathrm{d}t,$$

one can show (see [82], Chap. 2) that for any $\gamma \geqslant 0$ the norm $\|u\|_{H_\gamma^{m,m/2}(D^3_\infty)}$ is equivalent to the norm whose square is expressed by the formula

$$\|u\|^2_{m,\gamma,D^3_\infty} = \int_{\mathbb{R}^2}\int_{-\infty}^{\infty}\Big\{\sum_{j<m}\Big(\Big\|\frac{\partial^j}{\partial x_3^j}\widetilde{u}(\xi,x_3,s)\Big\|^2_{\mathbb{R}_+} |r^+|^{2m-2j} \\ + \Big\|\frac{\partial^j}{\partial x_3^j}\widetilde{u}(\xi,x_3,s)\Big\|^2_{\mathbb{R}_-} |r^-|^{2m-2j}\Big) + \|\widetilde{u}(\xi,\cdot,s)\|^2_{\dot{W}_2^m(\mathbb{R}_+\cup\mathbb{R}_-)}\Big\}\mathrm{d}\xi_0\,\mathrm{d}\xi.$$

Here $s = \gamma + \mathrm{i}\xi_0$, $r^\pm = \sqrt{\frac{s}{\nu^\pm} + \xi^2}$, $\mathbb{R}_\pm = \{y \in \mathbb{R} |\pm y > 0\}$.

The norm $\|u\|_{H_\gamma^{m,m/2}(\mathbb{R}^2_\infty)}$ of the trace of a function u on the plane $\mathbb{R}^2$ is equivalent to the norm

$$\|u\|_{m,\gamma,\mathbb{R}^2_\infty} = \left(\int\limits_{\mathbb{R}^2}\int\limits_{-\infty}^{\infty} |\widetilde{u}(\xi,0,s)|^2|r|^{2m}\,\mathrm{d}\xi_0\,\mathrm{d}\xi\right)^{1/2},$$

and the norm $\|u\|_{H_\gamma^{m+1/2,1/2,m/2}(\mathbb{R}^2_\infty)}$ does to the norm

$$\|u\|_{m,\gamma,\mathbb{R}^2_\infty} = \left(\int_{\mathbb{R}^2} |\xi| \int_{-\infty}^{\infty} |\widetilde{u}(\xi,0,s)|^2 |r|^{2m}\, \mathrm{d}\xi_0\, \mathrm{d}\xi \right)^{1/2},$$

where $r = \sqrt{s+\xi^2}$, $s = \gamma + \mathrm{i}\xi_0$. Substituting r^+ or r^- instead of r leaves the norm to be equivalent.

In addition to ones introduced earlier, for α, $\beta \in (0,1)$, we will use the following Hölder norms of u in Q_T:

$$|u|^{(0,\alpha)}_{Q_T} = \sup_{Q_T} |u| + \sup_{x\in\Omega} \sup_{t,\tau\leqslant T} \frac{|u(x,t)-u(x,\tau)|}{|t-\tau|^\alpha},$$

$$|u|^{(1,\beta)}_{Q_T} = \sup_{Q_T} |u| + \sup_{0\leqslant t\leqslant T} \sup_{x,y\in\Omega} \frac{|u(x,t)-u(y,t)|}{|x-y|} + \sup_{x\in\Omega} \sup_{t,\tau\leqslant T} \frac{|u(x,t)-u(x,\tau)|}{|t-\tau|^\beta}.$$

10.2 L_2-estimates of a Solution of the Model Problem with Plane Interface Between the Fluids

10.2.1 The Homogeneous Problem

We begin, as in the case of the Hölder spaces, with the analysis of problem (2.2.1). We estimate its solution in the Sobolev-Slobodetskiĭ spaces $W_2^{m,m/2}(Q_T)$. After the Fourier-Laplace transform (2.2.2), it can be represented in form (2.2.6), (2.2.7). The functions $e_1^\pm$ and $e_0^\pm = \mathrm{e}^{\mp r^\pm x_3}$ satisfy the following Lemma.

Lemma 10.2.1 *For $\xi \in \mathbb{R}^2$, $s = \gamma + \mathrm{i}\xi_0$, $\gamma > 0$, the inequalities*

$$\begin{aligned} &\pm \int_0^{\pm\infty} \left| \frac{\mathrm{d}^j e_0^\pm(x_3)}{\mathrm{d}x_3^j} \right|^2 \mathrm{d}x_3 \leqslant \frac{|r^\pm|^{2j-1}}{\sqrt{2}}, \\ &\pm \int_0^{\pm\infty} \int_0^{\infty} \left| \frac{\mathrm{d}^j e_0^\pm(x_3 \pm z)}{\mathrm{d}x_3^j} - \frac{\mathrm{d}^j e_0^\pm(x_3)}{\mathrm{d}x_3^j} \right|^2 \frac{\mathrm{d}z\, \mathrm{d}x_3}{z^{1+2\alpha}} \leqslant c_1^\pm |r^\pm|^{2(j+\alpha)-1}, \end{aligned} \tag{10.2.1}$$

$$\pm \int_0^{\pm\infty} \left| \frac{\mathrm{d}^j e_1^\pm(x_3)}{\mathrm{d}x_3} \right|^2 \mathrm{d}x_3 \leqslant c_2^\pm \frac{|r^\pm|^{2j-1} + |\xi|^{2j-1}}{|r^\pm|^2}, \tag{10.2.2}$$

$$\pm\int\limits_0^{\pm\infty}\int\limits_0^{\infty}\left|\frac{\mathrm{d}^j e_1^{\pm}(x_3\pm z)}{\mathrm{d}x_3^j}-\frac{\mathrm{d}^j e_1^{\pm}(x_3)}{\mathrm{d}x_3^j}\right|^2\frac{\mathrm{d}z\,\mathrm{d}x_3}{z^{1+2\alpha}}\leqslant c_3^{\pm}\frac{|r^{\pm}|^{2(j+\alpha)-1}+|\xi|^{2(j+\alpha)-1}}{|r^{\pm}|^2}$$

hold, where $j\in\mathbb{N}\cup\{0\}$, $\alpha\in(0,1)$, $c_1^{\pm}$, $c_2^{\pm}$, $c_3^{\pm}$ *are independent of* $|r^{\pm}|$ *and* $|\xi|$.

Theorem 10.2.1 ($\mathbf{L_2}$-estimate of the Solution of the Homogeneous Model Problem) *Let* b_1, $b_2\in H_{\gamma}^{l+\frac{1}{2},\frac{l}{2}+\frac{1}{4}}(\mathbb{R}^2_{\infty})$, $b_3\in H_{\gamma}^{l+\frac{1}{2},\frac{1}{2},\frac{l}{2}}(\mathbb{R}^2_{\infty})$, $B\in H_{\gamma}^{l-\frac{1}{2},\frac{l}{2}-\frac{1}{4}}(\mathbb{R}^2_{\infty})$ *with some* $l>1/2$. *Then the solution of problem* (2.2.1) *satisfies the estimate*

$$\|\boldsymbol{v}\|^2_{2+l,\gamma,D^3_{\infty}}+\|\nabla p\|^2_{l,\gamma,D^3_{\infty}}$$
$$\leqslant c_4(\gamma)\Big\{\sum_{\beta=1}^{2}\|b_\beta\|^2_{l+1/2,\gamma,\mathbb{R}^2_{\infty}}+\|b_3\|^2_{l,\gamma,\mathbb{R}^2_{\infty}}+\sigma^2\|B\|^2_{l-1/2,\gamma,\mathbb{R}^2_{\infty}}\Big\},\qquad(10.2.3)$$

where $c_4(\gamma)\leqslant c_4(\gamma_0)$ *for* $\gamma\geqslant\gamma_0>0$.

Proof First, we estimate velocity vector field $\boldsymbol{v}$. The Fourier-Laplace transform of it, given by (2.2.6), (2.2.7), can be written as follows:

$$\widetilde{\boldsymbol{v}}^{\pm}=\boldsymbol{\omega}e_0^{\pm}+(\mathrm{i}\xi_1,\mathrm{i}\xi_2,\mp|\xi|)\Psi^{\pm}e_1^{\pm},\quad \pm x_3>0.$$

If l is non-integral, the squared norm of $\boldsymbol{v}$ satisfies the inequality

$$\|\boldsymbol{v}^+\|^2_{2+l,\gamma,D^{+}_{\infty}}\leqslant\int\limits_{\mathbb{R}^2}\int\limits_{-\infty}^{\infty}\Big\{|\boldsymbol{\omega}|^2\Big(\sum_{j=0}^{m}|r^+|^{2l+4-2j}\int\limits_0^{\infty}\left|\frac{\mathrm{d}^j e_0^+(x_3)}{\mathrm{d}x_3^j}\right|^2\mathrm{d}x_3$$
$$+\int\limits_0^{\infty}\int\limits_0^{\infty}\left|\frac{\mathrm{d}^m e_0^+(x_3+z)}{\mathrm{d}x_3^m}-\frac{\mathrm{d}^m e_0^+(x_3)}{\mathrm{d}x_3^m}\right|^2\frac{\mathrm{d}z\,\mathrm{d}x_3}{z^{1+2(l-[l])}}\Big)$$
$$+2|\xi|^2|\Psi^+|^2\Big(\sum_{j=0}^{m}|r^+|^{2l+4-2j}\int\limits_0^{\infty}\left|\frac{\mathrm{d}^j e_1^+(x_3)}{\mathrm{d}x_3^j}\right|^2\mathrm{d}x_3$$
$$+\int\limits_0^{\infty}\int\limits_0^{\infty}\left|\frac{\mathrm{d}^m e_1^+(x_3+z)}{\mathrm{d}x_3^m}-\frac{\mathrm{d}^m e_1^+(x_3)}{\mathrm{d}x_3^m}\right|^2\frac{\mathrm{d}z\,\mathrm{d}x_3}{z^{1+2(l-[l])}}\Big)\Big\}\,\mathrm{d}\xi_0\,\mathrm{d}\xi,$$

where $m=[l]+2$. If l is an integer, then one should omit the terms with increments of mth derivatives of e_0^+ and e_1^+. Due to (10.2.1), (10.2.2), we conclude that

$$\|\boldsymbol{v}^+\|^2_{2+l,\gamma,D^+_\infty} \leqslant c^+ \int\limits_{\mathbb{R}^2}\int\limits_{-\infty}^{\infty} \{|\boldsymbol{\omega}|^2|r^+|^{2l+3} + |\Psi^+|^2|\xi||r^+|^{2l+2}\}\,\mathrm{d}\xi_0\,\mathrm{d}\xi. \tag{10.2.4}$$

A similar inequality holds for $\boldsymbol{v}^-$:

$$\|\boldsymbol{v}^-\|^2_{2+l,\gamma,D^-_\infty} \leqslant c^- \int\limits_{\mathbb{R}^2}\int\limits_{-\infty}^{\infty} \{|\boldsymbol{\omega}|^2|r^-|^{2l+3} + |\Psi^-|^2|\xi||r^-|^{2l+2}\}\,\mathrm{d}\xi_0\,\mathrm{d}\xi. \tag{10.2.5}$$

In the first line, we get an estimate of ω_3 and Ψ^-, since these functions are explicitly expressed in terms of problem data (see (2.2.7)):

$$|\omega_3|^2 \leqslant c\xi^2|P|^{-2}\Big\{\sum_{k=1}^{3}|\widetilde{b}_k|^2\big(|r^+|^2+|r^-|^2\big)|s|^2|q|^{-2} + \sigma^2|\widetilde{B}|^2\Big\};$$

$$|\Psi^-|^2|\xi| \leqslant c|\xi||P|^{-2}|q|^{-2}\Big\{\sum_{k=1}^{3}|\widetilde{b}_k|^2\big(|r^-|^4|s|^2+\sigma^2\xi^6\big) + \sigma^2|\widetilde{B}|^2|r^-|^4\Big\}.$$

We note that $|q| \geqslant (\mu^+|r^+|+\mu^-|r^-|)/\sqrt{2}$. Since $|r^\pm|^2$ is equivalent to $|s|+\xi^2$, we use Lemma 2.2.1 for obtaining a lower bound of $|P|$:

$$|P| \geqslant c^\pm(\sqrt{\gamma})\Big(|s||r^\pm|^2 + |r^\pm|^3\Big),$$

and arrive at the inequalities

$$|\omega_3|^2 \leqslant c_5(\gamma)\Big\{\sum_{k=1}^{3}\frac{|\widetilde{b}_k|^2\xi^2}{|r^-|^4} + \frac{\sigma^2|\widetilde{B}|^2|\xi|}{|r^-|^5}\Big\}, \tag{10.2.6}$$

$$|\Psi^-|^2|\xi| \leqslant c_6(\gamma)\Big\{\sum_{k=1}^{3}\frac{|\widetilde{b}_k|^2|\xi|}{|r^-|^2} + \frac{\sigma^2|\widetilde{B}|^2}{|r^-|^3}\Big\}, \tag{10.2.7}$$

where $c_5(\gamma)$, $c_6(\gamma)$ are bounded above by a positive constant if $\gamma \geqslant \gamma_0$. We estimate now the remaining functions in (2.2.7). Clearly, $|\Psi^+|^2|\xi|$ satisfies an inequality similar to (2.2.7). For ω_α, $\alpha = 1, 2$, we have

$$|\omega_\alpha|^2 \leqslant c_7\frac{|\widetilde{b}_\alpha|^2 + \xi_\alpha^2\big(|\Psi^+|^2+|\Psi^-|^2+|\omega_3|^2\big)}{|\mu^+r^+ + \mu^-r^-|^2}. \tag{10.2.8}$$

Since $|\mu^+r^+ + \mu^-r^-| \geqslant (\mu^+|r^+| + \mu^-|r^-|)/\sqrt{2}$, from inequalities (10.2.4)–(10.2.8), we obtain the required estimate

$$\|\boldsymbol{v}\|^2_{2+l,\gamma,D^3_\infty} \leqslant c_8(\gamma)\Big\{\sum_{\alpha=1}^{2} \|b_\alpha\|^2_{l+1/2,\gamma,\mathbb{R}^2_\infty} + \|b_3\|^2_{l,\gamma,\mathbb{R}^2_\infty} + \sigma^2\|B\|^2_{l-1/2,\gamma,\mathbb{R}^2_\infty}\Big\}$$

with a constant $c_8(\gamma)$ bounded for $\gamma \geqslant \gamma_0 > 0$.

The Fourier transform of pressure gradient is given by the vector

$$\widetilde{\nabla p} = (\mathrm{i}\xi_1, \mathrm{i}\xi_2, \mp|\xi|)\mu^\pm\Psi^\pm(r^\pm + |\xi|)\mathrm{e}^{\mp|\xi|x_3},$$

whose modulus can be estimated above in terms of $\sqrt{2}\mu^\pm|\xi||\Psi^\pm||r^\pm + |\xi||\mathrm{e}^{\mp|\xi|x_3}$. (The upper index corresponds to the half-space $\{x_3 > 0\}$, while the lower one corresponds to the half-space $\{x_3 < 0\}$.) For $\mathrm{e}^{\mp|\xi|x_3}$, inequalities (10.2.1) hold with $|\xi|$ in place of $|r^\pm|$; therefore,

$$\|\nabla p\|^2_{l,\gamma,D^3_\infty} \leqslant c \int\limits_{\mathbb{R}^2}\int\limits_{-\infty}^{\infty} (|\Psi^+|^2|r^+|^2 + |\Psi^-|^2|r^-|^2)|r|^{2l}|\xi|\,\mathrm{d}\xi_0\,\mathrm{d}\xi.$$

Applying (10.2.7), we arrive at estimate (10.2.3) which completes the proof of the theorem. □

10.2.2 The Nonhomogeneous Problem with Plane Interface

In this subsection, we investigate problem (2.2.1) with nonhomogeneous equations in the interval $(0, T)$, $T \leqslant \infty$. We notice that for functions in $H^{l,l/2}_\gamma(\mathbb{R}^3_+ \times (0,T))$, $T < \infty$, continuations are possible with preservation of class both into $\{x_3 < 0\}$ and $\{t > T\}$, i. e., into the domain $\mathbb{R}^3 \times (0,\infty) \equiv \mathbb{R}^3_\infty$ (in this regard, see [82]).

We introduce the notation:

$$\mathbb{R}^2_T \equiv \mathbb{R}^2 \times (0,T), \quad D^3 \equiv \mathbb{R}^3_- \cup \mathbb{R}^3_+, \quad D^\pm_T \equiv \mathbb{R}^3_\pm \times (0,T), \quad D^3_T \equiv D^-_T \cup D^+_T.$$

We consider problem (2.5.1) with $\boldsymbol{v}_0 = 0$ and given functions $\boldsymbol{f}$, g, $\boldsymbol{R}$, b_i, B which vanish at $t = 0$ together with some of their derivatives. In order to analyse this problem, one needs, in addition to homogeneous problem (2.2.1), to be able to estimate solutions $\boldsymbol{w}$ and Φ to auxiliary problems (2.5.13) with $\boldsymbol{F} = \boldsymbol{f}$ and (2.5.14) with $r = g$, respectively. We represent a solution of (2.5.1) in the form: $\boldsymbol{v} = \boldsymbol{w} + \boldsymbol{w}' + \boldsymbol{u}$, $p = q + \mu^\pm g' - \rho^\pm\mathcal{D}_t\Phi$, where $\boldsymbol{w}' = \nabla\Phi$, $g' = g - \nabla\cdot\boldsymbol{w}$, and $(\boldsymbol{u}, q)$ is a solution of problem (2.2.1) with

$$b'_\beta = b_\beta + \left[\mu^\pm\left(\frac{\partial(w_\beta + w'_\beta)}{\partial x_3} + \frac{\partial(w_3 + w'_3)}{\partial x_\beta}\right)\right]\Big|_{x_3=0}, \quad \beta = 1, 2;$$
$$b'_3 = b_3 - \left[2\mu^\pm \frac{\partial(w_3 + w'_3)}{\partial x_3} - \rho^\pm \mathcal{D}_t \Phi + \mu^\pm g'\right]\Big|_{x_3=0}, \tag{10.2.9}$$
$$B' = B - \Delta'(w_3 + w'_3)\big|_{x_3=0}.$$

Lemma 10.2.2 *If $\boldsymbol{f} \in \boldsymbol{H}_\gamma^{l,l/2}(D_T^3)$, $\gamma > 0$, then there exists a solution $\boldsymbol{w} \in \boldsymbol{H}_\gamma^{2+l,1+l/2}(D_T^3)$ of problem (2.5.13) with $\boldsymbol{F} = \boldsymbol{f}$, and for it, there holds the estimate*

$$\|\boldsymbol{w}\|_{\boldsymbol{H}_\gamma^{2+l,1+l/2}(D_T^3)} \leqslant c_1 \|\boldsymbol{f}\|_{\boldsymbol{H}_\gamma^{l,l/2}(D_T^3)}. \tag{10.2.10}$$

Proof We break (2.5.13) into three subproblems.

(a) It is required to find a vector $\boldsymbol{w}^+$ in D_T^+ such that

$$\mathcal{D}_t \boldsymbol{w}^+ - \nu^+ \Delta \boldsymbol{w}^+ = \boldsymbol{f}^+ \text{ in } D_T^+, \quad \boldsymbol{w}^+|_{t=0} = 0, \tag{10.2.11}$$

where $\boldsymbol{f}^+ = \boldsymbol{f}|_{D_T^+}$. We extend the vector $\boldsymbol{f}^+$ from the set D_T^+ into $\mathbb{R}^3_\infty$ first in $x_3 < 0$, and then in $t > T$, so that

$$\|\boldsymbol{f}^+\|_{\boldsymbol{H}_\gamma^{l,l/2}(\mathbb{R}^3_\infty)} \leqslant c\|\boldsymbol{f}\|_{\boldsymbol{H}_\gamma^{l,l/2}(D_T^+)} \tag{10.2.12}$$

(see [82]). We take the Fourier transform of the obtained functions $\boldsymbol{f}^+(x,t)$ in the variables x_1, x_2, x_3 and the Laplace transform in t. The vector

$$\widetilde{\boldsymbol{w}}^+(\xi, s) = \frac{\widetilde{\boldsymbol{f}}^+(\xi, s)}{s + \nu^+(\xi_1^2 + \xi_2^2 + \xi_3^2)}$$

after the inverse transformation satisfies system (10.2.11). (Here $\xi = (\xi_1, \xi_2, \xi_3)$ is the variable dual to $x = (x_1, x_2, x_3)$, and $s = \gamma + \mathrm{i}\xi_0$ is the one dual to t.) It is obvious that

$$\int\limits_{\mathbb{R}^3}\int\limits_{-\infty}^{\infty} \left|\widetilde{\boldsymbol{w}}^+(\xi, s)\right|^2 \left(|r^+|^2 + \xi_3^2\right)^{2+l} \mathrm{d}\xi_0\, \mathrm{d}\xi \leqslant c \int\limits_{\mathbb{R}^3}\int\limits_{-\infty}^{\infty} \left|\widetilde{\boldsymbol{f}}^+(\xi, s)\right|^2 \left(|r^+|^2 + \xi_3^2\right)^l \mathrm{d}\xi_0 \mathrm{d}\xi.$$

From the Parseval equality for the complete Fourier-Laplace transform and from inequality (10.2.12) it follows that

$$\|\boldsymbol{w}^+\|^2_{\boldsymbol{H}_\gamma^{2+l,1+l/2}(D_T^+)} \leqslant c\|\boldsymbol{f}\|^2_{\boldsymbol{H}_\gamma^{l,l/2}(D_T^+)}. \tag{10.2.13}$$

(b) In a similar way, we estimate the vector $\boldsymbol{w}_1^-$ satisfying in D_T^- the system of equations

$$\mathcal{D}_t \boldsymbol{w}_1^- - \nu^- \Delta \boldsymbol{w}_1^- = \boldsymbol{f}^- \text{ in } D_T^-, \quad \boldsymbol{w}_1^-|_{t=0} = 0 \qquad (\boldsymbol{f}^- = \boldsymbol{f}|_{D_T^-}),$$

namely,

$$\|\boldsymbol{w}_1^-\|^2_{\boldsymbol{H}_\gamma^{2+l,1+l/2}(D_T^-)} \leqslant c\|\boldsymbol{f}\|^2_{\boldsymbol{H}_\gamma^{l,l/2}(D_T^-)}. \tag{10.2.14}$$

(c) It is necessary to find a vector $\boldsymbol{w}_2^-$ satisfying the homogeneous vector-valued heat equation

$$\mathcal{D}_t \boldsymbol{w}_2^- - \nu^- \Delta \boldsymbol{w}_2^- = 0 \quad \text{in } D_T^-,$$

and also the initial and boundary conditions

$$\boldsymbol{w}_2^-|_{x_3=0} = (\boldsymbol{w}^+ - \boldsymbol{w}_1^-)|_{x_3=0} \equiv \boldsymbol{\varphi}(x_1, x_2, t),$$
$$\boldsymbol{w}_2^-|_{t=0} = 0.$$

Taking partial Fourier-Laplace transform (2.2.2) and solving the ordinary differential equation with respect to the variable x_3, we obtain

$$\widetilde{\boldsymbol{w}}_2^-(\xi_1, \xi_2, x_3, s) = \widetilde{\boldsymbol{\varphi}}(\xi_1, \xi_2, s) e_0^-(x_3), \qquad x_3 < 0.$$

From the imbedding theorem for the space $H_\gamma^{l,l/2}$ (see [82], Lemma 2.4) it follows that $\boldsymbol{\varphi} = (\boldsymbol{w}^+ - \boldsymbol{w}_1^-)|_{x_3=0} \in \boldsymbol{H}_\gamma^{3/2+l,3/4+l/2}(\mathbb{R}_T^2)$. Next, in view of (10.2.1), (10.2.13), and (10.2.14)

$$\begin{aligned}\|\boldsymbol{w}_2^-\|^2_{2+l,\gamma,D_\infty^\pm} &= \iint\limits_{\mathbb{R}^2 -\infty}^{\infty} \Big(\sum_{j=0}^{[2+l]} \Big\| \frac{d^j e_0^-(x_3)}{dx_3^j} \Big\|^2_{\mathbb{R}_-} |r^-|^{2l+4-2j} |\widetilde{\boldsymbol{\varphi}}|^2 \\ &+ \|\widetilde{\boldsymbol{\varphi}} e_0^-\|^2_{\dot{W}_2^{2+l}(\mathbb{R}_-)} \Big) d\xi_0\, d\xi \leqslant c \int\limits_{\mathbb{R}^3} |r^-|^{2l+3} |\widetilde{\boldsymbol{\varphi}}|^2\, d\xi_0\, d\xi \\ &\leqslant c\|\boldsymbol{\varphi}\|^2_{3/2+l,\gamma,\mathbb{R}^2_\infty} \leqslant c\|\boldsymbol{f}\|^2_{\boldsymbol{H}_\gamma^{l,l/2}(D_T^3)}.\end{aligned} \tag{10.2.15}$$

Thus, the vector $\boldsymbol{w} = \boldsymbol{w}^+$ for $x_3 > 0$ and $\boldsymbol{w} = \boldsymbol{w}^- + \boldsymbol{w}_2^-$ for $x_3 < 0$, being a solution of problem (2.5.13) with the right-hand side $\boldsymbol{F} = \boldsymbol{f}$, belongs to the space $\boldsymbol{H}_\gamma^{2+l,1+l/2}(D_T^3)$ and, in view of (10.2.13)–(10.2.15), satisfies estimate (10.2.10).

□

Lemma 10.2.3 *Let in equation* (2.5.14) *the function* $r = g \in H_\gamma^{1+l,1/2+l/2}(D_T^3)$, $\gamma > 0$, *can be represented in the form* $g = \nabla \cdot R$, *where* $\boldsymbol{R} \in \boldsymbol{H}_\gamma^{0,1+l/2}(D_T^3)$, $[R_3]|_{x_3=0} = 0$.[1] *Then equation* (2.5.14) *with* $r' = g' \equiv g - \nabla \cdot \boldsymbol{w}$ *is solvable and for the gradient of the solution* $\boldsymbol{w}' = \nabla\Phi$ *there holds the inequality*

$$\|\boldsymbol{w}'\|_{\boldsymbol{H}_\gamma^{2+l,1+l/2}(D_T^3)} \leqslant c\{\|g\|_{H_\gamma^{1+l,0}(D_T^3)} + \|\boldsymbol{R}\|_{\boldsymbol{H}_\gamma^{0,1+l/2}(D_T^3)} + \|\boldsymbol{w}\|_{\boldsymbol{H}_\gamma^{2+l,1+l/2}(D_T^3)}\}. \tag{10.2.16}$$

Proof Lemma 10.2.2 and trace theorem for the Sobolev-Slobodetskiĭ spaces imply that for a. e. $t \in (0,T)$ $\boldsymbol{w} \in \boldsymbol{W}_2^1(D^3)$, and since $[\boldsymbol{w}]|_{x_3=0} = 0$, it follows that $\boldsymbol{w} \in \boldsymbol{W}_2^1(\mathbb{R}^3)$. Next, under the assumptions of the lemma, $g' \in L_2(\mathbb{R}^3)$ for a. e. $t < T$. It is known from the theory of the elliptic equations that in this case a solution of equation (2.5.14) $\Phi \in \boldsymbol{W}_{2,loc}^2(\mathbb{R}^3)$ and is given by formula (2.5.17) due to the weak continuity of R_3 and w_3 when crossing the boundary $\{x_3 = 0\}$.

By differentiating (2.5.17) with respect to x and t and using the Calderón-Zygmund theorem, it can be found that for

$$\boldsymbol{w}' \equiv \nabla\Phi(x,t) = \int_{\mathbb{R}^3} \nabla_y \nabla_y \mathcal{E}(x,y) \cdot (\boldsymbol{R} - \boldsymbol{w})\mathrm{d}y,$$

the following inequalities

$$\gamma^{2+l}\|\boldsymbol{w}'\|_{\mathbb{R}^3}^2 \leqslant c\gamma^{2+l}\{\|\boldsymbol{R}\|_{\mathbb{R}^3}^2 + \|\boldsymbol{w}\|_{\mathbb{R}^3}^2\}, \tag{10.2.17}$$

$$\left\|\mathcal{D}_t^k \boldsymbol{w}'\right\|_{\mathbb{R}^3}^2 \leqslant c\left\{\left\|\mathcal{D}_t^k \boldsymbol{R}\right\|_{\mathbb{R}^3}^2 + \left\|\mathcal{D}_t^k \boldsymbol{w}\right\|_{\mathbb{R}^3}^2\right\}, \tag{10.2.18}$$

hold for integral $l/2$, $k = l/2 + 1$, or

$$\begin{aligned} &\int_0^\infty \left\|\mathcal{D}_t^k \boldsymbol{w}_0'(\cdot,t) - \mathcal{D}_t^k \boldsymbol{w}_0'(\cdot,t-\tau)\right\|_{\mathbb{R}^3}^2 \frac{\mathrm{d}\tau}{\tau^{3+l-2k}} \\ &\leqslant c\int_0^\infty \Big\{ \left\|\mathcal{D}_t^k \boldsymbol{R}_0(\cdot,t) - \mathcal{D}_t^k \boldsymbol{R}_0(\cdot,t-\tau)\right\|_{\mathbb{R}^3}^2 \\ &\quad + \left\|\mathcal{D}_t^k \boldsymbol{w}_0(\cdot,t) - \mathcal{D}_t^k \boldsymbol{w}_0(\cdot,t-\tau)\right\|_{\mathbb{R}^3}^2 \Big\} \frac{\mathrm{d}\tau}{\tau^{3+l-2k}}, \end{aligned} \tag{10.2.19}$$

[1] This condition is understood in the weak sense, i. e.,

$$\int_{\mathbb{R}^3} g(x,t)\varphi(x)\,\mathrm{d}x = -\int_{\mathbb{R}^3} \boldsymbol{R} \cdot \nabla\varphi(x)\,\mathrm{d}x, \quad \forall\varphi \in C_0^\infty(\mathbb{R}^3), \quad \text{a. e. } t \in (0,T).$$

for nonintegral $l/2$, $k = [l/2] + 1$, where $\boldsymbol{w}'_0, \boldsymbol{R}_0, \boldsymbol{w}_0$ are respective continuations by zero of the vectors $\boldsymbol{w}', \boldsymbol{R}, \boldsymbol{w}$ into the interval $t < 0$.

We now consider Φ as a solution of diffraction problem for equation (2.5.14) for $r = g$ with jump conditions

$$[\Phi]|_{x_3=0} = [\mathcal{D}_{x_3}\Phi]\Big|_{x_3=0} = 0.$$

By introducing the new unknown functions $\Psi(x_1, x_2, x_3) = \Phi(x_1, x_2, x_3) - \Phi(x_1, x_2, -x_3)$ and $\chi(x_1, x_2, x_3) = \Phi(x_1, x_2, x_3) + \Phi(x_1, x_2, -x_3)$, diffraction problem can be reduced to two problems in the half-space $\mathbb{R}^3_+ = \{x_3 > 0\}$: the Dirichlet problem

$$\Delta\Psi = g'^+ - g'^- \quad \text{in} \quad \mathbb{R}^3_+,$$
$$\Psi|_{x_3=0} = 0$$

and the Neumann problem

$$\Delta\chi = g'^+ + g'^- \quad \text{in} \quad \mathbb{R}^3_+,$$
$$\mathcal{D}_{x_3}\chi\Big|_{x_3=0} = 0.$$

The solutions of these problems satisfy the known estimates [74]

$$\|\Psi\|^2_{\dot{W}_2^{3+l}(\mathbb{R}^3_+)} \leqslant c\|g'^+ - g'^-\|^2_{\dot{W}_2^{1+l}(\mathbb{R}^3_+)},$$
$$\|\chi\|^2_{\dot{W}_2^{3+l}(\mathbb{R}^3_+)} \leqslant c\|g'^+ + g'^-\|^2_{\dot{W}_2^{1+l}(\mathbb{R}^3_+)},$$

which imply, as is easily seen, the inequality

$$\|\Phi\|^2_{\dot{W}_2^{3+l}(D^3)} \leqslant c\|g'\|^2_{\dot{W}_2^{1+l}(D^3)},$$

whence for the vector $\boldsymbol{w}'$ we obtain

$$\|\boldsymbol{w}'\|^2_{\dot{W}_2^{2+l}(D^3)} \leqslant c\big\{\|g\|^2_{\dot{W}_2^{1+l}(D^3)} + \|\boldsymbol{w}\|^2_{\dot{W}_2^{2+l}(D^3)}\big\}. \tag{10.2.20}$$

Multiplying inequalities (10.2.17)–(10.2.20) by $e^{-2\gamma t}$, integrating the estimates obtained from 0 to T and adding, we arrive at (10.2.16), which completes the proof of the lemma. □

Theorem 10.2.2 (The $\mathbf{L_2}$-estimate for Nonhomogeneous Model Problem) *Let $\boldsymbol{f} \in \boldsymbol{H}_\gamma^{l,l/2}(D^3_T)$ and let $g \in H_\gamma^{1+l,1/2+l/2}(D^3_T)$ can be* *represented in the form* $g = \nabla \cdot \boldsymbol{R}$, *where* $\boldsymbol{R} \in \boldsymbol{H}_\gamma^{0,1+l/2}(D_T^3)$, $[R_3]|_{x_3=0} = 0$[2]. *Moreover, we assume the functions* b_1, $b_2 \in H_\gamma^{l+1/2,l/2+1/4}(\mathbb{R}_T^2)$, $b_3 \in H_\gamma^{l+1/2,1/2,l/2}(\mathbb{R}_T^2)$, $B \in H_\gamma^{l-1/2,l/2-1/4}(\mathbb{R}_T^2)$ *with some* $l > 1/2$, $T < \infty$ *and* $\gamma \geqslant \gamma_0 > 0$. *Then there exists a unique solution* $(\boldsymbol{v}, p)$ *of problem* (2.5.1) *with* $\boldsymbol{v}_0 = 0$ *and* $p \xrightarrow[|x|\to\infty]{} 0$ *such that* $\boldsymbol{v} \in \boldsymbol{H}_\gamma^{2+l,1+l/2}(D_T^3)$, $\nabla p \in \boldsymbol{H}_\gamma^{l,l/2}(D_T^3)$, *and for it the estimate*

$$\begin{aligned}
&\|\boldsymbol{v}\|^2_{\boldsymbol{H}_\gamma^{2+l,1+l/2}(D_T^3)} + \|\nabla p\|^2_{\boldsymbol{H}_\gamma^{l,l/2}(D_T^3)} \leqslant c\Big\{\|\boldsymbol{f}\|^2_{\boldsymbol{H}_\gamma^{l,l/2}(D_T^3)} + \|g\|^2_{H_\gamma^{1+l,(1+l)/2}(D_T^3)} \\
&\quad + \|\boldsymbol{R}\|^2_{\boldsymbol{H}_\gamma^{0,1+l/2}(D_T^3)} + \|b_3\|^2_{H_\gamma^{l+1/2,1/2,l/2}(\mathbb{R}_T^2)} + \sum_{\beta=1}^{2} \|b_\beta\|^2_{H_\gamma^{l+1/2,l/2+1/4}(\mathbb{R}_T^2)} \\
&\quad + \sigma^2 \|B\|^2_{H_\gamma^{l-1/2,l/2-1/4}(\mathbb{R}_T^2)}\Big\}
\end{aligned} \tag{10.2.21}$$

holds.

Proof As already noted, a solution of problem (2.5.1) with $\boldsymbol{v}_0 = 0$ can be written in the form:

$$\boldsymbol{v} = \boldsymbol{w} + \boldsymbol{w}' + \boldsymbol{u}, \quad p = q + \rho^\pm \mathcal{D}_t \Phi + \mu^\pm g' \qquad g' = g - \nabla \cdot \boldsymbol{w}.$$

To estimate $(\boldsymbol{u}, q)$, we use Theorem 10.2.1 applied to problem (2.2.1), (10.2.9). The remaining terms $\boldsymbol{w}$, $\boldsymbol{w}'$ in the expression for velocity vector field can be estimated on the basis of Lemmas 10.2.2, 10.2.3.

We consider the norms on the right-hand side of inequality (10.2.3) with the functions b_i', B' defined in (10.2.9) instead of b_i, B. They are subjected to the inequality

$$\begin{aligned}
&\|b_\beta'\|^2_{l+1/2,\gamma,\mathbb{R}^2_\infty} \leqslant c\{\|b_\beta\|^2_{l+1/2,\gamma,\mathbb{R}^2_\infty} + \|\boldsymbol{w} + \boldsymbol{w}'\|^2_{l+3/2,\gamma,\mathbb{R}^2_\infty}\}, \qquad \beta = 1, 2, \\
&\quad \|b_3'\|^2_{l,\gamma,\mathbb{R}^2_\infty} \leqslant c\{\|b_3\|^2_{l,\gamma,\mathbb{R}^2_\infty} + \|w_3 + w_3'\|^2_{l+3/2,\gamma,\mathbb{R}^2_\infty} + \|[\rho^\pm \mathcal{D}_t \Phi]|_{x_3=0}\|^2_{l,\gamma,\mathbb{R}^2_\infty} \\
&\qquad\qquad + \|[\mu^\pm g']|_{x_3=0}\|^2_{l,\gamma,\mathbb{R}^2_\infty}\}, \qquad\qquad (10.2.22) \\
&\|B'\|^2_{l-1/2,\gamma,\mathbb{R}^2_\infty} \leqslant c\{\|B\|^2_{l-1/2,\gamma,\mathbb{R}^2_\infty} + \|w_3 + w_3'\|^2_{l+3/2,\gamma,\mathbb{R}^2_\infty}\}.
\end{aligned}$$

From trace theorem for the space $H_\gamma^{l,l/2}$ (see [82], Lemma 2.4) it follows that

$$\|\boldsymbol{w} + \boldsymbol{w}'\|^2_{l+3/2,\gamma,\mathbb{R}^2_\infty} \leqslant c\{\|\boldsymbol{w}\|^2_{2+l,\gamma,D_+^3} + \|\boldsymbol{w}'\|^2_{2+l,\gamma,D_+^3}\},$$

[2]See the footnote against Lemma 10.2.3.

$$\|[\rho^{\pm}\mathcal{D}_t\Phi]|_{x_3=0}\|^2_{l,\gamma,\mathbb{R}^2_\infty} \leqslant c\|\nabla\mathcal{D}_t\Phi\|^2_{l,\gamma,D^3_+} \leqslant c\|\boldsymbol{w}'\|^2_{2+l,\gamma,D^3_+}, \tag{10.2.23}$$

$$\|[\mu^{\pm}g']|_{x_3=0}\|^2_{l,\gamma,\mathbb{R}^2_\infty} \leqslant c\|\nabla g'\|^2_{l,\gamma,D^3_+} \leqslant c\|g'\|^2_{l+1,\gamma,D^3_+}. \tag{10.2.24}$$

Extending the functions in (10.2.9) in $t > T$ with preservation of class and using Theorem 10.2.1, we find that the solution $(\boldsymbol{u}, q)$ of system (2.2.1), (10.2.9) exists and estimate (10.2.3) for it implies, in view of (10.2.22)–(10.2.24), the inequality

$$\|\boldsymbol{u}\|^2_{\boldsymbol{H}^{2+l,1+l/2}_\gamma(D^3_T)} + \|\nabla q\|^2_{\boldsymbol{H}^{l,l/2}_\gamma(D^3_T)}$$

$$\leqslant c\Big\{\|g\|^2_{H^{1+l,(1+l)/2}_\gamma(D^3_T)} + \sum_{\beta=1}^{2}\|b_\beta\|^2_{H^{l+1/2,l/2+1/4}_\gamma(\mathbb{R}^2_T)} + \|b_3\|^2_{H^{l+1/2,1/2,l/2}_\gamma(\mathbb{R}^2_T)}$$

$$+\sigma^2\|B\|^2_{H^{l-1/2,l/2-1/4}_\gamma(\mathbb{R}^2_T)} + \|\boldsymbol{w}\|^2_{\boldsymbol{H}^{2+l,1+l/2}_\gamma(D^3_T)} + \|\boldsymbol{w}'\|^2_{\boldsymbol{H}^{2+l,1+l/2}_\gamma(D^3_T)}\Big\}. \tag{10.2.25}$$

Thus, we see that a solution $(\boldsymbol{v}, p)$ exists and estimate (10.2.21) for it follows from inequalities (10.2.10), (10.2.16), (10.2.23), and (10.2.25).

We verify solution uniqueness in these classes. Let $\boldsymbol{f} = 0$, $g = 0$, $b_i = 0$, $i = 1, 2, 3$, $B = 0$. Then problem (2.5.1) with $\boldsymbol{v}_0 = 0$ goes over into (2.2.1) whose unique solution given by formulas (2.2.6), (2.2.7) vanishes if $b_i = 0$, $i = 1, 2, 3$, and $B = 0$. This follows from the fact that, for $\gamma \geqslant \gamma_0 > 0$, the denominator of P (2.2.8) is bounded away from zero due to estimate (2.2.9). The theorem is proved. □

10.3 A Priori Estimates of a Solution of the Problem with a Closed Interface

Now we estimate, in the weighted Sobolev spaces, a solution to problem (1.1.7) which, in the linear approximation, governs the motion of a drop Ω^+ in a fluid Ω^- with zero initial velocity distribution.

We assume that the boundary of the drop $\Gamma \in W^{2+l}_2$, i. e., in a neighborhood of arbitrary its point z in a local coordinate system $\{y\}$, it is given by the equation $y_3 = \varphi_z(y_1, y_2)$, where $\varphi_z \in W^{2+l}_2(B^2_\rho)$ with some $l > 1/2$ ($B^2_\rho = \{(y_1, y_2)|\, y_1^2 + y_2^2 < \rho^2\}$; the axis y_3 is directed into Ω^+); $\varphi_z(0) = 0$, $\nabla\varphi_z(0) = 0$ and there holds a uniform estimate of $\|\varphi_z\|_{W^{2+l}_2(B^2_\rho)}$ on the entire surface Γ.

We denote by B_d the ball $\{x : |x| < d\}$. We choose a coordinate system $\{x\}$ so that Ω^+ is contained in the ball B_{d-4} for some $d > 4$. We set $B^+_d \equiv \Omega^+$, $B^-_d \equiv B_d \setminus \Omega^+ \setminus \Gamma$, $B_{dT} = B_d \times (0, T)$, $D_T = Q^+_T \cup Q^-_T$, $Q^{\pm}_T = \Omega \times (0, T)$.

Theorem 10.3.1 (A Priori Estimate of a Solution to the Linear Problem) *Let* $l > 1/2$, $\Gamma \in W_2^{2+l}$, $\boldsymbol{f} \in \boldsymbol{H}_\gamma^{l,l/2}(D_T)$, $g \in H_\gamma^{1+l,1/2+l/2}(D_T)$, $g = \nabla \cdot \boldsymbol{R}$, $\boldsymbol{R} \in \boldsymbol{H}_\gamma^{0,1+l/2}(\mathbb{R}_T^3)$, $[\boldsymbol{R} \cdot \boldsymbol{n}]|_{G_T} = 0$ *(this equality is understood in the weak sense),* $\boldsymbol{b} \in \boldsymbol{H}_\gamma^{l+1/2,l/2+1/4}(G_T)$, $\boldsymbol{b} \cdot \boldsymbol{n} = 0$, $b' \in H_\gamma^{l+1/2,1/2,l/2}(G_T)$, $B \in H_\gamma^{l-1/2,l/2-1/4}(G_T)$ *for* $T \leqslant \infty$ *and* $\gamma \geqslant \gamma_0 \gg 1$. *Then the vector* $\boldsymbol{v} \in \boldsymbol{H}_\gamma^{2+l,1+l/2}(D_T)$ *and the function* p *such that* $\nabla p \in \boldsymbol{H}_\gamma^{l,l/2}(D_T)$, $p \xrightarrow[|x|\to\infty]{} 0$, *solving system* (1.1.7) *with* $\boldsymbol{v}_0 = 0$, *satisfy the inequality*

$$\|\boldsymbol{v}\|^2_{\boldsymbol{H}_\gamma^{2+l,1+l/2}(D_T)} + \|\nabla p\|^2_{\boldsymbol{H}_\gamma^{l,l/2}(D_T)} + \|p\|^2_{H_\gamma^{0,l/2}(B_{dT})} \leqslant c_1\Big\{\|\boldsymbol{f}\|^2_{\boldsymbol{H}_\gamma^{l,l/2}(D_T)}$$
$$+ \|g\|^2_{H_\gamma^{1+l,(1+l)/2}(D_T)} + \|\boldsymbol{R}\|^2_{\boldsymbol{H}_\gamma^{0,1+l/2}(D_T)} + \|b'\|^2_{H_\gamma^{l+1/2,1/2,l/2}(G_T)}$$
$$+ \|\boldsymbol{b}\|^2_{\boldsymbol{H}_\gamma^{l+1/2,l/2+1/4}(G_T)} + \sigma^2\|B\|^2_{H_\gamma^{l-1/2,l/2-1/4}(G_T)}\Big\} \equiv c_1 M. \qquad (10.3.1)$$

Proof In order to prove inequality (10.3.1), we first estimate the solution outside some ball containing Ω^+. We multiply equations in (1.1.7) by a smooth function η such that $\eta = 0$ in the ball B_{d-3} and $\eta = 1$ outside the ball B_{d-2}. We introduce the new unknown functions $\boldsymbol{u} = \boldsymbol{v}\eta$ and $q = p\eta$. They are the solution of the following Cauchy problem:

$$\mathcal{D}_t \boldsymbol{u} - \nu^- \nabla^2 \boldsymbol{u} + \frac{1}{\rho^-}\nabla q = \boldsymbol{f}\eta - \nu^-\big(\boldsymbol{v}\nabla^2\eta + 2(\nabla\eta\cdot\nabla)\boldsymbol{v}\big) + \frac{1}{\rho^-}p\nabla\eta \equiv \boldsymbol{f}\eta + \boldsymbol{\Psi},$$
$$\nabla\cdot\boldsymbol{u} = g\eta + \boldsymbol{v}\cdot\nabla\eta \equiv h \qquad \text{in } \mathbb{R}_T^3 \equiv \mathbb{R}^3\times(0,T),$$
$$\boldsymbol{u}|_{t=0} = 0, \qquad \boldsymbol{u}\xrightarrow[|x|\to\infty]{} 0, \qquad q\xrightarrow[|x|\to\infty]{} 0.$$

We apply Theorem 3.3 [82] to this problem in the domain $\mathbb{R}_T^3$ which yields the estimate

$$\|\boldsymbol{u}\|^2_{\boldsymbol{H}_\gamma^{2+l,1+l/2}(\mathbb{R}_T^3)} + \|\nabla q\|^2_{\boldsymbol{H}_\gamma^{l,l/2}(\mathbb{R}_T^3)}$$
$$\leqslant c\Big\{\|\boldsymbol{f}\eta + \boldsymbol{\Psi}\|^2_{\boldsymbol{H}_\gamma^{l,l/2}(\mathbb{R}_T^3)} + \|h\|^2_{H_\gamma^{1+l,0}(\mathbb{R}_T^3)} + \|\boldsymbol{H}\|^2_{\boldsymbol{H}_\gamma^{0,1+l/2}(\mathbb{R}_T^3)}\Big\}, \qquad (10.3.2)$$

where the vector $\boldsymbol{H}$ satisfies the equality $h = \nabla\cdot\boldsymbol{H}$ and is given by the formula

$$\boldsymbol{H}(x,t) = \boldsymbol{R}(x,t)\eta - \frac{1}{4\pi}\int_{\mathbb{R}^3}\big((\boldsymbol{v}-\boldsymbol{R})\cdot\nabla\eta\big)\nabla_x\frac{1}{|x-y|}dy. \qquad (10.3.3)$$

Since $\nabla\eta \neq 0$ only in the ring $K_{d-2} = B_{d-2} \setminus B_{d-3}$, in view of the Hardy-Littlewood-Sobolev inequality [9], we have

$$\|\boldsymbol{H}\|_{\mathbb{R}^3} \leqslant c\Big\{\|\boldsymbol{R}\|_{\Omega^-} + \|(\boldsymbol{v}-\boldsymbol{R})\cdot\nabla\eta\|_{L_{6/5}(K_{d-2})}\Big\} \leqslant c\{\|\boldsymbol{v}\|_{K_{d-2}} + \|\boldsymbol{R}\|_{\Omega^-}\}.$$

Similar inequalities can be obtained by differentiating (10.3.3) with respect to t for the derivatives $\mathcal{D}_t^k\boldsymbol{H}$, if $l/2$ is an integer, $k = l/2+1$, and for the differences

$$\mathcal{D}_t^k\boldsymbol{H}(x,t) - \mathcal{D}_t^k\boldsymbol{H}(x,t-\tau) \text{ if } k = [l/2]+1 < l/2+1.$$

Thus, we see that

$$\|\boldsymbol{H}\|^2_{\boldsymbol{H}_\gamma^{0,1+l/2}(\mathbb{R}^3_T)} \leqslant c\Big\{\|\boldsymbol{v}\|^2_{\boldsymbol{H}_\gamma^{0,1+l/2}(K_{(d-2)T})} + \|\boldsymbol{R}\|^2_{\boldsymbol{H}_\gamma^{0,1+l/2}(Q_T^-)}\Big\}.$$

This inequality together with (10.3.2) gives us the estimate

$$\begin{aligned}
&\|\boldsymbol{v}\|^2_{\boldsymbol{H}_\gamma^{2+l,1+l/2}(\mathbb{R}^3_T\setminus B_{(d-2)T})} + \|\nabla p\|^2_{\boldsymbol{H}_\gamma^{l,l/2}(\mathbb{R}^3_T\setminus B_{(d-2)T})} \\
&\quad\leqslant c\Big\{\|\boldsymbol{f}\|^2_{\boldsymbol{H}_\gamma^{l,l/2}(Q_T^-)} + \|g\|^2_{H_\gamma^{1+l,0}(Q_T^-)} + \|\boldsymbol{R}\|^2_{\boldsymbol{H}_\gamma^{0,1+l/2}(Q_T^-)} \\
&\qquad + \|\boldsymbol{v}\|^2_{\boldsymbol{H}_\gamma^{2+l,1+l/2}(K_{(d-2)T})} + \|p\|^2_{\boldsymbol{H}_\gamma^{l,l/2}(K_{(d-2)T})}\Big\}. \qquad (10.3.4)
\end{aligned}$$

We have here used the fact that $\boldsymbol{\Psi}$ vanishes outside the cylinder $K_{(d-2)T} = K_{d-2}\times(0,T)$.

By applying the partition of unity and following the estimates for interior and boundary small domains carried out in [82] for a single fluid of finite volume, one can arrive at the inequality

$$\begin{aligned}
&\|\boldsymbol{v}\|^2_{\boldsymbol{H}_\gamma^{2+l,1+l/2}(B^-_{(d-1)T}\cup B^+_{d-1T})} + \|\nabla p\|^2_{\boldsymbol{H}_\gamma^{l,l/2}(B^-_{(d-1)T}\cup B^+_{(d-1)T})} \\
&\leqslant c\Big\{\|\boldsymbol{v}\|^2_{\boldsymbol{H}_\gamma^{1+l,(1+l)/2}(B^-_{dT}\cup B^+_{dT})} + \|p\|^2_{H_\gamma^{l,l/2}(B^-_{dT}\cup B^+_{dT})} + \|p\|^2_{H_\gamma^{0,l/2}(G_T\cap B_{dT})}\Big\} + cM,
\end{aligned}$$

which together with (10.3.4) leads to the estimate

$$\begin{aligned}
&\|\boldsymbol{v}\|^2_{\boldsymbol{H}_\gamma^{2+l,1+l/2}(D_T)} + \|\nabla p\|^2_{\boldsymbol{H}_\gamma^{l,l/2}(D_T)} \\
&\quad\leqslant c\Big\{M + \|\boldsymbol{v}\|^2_{\boldsymbol{H}_\gamma^{1+l,(1+l)/2}(\cup B^\pm_{dT})} + \|p\|^2_{H_\gamma^{l,l/2}(\cup B^\pm_{dT})} + \|p\|^2_{H_\gamma^{0,l/2}(G_T\cap B_{dT})}\Big\}.
\end{aligned} \qquad (10.3.5)$$

We shall estimate the pressure p. The interpolation inequalities

$$\|p\|^2_{W^l_2(B^-_d\cup B^+_d)} \leqslant \varepsilon\|\nabla p\|^2_{\boldsymbol{W}^l_2(B^-_d\cup B^+_d)} + c(\varepsilon)\|p\|^2_{B_d},$$
$$\|p\|^2_{\Gamma\cap B_d} \leqslant \varepsilon\|\nabla p\|^2_{B_d} + c(\varepsilon)\|p\|^2_{B_d}$$

imply that

$$\|p\|^2_{H^{l,0}_\gamma(\cup B^\pm_{dT})} + \|p\|^2_{H^{0,l/2}_\gamma(G_T\cap B_{dT})} \leqslant \varepsilon_1\|\nabla p\|^2_{\boldsymbol{H}^{l,l/2}_\gamma(\cup B^\pm_{dT})} + c(\varepsilon_1)\|p\|^2_{H^{0,l/2}_\gamma(B_{dT})}.$$

This means that it is necessary to estimate only the norm $\|p\|^2_{H^{0,l/2}_\gamma(B_{dT})}$. To this end, we consider diffraction problem

$$\frac{1}{\rho^\pm}\Delta p = \nabla\cdot\big(\boldsymbol{f} - \mathcal{D}_t\boldsymbol{v} + \nu^\pm\Delta\boldsymbol{v}\big) = \nabla\cdot\big(\boldsymbol{f} - \mathcal{D}_t\boldsymbol{R} + \nu^\pm\nabla g\big) \equiv \nabla\cdot\boldsymbol{F} \quad \text{in} \quad \Omega^+\cup\Omega^-,$$

$$[p]|_\Gamma = \Big[2\mu^\pm\Big(\frac{\partial\boldsymbol{v}}{\partial\boldsymbol{n}}\Big)_{\boldsymbol{n}}\Big]\Big|_\Gamma - \sigma\boldsymbol{n}\cdot\Delta_\Gamma\int_0^t \boldsymbol{v}\,\mathrm{d}t' - b' - \sigma\int_0^t B\,\mathrm{d}t' \equiv p_0,$$

$$\Big[\frac{1}{\rho^\pm}\frac{\partial p}{\partial\boldsymbol{n}}\Big]\Big|_\Gamma = [f_{\boldsymbol{n}} + \nu^\pm(\Delta\boldsymbol{v})_{\boldsymbol{n}}]|_\Gamma \equiv p_1, \quad p\xrightarrow[|x|\to\infty]{}0,$$

which is a consequence of system (1.1.7) with $\boldsymbol{v}_0 = 0$. The symbol $b_{\boldsymbol{n}}$ denotes the normal component of the vector $\boldsymbol{b}$, i. e., $b_{\boldsymbol{n}} = \boldsymbol{b}\cdot\boldsymbol{n}$.

We extend the functions $\boldsymbol{f}, g, \boldsymbol{R}, \boldsymbol{v}, b', B$ with preservation of class in the interval $t > T$ when $T < \infty$. Taking the Laplace transform

$$\hat{f}(x,s) = \int_0^\infty \mathrm{e}^{-st} f(x,t)\mathrm{d}t, \quad s = \gamma + \mathrm{i}\xi_0,\ \gamma > 0, \tag{10.3.6}$$

we arrive at the problem

$$\frac{1}{\rho^\pm}\Delta\hat{p} = \nabla\cdot\big(\hat{\boldsymbol{f}} - s\hat{\boldsymbol{R}} + \nu^\pm\nabla\hat{g}\big) \equiv \nabla\cdot\hat{\boldsymbol{F}} \quad \text{in} \quad \Omega^+\cup\Omega^-, \tag{10.3.7}$$

$$[\hat{p}]|_\Gamma = \Big[2\mu^\pm\Big(\frac{\partial\hat{\boldsymbol{v}}}{\partial\boldsymbol{n}}\Big)_{\boldsymbol{n}}\Big]\Big|_\Gamma - \frac{\sigma}{s}\boldsymbol{n}\cdot\Delta_\Gamma\hat{\boldsymbol{v}}|_\Gamma - \hat{b}' - \frac{\sigma}{s}\hat{B} \equiv \hat{p}_0, \tag{10.3.8}$$

$$\Big[\frac{1}{\rho^\pm}\frac{\partial\hat{p}}{\partial\boldsymbol{n}}\Big]\Big|_\Gamma = [\hat{f}_{\boldsymbol{n}} + \nu^\pm(\Delta\hat{\boldsymbol{v}})_{\boldsymbol{n}}]|_\Gamma \equiv \hat{p}_1, \quad \hat{p}\xrightarrow[|x|\to\infty]{}0.$$

Below we use the solution Φ of diffraction problem with homogeneous boundary conditions

$$\frac{1}{\rho^\pm}\Delta\Phi = \bar{\hat{p}}\zeta^2 \quad \text{in} \quad \Omega^+\cup\Omega^-,$$

$$[\Phi]|_\Gamma = 0, \tag{10.3.9}$$

$$\left[\frac{1}{\rho^\pm}\frac{\partial\Phi}{\partial \boldsymbol{n}}\right]\Big|_\Gamma = 0.$$

Here ζ is a smooth function with compact support equal to 1 in B_d and to 0 outside B_{d+1}. The bar means complex conjugation.

We multiply equation (10.3.7) by Φ and integrate by parts twice. In view of (10.3.8), (10.3.9) we then obtain

$$\int\limits_{\Omega^+\cup\Omega^-} |\hat{p}|^2\zeta^2\,\mathrm{d}x = \int\limits_\Gamma \frac{\hat{p}_0}{\rho^+}\frac{\partial\Phi^+}{\partial\boldsymbol{n}}\,\mathrm{d}\Gamma + \int\limits_\Gamma \big([\hat{F_{\boldsymbol{n}}}]|_\Gamma - \hat{p}_1\big)\Phi\,\mathrm{d}\Gamma - \int\limits_{\Omega^+\cup\Omega^-} \hat{\boldsymbol{F}}\cdot\nabla\Phi\,\mathrm{d}x. \tag{10.3.10}$$

By virtue of the fact that $\Delta\hat{\boldsymbol{v}} = \nabla(\nabla\cdot\hat{\boldsymbol{v}}) - \nabla\times(\nabla\times\hat{\boldsymbol{v}})$ and $[\hat{R}_{\boldsymbol{n}}]|_\Gamma = 0$, we have

$$\begin{aligned}[\hat{F_{\boldsymbol{n}}}]|_\Gamma - \hat{p}_1 &= \left[\hat{f}_{\boldsymbol{n}} - s\hat{R}_{\boldsymbol{n}} + \nu^\pm\frac{\partial\hat{g}}{\partial\boldsymbol{n}}\right]\Big|_\Gamma - \left[\hat{f}_{\boldsymbol{n}} + \nu^\pm\Big(\frac{\partial\hat{g}}{\partial\boldsymbol{n}} - \big(\nabla\times(\nabla\times\hat{\boldsymbol{v}})\big)_{\boldsymbol{n}}\Big)\right]\Big|_\Gamma \\ &= \big[\nu^\pm\big(\nabla\times(\nabla\times\hat{\boldsymbol{v}})\big)_{\boldsymbol{n}}\big]\big|_\Gamma.\end{aligned}$$

The symbol $\times$ denotes here the vector product. Further, since for an arbitrary vector $\boldsymbol{a}$ and for a closed surface S, the equality $\int\limits_S(\nabla\times\boldsymbol{a})_{\boldsymbol{n}}\,\mathrm{d}S = 0$ holds, we have

$$\int\limits_S \psi(\nabla\times\boldsymbol{a})_{\boldsymbol{n}}\,\mathrm{d}S = \int\limits_S\{(\nabla\times\psi\boldsymbol{a})_{\boldsymbol{n}} - (\nabla\psi\times\boldsymbol{a})_{\boldsymbol{n}}\}\,\mathrm{d}S = -\int\limits_S(\nabla\psi\times\boldsymbol{a})_{\boldsymbol{n}}\,\mathrm{d}S,$$

where ψ is an arbitrary scalar. Because of this remark and continuity of Φ when passing across Γ,

$$\int\limits_\Gamma \Phi\big[\nu^\pm\big(\nabla\times(\nabla\times\hat{\boldsymbol{v}})\big)_{\boldsymbol{n}}\big]\big|_\Gamma\,\mathrm{d}\Gamma = \int\limits_\Gamma \big(\nabla\Phi^+\times[\nu^\pm\nabla\times\hat{\boldsymbol{v}}]|_\Gamma\big)_{\boldsymbol{n}}\,\mathrm{d}\Gamma.$$

Therefore equality (10.3.10) implies that

$$\begin{aligned}\int\limits_{\mathbb{R}^3}|\hat{p}|^2\zeta^2\,\mathrm{d}x &\leqslant \varepsilon_1\Big\|\frac{\partial\Phi^+}{\partial\boldsymbol{n}}\Big\|_\Gamma^2 + c(\varepsilon_1)(\rho^+)^{-2}\|\hat{p}_0\|_\Gamma^2 + \varepsilon_2\|\nabla\Phi\|_{\mathbb{R}^3}^2 + c(\varepsilon_2)\|\hat{\boldsymbol{F}}\|_{\mathbb{R}^3}^2 \\ &+ \varepsilon_3\|\nabla\Phi^+\|_\Gamma^2 + c(\varepsilon_3)\max(\nu^\pm)\|\nabla\hat{\boldsymbol{v}}\|_\Gamma^2, \quad c(\varepsilon_i) = \frac{1}{4\varepsilon_i},\ i=1,2,3.\end{aligned} \tag{10.3.11}$$

To evaluate $\nabla\Phi^+$ on Γ, we use the well-known inequality for the norm on a boundary:

$$\|\nabla\Phi^+\|_\Gamma^2 \leqslant c_2\big(\|\nabla\nabla\Phi\|_{\Omega^+}^2 + \|\nabla\Phi\|_{\Omega^+}^2\big).$$

For the solution Φ^+ of diffraction problem (10.3.9) there hold local estimates of the second derivatives (see [43]):

$$\int\limits_{B_d^+\cup B_d^-} |\nabla\nabla\Phi|^2\,\mathrm{d}x \leqslant c_3\big(\|\hat{p}\zeta\|_{\mathbb{R}^3}^2 + \|\nabla\Phi\|_{\mathbb{R}^3}^2\big).$$

Thus, we see that one only needs to estimate $\|\nabla\Phi\|_{\mathbb{R}^3}^2$. For this we multiply the equation in (10.3.9) by $\bar{\Phi}$ and integrate by parts over $\Omega^+\cup\Omega^-$. Then we obtain the equality $\int\limits_{\Omega^+\cup\Omega^-} \frac{|\nabla\Phi|^2}{\rho^\pm}\,\mathrm{d}x = -\int\limits_{\Omega^+\cup\Omega^-} \bar{\Phi}\bar{\hat{p}}\zeta^2\,\mathrm{d}x$ which implies that

$$\int\limits_{\Omega^+\cup\Omega^-} \frac{|\nabla\Phi|^2}{\rho^\pm}\,\mathrm{d}x \leqslant \varepsilon_4 \int\limits_{\Omega^+\cup\Omega^-} |\Phi|^2\zeta^2\,\mathrm{d}x + c(\varepsilon_4)\int\limits_{\Omega^+\cup\Omega^-} |\hat{p}|^2\zeta^2\,\mathrm{d}x$$
$$\leqslant \varepsilon_4 c_4\|\Phi\|_{L_6(B_{d+1})}^2 + c(\varepsilon_4)\int\limits_{\mathbb{R}^3} |\hat{p}|^2\zeta^2\,\mathrm{d}x \leqslant \varepsilon_4 c_5\|\nabla\Phi\|_{\mathbb{R}^3}^2 + c(\varepsilon_4)\|\hat{p}\zeta\|_{\mathbb{R}^3}^2.$$

We choose $\varepsilon_4 = 1/(2c_5\max(\rho^+,\rho^-))$, then

$$\|\nabla\Phi\|_{\mathbb{R}^3}^2 \leqslant c_6\|\hat{p}\zeta\|_{\mathbb{R}^3}^2, \qquad c_6 = 2\max(\rho^+,\rho^-)c(\varepsilon_4).$$

Now, setting in (10.3.11)

$$\varepsilon_1 = \varepsilon_3 = 1/\big(6c_2(c_3(1+c_6)+c_6)\big), \quad \varepsilon_2 = 1/(6c_6),$$

we arrive at the required inequality

$$\|\hat{p}\zeta\|_{\mathbb{R}^3}^2 \leqslant c_7\big\{\|\hat{p}_0\|_\Gamma^2 + \|\hat{\boldsymbol{F}}\|_{\mathbb{R}^3}^2 + \|\nabla\hat{\boldsymbol{v}}\|_\Gamma^2\big\}.$$

Multiplying it by $|s|^l$, integrating with respect to ξ_0, and using the Parseval equality, we derive the estimate

$$\|p\|_{H_\gamma^{0,l/2}(B_{d\infty})}^2 \leqslant c\Big\{\|\boldsymbol{f}\|_{\boldsymbol{H}_\gamma^{0,l/2}(\mathbb{R}_\infty^3)}^2 + \|\boldsymbol{R}\|_{\boldsymbol{H}_\gamma^{0,l/2+1}(\mathbb{R}_\infty^3)}^2 + \|g\|_{H_\gamma^{1+l,1/2+l/2}(D_\infty)}^2$$
$$+ \|b'\|_{H_\gamma^{0,l/2}(G_\infty)}^2 + \sigma^2\|B\|_{H_\gamma^{0,l/2-1}(G_\infty)}^2 + \sigma^2\|(\Delta_\Gamma\boldsymbol{v})_{\boldsymbol{n}}\|_{H_\gamma^{0,l/2-1}(G_\infty)}^2$$
$$+ \Big\|\Big[\mu^\pm\Big(\frac{\partial\boldsymbol{v}}{\partial\boldsymbol{n}}\Big)_{\boldsymbol{n}}\Big]\Big|_{G_\infty}\Big\|_{H_\gamma^{0,l/2}(G_\infty)}^2 + \|\nabla\boldsymbol{v}\|_{\boldsymbol{H}_\gamma^{0,l/2}(G_\infty)}^2\Big\}. \tag{10.3.12}$$

If $l < 2$, the sum $\sigma^2 \left(\|B\|^2_{H^{0,l/2-1}_\gamma(G_\infty)} + \|(\Delta_\Gamma \boldsymbol{v})_{\boldsymbol{n}}\|^2_{H^{0,l/2-1}_\gamma(G_\infty)}\right)$ in (10.3.12) must be replaced by $\sigma^2\gamma^{l-2}\left(\|\mathrm{e}^{-\gamma t}B\|^2_{G_\infty} + \|\mathrm{e}^{-\gamma t}(\Delta_\Gamma \boldsymbol{v})_{\boldsymbol{n}}\|^2_{G_\infty}\right)$. Trace theorem implies

$$\gamma^{l-2}\|\mathrm{e}^{-\gamma t}(\Delta_\Gamma \boldsymbol{v})_{\boldsymbol{n}}\|^2_{G_\infty} \leqslant \gamma_0^{l-2}\|\boldsymbol{v}^+\|^2_{\boldsymbol{H}^{2,1}_\gamma(G_\infty)} \leqslant \gamma_0^{l-2}\|\boldsymbol{v}\|^2_{\boldsymbol{H}^{5/2,5/4}_\gamma(Q^\pm_\infty)}$$

and

$$\|\nabla \boldsymbol{v}\|^2_{\boldsymbol{H}^{0,l/2}_\gamma(G_\infty)} \leqslant \|\boldsymbol{v}\|^2_{H^{3/2+l,3/4+l/2}_\gamma(\cup B^\pm_{d\infty})}.$$

Finally, using extension theorem for functions in the weighted spaces $H^{\alpha,\beta}_\gamma$ (see [82], Chap. 2), we deduce from (10.3.5), (10.3.12) the inequality

$$\begin{aligned}\|\boldsymbol{v}\|^2_{\boldsymbol{H}^{2+l,1+l/2}_\gamma(D_T)} &+ \|\nabla p\|^2_{\boldsymbol{H}^{l,l/2}_\gamma(D_T)} + \|p\|^2_{H^{0,l/2}_\gamma(B_{dT})}\\ &\leqslant c\big\{M + \|\boldsymbol{v}\|^2_{\boldsymbol{H}^{3/2+j,3/4+j/2}_\gamma(\cup B^\pm_{dT})}\big\}.\end{aligned}$$

Here $j = \max(1, l)$, which is always less than $1/2 + l$, since $l > 1/2$. Hence, the norm of $\boldsymbol{v}$ on the right-hand side is weak. It can be estimated in terms of the norm of $\boldsymbol{v}$ in $\boldsymbol{H}^{2+l,1+l/2}_\gamma(\cup B^\pm_{dT})$ with a small coefficient for a sufficiently large value of the parameter γ ([82], Chap. 4). Thus, we arrive at inequality (10.3.1). □

10.4 L_2-Solvability of the Linearized Problem with Closed Interface in $\mathbb{R}^3$

In this section, we prove the solvability of problem (1.1.7). The proof is based on the above a priori estimates of the solution in the weighted spaces (Theorem 10.3.1). The results were first published in [27] (for detail see also [14]).

As before, we assume that the boundary $\Gamma \in W^{2+l}_2$, i. e., in the neighborhood of any of its points, one can introduce local coordinates $\{y\}$ so that Γ is given by the equation $\{y_3 = \varphi(y_1, y_2)\}$, where $\varphi \in W^{2+l}_2$, $\varphi(0) = 0$, $\nabla\varphi(0) = 0$.

The Laplace–Beltrami operator on Γ is defined by formula (1.1.4):

$$\Delta_\Gamma u = \frac{1}{\sqrt{g}}\frac{\partial}{\partial y_\alpha}\Big(\sqrt{g}g^{\alpha\beta}\frac{\partial u}{\partial y_\beta}\Big),$$

where $\{g^{\alpha\beta}\}^2_{\alpha,\beta=1}$ is the inverse of metric tensor matrix $\{g_{\alpha\beta}\}^2_{\alpha,\beta=1}$. It is given in the form:

$$\{g^{\alpha\beta}\} = \frac{1}{1+|\nabla'\varphi|^2}\begin{pmatrix} 1+\varphi'^{\,2}_{y_2} & -\varphi'_{y_1}\varphi'_{y_2} \\ -\varphi'_{y_1}\varphi'_{y_2} & 1+\varphi'^{\,2}_{y_1} \end{pmatrix},$$

where $\varphi'_{y_\alpha} = \frac{\partial\varphi}{\partial y_\alpha}$, $\nabla'\varphi = (\varphi'_{y_1}, \varphi'_{y_2})$, $g = \det\{g_{\alpha\beta}\} = 1+|\nabla'\varphi|^2$. The matrix $\{g^{\alpha\beta}\}$ is positive definite: $g^{\alpha\beta}\xi_\alpha\overline{\xi}_\beta \geqslant \frac{1}{g}\big(|\xi_1|^2+|\xi_2|^2\big)$ for arbitrary ξ_1, $\xi_2 \in \mathbb{C}$. The bar denotes complex conjugation. The notation $\langle\nabla'\zeta\cdot\nabla'\eta\rangle_\Gamma$ means the Hermitian scalar product of the vectors $\nabla'\zeta$ and $\nabla'\eta$ on the surface Γ; in the local coordinates

$$\langle\nabla'\zeta\cdot\nabla'\eta\rangle_\Gamma = g^{\alpha\beta}\frac{\partial\zeta}{\partial y_\alpha}\overline{\frac{\partial\eta}{\partial y_\beta}}.$$

The outward normal $\boldsymbol{n}$ to the surface Γ near 0 has the coordinates $\frac{1}{\sqrt{g}}\big(\varphi'_{y_1}, \varphi'_{y_2}, -1\big)$.

10.4.1 The Solvability of Problem (1.1.7) in a Simplified Case

Theorem 10.4.1 *We assume that* $\boldsymbol{v}_0 = 0$, $r = 0$, $b' = 0$ *and that, for* $l \in (1/2, 1)$, $0 < T < \infty$ *and* $\gamma \geqslant \gamma_1 \gg 1$, $\Gamma \in W_2^{2+l}$, $\boldsymbol{f} \in \boldsymbol{H}_\gamma^{l,l/2}(D_T)$, $\boldsymbol{b} \in \boldsymbol{H}_\gamma^{l+1/2,l/2+1/4}(G_T)$, $\boldsymbol{b}\cdot\boldsymbol{n} = 0$, $B \in H_\gamma^{l-1/2,l/2-1/4}(G_T)$.

Then there exist unique $\boldsymbol{v} \in \boldsymbol{H}_\gamma^{2+l,1+l/2}(D_T)$ *and* $p \xrightarrow[|x|\to\infty]{} 0$ *such that* $(\boldsymbol{v}, p)$ *is a solution to problem* (1.1.7), $\nabla p \in \boldsymbol{H}_\gamma^{l,l/2}(D_T)$ *and*

$$\begin{aligned} &\|\boldsymbol{v}\|^2_{\boldsymbol{H}_\gamma^{2+l,1+l/2}(D_T)} + \|\nabla p\|^2_{\boldsymbol{H}_\gamma^{l,l/2}(D_T)} + \|p\|^2_{H_\gamma^{0,l/2}(B_{dT})} \\ &\quad\leqslant c_1\Big\{\|\boldsymbol{f}\|^2_{\boldsymbol{H}_\gamma^{l,l/2}(D_T)} + \|\boldsymbol{b}\|^2_{\boldsymbol{H}_\gamma^{l+1/2,l+1/4}(G_T)} + \sigma^2\|B\|^2_{H_\gamma^{l-1/2,l-1/4}(G_T)}\Big\}. \end{aligned} \tag{10.4.1}$$

The proof of this theorem is based on the analysis of a generalized solution to a certain problem, close to (1.1.7). After establishing the existence and smoothness of this solution, one passes to a solution to system (1.1.7).

Namely, we consider the problem:

$$\begin{aligned} &\mathcal{D}_t\boldsymbol{u} - \nu^\pm\nabla^2\boldsymbol{u} + \frac{1}{\rho^\pm}\nabla q = \boldsymbol{f}, \qquad \nabla\cdot\boldsymbol{u} = 0 \ \text{ in } Q_T^\pm, \\ &\boldsymbol{u}|_{t=0} = 0, \qquad [\boldsymbol{u}]|_{G_T} = 0, \qquad \boldsymbol{u}\xrightarrow[|x|\to\infty]{}0, \quad q\xrightarrow[|x|\to\infty]{}0, \\ &[\mathbb{T}\boldsymbol{n} - (\boldsymbol{n}\cdot\mathbb{T}\boldsymbol{n})\boldsymbol{n}]|_{G_T} = \boldsymbol{b}, \quad \boldsymbol{b}\cdot\boldsymbol{n} = 0, \end{aligned} \tag{10.4.2}$$

$$[\boldsymbol{n}\cdot\mathbb{T}\boldsymbol{n}]|_{G_T}=\sigma\int_0^t\Big(n_i\Delta_\Gamma u_i+\langle\nabla' n_i\cdot\nabla' u_i\rangle_\Gamma+B\Big)\,\mathrm{d}t'. \tag{10.4.3}$$

Theorem 10.4.2 (The Solvability of the Auxiliary Problem) *Let the assumptions of Theorem* 10.4.1 *hold. Then problem* (10.4.2), (10.4.3) *has a unique solution* $(\boldsymbol{u},q)$: $\boldsymbol{u}\in\boldsymbol{H}_\gamma^{2+l,1+\frac{l}{2}}(D_T)$, $\nabla q\in\boldsymbol{H}_\gamma^{l,\frac{l}{2}}(D_T)$, *and*

$$\|\boldsymbol{u}\|_{\boldsymbol{H}_\gamma^{2+l,1+l/2}(D_T)}+\|\nabla q\|_{\boldsymbol{H}_\gamma^{l,l/2}(D_T)}+\|q\|_{H_\gamma^{0,l/2}(B_{dT})}\leqslant c(T)\Big\{\|\boldsymbol{f}\|_{\boldsymbol{H}_\gamma^{l,l/2}(D_T)}$$
$$+\|\boldsymbol{b}\|_{\boldsymbol{H}_\gamma^{l+1/2,l/2+1/4}(G_T)}+\sigma\|B\|_{H_\gamma^{l-1/2,l/2-1/4}(G_T)}\Big\},\qquad \gamma\geqslant\gamma_1\gg 1, \tag{10.4.4}$$

with the nondecreasing function $c(T)$.

We give the basic steps of the proof of Theorem 10.4.2.

I. We extend the given functions $\boldsymbol{f}$, $\boldsymbol{b}$, B in t into $(0,\infty)$ and consider system (10.4.2), (10.4.3) in the infinite time interval. We can do this with preservation of class as was shown in [82] for the functions belonging to $H_\gamma^{l,l/2}(Q_T)$.

The Laplace transform (10.3.6) in t converts (10.4.2), (10.4.3) into the stationary problem

$$\begin{aligned}
&s\hat{\boldsymbol{u}}-\nu^\pm\nabla^2\hat{\boldsymbol{u}}+\frac{1}{\rho^\pm}\nabla\hat{q}=\hat{\boldsymbol{f}},\qquad \nabla\cdot\hat{\boldsymbol{u}}=0\ \text{ in }\Omega^\pm,\\
&[\hat{\boldsymbol{u}}]|_\Gamma=0,\quad \hat{\boldsymbol{u}}\xrightarrow[|x|\to\infty]{}0,\quad \hat{q}\xrightarrow[|x|\to\infty]{}0,\\
&[\hat{\mathbb{T}}\boldsymbol{n}-(\boldsymbol{n}\cdot\hat{\mathbb{T}}\boldsymbol{n})\boldsymbol{n}]|_\Gamma=\hat{\boldsymbol{b}},\quad \hat{\boldsymbol{b}}\cdot\boldsymbol{n}=0,
\end{aligned} \tag{10.4.5}$$

$$[\boldsymbol{n}\cdot\hat{\mathbb{T}}\boldsymbol{n}]|_\Gamma=\frac{\sigma}{s}\Big(n_i\Delta_\Gamma\hat{u}_i+\langle\nabla' n_i\cdot\nabla'\hat{u}_i\rangle_\Gamma+\hat{B}\Big). \tag{10.4.6}$$

By a generalized solution of system (10.4.5), (10.4.6), we denote a vector-valued function $\hat{\boldsymbol{u}}\in\boldsymbol{J}^1(\mathbb{R}^3)\equiv\{\boldsymbol{v}\in\boldsymbol{W}_2^1(\mathbb{R}^3):\ \nabla\cdot\boldsymbol{v}=0,\ \boldsymbol{v}\cdot\boldsymbol{n}\in W_2^1(\Gamma)\}$ satisfying the identity

$$Q_s[\hat{\boldsymbol{u}},\boldsymbol{\eta}]\equiv s(\rho^\pm\hat{\boldsymbol{u}},\boldsymbol{\eta})+\frac{1}{2}E[\hat{\boldsymbol{u}},\boldsymbol{\eta}]+\frac{\sigma}{s}\int_\Gamma\Big(\langle\nabla'\hat{u}_{\boldsymbol{n}}\cdot\nabla'\eta_{\boldsymbol{n}}\rangle_\Gamma-\hat{u}_i\langle\nabla' n_i\cdot\nabla'\eta_{\boldsymbol{n}}\rangle_\Gamma\Big)\,\mathrm{d}\Gamma$$
$$=(\rho^\pm\hat{\boldsymbol{f}},\boldsymbol{\eta})+\int_\Gamma\Big(\hat{\boldsymbol{b}}\cdot\bar{\boldsymbol{\eta}}+\frac{\sigma}{s}\hat{B}\bar{\eta}_{\boldsymbol{n}}\Big)\,\mathrm{d}\Gamma \tag{10.4.7}$$

for any $\boldsymbol{\eta}\in\boldsymbol{J}^1(\mathbb{R}^3)$. Here $\eta_{\boldsymbol{n}}=\boldsymbol{\eta}\cdot\boldsymbol{n}$, $(\rho^\pm\boldsymbol{v},\boldsymbol{\eta})=\rho^+\int_{\Omega^+}\boldsymbol{v}\cdot\bar{\boldsymbol{\eta}}\,\mathrm{d}x+\rho^-\int_{\Omega^-}\boldsymbol{v}\cdot\bar{\boldsymbol{\eta}}\,\mathrm{d}x$,

$$E[\boldsymbol{v},\boldsymbol{\eta}] = \int_{\Omega^+\cup\Omega^-} \mu^\pm \left(\frac{\partial v_i}{\partial x_k} + \frac{\partial v_k}{\partial x_i}\right)\left(\overline{\frac{\partial \eta_i}{\partial x_k}} + \overline{\frac{\partial \eta_k}{\partial x_i}}\right) \mathrm{d}x.$$

We consider the expression

$$|\boldsymbol{v}| \equiv (\mathrm{Re}\, Q_s[\boldsymbol{v},\boldsymbol{v}])^{1/2}.$$

For sufficiently large $\gamma = \mathrm{Re}\, s \geqslant \gamma_2 \gg 1$, this norm is equivalent to the norm in $\boldsymbol{J}^1(\mathbb{R}^3)$: $\|\boldsymbol{v}\|_{J^1(\mathbb{R}^3)} = \left(\|\boldsymbol{v}\|^2_{\boldsymbol{W}^1_2(\mathbb{R}^3)} + \|v_{\boldsymbol{n}}\|^2_{W^1_2(\Gamma)}\right)^{1/2}$. This follows from the Korn inequality [80] and the estimates

$$\begin{aligned}
\frac{\sigma}{|s|}\left|\int_\Gamma g^{\alpha\beta} v_i \frac{\partial n_i}{\partial x_\alpha}\overline{\frac{\partial v_{\boldsymbol{n}}}{\partial x_\beta}}\, \mathrm{d}\Gamma\right| &\leqslant \frac{\varepsilon}{|s|^2}\sum_{\alpha=1}^2 \left\|\frac{\partial v_{\boldsymbol{n}}}{\partial x_\alpha}\right\|^2_\Gamma + c_2(\varepsilon)\|v_i\nabla' n_i\|^2_\Gamma \\
&\leqslant \frac{\varepsilon}{|s|^2}\sum_{\alpha=1}^2 \left\|\frac{\partial v_{\boldsymbol{n}}}{\partial x_\alpha}\right\|^2_\Gamma + c_3\|v_i\|^2_{L_r(\Gamma)}\|\nabla'\boldsymbol{n}\|^2_{L_m(\Gamma)} \\
&\leqslant \frac{\varepsilon}{|s|^2}\sum_{\alpha=1}^2 \left\|\frac{\partial v_{\boldsymbol{n}}}{\partial x_\alpha}\right\|^2_\Gamma + \delta\sum_{i=1}^3\left\|\frac{\partial \boldsymbol{v}}{\partial x_i}\right\|^2_{\Omega^+} + c_4(\varepsilon,\delta)\|\boldsymbol{v}\|^2_{\Omega^+},
\end{aligned}$$

where $2 < r < 4$, $m > 4$ are such that $1/r + 1/m = 1/2$. We have used here the Hölder and interpolation inequalities, trace theorem and observed that for $m \leqslant 2/(1-l)$, the norm $\|\nabla'\boldsymbol{n}\|^2_{L_m(\Gamma)}$ is finite by embedding theorem because $\Gamma \in W^{2+l}_2$, $l > 1/2$.

Since the right-hand side of (10.4.7) defines a linear continuous function with respect to the norm $|\cdot|$, the Lax-Milgram theorem [51] implies the existence of a unique generalized solution $\hat{\boldsymbol{u}} \in \boldsymbol{J}^1(\mathbb{R}^3)$. Setting $\boldsymbol{\eta} = \boldsymbol{u}$ in (10.4.7), we obtain an energy estimate

$$\begin{aligned}
&\gamma\min(\rho^\pm)\|\hat{\boldsymbol{u}}\|^2_{\mathbb{R}^3} + \frac{1}{2}E[\hat{\boldsymbol{u}},\hat{\boldsymbol{u}}] + \frac{\sigma\gamma}{|s|^2}\int_\Gamma\sum_{\alpha=1}^2\left|\frac{\partial \hat{u}_{\boldsymbol{n}}}{\partial x_\alpha}\right|^2 \mathrm{d}\Gamma \\
&\quad \leqslant |\hat{\boldsymbol{u}}|^2 \leqslant c_0(\gamma_2)\left\{\|\hat{\boldsymbol{f}}\|^2_{\mathbb{R}^3} + \|\hat{\boldsymbol{b}}\|^2_\Gamma + \frac{\sigma^2}{|s|^2}\|\hat{B}\|^2_\Gamma\right\}, \qquad \gamma \geqslant \gamma_2 \gg 1.
\end{aligned}$$

II. The main technical part of the proof consists in establishing the fact that $\hat{\boldsymbol{u}} \in \boldsymbol{W}^{2+l}_2(\cup\Omega^\pm)$, at least for a dense set of problem data $\hat{\boldsymbol{f}}$, $\hat{\boldsymbol{b}}$, $\hat{B}$, smoother than in the statement of the theorem. By the well-known procedure (see, for example, [42], Chap. III, Sect. 5), one can show that $\hat{\boldsymbol{u}} \in \boldsymbol{W}^{2+l}_2(\cup\Omega^\pm_1)$ for any strictly interior subdomains $\Omega^\pm_1 \subset \Omega^\pm$(i. e., such that $\overline{\Omega^\pm_1} \cap \Gamma = \emptyset$, moreover, Ω^-_1 may be also an unbounded "exterior" domain). In $\Omega^-_1 \cup \Omega^+_1$, $\hat{\boldsymbol{u}}$ satisfies equation (10.4.5) together with some $\nabla\hat{q}^\pm \in \boldsymbol{W}^l_2(\cup\Omega^\pm_1)$.

Now one needs to prove that $\hat{\boldsymbol{u}} \in \boldsymbol{W}_2^{2+l}(\cup U^\pm)$, where $U^\pm = U \cap \Omega^\pm$, U is some neighborhood of an arbitrary point $x^0 \in \Gamma$. We change the local coordinates $\{y\}$ near x^0 according to the formulas $z_1 = y_1,\ z_2 = y_2,\ z_3 = y_3 - \varphi(y_1, y_2)$. Let us assume that the transformation $z = \mathcal{T}y$ maps $U^\pm$ into $B_{d_0}^\pm = \{z \in \mathbb{R}^3 | |z| < d_0, \pm z_3 > 0\}$ and $\Gamma \cap U$ into $K_{d_0} = \{(z_1, z_2) \equiv z' \subset \mathbb{R}^2 | |z'| < d_0\}$. In identity (10.4.7), we pass to the coordinates z, replacing the vector field $\hat{\boldsymbol{u}}$ by $\hat{\boldsymbol{w}} = (\hat{u}_1, \hat{u}_2, \hat{u}_3 - \varphi'_{z_1}\hat{u}_1 - \varphi'_{z_2}\hat{u}_2)$, i. e., $\hat{\boldsymbol{u}} = \mathcal{A}\hat{\boldsymbol{w}}$, where

$$\mathcal{A} = \begin{pmatrix} 1 & 0 & 0 \\ 0 & 1 & 0 \\ \varphi'_{z_1} & \varphi'_{z_2} & 1 \end{pmatrix} \tag{10.4.8}$$

is the Jacobi matrix of the transformation $\overset{-1}{\mathcal{T}}$. Since

$$\frac{\partial}{\partial y_\alpha} \equiv \frac{\partial}{\partial z_\alpha} - \varphi'_{z_\alpha}\frac{\partial}{\partial z_3}, \quad \alpha = 1, 2, \qquad \frac{\partial}{\partial y_3} \equiv \frac{\partial}{\partial z_3}, \tag{10.4.9}$$

the solenoidality of $\hat{\boldsymbol{u}}$ implies the solenoidality of $\hat{\boldsymbol{w}}$. As test vector-functions, we take $\boldsymbol{\xi} = \overset{-1}{\mathcal{A}}\boldsymbol{\eta},\ \ \boldsymbol{\eta} \in \boldsymbol{J}^1(\mathbb{R}^3)$, $\boldsymbol{\eta} = 0$ outside U. As a result, instead of (10.4.7), we obtain

$$Q_s^1[\hat{\boldsymbol{w}}, \boldsymbol{\xi}] = l[\boldsymbol{\xi}] \equiv (\rho^\pm \mathcal{A}^T \hat{\boldsymbol{f}}, \boldsymbol{\eta}) + \int_{K_{d_0}} \hat{b}_\tau \bar{\xi}_\tau \sqrt{g}\,\mathrm{d}z' - \frac{\sigma}{s}\int_{K_{d_0}} \hat{B}\bar{\xi}_3\,\mathrm{d}z' \tag{10.4.10}$$

for arbitrary $\boldsymbol{\xi} \in \overset{\circ}{\boldsymbol{J}}{}^1(B_{d_0}) = \{\boldsymbol{\xi} \in \overset{\circ}{\boldsymbol{W}}{}^1_2(B_{d_0}),\ \xi_3 \in W_2^1(K_{d_0}),\ \nabla \cdot \boldsymbol{\xi} = 0\}$, where $B_{d_0} = B_{d_0}^+ \cup B_{d_0}^-$, $\overset{\circ}{\boldsymbol{W}}{}^1_2(B_{d_0})$ is the closure of the set of smooth vectors $\boldsymbol{C}_0^\infty(B_{d_0})$ in the norm of $\boldsymbol{W}_2^1$. Here

$$Q_s^1[\hat{\boldsymbol{w}}, \boldsymbol{\xi}] \equiv s(\rho^\pm \mathcal{A}\hat{\boldsymbol{w}}, \mathcal{A}\boldsymbol{\xi}) + \frac{1}{2}E^1[\mathcal{A}\hat{\boldsymbol{w}}, \mathcal{A}\boldsymbol{\xi}]$$
$$+\frac{\sigma}{s}\int_{K_{d_0}} g^{\alpha\beta}\left\{\frac{\partial}{\partial z_\alpha}\left(\frac{\hat{w}_3}{\sqrt{g}}\right)\overline{\frac{\partial}{\partial z_\beta}\left(\frac{\xi_3}{\sqrt{g}}\right)} + (\mathcal{A}\hat{\boldsymbol{w}})_i\frac{\partial n_i}{\partial z_\alpha}\overline{\frac{\partial}{\partial z_\beta}\left(\frac{\xi_3}{\sqrt{g}}\right)}\right\}\sqrt{g}\,\mathrm{d}z',$$

$E^1[\boldsymbol{v}, \boldsymbol{\eta}]$ is obtained from $E[\boldsymbol{v}, \boldsymbol{\eta}]$ by replacing the derivatives according to formulas (10.4.9). We note that $E^1[\mathcal{A}\hat{\boldsymbol{w}}, \mathcal{A}\boldsymbol{\xi}] = E[\hat{\boldsymbol{w}}, \boldsymbol{\xi}] + \varepsilon[\hat{\boldsymbol{w}}, \boldsymbol{\xi}]$, where $\varepsilon[\hat{\boldsymbol{w}}, \boldsymbol{\xi}]$ is a bilinear form whose coefficients depend on $\nabla'\varphi$ and are small near zero. Such terms do not appear in the boundary integral, since the functions on the plane K_{d_0} are independent of z_3.

By analogy with [94], one can show that the following lemma is valid.

Lemma 10.4.1 *There exists a function $\hat{q} \in L_2(B_{d_0})$ such that $\int_{B_{d_0}} \hat{q}\,\mathrm{d}z = 0$ and the integral identity*

$$Q_s^1[\hat{\boldsymbol{w}}, \boldsymbol{\psi}] - \int_{B_{d_0}} \hat{q}\ \overline{\nabla \cdot \boldsymbol{\psi}}\,\mathrm{d}z = l[\boldsymbol{\psi}] \tag{10.4.11}$$

holds for every $\boldsymbol{\psi} \in \mathring{\boldsymbol{H}}(B_{d_0}) \equiv \{\boldsymbol{\psi} \in \mathring{\boldsymbol{W}}{}_2^1(B_{d_0}),\ \psi_3 \in W_2^1(K_{d_0})\}$*. For* s *with a sufficiently large real part* $\gamma \geqslant \gamma_2 > 0$, $\hat{q}(s,z)$ *satisfies the inequality*

$$\|\hat{q}\|_{B_{d_0}} \leqslant c_5\Big\{\|\hat{\boldsymbol{w}}\|_{\boldsymbol{W}_2^1(B_{d_0})} + \|\hat{w}_3\|_{W_2^1(K_{d_0})} + \|\hat{\boldsymbol{f}}\|_{B_{d_0}} + \|\hat{\boldsymbol{b}}\|_{K_{d_0}} + \frac{\sigma}{|s|}\|\hat{B}\|_{K_{d_0}}\Big\}. \tag{10.4.12}$$

Proof The bilinear form $Q_s^1[\boldsymbol{v}, \boldsymbol{\eta}]$ is nonsingular, and the norm determined by it $|\boldsymbol{v}|_{B_{d_0}} \equiv (\mathrm{Re}\ Q_s^1[\boldsymbol{v}, \boldsymbol{v}])^{1/2}$ is equivalent to $|\boldsymbol{v}|$ for the vectors $\boldsymbol{v} \in \mathring{\boldsymbol{H}}(B_{d_0})$ defined in the ball B_{d_0} with a small enough radius d_0.

Since the difference of the right and left hand sides in (10.4.10) gives a linear functional continuous with respect to $|\cdot|_{B_{d_0}}$ and equal to 0 on $\mathring{\boldsymbol{J}}{}^1(B_{d_0})$, then, by the Lax–Milgram theorem, there exists $\boldsymbol{\chi} \in \mathring{\boldsymbol{H}}(B_{d_0}) \ominus \mathring{\boldsymbol{J}}{}^1(B_{d_0})$ such that

$$Q_s^1[\hat{\boldsymbol{w}}, \boldsymbol{\psi}] - l[\boldsymbol{\psi}] = Q_s^1[\boldsymbol{\chi}, \boldsymbol{\psi}], \qquad \forall \boldsymbol{\psi} \in \mathring{\boldsymbol{H}}(B_{d_0}).$$

Now we show that $Q_s^1[\boldsymbol{\chi}, \boldsymbol{\psi}] = \int_{B_{d_0}} \hat{q}\ \overline{\nabla \cdot \boldsymbol{\psi}}\,\mathrm{d}z$, where $\hat{q} \in D(B_{d_0}) \equiv \{p \in L_2(B_{d_0}) |\ \int_{B_{d_0}} p\,\mathrm{d}z = 0\}$.

Let $p \in D(B_{d_0})$. Then $\int_{B_{d_0}} p\ \overline{\nabla \cdot \boldsymbol{\psi}}\,\mathrm{d}z$ is a linear functional continuous with respect to $|\cdot|_{B_{d_0}}$. Consequently, there exists $\mathcal{A}p \in \mathring{\boldsymbol{H}}(B_{d_0})$ such that

$$\int_{B_{d_0}} p\ \overline{\nabla \cdot \boldsymbol{\psi}}\,\mathrm{d}z = Q_s^1[\mathcal{A}p, \boldsymbol{\psi}], \qquad \forall \boldsymbol{\psi} \in \mathring{\boldsymbol{H}}(B_{d_0}).$$

The operator $\mathcal{A}$ possesses the properties as follows:

1. $\mathcal{A}$ is bounded because

$$|\mathcal{A}p|_{B_{d_0}}^2 \leqslant \Big|\int_{B_{d_0}} p\ \overline{\nabla \cdot \mathcal{A}p}\,\mathrm{d}z\Big| \leqslant c\|p\|_{B_{d_0}} |\mathcal{A}p|_{B_{d_0}}.$$

2. $\mathcal{A}$ has the bounded inverse operator $\mathcal{A}^{-1}$.
   We are going to show below that, for any $p \in D(B_{d_0})$, there exists $\boldsymbol{P}^*$: $\nabla \cdot \boldsymbol{P}^* = p$, $\boldsymbol{P}^* \in \mathring{\boldsymbol{H}}(B_{d_0})$, $|\boldsymbol{P}^*|_{B_{d_0}} \leqslant \|p\|_{B_{d_0}}$. Then we will have

$$\|p\|_{B_{d_0}}^2 = Q_s^1[\mathcal{A}p, \boldsymbol{P}^*] \leqslant |\mathcal{A}p|_{B_{d_0}} |\boldsymbol{P}^*|_{B_{d_0}} \leqslant c|\mathcal{A}p|_{B_{d_0}} \|p\|_{B_{d_0}}.$$

3. $\overset{\circ}{\boldsymbol{H}}(B_{d_0}) \ominus \overset{\circ}{\boldsymbol{J}}{}^1(B_{d_0}) = \boldsymbol{R}(\mathcal{A})$ is the range of values of the operator $\mathcal{A}$.

We verify that $\overset{\circ}{\boldsymbol{H}}(B_{d_0}) \ominus \boldsymbol{R}(\mathcal{A}) = \overset{\circ}{\boldsymbol{J}}{}^1(B_{d_0})$. Let $\eta_0 \in \overset{\circ}{\boldsymbol{H}}(B_{d_0}) \ominus (B_{d_0})$. Then, for any $p \in D(B_{d_0})$, $\int_{B_{d_0}} p\, \overline{\nabla \cdot \mathcal{A}p}\, \mathrm{d}z \equiv Q_s^1[\mathcal{A}p, \boldsymbol{\eta}_0] = 0$. Consequently, $\nabla \cdot \eta_0 = c_7$. But $c_7|B_{d_0}| = \int_{B_{d_0}} \overline{\nabla \cdot \eta_0}\, \mathrm{d}z = \int_{S_{d_0}} \overline{\mathcal{A}p} \cdot \boldsymbol{n}_{S_{d_0}}\, \mathrm{d}S_{d_0} = 0$, where $S_{d_0} = \partial B_{d_0}$, $\boldsymbol{n}_{S_{d_0}}$ is the outward normal to S_{d_0}. Hence, $\eta_0 \in \overset{\circ}{\boldsymbol{J}}{}^1(B_{d_0})$.

Now we prove the existence of $\boldsymbol{P}^*$ in point 2. We consider some $p \in D(B_{d_0})$.

First, we admit that $\int_{B_{d_0}^-} p\, \mathrm{d}z = 0$ and $\int_{B_{d_0}^+} p\, \mathrm{d}z = 0$. Applying Lemma 2.5 in [45], we conclude that there exists $\boldsymbol{P}_1 \in \overset{\circ}{\boldsymbol{W}}{}_2^1(\cup B_{d_0}^{\pm})$ such that $\nabla \cdot \boldsymbol{P}_1 = p$ and

$$|\boldsymbol{P}_1|_{B_{d_0}} \leqslant c\|P_1\|_{\boldsymbol{W}_2^1(B_{d_0})} \leqslant c\|p\|_{B_{d_0}}. \tag{10.4.13}$$

The general case is deduced to this one by the introduction of the auxiliary function $\zeta' \in C_0^\infty(\cup B_{d_0}^{\pm})$ such that $\zeta'(z', z_3) = -\zeta'(z', -z_3)$, $\int_{B_{d_0}^+} \zeta'\, \mathrm{d}z = 1$. Next, we construct the smooth vector $\boldsymbol{P}_2$ equal to 0 on the boundary S_{d_0} and such that $\nabla \cdot \boldsymbol{P}_2 = \zeta'$ in B_{d_0}. It satisfies the inequality

$$|\boldsymbol{P}_2|_{B_{d_0}} \leqslant c\{\max_{B_{d_0}} |\zeta'| + \max_{B_{d_0}} |\nabla\zeta'|\} \leqslant c_8. \tag{10.4.14}$$

Then the desired vector $\boldsymbol{P}^* = \boldsymbol{P}_1 + a\boldsymbol{P}_2$, where $a = \int_{B_{d_0}^+} p\, \mathrm{d}z$. The norm of $\boldsymbol{P}^*$ is estimated in view of (10.4.13) and (10.4.14) as follows:

$$|\boldsymbol{P}^*|_{B_{d_0}} \leqslant |\boldsymbol{P}_1|_{B_{d_0}} + |a||\boldsymbol{P}_2|_{B_{d_0}} \leqslant (c_9 + c_{10}|B_{d_0}^+|^{1/2})\|p\|_{B_{d_0}}.$$

Thus, the properties of the operator $\mathcal{A}$ imply that $\boldsymbol{\chi} = \mathcal{A}\hat{q}$, where $\hat{q} \in D(B_{d_0})$, and

$$\|\hat{q}\|_{B_{d_0}} \leqslant c|\boldsymbol{\chi}|_{B_{d_0}} \leqslant c\Big\{|\hat{\boldsymbol{w}}|_{B_{d_0}} + |\boldsymbol{F}|_{B_{d_0}}\Big\}.$$

Here $\boldsymbol{F}$ is the element of $\overset{\circ}{\boldsymbol{H}}(B_{d_0})$ such that $l(\boldsymbol{\psi}) = Q_s^1(\boldsymbol{F}, \boldsymbol{\psi})$ for any $\boldsymbol{\psi} \in \overset{\circ}{\boldsymbol{H}}(B_{d_0})$. The last inequality implies the estimate (10.4.12) □

III. Let us continue the proof of Theorem 10.4.2.

We replace $\boldsymbol{\psi} \in \overset{\circ}{\boldsymbol{H}}(B_{d_0})$ by $\boldsymbol{\psi}\zeta$ in (10.4.11), where $\zeta \in C_0^\infty(B_{d_0})$, $\zeta(z) = 1$ for $|z| < \frac{d_0}{2}$, and rewrite (10.4.11) in the form:

$$Q_s^1[\boldsymbol{w}', \boldsymbol{\psi}] - \int_{B_{d_0}} q'\, \overline{\nabla \cdot \boldsymbol{\psi}}\, \mathrm{d}z = l[\boldsymbol{\psi}\zeta] + M[\boldsymbol{\psi}], \tag{10.4.15}$$

where $\boldsymbol{w}' = \hat{\boldsymbol{w}}\zeta$, $q' = \hat{q}\zeta - \frac{1}{|B_{d_0}|}\int_{B_{d_0}} \hat{q}\zeta\, \mathrm{d}z$,

$$M[\boldsymbol{\psi}] = Q_s^1[\hat{\boldsymbol{w}}\zeta, \boldsymbol{\psi}] - Q_s^1[\hat{\boldsymbol{w}}, \boldsymbol{\psi}\zeta] + \int_{B_{d_0}} \hat{q}\, \overline{\boldsymbol{\psi}\cdot\nabla\zeta}\, \mathrm{d}z$$

contains derivatives of the function ζ as multipliers in the integrands. Therefore $M[\boldsymbol{\psi}] = 0$ if $\operatorname{supp}\boldsymbol{\psi} \subset B_d$, $d < \frac{d_0}{2}$. Thus, $\boldsymbol{w}' \in \overset{\circ}{\boldsymbol{H}}(B_{d_0})$, $q' \in L_2(B_{d_0})$, $\int_{B_{d_0}} q'\, \mathrm{d}z = 0$, satisfy integral identity (10.4.15) and the equation $\nabla\cdot\boldsymbol{w}' = \hat{\boldsymbol{w}}\cdot\nabla\zeta$.

IV. In the process of proving the smoothness of $\boldsymbol{w}'$ and q', one should require that $\Gamma \in C^{3+l}$. Therefore, we approximate the function $\varphi \in W^{2+l}(K_{d_0})$ by functions $\varphi^m \in C^{3+l}(K_{d_0})$ satisfying the conditions $\varphi^m(0) = 0$, $\nabla'\varphi^m(0) = 0$. We denote by $Q_s^{(m)}[\boldsymbol{v}, \boldsymbol{\psi}]$ the bilinear form obtained from $Q_s^1[\boldsymbol{v}, \boldsymbol{\psi}]$ by replacing the function φ by φ^m in all the integrals, the functionals $l[\boldsymbol{\psi}\zeta]$ and $M[\boldsymbol{\psi}]$ going over in $l^{(m)}[\boldsymbol{\psi}\zeta]$ and $M^{(m)}[\boldsymbol{\psi}]$, respectively.

We define $\hat{\boldsymbol{w}}^{(m)} \in \overset{\circ}{\boldsymbol{H}}(B_{d_0})$, $\hat{q}^{(m)} \in L_2(B_{d_0})$, $\int_{B_{d_0}} \hat{q}^{(m)}\, \mathrm{d}z = 0$, as a vector field and a function satisfying the equation $\nabla\cdot\hat{\boldsymbol{w}}^{(m)} = \hat{\boldsymbol{w}}\cdot\nabla\zeta$ and the identity

$$Q_s^{(m)}[\hat{\boldsymbol{w}}^{(m)}, \boldsymbol{\psi}] - \int_{B_{d_0}} \hat{q}^{(m)}\, \overline{\nabla\cdot\boldsymbol{\psi}}\, \mathrm{d}z = l^{(m)}[\boldsymbol{\psi}\zeta] + M^{(m)}[\boldsymbol{\psi}], \quad \forall\boldsymbol{\psi} \in \overset{\circ}{\boldsymbol{H}}(B_{d_0}).$$

From here $\hat{\boldsymbol{w}}^{(m)}$, $\hat{q}^{(m)}$ are defined uniquely and, moreover,

$$\|\hat{\boldsymbol{w}}^{(m)}\|^2_{\boldsymbol{W}_2^1(B_{d_0})} + \|\hat{w}_3^{(m)}\|^2_{W_2^1(K_{d_0})} + \|\hat{q}^{(m)}\|^2_{B_{d_0}} \leqslant c_{11}\Big\{\|\hat{\boldsymbol{f}}\|^2_{B_{d_0}} + \|\hat{\boldsymbol{b}}\|^2_{K_{d_0}} + \frac{\sigma^2}{|s|^2}\|\hat{B}\|^2_{K_{d_0}}\Big\}$$

for the ball B_{d_0} with a sufficiently small radius d_0. The last inequality implies that $\{\hat{\boldsymbol{w}}^{(m)}\}_{m=1}^{\infty}$, $\{\hat{q}^{(m)}\}_{m=1}^{\infty}$ contain weakly convergent subsequences: $\hat{\boldsymbol{w}}^{(m)} \rightharpoonup \boldsymbol{w}'$ in $\overset{\circ}{\boldsymbol{H}}(B_{d_0})$, $\hat{q}^{(m)} \rightharpoonup q'$ in $L_2(B_{d_0})$ as $m \to \infty$. Making use of the method developed in [94] and also of inequalities from Chap. 2 of [75], one can show that $\hat{\boldsymbol{w}}^{(m)} \in \boldsymbol{W}_2^{2+l}(\cup B_{d_1}^{\pm})$, $\nabla\hat{q}^{(m)} \in W_2^l(\cup B_{d_1}^{\pm})$ for $d_1 < \frac{d_0}{4}$ and the estimate

$$\begin{aligned}
&|s|^{2+l}\|\hat{\boldsymbol{w}}^{(m)}\|^2_{B_{d_1}} + \|\hat{\boldsymbol{w}}^{(m)}\|^2_{\boldsymbol{W}_2^{2+l}(\cup B_{d_1}^{\pm})} + \|\nabla\hat{q}^{(m)}\|^2_{\boldsymbol{W}_2^l(\cup B_{d_1}^{\pm})} + |s|^l\|\nabla\hat{q}^{(m)}\|^2_{\cup B_{d_1}^{\pm}} \\
&+ |s|^l\|\hat{q}^{(m)}\|^2_{B_{d_1}} \leqslant c_m\Big\{|s|^{2+l}\|\hat{\boldsymbol{f}}\|^2_{B_{d_0}} + \|\hat{\boldsymbol{f}}\|^2_{\boldsymbol{W}_2^l(\cup B_{d_0}^{\pm})} + |s|^{2+l}\|\hat{\boldsymbol{b}}\|^2_{K_{d_0}} + |s|\|\hat{\boldsymbol{b}}\|^2_{\boldsymbol{W}_2^l(K_{d_0})} + \\
&\quad + \|\hat{\boldsymbol{b}}\|^2_{\boldsymbol{W}_2^{1+l}(K_{d_0})} + \sigma^2\big(|s|^l\|\hat{B}\|^2_{K_{d_0}} + \|\hat{B}\|^2_{W_2^l(K_{d_0})}\big)\Big\}, \quad \gamma \geqslant \gamma_2, \qquad (10.4.16)
\end{aligned}$$

holds (in general, the constant c_m increases unboundedly as $m \to \infty$).

We take the inverse Laplace transform and introduce the functions

$$\boldsymbol{w}^{(m)}(z,t)=\frac{1}{2\pi \mathrm{i}}\int_{\gamma-\mathrm{i}\infty}^{\gamma+\mathrm{i}\infty}\hat{\boldsymbol{w}}^{(m)}\mathrm{e}^{st}\,\mathrm{d}s,\quad q^{(m)}(z,t)=\frac{1}{2\pi \mathrm{i}}\int_{\gamma-\mathrm{i}\infty}^{\gamma+\mathrm{i}\infty}\hat{q}^{(m)}\mathrm{e}^{st}\,\mathrm{d}s.$$

Since the given functions on the right-hand sides of system (10.4.2), (10.4.3) belong to the spaces $H_\gamma^{\beta,\beta/2}$, it follows that these functions admit zero extensions with preservation of class into the interval $t<0$. Consequently, the Laplace transform of them are analytic with respect to s, $\gamma>0$. The arguments in [3] allow us to conclude that $\hat{\boldsymbol{w}}^{(m)}$, $\hat{q}^{(m)}$ obtained are analytic too. Then the Paley-Wiener theorem implies that $\boldsymbol{w}^{(m)}$, $q^{(m)}$ vanish for $t\leqslant 0$. Therefore, integrating (10.4.16) along the line $\mathrm{Re}s=\gamma\geqslant\gamma_2$, we can deduce that $\boldsymbol{w}^{(m)}\in \boldsymbol{H}_\gamma^{2+l,1+l/2}(\cup B_{d\infty}^{\pm})$, $\nabla q^{(m)}\in H_\gamma^{l,l/2}(\cup B_{d\infty}^{\pm})$, $B_{d\infty}^{\pm}=B_d^{\pm}\times(0,\infty)$, for a dense set of problem data $\boldsymbol{f}$, $\boldsymbol{b}$ and B, smoother than ones in theorem statement.

V. We pass from the vector-valued functions $\boldsymbol{w}^{(m)}$ to $\boldsymbol{u}^{(m)}$ by formula $\boldsymbol{u}^{(m)}=\mathcal{A}_m\boldsymbol{w}^{(m)}$, where the matrix $\mathcal{A}_m$ is obtained from $\mathcal{A}$, defined by (10.4.8), by replacing φ by $\varphi^{(m)}$. The functions $\boldsymbol{u}^{(m)}$ satisfy the equations

$$\rho^{\pm}\mathcal{D}_t u_i^{(m)}-\mu^{\pm}\nabla^2 u_i^{(m)}+\frac{\partial q^{(m)}}{\partial z_i}=\rho^{\pm}f_i-\mu^{\pm}\Big((\nabla'\varphi_i^{(m)})^2\frac{\partial^2 u_i^{(m)}}{\partial z_3^2}-2\varphi_{z_\alpha}^{(m)\prime}\frac{\partial^2 u_i^{(m)}}{\partial z_\alpha\partial z_3}$$
$$-\Delta'\varphi^{(m)}\frac{\partial u_i^{(m)}}{\partial z_3}\Big)+\delta_i^\beta\varphi_{z_\beta}^{(m)\prime}\frac{\partial q^{(m)}}{\partial z_3},\quad i=1,2,3,\tag{10.4.17}$$

$$\nabla\cdot\boldsymbol{u}^{(m)}=\varphi_{z_\alpha}^{(m)\prime}\frac{\partial u_\alpha^{(m)}}{\partial z_3}\quad\text{in } B^-_{\frac{d_0}{4}\infty}\cup B^+_{\frac{d_0}{4}\infty}$$

and also the initial and boundary conditions

$$\boldsymbol{u}^{(m)}\big|_{t=0}=0\ \text{ in } B^-_{d_0/4}\cup B^+_{d_0/4},\quad [\boldsymbol{u}^{(m)}]\big|_{K_{d_0/4\infty}}=0,$$

$$-\Big[\mu^{\pm}\Big(\frac{\partial u_\alpha^{(m)}}{\partial z_3}+\frac{\partial u_3^{(m)}}{\partial z_\alpha}\Big)\Big]\Big|_{K_{d_0/4\infty}}=-\Big[\mu^{\pm}\Big\{\varphi_{z_\beta}^{(m)\prime}\Big(\frac{\partial u_\alpha^{(m)}}{\partial z_\beta}+\frac{\partial u_\beta^{(m)}}{\partial z_\alpha}\Big)$$
$$+\varphi_{z_\alpha}^{(m)\prime}\Big(2\frac{\partial u_\alpha^{(m)}}{\partial z_\alpha}-g_m\frac{\partial u_3^{(m)}}{\partial z_3}\Big)+\varphi_{z_\beta}^{(m)\prime}\Big(\varphi_{z_\alpha}^{(m)\prime}\frac{\partial u_3^{(m)}}{\partial z_\beta}-\varphi_{z_\beta}^{(m)\prime}\frac{\partial u_\alpha^{(m)}}{\partial z_3}\Big)\tag{10.4.18}$$
$$+(\varphi_{z_\alpha}^{(m)\prime})^2\Big(\frac{\partial u_3^{(m)}}{\partial z_\alpha}-\frac{\partial u_\alpha^{(m)}}{\partial z_3}\Big)\Big\}\Big]\Big|_{K_{d_0/4\infty}}+b_\alpha\sqrt{(1+(\varphi_{z_\alpha}^{(m)\prime})^2)g_m},\ \alpha,\beta=1,2,\ \alpha\neq\beta,$$

$$\Big[-q^{(m)}+2\mu^{\pm}\frac{\partial u_3^{(m)}}{\partial z_3}\Big]\Big|_{K_{d_0/4\infty}}+\sigma\int_0^t\Delta' u_3^{(m)}\big|_{K_{d_0/4\infty}}\,\mathrm{d}\tau=2\Big[\frac{\mu^{\pm}}{g_m}\sum_\beta\varphi_{z_\beta}^{(m)\prime}\Big(g_m\frac{\partial u_\beta^{(m)}}{\partial z_3}$$

$$+\frac{\partial u_3^{(m)}}{\partial z_\beta}-\sum_\alpha \varphi_{z_\alpha}^{(m)\prime}\frac{\partial u_\beta^{(m)}}{\partial z_\alpha}\Big)\Big]\Big|_{K_{d_0/4\infty}}+\sigma\int\limits_0^t\Big\{B+\frac{1}{g_m}\Big\{\frac{\partial^2 u_3^{(m)}}{\partial z_\alpha\partial z_\beta}\varphi_{z_\alpha}^{(m)\prime}\varphi_{z_\beta}^{(m)\prime}$$

$$-2\frac{\partial^2 u_3^{(m)}}{\partial z_3\partial z_\alpha}\varphi_{z_\alpha}^{(m)\prime}\big((\varphi_{z_\alpha}^{(m)\prime})^2+\varphi_{z_1}^{(m)\prime}\varphi_{z_2}^{(m)\prime}\big)+\frac{\partial^2 u_3^{(m)}}{\partial z_3^2}\sum_{\alpha,\beta}(\varphi_{z_\alpha}^{(m)\prime})^2(\varphi_{z_\beta}^{(m)\prime})^2$$

$$-\frac{\partial u_3^{(m)}}{\partial z_3}\varphi_{z_\alpha z_\beta}^{(m)\prime\prime}\varphi_{z_\alpha}^{(m)\prime}\varphi_{z_\beta}^{(m)\prime}-\big(\sqrt{g_m}g_m^{\alpha\beta}\big)'_{z_\alpha}\Big(\frac{\partial u_3^{(m)}}{\partial z_\beta}-\varphi_{z_\beta}^{(m)\prime}\frac{\partial u_3^{(m)}}{\partial z_3}\Big)$$

$$+\varphi_{z_\beta}^{(m)\prime}\Big[\frac{\partial}{\partial z_\alpha}\Big(\sqrt{g_m}g_m^{\alpha\mu}\Big(\frac{\partial u_\beta^{(m)}}{\partial z_\mu}-\varphi_{z_\mu}^{(m)\prime}\frac{\partial u_\beta^{(m)}}{\partial z_3}\Big)\Big)$$

$$-\varphi_{z_\beta}^{(m)\prime}\sqrt{g_m}g_m^{\alpha\mu}\Big(\frac{\partial^2 u_\beta^{(m)}}{\partial z_3\partial z_\mu}-\varphi_{z_\mu}^{(m)\prime}\frac{\partial^2 u_\beta^{(m)}}{\partial z_3^2}\Big)\Big]\Big\}$$

$$+\frac{|\nabla'\varphi^{(m)}|^2 g_m^{\alpha\beta}}{g_m+\sqrt{g_m}}\Big(\frac{\partial^2 u_3^{(m)}}{\partial z_\alpha\partial z_\beta}-\varphi_{z_\alpha}^{(m)\prime}\frac{\partial}{\partial z_3}\Big(\frac{\partial u_3^{(m)}}{\partial z_\beta}-\varphi_{z_\beta}^{(m)\prime}\frac{\partial u_3^{(m)}}{\partial z_3}\Big)\Big)$$

$$+g_m^{\alpha\beta}\Big(\frac{\partial u_i^{(m)}}{\partial z_\alpha}-\varphi_{z_\alpha}^{(m)\prime}\frac{\partial u_i^{(m)}}{\partial z_3}\Big)\frac{\partial n_i^{(m)}}{\partial z_\beta}\Big\}\Big|_{K_{(d_0/4)\infty}}\mathrm{d}\tau, \tag{10.4.19}$$

here $K_{d\infty}=K_d\times(0,\infty)$, $\Delta'=\frac{\partial^2}{\partial z_1^2}+\frac{\partial^2}{\partial z_2^2}$. In addition, we have used the relation

$$\Delta' u_3+\boldsymbol{n}\cdot\Delta_\Gamma\boldsymbol{u}+\langle\nabla' n_i\cdot\nabla' u_i\rangle_\Gamma=(\Delta'-\Delta_\Gamma)u_3+(1+n_3)\Delta_\Gamma u_3+n_\beta\Delta_\Gamma u_\beta+g^{\alpha\beta}\frac{\partial u_i}{\partial y_\alpha}\frac{\partial n_i}{\partial y_\beta}.$$

We note also that we write only two tangential boundary conditions from three ones since the third condition is a linear combination of two first ones.

Multiplying system (10.4.17), (10.4.18), and (10.4.19) by a cutoff function and extending it by zero into the entire space, we arrive at a problem in $\mathbb{R}^3_\infty$ with a plane interface between two media. To it, we can apply the results of § 10.2, in particular Theorem 10.2.2. We estimate the norms of function products in $H_\gamma^{l,l/2}$ by the inequalities proved in [82] (the Corollary to Lemma 4.1). We state them in lemma form.

Lemma 10.4.2 *For any $a\in W_2^{1+l}(\Omega)$, $b\in W_2^l(\Omega)$, $f\in H_\gamma^{l,l/2}(Q_T)$, $g\in H_\gamma^{1+l,(1+l)/2}(Q_T)$, where Ω is a bounded domain, $Q_T\equiv\Omega\times(0,T)$, $l>1/2$, thc incqualities*

$$\|af\|_{H_\gamma^{l,l/2}(Q_T)}\leqslant\big\{\tilde c_1\sup_\Omega|a(x)|+(\varepsilon+\tilde c_2(\varepsilon)\gamma^{-l/2})\|a\|_{W_2^{1+l}(\Omega)}\big\}\|f\|_{H_\gamma^{l,l/2}(Q_T)},$$

$$\|bg\|_{H_\gamma^{l,l/2}(Q_T)}\leqslant\tilde c_3(\varepsilon+\tilde c_4(\varepsilon)\gamma^{-1/2})\|b\|_{W_2^l(\Omega)}\|g\|_{H_\gamma^{1+l,\frac{1+l}{2}}(Q_T)},$$

$$\|ag\|_{H_\gamma^{1+l,\frac{1+l}{2}}(Q_T)}\leqslant\big\{\tilde c_5\sup_\Omega|a(x)|+(\varepsilon+\tilde c_6(\varepsilon)\gamma^{-\frac{1+l}{2}})\|a\|_{W_2^{1+l}(\Omega)}\big\}\|g\|_{H_\gamma^{1+l,\frac{1+l}{2}}(Q_T)}$$

hold.

We choose the ball B_{d_0} with a sufficiently small radius. Then the method described in [47] (Chap. IV, § 10) (see also the proof to Prop. 6.4.2 where a similar procedure has been applied with respect to time variable) allows us to obtain the estimate for $\gamma \geqslant \gamma_2 \geqslant \gamma_1$:

$$\|\boldsymbol{u}^{(m)}\|_{\boldsymbol{H}_\gamma^{2+l,1+l/2}(\cup B^{\pm}_{\frac{d_0}{4}\infty})} + \|\nabla q^{(m)}\|_{\boldsymbol{H}_\gamma^{l,l/2}(\cup B^{\pm}_{\frac{d_0}{4}\infty})} + \|q^{(m)}\|_{H_\gamma^{0,l/2}(B_{\frac{d_0}{4}\infty})} \leqslant$$
$$\leqslant c_{12}(\gamma_1, \|\varphi\|_{W_2^{2+l}}) \Big\{ \|\boldsymbol{f}\|_{\boldsymbol{H}_\gamma^{l,l/2}(\cup U_T^{\pm})} + \|\boldsymbol{b}\|_{\boldsymbol{H}_\gamma^{l+1/2,l/2+1/4}(G_T \cap U_T)} +$$
$$+ \sigma \|B\|_{H_\gamma^{l-1/2,l/2-1/4}(G_T \cap U_T)} \Big\}, \qquad (10.4.20)$$

in which the constant c_{12} does not depend on m, $U_T = U \times (0,T)$.

VI. Passing to the limit as $m \to \infty$ in system (10.4.17)–(10.4.19) and in inequality (10.4.20), we can see that the solution $(\boldsymbol{u}, q)$ of problem (10.4.2), (10.4.3) is smooth near an arbitrary point $x^0 \in \Gamma$, i.e., $\boldsymbol{u} \in H_\gamma^{2+l,1+l/2}(\cup B^{\pm}_{dT})$, $\nabla q \in H_\gamma^{l,l/2}(\cup B^{\pm}_{dT})$, $d \leqslant d_0/4$, and for it estimate (10.4.20) holds. Similar estimates for domains not intersecting the boundary Γ follows from [82].

As mentioned in paragraph II, there are $\nabla \hat{q}^+ \in \boldsymbol{L}_2(\Omega^{+\prime})$ and $\nabla \hat{q}^- \in \boldsymbol{L}_2(\Omega^{-\prime})$ separately for the inner and outer subregions of $\Omega^{\pm\prime} \subset \Omega^{\pm}$. After the smoothness of $\hat{\boldsymbol{u}}$ is proved, this can be deduced from integral identity (10.4.7) for $\boldsymbol{\eta} \in \boldsymbol{J}^1(\Omega^{\pm\prime})$ using the orthogonal decomposition of $\boldsymbol{L}_2(\Omega^{\pm\prime})$. The condition $\hat{q} \xrightarrow[|x| \to \infty]{} 0$ implies that $\hat{q}^-$ is unique.

Since the pressure functions $\hat{q}$ for a fixed s, corresponding to different regions U, can differ at their intersection only by a constant, we can construct a single pressure $\hat{q}$ for the whole space. We choose $\hat{q}^+$ so that $\hat{q}^+ - \hat{q}' = \hat{q}^- - \hat{q}'$ for $\hat{q}'(\mathcal{T}y)$ constructed in Lemma 10.4.1 and corresponding to a certain region U', $U' \cap \Gamma \neq \emptyset$. Then, choosing a finite covering of Γ by bounded domains U_k, $k = 1, \dots, N$, $U_k \cap \Omega^{\pm\prime} \neq \emptyset$, we build $\hat{q}$ as follows: $\hat{q} = \hat{q}^+$ in $\Omega^{+\prime}$, $\hat{q} = \hat{q}^-$ in $\Omega^{-\prime}$ and $\hat{q} = \hat{q}_k + \hat{c}_k$ in U_k, where $\hat{c}_k = \hat{q}^- - \hat{q}_k$ in $U_k \cap \Omega^{-\prime}$, $\hat{q}_k$ is the pressure corresponding to the region U_k, $\int_{U_k} \hat{q}_k \, dy = 0$. Such a function $\hat{q}$ is unique, it does not depend on the choice of the region U'.

After the inverse Laplace transform, estimates (10.4.20) for U_k, $k = 1, \dots, N$, and the results in [82] for the Cauchy problem imply that $q \in H_{\gamma,loc}^{0,l/2}(\mathbb{R}_T^3)$, $\nabla q \in \boldsymbol{H}_\gamma^{l,l/2}(\cup Q_T^{\pm})$ and inequality (10.4.4) holds. An estimate of the functions $c_k(t)$ in $H_\gamma^{0,l/2}(U_{k,T})$ in terms of the norms of the given functions follows from inequality (10.4.20) for q_k and q^- and the equalities

$$\int_{U_k \cap \Omega^{-\prime}} c_k^2 \, dy = \int_{U_k \cap \Omega^{-\prime}} (q^- - q_k)^2 \, dy.$$

This remark completes the proof of Theorem 10.4.2. □

Proof of Theorem 10.4.1 Now, using iterations, we construct a solution $(\boldsymbol{v}, p)$ of original problem (1.1.7), and as the zero approximation we take the solution of auxiliary problem (10.4.2), (10.4.3): $\boldsymbol{v}^0 = \boldsymbol{u}$, $p^0 = q$. The successive approximations $(\boldsymbol{v}^{k+1}, p^{k+1})$, $k = 0, 1, \dots,$ are constructed as satisfying system (10.4.2) and the boundary condition

$$[\boldsymbol{n} \cdot \mathbb{T}(\boldsymbol{v}^{k+1}, p^{k+1})\boldsymbol{n}]\big|_{G_T} = \sigma \int_0^t \Big(n_i \Delta_\Gamma \boldsymbol{v}^{k+1} + \langle \nabla' n_i \cdot \nabla' v_i^{k+1} \rangle_\Gamma - \langle \nabla' n_i \cdot \nabla' v_i^k \rangle_\Gamma + B\Big)\, dt'. \qquad (10.4.21)$$

On the basis of Theorem 10.4.2 and embedding theorems for the spaces $H_\gamma^{\beta,\beta/2}$ [82], we conclude that $\nabla' v_i^0 \in \boldsymbol{H}_\gamma^{l+1/2,l/2+1/4}(G_T)$, $i = 1, 2, 3$. Using the theorems on the continuation of functions from the boundary into the domain with preservation of the class, as well as Lemma 10.4.2, we verify that $\langle \nabla' n_i \cdot \nabla' v_i^0 \rangle_\Gamma \in H_\gamma^{l-1/2,l/2-1/4}(G_T)$, because $\nabla' n_i \in \boldsymbol{W}_2^l$. Therefore, from Theorem 10.4.2 we deduce by induction that $(\boldsymbol{v}^{k+1}, p^{k+1})$ exist, $\boldsymbol{v}^{k+1} \in H_\gamma^{2+l,1+l/2}(D_T)$, $\nabla p^{k+1} \in H_\gamma^{l,l/2}(D_T)$, and they satisfy inequality (10.4.4) with constant $c(k)$.

Next, considering the differences $\boldsymbol{v}^{k+1} - \boldsymbol{v}^k$, we arrive at inequality chain

$$\begin{aligned} &\|\boldsymbol{v}^{k+1} - \boldsymbol{v}^k\|_{\boldsymbol{H}_\gamma^{2+l,1+l/2}(D_T)} + \|\nabla(p^{k+1} - p^k)\|_{\boldsymbol{H}_\gamma^{l,l/2}(D_T)} \\ &\leqslant c\sigma \|\langle \nabla' n_i \cdot \nabla'(v_i^k - v_i^{k-1})\rangle_\Gamma\|_{\boldsymbol{H}_\gamma^{l-1/2,l/2-1/4}(G_T)} \\ &\leqslant c_{13} \|\langle \nabla' n_i \cdot \nabla'(v_i^k - v_i^{k-1})\rangle_\Gamma\|_{\boldsymbol{H}_\gamma^{l,l/2}(B_{dT})} \leqslant c_{14}(\gamma) \|\boldsymbol{v}^k - \boldsymbol{v}^{k-1}\|_{\boldsymbol{H}_\gamma^{2+l,1+l/2}(D_T)}, \end{aligned}$$

where $c_{14}(\gamma)$ is independent of k and can be chosen less than 1 for sufficiently large $\gamma \geqslant \gamma_3 > \gamma_2$ (see Lemma 10.4.2). We have used here the fact that $\nabla' n_i$ depends only the tangential variables and admits the continuation into B_{dT} in the class W_2^l. Due to the completeness of the space $H_\gamma^{l,l/2}$, the Cauchy convergence of the consequences $\{\boldsymbol{v}^k\}_{k=0}^\infty$, $\{\nabla p^k\}_{k=0}^\infty$ implies the existence of $\boldsymbol{v} = \lim_{k\to\infty} \boldsymbol{v}^k$, $\boldsymbol{P} = \lim_{k\to\infty} \nabla p^k$.

The vector space $\boldsymbol{L}_2$ is decomposable into the direct sum $\mathring{\boldsymbol{J}} \oplus \boldsymbol{G}$, where $\mathring{\boldsymbol{J}}$ is the closure of compactly supported infinitely differentiable solenoid vectors, and $\boldsymbol{G}$ consists of $\nabla\eta$, where η is a function whose first derivatives belong to L_2. It follows from this orthogonal expansion that there exists a function $p \in L_{2,loc}$ such that $\nabla p = \boldsymbol{P}$ (see [42], Ch. I, § 2). (For $T < \infty$, $\boldsymbol{H}_\gamma^{l,l/2}(Q_T^\pm) \subset \boldsymbol{L}_2(Q_T^\pm)$.)

Thus, $\boldsymbol{v} \in H_\gamma^{2+l,1+l/2}(D_T)$, $\nabla p \in H_\gamma^{l,l/2}(D_T)$. Passing to the limit as $k \to \infty$ in system (10.4.2), (10.4.21), we verify that $(\boldsymbol{v}, p)$ is a solution to problem (1.1.7). Inequality (10.4.1) for $(\boldsymbol{v}, p)$ follows from a priori estimate of a solution (10.3.1). Theorem 10.4.1 is proved. □

10.4.2 The Case of Nonzero r, b' and $\boldsymbol{v}_0 = 0$

In case of $r \neq 0$ and $b' \neq 0$, a solution to (1.1.7) with $\boldsymbol{v}_0 = 0$ can be found in the form $\boldsymbol{v} = \boldsymbol{v}^{(1)} + \boldsymbol{v}^{(2)}$, $p = p^{(1)} + p^{(2)}$, where $\boldsymbol{v}^{(1)} = \nabla\Phi$, Φ is a solution of diffraction problem

$$\begin{aligned} &\Delta\Phi = r \ \text{ in } \Omega^{\pm}, \\ &[\Phi]|_{\Gamma} = \Big[\frac{\partial\Phi}{\partial\boldsymbol{n}}\Big]\Big|_{\Gamma} = 0, \end{aligned} \tag{10.4.22}$$

$p^{(1)}$ is a solution of the Dirichlet problem in two domains

$$\begin{aligned} &\Delta p_+^{(1)} = 0 \ \text{ in } \Omega^+, \qquad p_+^{(1)}|_{\Gamma} = -b' + 2\mu^+\boldsymbol{n}\cdot\frac{\partial\boldsymbol{v}^{(1)+}}{\partial\boldsymbol{n}}\Big|_{\Gamma}, \quad \text{and} \\ &\Delta p_-^{(1)} = 0 \ \text{ in } \Omega^-, \qquad p_-^{(1)}|_{\Gamma} = 2\mu^-\boldsymbol{n}\cdot\frac{\partial\boldsymbol{v}^{(1)-}}{\partial\boldsymbol{n}}\Big|_{\Gamma}, \quad p_-^{(1)}\xrightarrow[|x|\to\infty]{}0, \end{aligned} \tag{10.4.23}$$

and $(\boldsymbol{v}^{(2)}, p^{(2)})$ is a solution to the system

$$\begin{aligned} &\mathcal{D}_t\boldsymbol{v}^{(2)} - \nu^{\pm}\nabla^2\boldsymbol{v}^{(2)} + \frac{1}{\rho^{\pm}}\nabla p^{(2)} = \boldsymbol{f} - \mathcal{D}_t\boldsymbol{v}^{(1)} + \nu^{\pm}\nabla^2\boldsymbol{v}^{(1)} - \frac{1}{\rho^{\pm}}\nabla p_{\pm}^{(1)}, \\ &\qquad \nabla\cdot\boldsymbol{v}^{(2)} = 0 \ \text{ in } Q_T^{\pm}, \\ &\boldsymbol{v}^{(2)}|_{t=0} = 0, \qquad [\boldsymbol{v}^{(2)}]|_{G_T} = 0, \quad \boldsymbol{v}^{(2)}\xrightarrow[|x|\to\infty]{}0, \quad p^{(2)}\xrightarrow[|x|\to\infty]{}0, \\ &\big[\mathbb{T}(\boldsymbol{v}^{(2)})\boldsymbol{n} - (\boldsymbol{n}\cdot\mathbb{T}(\boldsymbol{v}^{(2)})\boldsymbol{n})\boldsymbol{n}\big]\big|_{G_T} = \boldsymbol{b} - \big[\mathbb{T}(\boldsymbol{v}^{(1)})\boldsymbol{n} - (\boldsymbol{n}\cdot\mathbb{T}(\boldsymbol{v}^{(1)})\boldsymbol{n})\boldsymbol{n}\big]\big|_{G_T}, \\ &\big[\boldsymbol{n}\cdot\mathbb{T}(\boldsymbol{v}^{(2)}, p^{(2)})\boldsymbol{n}\big]\big|_{G_T} - \sigma\boldsymbol{n}\cdot\int_0^t \Delta_{\Gamma}\boldsymbol{v}^{(2)}\,\mathrm{d}t' = \sigma\int_0^t \big(B + \boldsymbol{n}\cdot\Delta_{\Gamma}\boldsymbol{v}^{(1)}\big)\,\mathrm{d}t'. \end{aligned} \tag{10.4.24}$$

Theorem 10.4.3 (The Solvability of the Nonhomogeneous Problem in the Weighted Spaces) *Let the hypotheses of Theorem* 10.4.1 *hold, except the assumptions on r and b', and let, for $\gamma \geqslant \gamma_2$, $r \in H_{\gamma}^{1+l,1/2+l/2}(D_T)$, $r = \nabla\cdot\boldsymbol{R}$, $\boldsymbol{R} \in \boldsymbol{H}_{\gamma}^{0,1+l/2}(\mathbb{R}_T^3)$, $[\boldsymbol{R}\cdot\boldsymbol{n}]|_{G_T} = 0$ in the weak sense,*[3] *$b' \in H_{\gamma}^{l+1/2,1/2,l/2}(G_T)$.*

Then problem (1.1.7) *with $p\xrightarrow[|x|\to\infty]{}0$ uniquely solvable and the solution $(\boldsymbol{v}, p)$ satisfies the inequality*

[3] I. e., $\int_{\mathbb{R}^3} r(x,t)\eta(x)\,\mathrm{d}x = -\int_{\mathbb{R}^3}\boldsymbol{R}\cdot\nabla\eta\,\mathrm{d}x$ for any $\eta \in C_0^{\infty}(\mathbb{R}^3)$ and a.e. $t \in (0,T)$.

$$\|\boldsymbol{v}\|_{\boldsymbol{H}_\gamma^{2+l,1+\frac{l}{2}}(D_T)} + \|\nabla p\|_{\boldsymbol{H}_\gamma^{l,\frac{l}{2}}(D_T)} + \|p\|_{H_\gamma^{0,l/2}(B_{dT})}$$
$$\leqslant c\Big\{\|\boldsymbol{f}\|_{\boldsymbol{H}_\gamma^{l,\frac{l}{2}}(D_T)} + \|r\|_{\boldsymbol{H}_\gamma^{1+l,\frac{1+l}{2}}(D_T)} + \|\boldsymbol{R}\|_{\boldsymbol{H}_\gamma^{0,1+\frac{l}{2}}(\mathbb{R}^3_T)}$$
$$+ \|\boldsymbol{b}\|_{\boldsymbol{H}_\gamma^{l+\frac{1}{2},\frac{l}{2}+\frac{1}{4}}(G_T)} + \|b'\|_{H_\gamma^{l+\frac{1}{2},\frac{1}{2},\frac{l}{2}}(G_T)} + \sigma\|B\|_{H_\gamma^{l-\frac{1}{2},\frac{l}{2}-\frac{1}{4}}(G_T)}\Big\}.$$

The proof of this theorem reduces to using the known estimates for diffraction problem (10.4.22) [69] and for the Dirichlet one (10.4.23) in the spaces W_2^m. Multiplying these estimates by the weight $e^{-2\gamma t}$ and integrating with respect to t, we obtain that

$$\|\boldsymbol{v}^{(1)}\|^2_{\boldsymbol{H}_\gamma^{2+l,1+l/2}(D_T)} \leqslant c\Big\{\|r\|^2_{\boldsymbol{H}_\gamma^{1+l,1/2+l/2}(D_T)} + \|\boldsymbol{R}\|^2_{\boldsymbol{H}_\gamma^{0,1+l/2}(\mathbb{R}^3_T)}\Big\}, \tag{10.4.25}$$

$$\|\nabla p^{(1)}\|^2_{\boldsymbol{H}_\gamma^{l,l/2}(D_T)} \leqslant c\Big\{\|b'\|^2_{H_\gamma^{l+1/2,1/2,l/2}(G_T)} + \|\boldsymbol{v}^{(1)}\|^2_{\boldsymbol{H}_\gamma^{2+l,1+l/2}(D_T)}\Big\}. \tag{10.4.26}$$

In addition, according to Lemma 4.2 in [82] (inequality (4.21))

$$\|p_+^{(1)}\|^2_{\Omega+} \leqslant c\Big\{\|b'\|^2_\Gamma + \Big\|\frac{\partial \boldsymbol{v}^{(1)+}}{\partial \boldsymbol{n}}\Big\|^2_\Gamma\Big\}.$$

The same inequality is obviously true for the increments $p_+^{(1)}$ in t, therefore

$$\|p_+^{(1)}\|^2_{\boldsymbol{H}_\gamma^{0,l/2}(B^-_{dT})} \leqslant c\Big\{\|b'\|^2_{H_\gamma^{0,l/2}(G_T)} + \Big\|\frac{\partial \boldsymbol{v}^{(1)+}}{\partial \boldsymbol{n}}\Big\|^2_{\boldsymbol{H}_\gamma^{0,l/2}(G_T)}\Big\}. \tag{10.4.27}$$

Arguing in the same way as in the proof of the local quadratic summability of the pressure at the end of § 10.3, we can show that

$$\|p_-^{(1)}\|^2_{\boldsymbol{H}_\gamma^{0,l/2}(B^-_{dT})} \leqslant c\|p_-^{(1)}\|^2_{H_\gamma^{0,l/2}(G_T)} \leqslant c\|\boldsymbol{v}^{(1)}\|^2_{\boldsymbol{H}_\gamma^{2+l,1+l/2}(Q^-_T)}\Big\}. \tag{10.4.28}$$

Now we use Theorem 10.4.1 applied to problem (10.4.24). Taking inequalities (10.4.25) – (10.4.28) into account, we arrive at the necessary estimate. Theorem 10.4.3 is proved.

10.4.3 Problem (1.1.7) in the General Case

We consider now problem (1.1.7) with a nonzero vector $\boldsymbol{v}_0$ and estimate its solution $(\boldsymbol{v}, p)$ in the spaces $W_2^{l,l/2}$ (see also [82]).

Theorem 10.4.4 (Solvability of the Complete Linear Problem) *We assume that* $l \in (1/2, 1)$, $T < \infty$, $\Gamma \in W_2^{2+l}$, $\boldsymbol{v}_0 \in \boldsymbol{W}_2^{1+l}(\cup\Omega^\pm)$, $[\boldsymbol{v}_0]|_\Gamma = 0$, $\boldsymbol{f} \in \boldsymbol{W}_2^{l,l/2}(D_T)$, $r \in W_2^{1+l,1/2+l/2}(D_T)$, $r = \nabla\cdot\boldsymbol{R}$, $\boldsymbol{R} \in \boldsymbol{W}_2^{0,1+l/2}(\mathbb{R}_T^3)$, $[\boldsymbol{R}\cdot\boldsymbol{n}]|_{G_T} = 0$, $\boldsymbol{b} \in \boldsymbol{W}_2^{l+1/2,l/2+1/4}(G_T)$, $b' \in W_2^{l+1/2,l/2+1/4}(G_T)$, $B \in W_2^{l-1/2,l/2-1/4}(G_T)$ *and compatibility conditions*

$$\nabla\cdot\boldsymbol{v}_0(x) = r(x,0), \qquad [\mathbb{T}(\boldsymbol{v}_0)\boldsymbol{n} - (\boldsymbol{n}\cdot\mathbb{T}(\boldsymbol{v}_0)\boldsymbol{n})\boldsymbol{n}]|_\Gamma = \boldsymbol{b}(x,0)$$

hold. Then there are unique $\boldsymbol{v}$ *and* p *such that* $(\boldsymbol{v},p)$ *is a solution of system* (1.1.7) *with* $p \xrightarrow[|x|\to\infty]{} 0$, $\boldsymbol{v} \in \boldsymbol{W}_2^{2+l,1+l/2}(D_T)$, $\nabla p \in \boldsymbol{W}_2^{l,l/2}(D_T)$, $p \in W_2^{l,l/2}(\cup B_{dT}^\pm)$, $d < \infty$, $[p]|_\Gamma \in W_2^{l+1/2,l/2+1/4}(G_T)$ *and the inequality*

$$\begin{aligned} N_T[\boldsymbol{v},p] &\equiv \|\boldsymbol{v}\|_{D_T}^{(2+l,1+l/2)} + \|\nabla p\|_{D_T}^{(l,l/2)} + \|p\|_{\cup B_{dT}^\pm}^{(l,l/2)} + \|[p]|_\Gamma\|_{W_2^{l+1/2,l/2+1/4}(G_T)} \\ &\leqslant c_{15}(T)\Big\{\|\boldsymbol{f}\|_{D_T}^{(l,l/2)} + \|r\|_{\boldsymbol{W}_2^{1+l,1/2+l/2}(D_T)} + \|\mathcal{D}_t\boldsymbol{R}\|_{\mathbb{R}_T^3}^{(0,l/2)} + \|\boldsymbol{v}_0\|_{\boldsymbol{W}_2^{1+l}(\cup\Omega^\pm)} \\ &\quad + \|\boldsymbol{b}\|_{\boldsymbol{W}_2^{l+1/2,l/2+1/4}(G_T)} + \|b'\|_{W_2^{l+1/2,l/2+1/4}(G_T)} \\ &\quad + T^{-l/2}\|b'\|_{W_2^{1/2,0}(G_T)} + \sigma\|B\|_{G_T}^{(l-1/2,l/2-1/4)}\Big\} \end{aligned} \tag{10.4.29}$$

is satisfied.

The basic idea of the proof consists in the construction of a vector $\boldsymbol{V}$ such that $\boldsymbol{V} \in \boldsymbol{W}_2^{2+l,1+l/2}(D_T)$, $\boldsymbol{V}(x,0) = \boldsymbol{v}_0$, $[\boldsymbol{V}]|_{G_T} = 0$ and

$$\|\boldsymbol{V}\|_{\boldsymbol{W}_2^{2+l,1+l/2}(D_T)} \leqslant c\|\boldsymbol{v}_0\|_{\boldsymbol{W}_2^{1+l}(\cup\Omega^\pm)}. \tag{10.4.30}$$

Then the couple $(\boldsymbol{u} = \boldsymbol{v} - \boldsymbol{V}, p)$ satisfies the problem with zero initial data to which Theorem 10.4.3 is applied. Next, we take into account the fact that the norms in the spaces $H_0^{l,l/2}(D_T)$ and $H_\gamma^{l,l/2}(D_T)$ for $T < \infty$ are equivalent:

$$\mathrm{e}^{-2\gamma T}\|v\|_{H_0^{l,l/2}(D_T)}^2 \leqslant \|v\|_{H_\gamma^{l,l/2}(D_T)}^2 \leqslant (1 + T^{l/2}\gamma^l)\|v\|_{H_0^{l,l/2}(D_T)}^2$$

for any $v \in H_0^{l,l/2}(D_T)$, and that

$$\|b'\|_{H_0^{l+1/2,1/2,l/2}(G_T)}^2 \leqslant c\Big\{\|b'\|_{W_2^{l+1/2,l/2+1/4}(G_T)}^2 + T^{-l}\|b'\|_{W_2^{1/2,0}(G_T)}^2\Big\}.$$

Following [82], we arrive at the estimate

$$\|\boldsymbol{u}\|_{\boldsymbol{H}_0^{2+l,1+l/2}(D_T)}^2 + \|\nabla p\|_{\boldsymbol{H}_0^{l,l/2}(D_T)}^2 + \|p\|_{H_0^{l,l/2}(\cup B_{dT}^\pm)}^2 \leqslant$$

$$\leqslant \mathrm{e}^{2\gamma T}\Big\{\|\boldsymbol{u}\|^2_{\boldsymbol{H}^{2+l,1+l/2}_\gamma(D_T)} + \|\nabla p\|^2_{\boldsymbol{H}^{l,l/2}_\gamma(D_T)} + \|p\|^2_{H^{l,l/2}_\gamma(\cup B^\pm_{dT})}\Big\}$$

$$\leqslant c(T)\Big\{\big(\|\boldsymbol{f}\|^{(l,l/2)}_{D_T}\big)^2 + \|r\|^2_{\boldsymbol{W}^{1+l,1/2+l/2}_2(D_T)} + \|\boldsymbol{R}\|^2_{\boldsymbol{W}^{0,1+l/2}_2(\mathbb{R}^3_T)} + T^{-l}\|\mathcal{D}_t\boldsymbol{R}\|^2_{\mathbb{R}^3_T}$$
$$+ \|\boldsymbol{b}\|^2_{\boldsymbol{W}^{l+1/2,l/2+1/4}_2(G_T)} + \|b'\|^2_{W^{l+1/2,l/2+1/4}_2(G_T)} + T^{-l}\|b'\|^2_{W^{1/2,0}_2(G_T)}$$
$$+ \sigma\big(\|B\|^{(l-1/2,l/2-1/4)}_{G_T}\big)^2 + \|\boldsymbol{v}_0\|^2_{\boldsymbol{W}^{1+l}_2(\cup\Omega^\pm)}\Big\}. \qquad (10.4.31)$$

We estimate the function of pressure jump $[p]|_{G_T}$ using the boundary condition

$$\big[\boldsymbol{n}\cdot\mathbb{T}(\boldsymbol{v},p)\boldsymbol{n}\big]\big|_{G_T} - \sigma\boldsymbol{n}\cdot\int_0^t \Delta_\Gamma\boldsymbol{v}\,\mathrm{d}t' = b' + \sigma\int_0^t B\mathrm{d}t'.$$

We have

$$\|[p]|_{G_T}\|_{W^{0,l/2+1/4}_2(G_T)} \leqslant 2\Big\|\Big[\mu^\pm\frac{\partial\boldsymbol{v}}{\partial\boldsymbol{n}}\Big]\Big|_{G_T}\Big\|_{W^{0,l/2+1/4}_2(G_T)} + \|b'\|_{W^{0,l/2+1/4}_2(G_T)}$$
$$+\sigma\Big\|\int_0^t(\boldsymbol{n}\cdot\Delta_\Gamma\boldsymbol{v} + B)\,\mathrm{d}t'\Big\|_{W^{0,l/2+1/4}_2(G_T)}$$

Since for any $\Phi\in L_2(G_T)$

$$\int_0^T \mathrm{d}t\int_0^t\frac{\mathrm{d}\tau}{\tau^{3/2+l}}\Big(\int_{t-\tau}^t\|\Phi\|_\Gamma\,\mathrm{d}t'\Big)^2 \leqslant \int_0^T\mathrm{d}t\int_0^t\frac{\mathrm{d}\tau}{\tau^{1/2+l}}\int_{t-\tau}^t\|\Phi\|^2_\Gamma\,\mathrm{d}t'$$
$$\leqslant\frac{2}{2l-1}\int_0^T\mathrm{d}t\int_0^t\|\Phi\|^2_\Gamma\frac{\mathrm{d}t'}{(t-t')^{l-1/2}} \leqslant \frac{4T^{3/2-l}}{(2l-1)(3-2l)}\int_0^T\|\Phi\|^2_\Gamma\,\mathrm{d}t',$$

then

$$\|[p]|_{G_T}\|_{W^{0,l/2+1/4}_2(G_T)} \leqslant c\Big\{\|b'\|_{W^{0,l/2+1/4}_2(G_T)} + \|\boldsymbol{v}\|_{W^{l+2,l/2+1}_2(G_T)}$$
$$+\sigma T^{3/4-l/2}\Big(\|B\|_{G_T} + \|\Delta_\Gamma\boldsymbol{v}\|_{G_T}\Big)\Big\}. \qquad (10.4.32)$$

It follows from the embedding theorems that $\nabla p^\pm\in\boldsymbol{W}^{l-1/2,l/2-1/4}_2(G_T)$, $[p]|_\Gamma\in W^{l-1/2,l/2-1/4}_2(G_T)$, therefore, combining these facts with inequality (10.4.32), we obtain the estimate of $\|[p]|_\Gamma\|_{W^{l+1/2,l/2+1/4}_2(G_T)}$. And taking into account inequalities (10.4.30) and (10.4.31), we get (10.4.29).

10.5 Local Solvability of the Nonlinear Problem in the L_2-Setting

Now we study the system

$$
\begin{aligned}
&\mathcal{D}_t\boldsymbol{u} - \nu^{\pm}\nabla^2_{\boldsymbol{u}}\boldsymbol{u} + \frac{1}{\rho^{\pm}}\nabla_{\boldsymbol{u}}q = \boldsymbol{f}, \quad \nabla_{\boldsymbol{u}}\cdot\boldsymbol{u} = 0 \text{ in } D_T \equiv \cup Q^{\pm}_T,\\
&\boldsymbol{u}\big|_{t=0} = \boldsymbol{v}_0 \quad \text{in } \Omega^-_0 \cup \Omega^+_0, \quad \boldsymbol{u}\xrightarrow[|\xi|\to\infty]{} 0, \quad q \xrightarrow[|\xi|\to\infty]{} 0,\\
&[\boldsymbol{u}]\big|_{G_T} = 0, \quad [\mu^{\pm}\Pi_0\Pi\mathbb{S}_{\boldsymbol{u}}(\boldsymbol{u})\boldsymbol{n}]\big|_{G_T} = 0 \quad (G_T \equiv \Gamma\times(0,T)),\\
&[\boldsymbol{n}_0\cdot\mathbb{T}_{\boldsymbol{u}}(\boldsymbol{u},q)\boldsymbol{n}]\big|_{G_T} = \sigma H\boldsymbol{n}\cdot\boldsymbol{n}_0,
\end{aligned}
\tag{10.5.1}
$$

where surface tension coefficient $\sigma \geqslant 0$.

Theorem 10.5.1 (Local Existence Theorem for the Problem with the Surface Tension) *We assume that for some* $l \in (1/2, 1)$ $\Gamma \in W^{5/2+l}_2$, $\boldsymbol{f} \in \boldsymbol{W}^{l,l/2}_2(Q_T)$, $0 < T < \infty$, $\boldsymbol{f}(\cdot,t)$, $\nabla\boldsymbol{f}(\cdot,t) \in \mathrm{Lip}(\Omega_0)$ *for any* $t \in [0,T]$, $\boldsymbol{f}(\xi,\cdot)$, $\nabla\boldsymbol{f}(\xi,\cdot) \in \boldsymbol{C}^{\beta}(0,T)$ *for any* $\xi \in \Omega_0$ *with some* $\beta \in [1/2,1)$. *In addition, let the initial velocity vector field* $\boldsymbol{v}_0 \in \boldsymbol{W}^{1+l}_2(\cup\Omega^{\pm}_0)$ *satisfy compatibility conditions*

$$
\nabla\cdot\boldsymbol{v}_0 = 0 \text{ in } \Omega^-_0\cup\Omega^+_0, \quad [\boldsymbol{v}_0]\big|_{\Gamma} = 0, \quad [\mu^{\pm}\Pi_0\mathbb{S}(\boldsymbol{v}_0)\boldsymbol{n}_0]\big|_{\Gamma} = 0. \tag{10.5.2}
$$

Under this assumptions, there is a constant $T_0 \in (0,T]$ *such that problem* (10.5.1) *is solvable in the interval* $(0,T_0]$, *its solution* $(\boldsymbol{u},q)$ *has the following properties:* $\boldsymbol{u} \in \boldsymbol{W}^{2+l,1+l/2}_2(D_{T_0})$, $q \in W^{l,l/2}_2(\cup B^{\pm}_{dT_0})$, $\nabla q \in \boldsymbol{W}^{l,l/2}_2(D_{T_0})$, $[q]\big|_{\Gamma} \in W^{l+1/2,l/2+1/4}_2(G_{T_0})$, *and the inequality*

$$
\begin{aligned}
N_{T_0}[\boldsymbol{u},q] &\equiv \|\boldsymbol{u}\|^{(2+l,1+l/2)}_{D_{T_0}} + \|\nabla q\|^{(l,l/2)}_{D_{T_0}} + \|q\|^{(l,l/2)}_{\cup B^{\pm}_{dT_0}} + \|[q]|_{\Gamma}\|_{W^{l+1/2,l/2+1/4}_2(G_{T_0})}\\
&\leqslant c_{16}\Big(c + |\boldsymbol{f}|^{(1,\beta)}_{Q_{T_0}} + |\nabla\boldsymbol{f}|^{(0,\beta)}_{Q_{T_0}} + T_0^{\frac{1-l}{2}}\|\boldsymbol{v}_0\|_{\boldsymbol{W}^1_2(\Omega_0)}\Big)\\
&\quad\times\Big\{\|\boldsymbol{f}\|^{(l,l/2)}_{Q_{T_0}} + \|\boldsymbol{v}_0\|_{\boldsymbol{W}^{1+l}_2(\cup\Omega^{\pm}_0)} + \sigma\|H_0\|_{W^{l+\frac12}_2(\Gamma)}\Big\}
\end{aligned}
\tag{10.5.3}
$$

holds; here H_0 *denotes twice the mean curvature of* Γ. *The value* T_0 *depends on the norms of* $\boldsymbol{f}$, $\boldsymbol{v}_0$ *and* H_0.

Remark 10.5.1 *We note that this result is also valid when the domain* $\Omega^+_0 \cup \Omega^-_0$ *is bounded, decay condition for the velocity* $\boldsymbol{u}$ *at infinity being replaced by adhesion condition* $\boldsymbol{u}\big|_{\Upsilon_T} = 0$, *where* $\Upsilon_T \equiv \Sigma\times(0,T)$, Σ *is the outer boundary.*

Remark 10.5.2 *In the case of the bounded domain, the pressure is determined up to a function $c(t)$. To be definite, we can assume, for instance, that $\int_{\Omega_0^+} q(\xi,t)\,\mathrm{d}\xi = 0$ for each $t \leqslant T$. This assumption is sufficient to estimate the L_2-norm of q in $D_{\Sigma T} \equiv \cup\Omega_0^\pm \times (0,T)$. Using the Poincaré inequality, one can deduce that*

$$\|q\|_{\cup\Omega_0^\pm} \leqslant c\big\{\|\nabla q\|_{\cup\Omega_0^\pm} + \|[q]|_\Gamma\|_\Gamma\big\}$$

(for more detail, see § 12.1).

Theorem 10.5.1 is based on the unique solvability of a linearized problem in an arbitrary finite time interval. Therefore we apply relation (1.1.3) and linearize problem (10.5.1) on a given vector field $\boldsymbol{u}$ which leads to system (5.3.1).

The first step of the proof is the analysis of problem (5.3.1) with $\boldsymbol{u} = 0$. Unique solvability of this system has been obtained in § 10.4 in the whole space $\mathbb{R}^3 = \overline{\Omega_0^+ \cup \Omega_0^-}$ (Theorem 10.4.4). In order to prove this result for a bounded domain, we need an a priori estimate of a solution near the outer boundary which is obtained by estimating the solution of the Dirichlet problem for the Stokes system in a half-space [76].

Now we can state existence and uniqueness theorem for the problem in the bounded domain.

Theorem 10.5.2 (The Solvability of the Linear Problem in a Bounded Domain) *We assume that, for some $l \in (1/2,1)$, $T < \infty$, $\Gamma \in W_2^{2+l}$, $\Sigma \in W_2^{3/2+l}$, $\boldsymbol{f} \in \boldsymbol{W}_2^{l,l/2}(D_{\Sigma T})$, $r \in W_2^{1+l,1/2+l/2}(D_{\Sigma T})$, $r = \nabla\cdot\boldsymbol{R}$, $\boldsymbol{R} \in \boldsymbol{W}_2^{0,1+l/2}(D_{\Sigma T})$, $[\boldsymbol{R}\cdot\boldsymbol{n}]|_{G_T} = 0$, $\boldsymbol{w}_0 \in \boldsymbol{W}_2^{1+l}(\cup\Omega_0^\pm)$, $\boldsymbol{b} \in \boldsymbol{W}_2^{l+1/2,l/2+1/4}(G_T)$, $b \in W_2^{l+1/2,l/2+1/4}(G_T)$, and $B \in W_2^{l-1/2,l/2-1/4}(G_T)$. In addition, we suppose that compatibility conditions*

$$[\boldsymbol{w}_0]\big|_\Gamma = 0, \quad [\mu^\pm \Pi_0 \mathbb{S}(\boldsymbol{w}_0)\boldsymbol{n}_0]\big|_\Gamma = \Pi_0\boldsymbol{b}\big|_{t=0}, \quad \boldsymbol{w}_0\big|_\Sigma = 0,$$

$$\nabla\cdot\boldsymbol{w}_0 = r\big|_{t=0} \quad \text{in} \quad \Omega_0^- \cup \Omega_0^+$$

hold.

Then problem (7.3.1) with $\boldsymbol{u} = 0$ and $\int_{\Omega_0^+} s(x,t)\,\mathrm{d}x = 0$ is uniquely solvable for $t \in (0,T]$ and its solution $(\boldsymbol{w},s)$ has the properties: $\boldsymbol{w} \in \boldsymbol{W}_2^{2+l,1+l/2}(D_{\Sigma T})$, $s \in W_2^{l,l/2}(D_{\Sigma T})$, $\nabla s \in \boldsymbol{W}_2^{l,l/2}(D_{\Sigma T})$, $[s]\big|_{G_T} \in W_2^{l+1/2,l/2+1/4}(G_T)$, and

$$\|\boldsymbol{w}\|_{D_{\Sigma T}}^{(2+l,1+l/2)} + \|\nabla s\|_{D_{\Sigma T}}^{(l,l/2)} + \|s\|_{D_{\Sigma T}}^{(l,l/2)} + \|[s]|_\Gamma\|_{W_2^{l+1/2,l/2+1/4}(G_T)}$$
$$\leqslant c_{15}(T)\Big\{\|\boldsymbol{f}\|_{D_{\Sigma T}}^{(l,l/2)} + \|\boldsymbol{w}_0\|_{\boldsymbol{W}_2^{1+l}(\cup\Omega_0^\pm)} + \|r\|_{W_2^{1+l,0}(D_{\Sigma T})}$$

$$+ \|\boldsymbol{R}\|_{\boldsymbol{W}_2^{0,1+l/2}(D_{\Sigma T})} + T^{-l/2}\|\mathcal{D}_t\boldsymbol{R}\|_{D_{\Sigma T}} + \|\boldsymbol{b}\|_{\boldsymbol{W}_2^{l+1/2,l/2+1/4}(G_T)}$$

$$+ \|b\|_{W_2^{l+1/2,l/2+1/4}(G_T)} + T^{-l/2}\|b\|_{W_2^{1/2,0}(G_T)}$$

$$+ \sigma\|B\|_{G_T}^{(l-1/2,l/2-1/4)}\Big\} \equiv c_{15}(T)F, \tag{10.5.4}$$

where $c_{15}(T)$ is a nondecreasing function of T.

The second step is to prove the solvability of problem (5.3.1) in the general case. We give the statement of this result.

Theorem 10.5.3 (Existence Theorem for Linearized Problem (5.3.1)) *Let the hypotheses of Theorem* 10.4.4 *be satisfied and let, in addition, a vector field $\boldsymbol{u} \in \boldsymbol{W}_2^{2+l,1+l/2}(D_T)$ be continuous when passing across the boundary Γ and satisfy the inequality*

$$T^{1/2}\|\boldsymbol{u}\|_{D_T}^{(2+l,1+l/2)} \leqslant \delta \tag{10.5.5}$$

with small number δ for some $T < \infty$.

Then there exists a unique solution $(\boldsymbol{w}, s)$ of problem (5.3.1) *such that $\boldsymbol{w} \in \boldsymbol{W}_2^{2+l,1+l/2}(D_T)$, $s \in W_2^{l,l/2}(\cup B_{dT}^{\pm})$, $\nabla s \in \boldsymbol{W}_2^{l,l/2}(D_T)$, $[s]|_{G_T} \in W_2^{l+1/2,l/2+1/4}(G_T)$ and inequality* (10.4.29) *holds for it with $c_{15}(T) = \overline{c}_1 + \overline{c}_2 T^{\frac{1-l}{2}}\|\boldsymbol{u}(\cdot,0)\|_{\boldsymbol{W}_2^1(\Omega)}$, $\overline{c}_1, \overline{c}_2$ being nondecreasing functions of T.*

Remarks 10.5.1 and 10.5.2 are also true for the linearized problem. The proof of Theorem 10.5.3 follows the same scheme as that of Theorem 5.3.1 for the Hölder case, therefore, we will often refer to Chap. 5.

We solve problem (5.3.1) by successive approximations taking $\boldsymbol{w}^{(0)} = 0$, $s^{(0)} = 0$ and defining $(\boldsymbol{w}^{(m+1)}, s^{(m+1)})$, $m \geqslant 0$, as a solution to problem (5.3.25), where notation (5.3.10) is used. In addition, $l_2(\boldsymbol{w}^{(m)}) = \nabla \cdot \boldsymbol{\mathcal{L}}(\boldsymbol{w}) = (\mathbb{I} - \mathbb{A}^*)\boldsymbol{w}$.

The operators $\boldsymbol{l}_1, \ldots, l_5$ and $\boldsymbol{\mathcal{L}}$ were considered in [83]. The estimates obtained there imply the following lemma.

Lemma 10.5.1 *If $\boldsymbol{u}$ and $\boldsymbol{u}'$ satisfy inequality* (10.5.5) and $[\boldsymbol{u}]|_\Gamma = [\boldsymbol{u}']|_\Gamma = 0$, then

$$\|\boldsymbol{l}_1(\boldsymbol{w}, s) - \boldsymbol{l}_1'(\boldsymbol{w}, s)\|_{D_T}^{(l,l/2)} + \|l_2(\boldsymbol{w}) - l_2'(\boldsymbol{w})\|_{W_2^{1+l,\frac{1+l}{2}}(D_T)} +$$

$$+ \|\boldsymbol{l}_3(\boldsymbol{w}) - \boldsymbol{l}_3'(\boldsymbol{w})\|_{W_2^{1/2+l,1/4+l/2}(G_T)} + \|l_5(\boldsymbol{w}) - l_5'(\boldsymbol{w})\|_{G_T}^{(l-1/2,l/2-1/4)} \leqslant$$

$$\leqslant c_{17}\sqrt{T}\|\boldsymbol{u} - \boldsymbol{u}'\|_{D_T}^{(2+l,1+l/2)}\Big\{\|\boldsymbol{w}\|_{D_T}^{(2+l,1+l/2)} + \|\nabla s\|_{D_T}^{(l,l/2)}\Big\},$$

$$\|\mathcal{D}_t(\boldsymbol{\mathcal{L}}(\boldsymbol{w}) - \boldsymbol{\mathcal{L}}'(\boldsymbol{w}))\|_{D_T}^{(0,l/2)} \leqslant c_{18}\Big\{\sqrt{T}\|\boldsymbol{u} - \boldsymbol{u}'\|_{D_T}^{(2+l,1+l/2)} + \tag{10.5.6}$$

$$+ T^{\frac{1-l}{2}}\|\boldsymbol{u}(\cdot,0) - \boldsymbol{u}'(\cdot,0)\|_{\boldsymbol{W}_2^1(\Omega)}\Big\}\|\boldsymbol{w}\|_{D_T}^{(2+l,1+l/2)},$$

$$\|l_4(\boldsymbol{w},s) - l_4'(\boldsymbol{w},s)\|_{W^{1/2+l,1/4+l/2}(G_T)} \leqslant c_{19}\sqrt{T}\|\boldsymbol{u} - \boldsymbol{u}'\|_{D_T}^{(2+l,1+l/2)} \times$$

$$\times \Big\{\|\boldsymbol{w}\|_{D_T}^{(2+l,1+l/2)} + \|\nabla s\|_{D_T}^{(l,l/2)} + \|q\|_{\cup B_{dT}^{\pm}}^{(l,l/2)} + \|s\|_{W_2^{l+1/2,l/2+1/4}(G_T)}\Big\}.$$

Here the operators $\boldsymbol{l}_1', \ldots, l_5'$ and $\boldsymbol{\mathcal{L}}'$ are calculated according to formulas (5.3.10), where the vector $\boldsymbol{u}$ is replaced by $\boldsymbol{u}'$. If $\boldsymbol{w}\big|_{t=0} = 0$, then inequality (10.5.6) is valid without the term $T^{\frac{1-l}{2}}\|\boldsymbol{u}(\cdot,0) - \boldsymbol{u}'(\cdot,0)\|_{\boldsymbol{W}_2^1(\Omega)}$ on the right-hand side.

The following proposition is a consequence of Lemma 10.5.1.

Lemma 10.5.2 *If $\boldsymbol{u}$ satisfies inequality* (10.5.5) *and* $[\boldsymbol{u}]|_\Gamma = 0$, *then*

$$\|\boldsymbol{l}_1(\boldsymbol{w},s)\|_{D_T}^{(l,l/2)} + \|l_2(\boldsymbol{w})\|_{W_2^{1+l,(1+l)/2}(D_T)} + \|\boldsymbol{l}_3(\boldsymbol{w})\|_{W_2^{1/2+l,1/4+l/2}(G_T)} +$$

$$+ \|l_4(\boldsymbol{w},s)\|_{W^{1/2+l,1/4+l/2}(G_T)} + \|l_5(\boldsymbol{w})\|_{G_T}^{(l-1/2,l/2-1/4)} \leqslant$$

$$\leqslant c_{20}\delta\Big\{\|\boldsymbol{w}\|_{D_T}^{(2+l,1+l/2)} + \|\nabla s\|_{D_T}^{(l,l/2)} + \|q\|_{\cup B_{dT}^{\pm}}^{(l,l/2)} + \|s\|_{W_2^{l+1/2,l/2+1/4}(G_T)}\Big\},$$

$$\|\mathcal{D}_t\boldsymbol{\mathcal{L}}(\boldsymbol{w})\|_{D_T}^{(0,l/2)} \leqslant c_{21}\Big\{\delta + T^{\frac{1-l}{2}}\|\boldsymbol{u}(\cdot,0)\|_{\boldsymbol{W}_2^1(\Omega)}\Big\}\|\boldsymbol{w}\|_{D_T}^{(2+l,1+l/2)}. \tag{10.5.7}$$

If $\boldsymbol{w}(\cdot,0) = 0$ in Ω, then the term with $\|\boldsymbol{u}(\cdot,0)\|_{\boldsymbol{W}_2^1(\Omega)}$ may be dropped in the last inequality.

The difference

$$\boldsymbol{f}(X_{\boldsymbol{u}},t) - \boldsymbol{f}(X_{\boldsymbol{u}'},t) = \sum_{k=1}^{3}\int_0^1 \partial\boldsymbol{f}(X_{\boldsymbol{u}_s},t)/\partial x_k\,\mathrm{d}s\int_0^t (u_k - u_k')\,\mathrm{d}\tau,$$

where $\boldsymbol{u}_s = \boldsymbol{u}' + s\widetilde{\boldsymbol{u}}$ is a continuous transformation from $\boldsymbol{u}'$ into $\boldsymbol{u}$ with $\widetilde{\boldsymbol{u}} \equiv \boldsymbol{u} - \boldsymbol{u}'$. The following lemma was proved in [83].

Lemma 10.5.3 *Let $\boldsymbol{f}$ satisfy the assumption of Theorem* 10.5.1, *and let vectors $\boldsymbol{u}$, $\boldsymbol{u}' \in \boldsymbol{W}_2^{l,l/2}(D_T)$, $[\boldsymbol{u}]|_\Gamma = [\boldsymbol{u}']|_\Gamma = 0$, satisfy inequality* (10.5.5). *Then*

$$\|\boldsymbol{f}(X_{\boldsymbol{u}},t) - \boldsymbol{f}(X_{\boldsymbol{u}'},t)\|_{Q_T}^{(l,l/2)} \leqslant c(T)\int_0^T \|\boldsymbol{u} - \boldsymbol{u}'\|_{\boldsymbol{W}_2^l(\Omega)}\,\mathrm{d}t.$$

Here $c(T)$ is a power function of T.

Proof of Theorem 10.5.3 We return to problem (5.3.25). We observe that the vector $\boldsymbol{\mathcal{L}}(\boldsymbol{w}^{(m)}) = (\mathbb{I} - \mathbb{A}^T)\boldsymbol{w}^{(m)}$ is continuous when crossing Γ: $[\boldsymbol{\mathcal{L}}(\boldsymbol{w}^{(m)}) \cdot \boldsymbol{n}_0]|_\Gamma = [\boldsymbol{n}_0 \cdot (\mathbb{I} - \mathbb{A}^T)\boldsymbol{w}^{(m)}]_\Gamma = [\mathbb{A}\boldsymbol{n}_0]|_\Gamma \cdot \boldsymbol{w}^{(m)} = 0$. It follows from the formula for the cofactors A_{ij} to $a_{ij} = \partial x_i/\partial \xi_j$ due to the continuity of $\boldsymbol{x}$ and its tangent derivatives $\nabla_\Gamma \boldsymbol{x} = \Pi_0 \nabla \boldsymbol{x}$: for example, for A_{1j} we have

$$[A_{1j} n_{0j}]_\Gamma = [\boldsymbol{n}_0 \cdot (\nabla x_2 \times \nabla x_3)]_\Gamma = [\boldsymbol{n}_0 \cdot (\nabla_\Gamma x_2 \times \nabla_\Gamma x_3)]_\Gamma = 0.$$

Hence, we can apply Theorem 10.4.4 to (5.3.25) and conclude by Lemma 10.5.2 that $(\boldsymbol{w}^{(m+1)}, s^{(m+1)})$, $m \in \mathbb{N}$, are uniquely defined, moreover, $\boldsymbol{w}^{(0)} = 0$, $s^{(0)} = 0$, the functions $\boldsymbol{w}^{(1)}$, $s^{(1)}$ being a solution of problem (5.3.1) with $\boldsymbol{u} = 0$ and satisfying inequality (10.4.29).

Now we consider the differences $\boldsymbol{z}^{(m+1)} = \boldsymbol{w}^{(m+1)} - \boldsymbol{w}^{(m)}$, $g^{(m+1)} = s^{(m+1)} - s^{(m)}$, $m \in \mathbb{N} \cup \{0\}$. We have problem (5.3.28) for $m \in \mathbb{N}$. If $m > 1$, then $\boldsymbol{z}^{(m)}\big|_{t=0} = 0$, and we deduce from (10.4.29) and Lemma 10.5.2 that

$$N_T[\boldsymbol{z}^{(m+1)}, g^{(m+1)}] \leqslant c_{22}\delta N_T[\boldsymbol{z}^{(m)}, g^{(m)}], \qquad m = 2, 3, \ldots \tag{10.5.8}$$

If $m = 1$, then by virtue of (10.5.7) we obtain

$$N_T[\boldsymbol{z}^{(2)}, g^{(2)}] \leqslant (c_{22}\delta + c_{21}\delta_1) N_T[\boldsymbol{w}^{(1)}, s^{(1)}] \tag{10.5.9}$$

with $\delta_1 = T^{\frac{1-l}{2}} \|\boldsymbol{u}(\cdot, 0)\|_{\boldsymbol{W}_2^1(\Omega)}$, since $\boldsymbol{z}^{(1)}|_{t=0} \equiv \boldsymbol{w}^{(1)}|_{t=0} = \boldsymbol{w}_0 \neq 0$ in the general case.

Next, for $\Sigma_m = \sum_{j=2}^m N_T[\boldsymbol{z}^{(j)}, g^{(j)}]$, the following inequality

$$\Sigma_{m+1} \leqslant c_{22}\delta \Sigma_m + N_T[\boldsymbol{z}^{(2)}, g^{(2)}]$$

holds due to (10.5.8). Let us choose δ such that $c_{22}\delta < 1$. It is obvious that

$$\Sigma_{m+1} \leqslant (1 - c_{22}\delta)^{-1} N_T[\boldsymbol{z}^{(2)}, g^{(2)}].$$

In view of (10.4.29), (10.5.9), we have:

$$\begin{aligned} N_T[\boldsymbol{w}^{(m+1)}, s^{(m+1)}] &\leqslant \Sigma_{m+1} + N_T[\boldsymbol{w}^{(1)}, s^{(1)}] \\ &\leqslant \Big(\frac{1}{1 - c_{22}\delta} + \frac{c_{21}}{1 - c_{22}\delta} T^{\frac{1-l}{2}} \|\boldsymbol{u}(\cdot, 0)\|_{\boldsymbol{W}_2^1(\Omega)}\Big) c_{15} F, \end{aligned}$$

where F is the sum of right-hand side norms in (10.4.29) which are independent of m. Hence, the sequence $\{\boldsymbol{w}^{(m+1)}, s^{(m+1)}\}$ is convergent in the norm $N_T[\cdot, \cdot]$ and its limit $(\boldsymbol{w}, s)$ is a solution of (5.3.1) satisfying inequality (10.4.29) with

$$\widetilde{c_1}(T) \equiv \frac{c_{15}}{1-c_{22}\delta}\left(1 + c_{21}T^{\frac{1-l}{2}}\|\boldsymbol{u}(\cdot,0)\|_{\boldsymbol{W}_2^1(\Omega)}\right).$$

In a similar way, we can conclude that the difference ($\boldsymbol{z}=\boldsymbol{w}-\boldsymbol{w}', g=s-s'$) of two solutions of (5.3.1) satisfies the estimate

$$N_T[\boldsymbol{z},g] \leqslant c_{22}\delta N_T[\boldsymbol{z},g],$$

whence it follows $\boldsymbol{z}=0$, $g=0$. Thus, the uniqueness of the solution is also proved. □

Proof of Theorem 10.5.1 In order to verify the validity of Theorem 10.5.1, we apply again successive approximations, now for solving system (10.5.1), where we make use of formula (1.1.3) continued as follows:

$$H\boldsymbol{n} = \Delta(t)\boldsymbol{X}_{\boldsymbol{u}} \equiv \Delta(t)\boldsymbol{\xi}+\Delta(t)\int_0^t \boldsymbol{u}\,\mathrm{d}\tau = \Delta(0)\boldsymbol{\xi}+\int_0^t \dot{\Delta}(\tau)\boldsymbol{\xi}\,\mathrm{d}\tau+\Delta(t)\int_0^t \boldsymbol{u}\,\mathrm{d}\tau,$$

here $\dot{\Delta}(t) = \mathcal{D}_t\Delta(t)$.

We set $\boldsymbol{u}^{(0)}=0$, $q^{(0)}=0$ and define the first approximation ($\boldsymbol{u}^{(1)}$, $q^{(1)}$) as a solution to the problem

$$\begin{aligned}
&\mathcal{D}_t\boldsymbol{u}^{(1)} - \nu^{\pm}\nabla^2\boldsymbol{u}^{(1)} + \frac{1}{\rho^{\pm}}\nabla q^{(1)} = \boldsymbol{f}, \qquad \nabla\cdot\boldsymbol{u}^{(1)} = 0 \quad \text{in} \quad D_T,\\
&\boldsymbol{u}^{(1)}\big|_{t=0} = \boldsymbol{v}_0 \quad \text{in} \quad \Omega_0^-\cup\Omega_0^+, \quad \boldsymbol{u}^{(1)}\xrightarrow[|x|\to\infty]{}0, \quad p\xrightarrow[|x|\to\infty]{}0, \qquad (10.5.10)\\
&[\boldsymbol{u}^{(1)}]\big|_{G_T} = 0, \quad [\mu^{\pm}\Pi_0\mathbb{S}(\boldsymbol{u}^{(1)})\boldsymbol{n}_0]\big|_{G_T} = 0,\\
&[\boldsymbol{n}_0\cdot\mathbb{T}(\boldsymbol{u}^{(1)},q^{(1)})\boldsymbol{n}_0]\big|_\Gamma - \sigma\boldsymbol{n}_0\cdot\Delta(0)\int_0^t \boldsymbol{u}^{(1)}\,\mathrm{d}\tau\big|_\Gamma = \sigma H_0, \qquad t\in(0,T),
\end{aligned}$$

here $H_0(\xi) = \boldsymbol{n}_0\cdot\Delta(0)\boldsymbol{\xi}$ is twice the mean curvature of Γ. As $H_0 \in W_2^{l+1/2}(\Gamma)$, problem (10.5.10) is solvable by Theorem 10.4.4 in the interval $(0,T_1)$, $T_1 = T$, and

$$N_{T_1}[\boldsymbol{u}^{(1)},q^{(1)}] \leqslant c(T_1)\Big\{\|\boldsymbol{f}\|_{Q_T}^{(l,l/2)} + \|\boldsymbol{v}_0\|_{\boldsymbol{W}_2^{1+l}(\cup\Omega^{\pm})} + \sigma\|H_0\|_{W_2^{l+1/2}(\Gamma)}\Big\}. \tag{10.5.11}$$

Let the functions $\boldsymbol{u}^{(m+1)}$, $q^{(m+1)}$, $m\in\mathbb{N}$, solve problems (5.4.4). We recall the notation: $\nabla_m = \nabla_{\boldsymbol{u}^{(m)}}$ and etc.; $\boldsymbol{n}_m$ is the outward normal to the surface $\Gamma_m(t) = \{x = X_m(\xi,t)|\xi\in\Gamma\}$, where X_m are calculated by formula (1.1.2) with $\boldsymbol{u}=\boldsymbol{u}^{(m)}$; Π_m is the projector onto the tangential plane to $\Gamma_m(t)$, $\Delta_m(t)$ is the Beltrami-Laplace operator on $\Gamma_m(t)$.

Since the vector $\boldsymbol{f}$ satisfies inequality

$$\|\boldsymbol{f}(X_m,t)\|^{(l,l/2)}_{Q_{T_m}} \leqslant c(T_m)\langle\!\langle \boldsymbol{f}\rangle\!\rangle^{(l,l/2)}_{T_m},$$

where $c(T)$ is a power function of T, and since $\boldsymbol{n}_0\cdot\dot{\Delta}_m(\tau)\boldsymbol{\xi} \in W_2^{l-\frac{1}{2},\frac{l}{2}-\frac{1}{4}}(G_{T_m})$ if $\boldsymbol{u}^{(m)} \in \boldsymbol{W}_2^{2+l,1+l/2}(D_{T_m})$ [83], $H_0 \in W_2^{l+\frac{1}{2},\frac{l}{2}+\frac{1}{4}}(G_{T_m})$, then, by Theorem 10.5.3, there exists a solution $(\boldsymbol{u}^{(m+1)}, q^{(m+1)})$ to problem (5.4.4) in an interval $(0,T_{m+1})$, provided that, in this interval, the approximation $(\boldsymbol{u}^{(m)}, q^{(m)})$ is also defined and condition (10.5.5) holds for $\boldsymbol{u}^{(m)}$ with sufficiently small $\delta > 0$. Thus, it should be $T_{m+1} \leqslant T_m$.

Now it is necessary to show that there exists T_0 such that $T_m \geqslant T_0 > 0$ for all $m \in \mathbb{N}$, $N_{T_0}[\boldsymbol{u}^{(m)}q^{(m)}]$ are uniformly bounded and that the sequence $\{\boldsymbol{u}^{(m)},\ q^{(m)}\}_{m=1}^{\infty}$ converges to a solution of problem (10.5.1). The proof of these facts is based on Lemmas 10.5.1 and 10.5.3 applied to the right-hand sides of system (5.4.6), where $\widetilde{\boldsymbol{w}}^{(j+1)}$, $\widetilde{s}^{(j+1)}$ mean the differences $\boldsymbol{u}^{(j+1)} - \boldsymbol{u}^{(j)}, q^{(j+1)} - q^{(j)}$, respectively, $j \leqslant m$.

The norms of the right-hand sides in (5.4.6) are estimated either by the lower order norms of $\widetilde{\boldsymbol{w}}^{(j)}$ and $\widetilde{s}^{(j)}$, or by the leading part of their norms but with small coefficients including δ from inequality (10.5.5). In particular,

$$\big\|\boldsymbol{n}_0\cdot\big(\dot{\Delta}_j(\tau)-\dot{\Delta}_{j-1}(\tau)\big)\boldsymbol{\xi}\big\|^{(l-\frac{1}{2},\frac{l}{2}-\frac{1}{4})}_{G_{T_{m+1}}} \leqslant c\big\|\nabla\big(\widetilde{\boldsymbol{w}}^{(j)}\big)\big\|^{(l-\frac{1}{2},\frac{l}{2}-\frac{1}{4})}_{G_{T_{m+1}}} \leqslant c\big\|\widetilde{\boldsymbol{w}}^{(j)}\big\|^{(1+l,\frac{1}{2}+\frac{l}{2})}_{D_{T_{m+1}}}.$$

In addition, by Lemma 10.5.3 we have

$$\|\boldsymbol{f}(X_j,t) - \boldsymbol{f}(X_{j-1},t)\|^{(l,l/2)}_{Q_{T_{m+1}}} \leqslant c(T_{m+1})\int_0^{T_{m+1}} \|\widetilde{\boldsymbol{w}}^{(j)}\|_{\boldsymbol{W}_2^l(\Omega)}\,\mathrm{d}t,$$

where $c(T)$ is a nondecreasing function of T, depending on the norms $|\boldsymbol{f}|^{(1,\beta)}_{Q_T}$ and $|\nabla \boldsymbol{f}|^{(0,\beta)}_{Q_T}$. Similarly to § 5.4, one can deduce from this the boundedness of the sum $\Sigma'_{m+1}(T') \equiv \sum_{j=2}^{m+1} N_{T'}[\widetilde{\boldsymbol{w}}^{(j)}, \widetilde{s}^{(j)}]$:

$$\Sigma'_{m+1}(T') \leqslant c_{23}\big(c_{24}(T',\delta)+T'^{\frac{1-l}{2}}\|\boldsymbol{v}_0\|_{\boldsymbol{W}_2^{\frac{1}{2}}(\Omega)}\big)N_{T'}[\boldsymbol{u}^{(1)},q^{(1)}], \quad T' \in (0,T_{m+1}],$$

which implies the Cauchy convergence of the sequence $\{\boldsymbol{u}^{(m)}, q^{(m)}\}_{m=1}^{\infty}$ and, due to (10.5.11), the estimate

$$\begin{aligned} N_{T'}[\boldsymbol{u}^{(m+1)},q^{(m+1)}] &\leqslant \Sigma'_{m+1}(T') + N_{T'}[\boldsymbol{u}^{(1)},q^{(1)}] \\ &\leqslant c_{25}\big(T',\|\boldsymbol{v}_0\|_{\boldsymbol{W}_2^{\frac{1}{2}}(\Omega)}\big)\Big\{\langle\!\langle \boldsymbol{f}\rangle\!\rangle^{(l,l/2)}_{T'} + \|\boldsymbol{v}_0\|_{\boldsymbol{W}_2^{1+l}(\cup\Omega_0^{\pm})} + \sigma\|H_0\|_{W_2^{l+\frac{1}{2}}(\Gamma)}\Big\}. \end{aligned} \tag{10.5.12}$$

Since the right-hand side of the last inequality in (10.5.12) is independent of m, and c_{25} is a nondecreasing function of T', one can find such $T_0 \in (0,T_{m+1}]$ that

$$T_0^{1/2} N_{T_0}[\boldsymbol{u}^{(j)}, q^{(j)}] \leqslant \delta, \qquad \forall j \in \mathbb{N}.$$

Hence, as follows from (10.5.12), $N_{T_0}[\boldsymbol{u}^{(j)}, q^{(j)}]$ are uniformly bounded and the sequence $\{\boldsymbol{u}^{(j)}, q^{(j)}\}_{j=1}^{\infty}$ is convergent. Passing to the limit in system (5.4.4), we make sure that the approximations $(\boldsymbol{u}^{(j)}, q^{(j)})$, $j \in \mathbb{N}$, converge to a solution of problem (10.5.1) for which the inequality (10.5.3) holds. □

A similar consideration for the case of a single fluid was presented in detail in [83].

Remark 10.5.3 *It remains to give equivalence condition for systems* (10.5.1) *and* (1.1.1). *These problems are equivalent if the angle between* $\boldsymbol{n}_0$ *and* $\boldsymbol{n}$ *is acute, that is, the condition* $\mathbb{A}\boldsymbol{n}_0 \cdot \boldsymbol{n}_0 > 0$ *must be satisfied which, in turn, means the positive definiteness of the matrix* $\mathbb{A}$. *From matrix presentation in the form* $\{A_{ij}\} = \{\delta^i_j + B_{ij}\}$, *where* $B_{ij} = -b_{ji} + b_{ki}b_{jk} - b_{ji}b_{kk}$, $B_{ii} = b_{jj} + b_{kk}(1 + b_{jj}) - b_{jk}b_{kj}$ *with* $i \neq j$, $j \neq k$, $k \neq i$, $b_{ij} = \int_0^t \frac{\partial u_i}{\partial \xi_j}\,\mathrm{d}\tau$, *and from estimate* (2.7) *in* [83] *for* B_{ij}, *it follows that the matrix* $\mathbb{A}$ *is close to unit one if* δ *in* (10.5.5) *is small. This can always be achieved for a sufficiently small time* T'.

By virtue of Remark 10.5.3 and Theorem 10.5.1, problem (1.1.1) is solvable on a sufficiently small time interval $(0, T')$.

Chapter 11
Global L_2-Solvability of the Problem Without Surface Tension

Abstract The chapter deals with an interface problem for the Navier-Stokes system governing the motion of two-phase fluid without surface tension in a container, one liquid being inside another one. We prove the unique solvability of the problem in an infinite time interval provided that the data are small enough and mass forces decrease exponentially in time. The norms of the solution are shown to decay in an exponential way at infinity with respect to time. The proof is based on an exponential estimate of generalized energy and on the local existence theorem of the problem in the Sobolev-Slobodetskiĭ spaces.

This material was published in [23].

11.1 The Statement of Global Existence Theorem

We consider again that $\cup\Omega_0^\pm$ is a bounded domain with outer boundary Σ and we study the initial-boundary value problem for the Navier-Stokes system with homogeneous boundary conditions:

$$\mathcal{D}_t \boldsymbol{v} + (\boldsymbol{v}\cdot\nabla)\boldsymbol{v} - \nu^\pm \nabla^2 \boldsymbol{v} + \frac{1}{\rho^\pm}\nabla p = \boldsymbol{f}, \quad \nabla\cdot\boldsymbol{v} = 0 \text{ in } \Omega_t^- \cup \Omega_t^+, \quad t > 0,$$

$$\boldsymbol{v}\big|_{t=0} = \boldsymbol{v}_0 \text{ in } \Omega_0^- \cup \Omega_0^+, \quad \boldsymbol{v}\big|_\Sigma = 0, \quad \int_{\Omega_0^+} p\,\mathrm{d}x = 0, \tag{11.1.1}$$

$$[\boldsymbol{v}]\big|_{\Gamma_t} = 0, \quad [\mathbb{T}\boldsymbol{n}]\big|_{\Gamma_t} = 0, \quad t > 0.$$

After the passage to the Lagrangian coordinates in problem (11.1.1), (1.0.3), we get system (6.1.1). We state the theorem on local solvability for this problem.

© The Author(s), under exclusive license to Springer Nature Switzerland AG 2021
I. V. Denisova, V. A. Solonnikov, *Motion of a Drop in an Incompressible Fluid*, Advances in Mathematical Fluid Mechanics, https://doi.org/10.1007/978-3-030-70053-9_11

Theorem 11.1.1 (Local L_2-Solvability of the Problem Without Surface Tension) *Let the hypotheses of Theorem* 10.5.1 *except the assumption on* Γ *hold. Moreover, we suppose that* $\boldsymbol{u}_0|_\Sigma = 0$ *and* $\Gamma, \Sigma \in W_2^{3/2+l}$, $l \in (1/2, 1)$.

Then there is a time moment $T_0 \in (0, T]$ *such that problem* (6.1.1) *is uniquely solvable in the interval* $(0, T_0]$, *and the solution* $(\boldsymbol{u}, q)$ *has the properties:* $\boldsymbol{u} \in \boldsymbol{W}_2^{2+l,1+l/2}(D_{\Sigma T_0})$, $q \in W_2^{l,l/2}(D_{\Sigma T_0})$, $\nabla q \in \boldsymbol{W}_2^{l,l/2}(D_{\Sigma T_0})$, $[q]\big|_\Gamma \in W_2^{l+1/2,l/2+1/4}(G_{T_0})$ *and*

$$\|\boldsymbol{u}\|_{D_{\Sigma T_0}}^{(2+l,1+l/2)} + \|\nabla q\|_{D_{\Sigma T_0}}^{(l,l/2)} + \|q\|_{D_{\Sigma T_0}}^{(l,l/2)} + \|[q]|_\Gamma\|_{W_2^{l+1/2,l/2+1/4}(G_{T_0})}$$
$$\leqslant c_1\Big(c_2 + |\nabla \boldsymbol{f}|_{Q_{T_0}}^{(0,\beta)} + T_0^{\frac{1-l}{2}}\|\boldsymbol{v}_0\|_{\boldsymbol{W}_2^1(\Omega)}\Big)\Big\{\|\boldsymbol{f}\|_{Q_{T_0}}^{(l,l/2)} + |\boldsymbol{f}|_{Q_{T_0}}^{(1,\beta)} + \|\boldsymbol{v}_0\|_{\boldsymbol{W}_2^{1+l}(\cup\Omega_0^\pm)}\Big\}. \tag{11.1.2}$$

The value of T_0 *depends on the norms of* $\boldsymbol{f}$ *and* $\boldsymbol{v}_0$.

Remark 10.5.2 is also valid for Theorem 11.1.1. See § 12.1 for more details about conditions for the pressure being unique.

Remark 11.1.1 *If* $\sigma = 0$, *the solvability of problem* (10.5.1) *requires lower smoothness of the initial interface* Γ *than in the case of positive* σ. *Indeed, in this case, we have the homogeneous boundary conditions in* (5.4.4) *and we do not need to calculate* H_0 *and* $\dot{\Delta}_m(\tau)\boldsymbol{\xi}$ *on* Γ. *It is the estimates of these functions that make us suppose* Γ *to belong to* $W_2^{5/2+l}$ *when* $\sigma > 0$. *Therefore for* $\sigma = 0$, *Theorem* 10.5.1 *holds with the initial interface* $\Gamma \in W_2^{3/2+l}$ *and the magnitude of* T_0 *does not depend on the curvature of* Γ.

In view of Remarks 11.1.1 and 10.5.2, Theorem 10.5.1 implies Theorem 11.1.1.

Let $T \in (0, \infty]$, $t, \tau > 0$. We introduce the notation:

$$Q^\pm_{(t,t+\tau)} = \Omega_t^\pm \times (t, t+\tau),\ D_{(t,t+\tau)} = \cup Q^\pm_{(t,t+\tau)},\ G_{(t,t+\tau)} = \Gamma_t \times (t, t+\tau).$$

The following theorem guarantees the existence of a global solution to system (11.1.1), (1.0.3).

Theorem 11.1.2 (Global Solvability of the Nonlinear Problem Without Surface Tension) *For some* $l \in (1/2, 1)$ *let the boundaries* Γ, $\Sigma \in W_2^{3/2+l}$, *vector field* $\boldsymbol{f} \in \boldsymbol{W}_2^{l,l/2}(Q_\infty)$, $\boldsymbol{f}(\cdot, t)$, $\nabla \boldsymbol{f}(\cdot, t) \in \mathrm{Lip}(\Omega)$ *for any* $t \in [0, \infty]$, $\boldsymbol{f}(\xi, \cdot)$, $\nabla \boldsymbol{f}(\xi, \cdot) \in \boldsymbol{C}^\beta(0, T)$ *for any* $\xi \in \Omega$ *with some* $\beta \in [1/2, 1)$. *We also suppose that the initial velocity vector field* $\boldsymbol{v}_0 \in \boldsymbol{W}_2^{1+l}(\cup\Omega_0^\pm)$ *satisfies conditions* (10.5.2) *and* $\boldsymbol{u}_0|_\Sigma = 0$, *in addition, it is sufficiently small, as well as mass forces, i.e.,*

$$\|\boldsymbol{v}_0\|_{\boldsymbol{W}_2^{1+l}(\cup\Omega_0^\pm)} + \|\mathrm{e}^{bt}\boldsymbol{f}\|_{\boldsymbol{W}_2^{l,l/2}(Q_\infty)} + \int\limits_0^\infty \mathrm{e}^{bt}\|\boldsymbol{f}\|_\Omega\,\mathrm{d}t + |\mathrm{e}^{bt}\boldsymbol{f}|_{Q_\infty}^{(1,\beta)} \leqslant \varepsilon \ll 1, \tag{11.1.3}$$

where $b = \min\{\nu^+, \nu^-\}/(2c_0)$, c_0 *is the constant in the Korn inequality.*

Then problem (11.1.1), (1.0.3) *is uniquely solvable for all positive moments of time* t, *and the solution* $(\boldsymbol{v}, p)$ *possesses the properties:* $\boldsymbol{v} \in \boldsymbol{W}_2^{2+l,1+l/2}$, $p \in W_2^{l,l/2}$, $\nabla p \in \boldsymbol{W}_2^{l,l/2}$, $[p]\big|_{\Gamma_t} \in W_2^{l+1/2,l/2+1/4}$, $\Gamma_t \in W_2^{3/2+l}$. *It means that for any* $t_0 \in (0,\infty)$, *the solution* $(\mathbf{u}, q)$ *and its derivatives written in the Lagrangian coordinates belong to the corresponding Sobolev spaces over* $D_{(t_0,t_0+\tau)}$ *for a sufficiently small time interval* $(t_0, t_0+\tau)$. *Moreover, there holds the estimate*

$$\|\boldsymbol{u}\|_{D_{(t_0,t_0+\tau)}}^{(2+l,1+\frac{l}{2})} + \|\nabla q\|_{D_{(t_0,t_0+\tau)}}^{(l,\frac{l}{2})} + \|q\|_{D_{(t_0,t_0+\tau)}}^{(l,\frac{l}{2})} + \|[q]|_\Gamma\|_{W_2^{l+\frac{1}{2},\frac{l}{2}+\frac{1}{4}}(G_{(t_0,t_0+\tau)})}$$
$$\leqslant c_3\mathrm{e}^{-bt_0}\Big\{\|\boldsymbol{v}_0\|_{\boldsymbol{W}_2^{1+l}(\cup\Omega_0^\pm)} + \|\mathrm{e}^{bt}\boldsymbol{f}\|_{\boldsymbol{W}_2^{l,l/2}(Q_\infty)} + \int\limits_0^\infty \mathrm{e}^{bt}\|\boldsymbol{f}\|_\Omega\,\mathrm{d}t + |\mathrm{e}^{bt}\boldsymbol{f}|_{Q_\infty}^{(1,\beta)}\Big\}, \tag{11.1.4}$$

where c_3 *is independent of* t_0.

One can conclude from this theorem that the trivial solution is unique when the initial velocity and mass forces vanish. The stability of this solution takes place in the sense that the solution differs a little from zero under a small deviation of the data from zero.

At the end of the chapter, we give a sufficient upper bound of the initial distance between fluid interface and the outer boundary.

11.2 Auxiliary Propositions

To prove the existence of a global solution to the nonlinear problem, we use the exponential L_2-estimates in Proposition 6.4.1, which are true for a solution to problem (11.1.1), (1.0.3), defined on $[0,T]$, with an initial velocity vector $\boldsymbol{v}_0$, satisfying compatibility conditions (10.5.2).

Below we use the following lemma.

Lemma 11.2.1 *Let* $v \in W_2^{2+l,1+l/2}(Q_T)$, $T > 0$, $l \in (0,1)$, $\theta > 0$. *Then the function* v *is subjected to the inequality*

$$\|v\|_{Q_T}^{(l,l/2)} \leqslant c\Big\{\theta\|v\|_{Q_T}^{(2+l,1+l/2)} + \Big(\frac{1}{\theta^{l/2}} + \frac{1}{T^{l/2}}\Big)\|v\|_{Q_T}\Big\}. \tag{11.2.1}$$

Proof We apply the well-known interpolation inequality (see, for example, [2]) for $v \in W_2^m(\Omega)$ with any $\varepsilon_1 > 0$:

$$\|v\|_{\dot{W}_2^j(\Omega)} \leqslant c\big(\varepsilon_1 \|v\|_{\dot{W}_2^m(\Omega)} + \varepsilon_1^{-\frac{j}{m-j}} \|v\|_\Omega\big), \quad 0 \leqslant j \leqslant m-1, \quad m \geqslant 1.$$

Inequality (11.2.1) follows from the estimates

$$\|v\|_{\dot{W}_2^{l,0}(Q_T)} \leqslant c\big(\theta \|v\|_{\dot{W}_2^{2+l,0}(Q_T)} + \theta^{-\frac{l}{2}} \|v\|_{Q_T}\big),$$

$$\|v\|_{\dot{W}_2^{0,l/2}(Q_T)} \leqslant c\big(\theta \|v\|_{\dot{W}_2^{0,1+l/2}(Q_T)} + \theta^{-\frac{l}{2}} \|v\|_{Q_T}\big).$$

□

Proposition 11.2.1 *Let a solution of problem* (11.1.1), (1.0.3) *be defined on an interval* $(0, T]$ *and let the estimate*

$$N_{(0,T)}[\boldsymbol{v}, p] \equiv \|\boldsymbol{u}\|_{D_T}^{(2+l,1+l/2)} + \|\nabla q\|_{D_T}^{(l,l/2)} + \|q\|_{D_T}^{(l,l/2)} + \|q\|_{W_2^{l+1/2,l/2+1/4}(G_T)} \leqslant \mu$$

hold, where the pair $(\boldsymbol{u}, q)$ *is a solution of problem* (11.1.1), (1.0.3), *written in the Lagrangian coordinates.*

Then, for any $t_0 \in (0, T]$, *there exists* $\tau_0 \in (0, t_0/2)$ *such that the norm*

$$N_{(t_0-2\tau_0+\gamma,t_0)}[\boldsymbol{v}, p] \equiv \|\boldsymbol{u}\|_{D'_\gamma}^{(2+l,1+\frac{l}{2})} + \|\nabla q\|_{D'_\gamma}^{(l,\frac{l}{2})}$$
$$+ \|q\|_{D'_\gamma}^{(l,\frac{l}{2})} + \|q\|_{W_2^{l+\frac{1}{2},\frac{l}{2}+\frac{1}{4}}(G_{(t_0-2\tau_0+\gamma,t_0)})}$$

satisfies the inequality

$$N_{(t_0-\tau_0,t_0)}[\boldsymbol{v}, p] \leqslant c(\delta, \tau_0)\Big\{\|\boldsymbol{f}\|_{Q'_0}^{(l,\frac{l}{2})} + |\boldsymbol{f}|_{Q'_0}^{(1,\beta)} + \|\boldsymbol{v}\|_{Q'_0}\Big\}, \tag{11.2.2}$$

where $Q'_\gamma \equiv \Omega \times (t_0 - 2\tau_0 + \gamma, t_0)$, $D'_\gamma \equiv D'_{(t_0-2\tau_0+\gamma,t_0)}$, $\gamma \geqslant 0$, τ_0 *depends on* μ *and on the constant* δ *in* (11.2.4), $c(\delta, \tau_0)$ *is a nondecreasing function.*

Proof We fix arbitrary $t_0 \in (0, T]$. Let $\tau_0 \in (0, t_0/2)$, and let $\eta_\lambda(t)$ be a smooth monotone function of t such that

$$\eta_\lambda(t) = \begin{cases} 0 & \text{if } t \leqslant t_0 - 2\tau_0 + \lambda/2, \\ 1 & \text{if } t \geqslant t_0 - 2\tau_0 + \lambda, \end{cases}$$

$\lambda \in (0, \tau_0]$, and for $\dot{\eta}_\lambda(t) \equiv \frac{\mathrm{d}\eta_\lambda(t)}{\mathrm{d}t}$ the inequalities

$$\sup_{\mathbb{R}} \left|\dot{\eta}_\lambda(t)\right| \leqslant c\lambda^{-1}, \quad \sup_{t,\tau\in\mathbb{R}} \frac{|\dot{\eta}_\lambda(t) - \dot{\eta}_\lambda(\tau)|}{|t-\tau|^{\,l/2}} \leqslant c\lambda^{-1-l/2}$$

hold.

The couple $\boldsymbol{w} = \boldsymbol{v}\eta_\lambda$, $s = p\eta_\lambda$ satisfies the system

$$\begin{aligned}
&\mathcal{D}_t\boldsymbol{w} + (\boldsymbol{v}\cdot\nabla)\boldsymbol{w} - \nu^\pm\nabla^2\boldsymbol{w} + \frac{1}{\rho^\pm}\nabla s = \boldsymbol{f}\eta_\lambda + \boldsymbol{v}\dot{\eta}_\lambda,\\
&\nabla\cdot\boldsymbol{w} = 0 \quad \text{in} \quad \Omega_t^+\cup\Omega_t^-, \quad t > t_0 - 2\tau_0,\\
&\boldsymbol{w}\big|_{t=t_0-2\tau_0} = 0 \quad \text{in} \quad \cup\,\Omega' \equiv \Omega^-_{t_0-2\tau_0}\cup\Omega^+_{t_0-2\tau_0},\\
&[\boldsymbol{w}]\big|_{\Gamma_t} = 0, \quad [\mathbb{T}(\boldsymbol{w},s)\boldsymbol{n}]\big|_{\Gamma_t} = 0, \quad \boldsymbol{w}\big|_\Sigma = 0, \quad t > t_0 - 2\tau_0.
\end{aligned}$$

We introduce the Lagrangian coordinates by formula (6.4.15), where $\boldsymbol{u}(\xi',t)$ means the vector $\boldsymbol{v}(X(\xi',t),t)$, $\xi'\in\cup\Omega^\pm_{t_0-2\tau_0}$. The functions $\boldsymbol{w}$ and s written in the Lagrangian coordinates will be denoted by the same symbols. They solve the problem

$$\begin{aligned}
&\mathcal{D}_t\boldsymbol{w} - \nu^\pm\nabla^2_{\boldsymbol{u}}\boldsymbol{w} + \frac{1}{\rho^\pm}\nabla_{\boldsymbol{u}}s = \boldsymbol{f}(X,t)\eta_\lambda + \boldsymbol{u}\dot{\eta}_\lambda, \qquad \nabla_{\boldsymbol{u}}\cdot\boldsymbol{w} = 0 \quad \text{in} \quad D_0',\\
&\boldsymbol{w}\big|_{t=t_0-2\tau_0} = 0 \quad \text{in} \quad \cup\,\Omega', \qquad\qquad (11.2.3)\\
&[\boldsymbol{w}]\big|_{\Gamma'} = 0, \quad [\mu^\pm\Pi_0'\Pi\mathbb{S}_{\boldsymbol{u}}(\boldsymbol{w})\boldsymbol{n}]\big|_{\Gamma'} = 0, \qquad \boldsymbol{w}|_\Sigma = 0,\\
&[\boldsymbol{n}_0'\cdot\mathbb{T}_{\boldsymbol{u}}(\boldsymbol{w},s)\boldsymbol{n}]\big|_{\Gamma'} = 0, \quad t > t_0 - 2\tau_0.
\end{aligned}$$

Here $\Gamma' = \Gamma_{t_0-2\tau_0}$, $\boldsymbol{n}_0'$ is the outward normal to Γ', Π_0' and Π are the projectors onto the tangent planes to Γ' and to Γ_t, respectively. The other notation, for instance, $\nabla_{\boldsymbol{u}}$, also corresponds to transformation (6.4.15).

We apply Theorem 10.5.3 to problem (11.2.3). To this end, we choose τ_0 so small that inequality (10.5.5) holds. It is sufficient to take τ_0 such that

$$(2\tau_0)^{1/2}\mu \leqslant \delta. \qquad (11.2.4)$$

The right-hand side of the first equation in (11.2.3) belongs to $\boldsymbol{W}_2^{l,l/2}(D_0')$. Hence, by (10.5.4)

$$\begin{aligned}
N_{(t_0-2\tau_0+\lambda,t_0)}[\boldsymbol{v},p] &\leqslant N_{(t_0-2\tau_0,t_0)}[\boldsymbol{w},s]\\
&\leqslant c_1(2\tau_0)\Big\{\|\boldsymbol{f}(X,t)\eta_\lambda\|^{(l,l/2)}_{Q_0'} + \|\boldsymbol{u}\dot{\eta}_\lambda\|^{(l,l/2)}_{D_0'}\Big\}.
\end{aligned}$$

We can estimate the Sobolev norm of the composite function $\boldsymbol{f}(X(\xi,t),t)$ as follows:

$$\|\boldsymbol{f}(X,t)\|_{Q_0'}^{(l,l/2)} \leqslant \|\boldsymbol{f}\|_{Q_0'}^{(l,l/2)} + c\Big(1 + (2\tau_0)^{1-l/2}\|\boldsymbol{u}\|_{\boldsymbol{W}_2^{l,0}(D_0')}\Big)|\boldsymbol{f}|_{Q_0'}^{(1,\beta)}.$$

By Lemma 11.2.1, we conclude for $\lambda \leqslant 1$ that

$$\begin{aligned} N_{(t_0-2\tau_0+\lambda,t_0)}[\boldsymbol{v},p] &\leqslant c_2\Big\{\|\boldsymbol{f}(X,t)\|_{Q_0'}^{(l,l/2)} + \frac{1}{\lambda^{\frac{l}{2}}}\|\boldsymbol{f}(X,t)\|_{Q_0'} + \frac{1}{\lambda}\|\boldsymbol{u}\|_{D'_{\lambda/2}}^{(l,l/2)} \\ &\qquad + \frac{1}{\lambda^{1+\frac{l}{2}}}\|\boldsymbol{u}\|_{D'_{\lambda/2}}\Big\} \\ &\leqslant c_3(1+\delta)\Big\{\frac{1}{\lambda^{\frac{l}{2}}}\|\boldsymbol{f}\|_{Q_0'}^{(l,l/2)} + |\boldsymbol{f}|_{Q_0'}^{(1,\beta)} + \frac{\theta}{\lambda}\|\boldsymbol{u}\|_{D'_{\lambda/2}}^{(2+l,1+l/2)} \\ &\qquad + \Big(\frac{1}{\lambda\theta^{\frac{l}{2}}} + \frac{1}{\lambda(2\tau_0)^{\frac{l}{2}}} + \frac{1}{\lambda^{1+\frac{l}{2}}}\Big)\|\boldsymbol{u}\|_{D'_{\lambda/2}}\Big\} \end{aligned} \tag{11.2.5}$$

We set now $\theta = \varepsilon\lambda$ in estimate (11.2.5). Then we have

$$\begin{aligned} N_{(t_0-2\tau_0+\lambda,t_0)}[\boldsymbol{v},p] \leqslant c_4(\delta)\Big\{&\varepsilon N_{(t_0-2\tau_0+\lambda/2,t_0)}[\boldsymbol{v},p] + \frac{1}{\lambda^{\frac{l}{2}}}\|\boldsymbol{f}\|_{Q_0'}^{(l,l/2)} + |\boldsymbol{f}|_{Q_0'}^{(1,\beta)} + \\ &+ \frac{1}{\lambda^{1+\frac{l}{2}}}\Big(\varepsilon^{-\frac{l}{2}} + 1\Big)\|\boldsymbol{v}\|_{Q'_{\lambda/2}}\Big\}. \end{aligned} \tag{11.2.6}$$

We employ again the technique described in § 7.4, and we introduce the function $\Phi(\lambda) = \lambda^{1+\frac{l}{2}} N_{(t_0-2\tau_0+\lambda,t_0)}[\boldsymbol{v},p]$. Then we can rewrite (11.2.6) in the form:

$$\Phi(\lambda) \leqslant c_5\varepsilon\Phi(\lambda/2) + K, \tag{11.2.7}$$

where $c_5 = c_4(\delta)2^{1+\frac{l}{2}}$,

$$K = c_4(\delta)\Big\{\|\boldsymbol{f}\|_{Q_0'}^{(l,l/2)} + |\boldsymbol{f}|_{Q_0'}^{(1,\beta)} + c(\varepsilon)\|\boldsymbol{v}\|_{Q_0'}\Big\}.$$

We put $\varepsilon = \frac{1}{2c_5}$ in (11.2.7). By iterations with $\lambda/2,\ldots,\lambda/2^k$, we deduce from inequality (11.2.7) in the limit as $k \to \infty$ that

$$\Phi(\lambda) \leqslant 2K.$$

This inequality with $\lambda = \tau_0$ implies (11.2.2). □

11.3 The Proof of the Existence of a Global Solution

Proof of Theorem 11.1.2 By Theorem 11.1.1, we have a solution $(\boldsymbol{v}, p)$ in the interval $(0, T_0]$. We can take ε in (11.1.3) so small that T_0 is greater than 1. Moreover, according to (11.1.2), the solution norm satisfies the inequality

$$N_{(0,T_0)}[\boldsymbol{v}, p] \leqslant \mu \tag{11.3.1}$$

with some $\mu > 0$. Then, due to Proposition 11.2.1, there exists $\tau_0 < T_0/2$ such that (11.2.4) is satisfied and estimate (11.2.2) holds for any $t_0 \in (T_0/2, T_0]$. Next, inequalities (6.4.4), (11.1.3) imply that

$$\|\boldsymbol{v}\|_{Q_0'} \leqslant \left\{ \int_{t_0-2\tau_0}^{t_0} \mathrm{e}^{-2bt} \Big(\|\boldsymbol{v}_0\|_\Omega + c_6 \int_0^t \mathrm{e}^{b\tau} \|\boldsymbol{f}\|_\Omega \mathrm{d}\tau \Big)^2 \mathrm{d}t \right\}^{\frac{1}{2}} \leqslant c\, \mathrm{e}^{-b(t_0-2\tau_0)} \sqrt{2\tau_0}\varepsilon. \tag{11.3.2}$$

Thus,

$$\begin{aligned} N_{(t_0-\tau_0,t_0)}[\boldsymbol{v}, p] &\leqslant c_7(\delta, \tau_0)\mathrm{e}^{-bt_0} \Big\{ \|\mathrm{e}^{bt}\boldsymbol{f}\|_{Q_0'}^{(l,\frac{l}{2})} + |\mathrm{e}^{bt}\boldsymbol{f}|_{Q_0'}^{(1,\beta)} + \mathrm{e}^{2b\tau_0}\sqrt{2\tau_0}\varepsilon \Big\} \\ &\leqslant c_8(\delta, \tau_0)\mathrm{e}^{-bt_0}\varepsilon \quad \text{for} \quad \forall t_0 \in (T_0/2, T_0], \end{aligned} \tag{11.3.3}$$

here $c_7(\delta, \tau_0)$, $c_8(\delta, \tau_0)$ are nondecreasing functions of τ_0. From trace theorem for $W_2^{2+l,1+l/2}(D^{\pm}_{(T_0-\tau_0,T_0)})$ and (11.3.3), it follows that

$$\|\boldsymbol{u}(\cdot, T_0)\|_{\boldsymbol{W}_2^{1+l}(\cup\Omega_0^{\pm})} \leqslant c_8(\delta, \tau_0)\mathrm{e}^{-bT_0}\varepsilon.$$

In addition, because of (6.4.4)

$$\|\boldsymbol{u}(\cdot, T_0)\|_\Omega = \|\boldsymbol{v}(\cdot, T_0)\|_\Omega \leqslant \mathrm{e}^{-bT_0} \Big\{ \|\boldsymbol{v}_0\|_\Omega + c_6 \int_0^{T_0} \|\mathrm{e}^{b\tau}\boldsymbol{f}(\cdot, \tau)\|_\Omega \mathrm{d}\tau \Big\} \leqslant (c_6+1)\varepsilon. \tag{11.3.4}$$

We apply Theorem 11.1.1 once more and obtain a solution on an interval $(T_0, T_0 + T_1]$ with $0 < T_1 \leqslant T_0$ corresponding to the initial data $\boldsymbol{v}(\cdot, T_0)$. Due to (11.1.2), we get

$$N_{(T_0,T_0+T_1)}[\boldsymbol{v}, p] \leqslant c(T_1)(\varepsilon + c_8(\delta, \tau_0)\mathrm{e}^{-bT_0}\varepsilon) \leqslant \mu,$$

where μ is the same as in (11.3.1) for sufficiently small ε. Then by Proposition 11.2.1 and in view of (11.3.2), (11.3.4), we have similarly to (11.3.3) that

$$\begin{aligned} N_{(T_0+T_1-\tau_1,T_0+T_1)}[\boldsymbol{v}, p] &\leqslant c_7(\delta, \tau_1)\mathrm{e}^{-b(T_0+T_1)} \Big\{ \|\mathrm{e}^{bt}\boldsymbol{f}\|_{Q_0''}^{(l,l/2)} + |\mathrm{e}^{bt}\boldsymbol{f}|_{Q_0''}^{(1,\beta)} + 2\mathrm{e}^{2b\tau_1}\sqrt{2\tau_1}\varepsilon \Big\} \\ &\leqslant c_8(\delta, \tau_0)\mathrm{e}^{-b(T_0+T_1)}\varepsilon, \end{aligned} \tag{11.3.5}$$

where $Q_0'' \equiv \Omega\times(T_0+T_1-2\tau_1, T_0+T_1)$ and $\tau_1\in(0,T_1/2)$, $\tau_1\leqslant\tau_0/4$. Hence, (11.3.5) gives us

$$\|\boldsymbol{u}(\cdot,T_0+T_1)\|_{\boldsymbol{W}_2^{1+l}(\cup\Omega_0^\pm)} \leqslant c_8(\delta,\tau_0)\mathrm{e}^{-b(T_0+T_1)}\varepsilon,$$

and in virtue of (6.4.4), (11.3.4),

$$\|\boldsymbol{u}(\cdot,T_0+T_1)\|_\Omega \leqslant \mathrm{e}^{-bT_1}\Big\{\|\boldsymbol{v}(\cdot,T_0)\|_\Omega + c_6\int_{T_0}^{T_0+T_1}\|\mathrm{e}^{b\tau}\boldsymbol{f}(\cdot,\tau)\|_\Omega \mathrm{d}\tau\Big\}$$

$$\leqslant \mathrm{e}^{-bT_1}\Big\{\mathrm{e}^{-bT_0}\|\boldsymbol{v}_0\|_\Omega + c_6\mathrm{e}^{-bT_0}\int_0^{T_0}\|\mathrm{e}^{b\tau}\boldsymbol{f}(\cdot,\tau)\|_\Omega\mathrm{d}\tau$$

$$+ c_6\int_{T_0}^{T_0+T_1}\|\mathrm{e}^{b\tau}\boldsymbol{f}(\cdot,\tau)\|_\Omega\mathrm{d}\tau\Big\} \leqslant (c_6+1)\varepsilon.$$

Since data norms are bounded above by the same constant as before, $(\boldsymbol{v},p)$ exists on $(T_0+T_1, T_0+2T_1]$ and

$$N_{(T_0+T_1,T_0+2T_1)}[\boldsymbol{v},p] \leqslant \mu.$$

Hence, inequality

$$N_{(t_0-\tau_1,t_0)}[\boldsymbol{v},p] \leqslant c_8(\delta,\tau_0)\mathrm{e}^{-bt_0}\varepsilon \qquad (11.3.6)$$

holds for any $t_0\in(T_0/2, T_0+2T_1]$ and so on. Thus, the solution of problem (11.1.1), (1.0.3) can be extended as far as one likes, estimate (11.1.4) holding for any positive t_0.

The uniqueness of a global solution follows from the uniqueness of local ones.

In conclusion, we estimate the expansion of the interface Γ_t. To this end, we need to evaluate the speed of interface displacement. As $l>1/2$, $\boldsymbol{W}_2^{1+l}(\Omega_0^+)$ is embedded in the space of the continuous functions. Consequently, by the embedding theorem, we can deduce from inequality (11.3.6) the estimate

$$\max_{\Omega_0^+}|\boldsymbol{u}(\cdot,t)| \leqslant c_9\mathrm{e}^{-bt}, \qquad t>T_0/2.$$

We integrate this inequality by t from $T_0/2$ till infinity, then we have

$$\int_{T_0/2}^\infty \max_{\Omega_0^+}|\boldsymbol{u}(\cdot,t)|\mathrm{d}t \leqslant c_{10}.$$

Thus, if the distance between the interface Γ and the solid boundary Σ at the initial moment is greater than $c_{11}=\frac{T_0}{2}\sup_{Q^+_{T_0/2}}|\boldsymbol{u}|+c_{10}$, we can guarantee that these surfaces never intersect. □

Chapter 12
L_2-Theory for Two-Phase Capillary Fluid

Abstract The chapter is devoted to the problem of the unsteady motion of a viscous drop in an incompressible fluid which is contained in a bounded vessel. It is assumed that the fluids are subjected to mass forces and capillary ones on the interface. We prove the stability of the rest state under the assumption that initial velocities are small, the drop is close to a ball at the initial instant, and mass forces decay as $t \to \infty$ but not necessarily in an exponential way.

This material is based on the article [32]. Its Russian version was published as a preprint [31].

Here we continue to study two-phase problem (7.1.1) in the Sobolev–Slobodetskiĭ spaces $W_2^{2+l,1+l/2}$, $l \in (1/2, 1)$, in the three-dimensional case. We concentrate on the proof of the stability of the rest state and construct a solution assuming that the data of the problem are close to this state, i.e., the initial velocities and mass forces are small, and the initial interface is close to a sphere S_{R_0} of the radius R_0 such that the ball bounded by this sphere has the same volume as the inner fluid. We place the center of this ball into the origin which coincides with the barycenter of the drop at the initial instant $t = 0$, the interface being defined as a normal perturbation of $S_{R_0}(0)$. We find it reasonable to consider also the unknown interface for $t > 0$ as a normal perturbation of the sphere $S_{R_0}(h)$ of the same radius R_0 and with the center placed at the barycenter $h(t)$ of the inner domain. Therefore, we add the term with the vector $\boldsymbol{h}(t)$ into the standard Hanzawa transformation of two-part domain with an unknown interface into the domain with the interface $S_{R_0}(0)$. In our opinion, this allows us to take interface evolution into account in the most precise way.

Next, we linearize the problem transformed. In Sect. 12.2 we analyze a linear problem in two domains separated by S_{R_0} and prove maximal regularity estimates for a solution of the problem first on an arbitrary finite time interval in the standard Sobolev–Slobodetskiĭ spaces and then, under some additional assumptions, on the infinite interval $t > 0$ in these spaces

© The Author(s), under exclusive license to Springer Nature Switzerland AG 2021

I. V. Denisova, V. A. Solonnikov, *Motion of a Drop in an Incompressible Fluid*, Advances in Mathematical Fluid Mechanics,
https://doi.org/10.1007/978-3-030-70053-9_12

with the exponential weight $e^{\beta t}$, $\beta > 0$. In Sect. 12.3, on the basis of these estimates and of inequalities for nonlinear terms, we construct a solution at first for $t \in (0, T_0)$ with appropriate $T_0 > 1$, and then we extend this solution into the interval $(T_0, 2T_0)$ and so forth, step by step, we get a solution for any $t > 0$. We show that the velocities and pressure gradient decay exponentially to zero as $t \to \infty$, and Γ_t tends to a sphere of radius R_0 centered at $h(\infty)$ close to $S_{R_0}(0)$ but, in general, different from it.

In addition, we admit here a more general than exponential decay of the vector field of mass forces. The proofs are constructed in the same manner as in Chap. 7, but the final estimate of the solution (see Theorem 12.3.2) is slightly different from those in the preceding chapters.

As before, the idea of constructing a function of generalized energy [58, 87] is used for obtaining an exponential estimate of the solution instead of an analysis of the spectrum of the linear problem. It is worth noting that our technique can be generalized to the case of a multi-phase fluid and that of the dimension $n > 3$.

12.1 The Statement of the Theorem on Global Solvability

We turn to problem (7.1.1), (7.1.2) again.

We recall that $\boldsymbol{n}(x,t)$ is the outward normal to Γ_t with respect to Ω_t^+, $[\boldsymbol{u}]|_{\Gamma_t} = \boldsymbol{u}^+ - \boldsymbol{u}^-$ is the jump of the vector $\boldsymbol{u}$ across Γ_t, V_n is evolution speed of Γ_t in the direction of $\boldsymbol{n}$, $R_0 = (3|\Omega_0^+|/4\pi)^{1/3}$, $|\Omega_0^+| = \text{mes}\, \Omega_0^+$.

We assume that the surface Γ_0 is close to the sphere S_{R_0} the center of which coincides with the barycenter of the domain Ω_0^+. Without the restriction of generality, we suppose that it is placed at the origin. Then Γ_0 can be viewed as a normal perturbation of S_{R_0}, i.e.,

$$\Gamma_0 = \{x \in \mathbb{R}^3 | x = y + r_0(y)\boldsymbol{N}(y)\}, \quad y \in S_{R_0},$$

where $\boldsymbol{N}(y) = \boldsymbol{y}/|y|$, $y \in S_{R_0}$, and r_0 is a small given function.

We use a similar representation formula for the unknown surface Γ_t, $t > 0$:

$$\Gamma_t = \{x \in \mathbb{R}^3 | x = y + \boldsymbol{h}(t) + r(y,t)\boldsymbol{N}(y)\},$$

where $r(y,t)$ is an unknown function on S_{R_0}. The coordinates of the barycenter of Ω_t^+ are given by (7.2.1).

We extend $\boldsymbol{N}$ in $\mathbb{R}^3$ by the formula $\boldsymbol{N}^*(y) = \omega(y)\boldsymbol{y}/|y|$, where $\omega(y)$ is a smooth function, equal to 1 for $|y| \geqslant 2R_0/3$ and to 0 for $|y| \leqslant R_0/3$. For r, we introduce the extension $r^*(y,t) = \phi(y)\big(\mathcal{E}r\big)(y,t)$, where $\phi(y)$ is a smooth cutoff function equal to 1 in the neighborhood of S_{R_0} and to 0 near the outer boundary Σ, while $\mathcal{E}$ is a fixed extension operator from S_{R_0} into $\mathbb{R}^3$. We also require that

$$\frac{\partial r^*}{\partial N}|_{S_{R_0}} = 0,$$
$$\|r^*\|_{W_2^{l'+1/2}(\mathbb{R}^3)} \leqslant c\|r\|_{W_2^{l'}(S_{R_0})}, \quad \forall l' \in (0, 2+l],$$

and $r^*(y,t) = 0$ for $\big||y| - R_0\big| \geqslant d_0$, in particular, for y close to Σ, where d_0 is a small positive number (W_2^m is a Sobolev–Slobodetskiĭ space, see Sect. 10.1.) If $r^*(y,t)$ is differentiable with respect to t, then

$$\|\mathcal{D}_t r^*\|_{W_2^{l'+1/2}(\mathbb{R}^3)} \leqslant c\|\mathcal{D}_t r\|_{W_2^{l'}(S_{R_0})}. \tag{12.1.1}$$

We introduce the modified Hanzawa transformation

$$x = y + r^*(y,t)\boldsymbol{N}^*(y) + \chi(y)\boldsymbol{h}(t) \equiv e_{r,\boldsymbol{h}}(y,t), \tag{12.1.2}$$

where $\chi(y)$ is a smooth cutoff function, equal to 1 for $\big||y| - R_0\big| \leqslant d_0/2$ and to 0 for $\big||y| - R_0\big| \geqslant d_0$. If r and $\boldsymbol{h}(t)$ are small, and d_0 is chosen in a proper way, then this mapping is invertible and it establishes one-to-one correspondences between the ball $B^+ \equiv \{|y| < R_0\}$ and Ω_t^+, between sphere S_{R_0} and Γ_t, as well as between $B^- \equiv \Omega \setminus \overline{B^+}$ and Ω_t^-.(This is obvious for $t = 0$ when $\boldsymbol{h}(0) = 0$, and it remains true for small $\boldsymbol{h}(t)$.) We denote by $\mathbb{L}$ the Jacobi matrix of transformation (12.1.2), and we set $L = \det\mathbb{L}$, $\widehat{\mathbb{L}} = L\mathbb{L}^{-1}$. It is evident that

$$\mathbb{L}(r,\boldsymbol{h}) = \Big\{\delta_j^i + \frac{\partial\big(r^*(y,t)N_i^*(y)\big)}{\partial y_j} + h_i(t)\frac{\partial\chi(y)}{\partial y_j}\Big\}_{i,j=1}^3.$$

For the points y located on the sphere S_{R_0} and near it, we have $\nabla\chi = 0$ and

$$\mathbb{L} = \mathbb{L}(r,\boldsymbol{0}) = \Big\{\delta_j^i + \frac{\partial\big(r^*(y,t)N_i^*(y)\big)}{\partial y_j}\Big\}_{i,j=1}^3.$$

Mapping (12.1.2) converts (7.1.3), (7.1.2) into the system

$$\mathcal{D}_t\boldsymbol{u} - \nu^{\pm}\widetilde{\nabla}^2\boldsymbol{u} - (\mathbb{L}^{-1}(\mathcal{D}_t r^*\boldsymbol{N}^* + \chi\dot{\boldsymbol{h}})\cdot\nabla)\boldsymbol{u} + (\mathbb{L}^{-1}\boldsymbol{u}\cdot\nabla)\boldsymbol{u} + \frac{1}{\rho^{\pm}}\widetilde{\nabla}q = \widehat{\boldsymbol{f}},$$
$$\widetilde{\nabla}\cdot\boldsymbol{u} = 0 \quad \text{in } B^{\pm},\ t > 0,$$
$$\boldsymbol{u}(y,0) = \boldsymbol{u}_0(y) \quad \text{in } B \equiv B^+ \cup B^-, \quad r(y,0) = r_0(y) \quad \text{on } S_{R_0},$$
$$[\boldsymbol{u}]\big|_{S_{R_0}} = 0, \quad [\mu^{\pm}\Pi\widetilde{\mathbb{S}}(\boldsymbol{u})\boldsymbol{n}]\big|_{S_{R_0}} = 0, \qquad \boldsymbol{u}|_{\Sigma} = 0, \tag{12.1.3}$$

$$[-q+\mu^{\pm}\boldsymbol{n}\cdot\widetilde{\mathbb{S}}(\boldsymbol{u})\boldsymbol{n}]\big|_{S_{R_0}}=\sigma\Big(H\big(e_{r,\mathbf{0}}(y,t),t\big)+\frac{2}{R_0}\Big)\Big|_{S_{R_0}},\qquad \int_{B^+}q(y,t)\,\mathrm{d}y=0,$$

$$\mathcal{D}_t r-\Big(\boldsymbol{u}-\frac{1}{|B^+|}\int_{B^+}\boldsymbol{u}L(r,\mathbf{0})\,\mathrm{d}y\Big)\cdot\frac{\boldsymbol{n}}{\boldsymbol{N}\cdot\boldsymbol{n}}=0 \text{ on } S_{R_0},\ t>0,$$

where $\boldsymbol{u}(y,t)=\boldsymbol{v}(e_{r,\boldsymbol{h}}(y,t),t)$, $q(y,t)=p(e_{r,\boldsymbol{h}}(y,t),t)$, $\widetilde{\nabla}=\mathbb{L}^{-T}\nabla=(\mathbb{L}^{-1})^T\nabla$ is the transformed gradient of ∇_x (the superscript "T" means transposition), $\widetilde{\mathbb{S}}(\boldsymbol{u})=\widetilde{\nabla}\boldsymbol{u}+(\widetilde{\nabla}\boldsymbol{u})^T$ is the transformed doubled rate-of-strain tensor, $\widehat{\boldsymbol{f}}(y,t)=\boldsymbol{f}(e_{r,\boldsymbol{h}}(y,t),t)$, $\boldsymbol{u}_0(y)=\boldsymbol{v}_0(e_{r_0,\mathbf{0}}(y,0))$, $\Pi\boldsymbol{g}=\boldsymbol{g}-\boldsymbol{n}(\boldsymbol{n}\cdot\boldsymbol{g})$.

In problem (12.1.3), we have added normalization condition $\int_{B^+}q\,\mathrm{d}y=0$ for the pressure. It can be also taken in another form, for instance,

$$\int_{B^-}q(y,t)\,\mathrm{d}y=0, \tag{12.1.4}$$

or $\int_\Sigma q(y,t)\,\mathrm{d}\Sigma=0$. Pressure functions satisfying different normalization conditions differ from each other by certain functions of time. If, say, $\int_{B^+}q\,\mathrm{d}y=0$, $\widehat{q}$ satisfies (12.1.4) in the domain B^-, and $\widetilde{q}$ does it on Σ, respectively, then

$$q(y,t)=\widehat{q}(y,t)+\widehat{c}(t)=\widetilde{q}(y,t)+\widetilde{c}(t)$$

with

$$\widehat{c}(t)=|B^-|^{-1}\int_{B^-}q(y,t)\,\mathrm{d}y,\quad \widetilde{c}(t)=|\Sigma|^{-1}\int_\Sigma q(y,t)\,\mathrm{d}\Sigma.$$

It is obvious that $[q]|_{S_{R_0}}=[\widehat{q}]|_{S_{R_0}}=[\widetilde{q}]|_{S_{R_0}}$.

The last equation in (12.1.3) for unknown function r on S_{R_0} describing the evolution of the interface Γ_t arises from the condition $V_n=\boldsymbol{v}\cdot\boldsymbol{n}$ on Γ_t in view of (12.1.2) and (7.2.1), since $V_n\equiv\mathcal{D}_t\boldsymbol{x}\cdot\boldsymbol{n}=\mathcal{D}_t r(\boldsymbol{N}\cdot\boldsymbol{n})+\dot{\boldsymbol{h}}\cdot\boldsymbol{n}$, $\dot{\boldsymbol{h}}\equiv\mathrm{d}\boldsymbol{h}/\mathrm{d}t$.

System (12.1.3) can be written in the form:

$$\begin{aligned}
&\mathcal{D}_t\boldsymbol{u}-\nu^{\pm}\nabla^2\boldsymbol{u}+\frac{1}{\rho^{\pm}}\nabla q=\boldsymbol{l}_1(\boldsymbol{u},q,r)+\widehat{\boldsymbol{f}},\qquad \nabla\cdot\boldsymbol{u}=l_2(\boldsymbol{u},r) \text{ in } B^{\pm},\ t>0,\\
&\boldsymbol{u}(y,0)=\boldsymbol{u}_0(y)\quad \text{in } B,\quad r(y,0)=r_0(y)\quad \text{on } S_{R_0},\\
&[\boldsymbol{u}]\big|_{S_{R_0}}=0,\quad [\mu^{\pm}\Pi_0\mathbb{S}(\boldsymbol{u})\boldsymbol{N}]\big|_{S_{R_0}}=\boldsymbol{l}_3(\boldsymbol{u},r),\qquad \boldsymbol{u}|_\Sigma=0,\\
&[-q+\mu^{\pm}\boldsymbol{N}\cdot\mathbb{S}(\boldsymbol{u})\boldsymbol{N}]\big|_{S_{R_0}}-\sigma\mathcal{B}_0 r=l_4(\boldsymbol{u},r)+\sigma l_5(r),\qquad \int_{B^+}q(y,t)\,\mathrm{d}y=0,\\
&\mathcal{D}_t r-\Big(\boldsymbol{u}-\frac{1}{|B^+|}\int_{B^+}\boldsymbol{u}\,\mathrm{d}y\Big)\cdot\boldsymbol{N}=l_6(\boldsymbol{u},r)\qquad \text{on } S_{R_0},\ t>0,
\end{aligned} \tag{12.1.5}$$

where $\mathcal{B}_0 r = \Delta_{S_{R_0}} r + 2R_0^{-2} r$, $\Delta_{S_{R_0}}$ is the Laplace–Beltrami operator on S_{R_0},

$$\begin{aligned}
\boldsymbol{l}_1(\boldsymbol{u}, q, r) &= \nu^{\pm}(\widetilde{\nabla}^2 - \nabla^2)\boldsymbol{u} + \frac{1}{\rho^{\pm}}(\nabla - \widetilde{\nabla})q + \Big(\mathbb{L}^{-1}(\mathcal{D}_t r^* \boldsymbol{N}^* + \chi \dot{\boldsymbol{h}}(t)) \cdot \nabla\Big)\boldsymbol{u} \\
&\quad - (\mathbb{L}^{-1}\boldsymbol{u} \cdot \nabla)\boldsymbol{u}, \\
l_2(\boldsymbol{u}, r) &= (\mathbb{I} - \widehat{\mathbb{L}}^T)\nabla \cdot \boldsymbol{u} = \nabla \cdot \boldsymbol{L}(\boldsymbol{u}, r), \quad \boldsymbol{L}(\boldsymbol{u}, r) = (\mathbb{I} - \widehat{\mathbb{L}})\boldsymbol{u}, \\
\boldsymbol{l}_3(\boldsymbol{u}, r) &= \big[\mu^{\pm}\Pi_0\big(\Pi_0 \mathbb{S}(\boldsymbol{u})\boldsymbol{N} - \Pi\widetilde{\mathbb{S}}(\boldsymbol{u})\boldsymbol{n}\big)\big]\big|_{S_{R_0}}, \\
l_4(\boldsymbol{u}, r) &= \big[\mu^{\pm}\big(\boldsymbol{N} \cdot \mathbb{S}(\boldsymbol{u})\boldsymbol{N} - \boldsymbol{n} \cdot \widetilde{\mathbb{S}}(\boldsymbol{u})\boldsymbol{n}\big)\big]\big|_{S_{R_0}}, \qquad (12.1.6) \\
l_5(r) &= -\int_0^1 (1-s)\frac{\mathrm{d}^2}{\mathrm{d}s^2}\widehat{\mathbb{L}}^T(sr, \mathbf{0})\nabla \cdot \boldsymbol{n}_s \,\mathrm{d}s, \quad \boldsymbol{n}_s = \frac{\widehat{\mathbb{L}}^T(sr, \mathbf{0})\boldsymbol{N}}{|\widehat{\mathbb{L}}^T(sr, \mathbf{0})\boldsymbol{N}|}, \\
l_6(\boldsymbol{u}, r) &= \Big(\boldsymbol{u} - \frac{1}{|B^+|}\int_{B^+} \boldsymbol{u}(y', t)\,\mathrm{d}y'\Big) \cdot \Big(\frac{\widehat{\mathbb{L}}^T(r, \mathbf{0})\boldsymbol{N}}{\boldsymbol{N} \cdot \widehat{\mathbb{L}}^T(r, \mathbf{0})\boldsymbol{N}} - \boldsymbol{N}\Big) \\
&\quad - \frac{1}{|B^+|}\int_{B^+} \big(L(r, \mathbf{0}) - 1\big)\boldsymbol{u}\,\mathrm{d}y \cdot \frac{\widehat{\mathbb{L}}^T(r, \mathbf{0})\boldsymbol{N}}{\boldsymbol{N} \cdot \widehat{\mathbb{L}}^T(r, \mathbf{0})\boldsymbol{N}}, \\
\Pi_0 \boldsymbol{g} &= \boldsymbol{g} - \boldsymbol{N}(\boldsymbol{N} \cdot \boldsymbol{g}), \qquad \boldsymbol{N}(y) \cdot \widehat{\mathbb{L}}^T(y, t)\boldsymbol{N}(y) = \boldsymbol{y} \cdot \widehat{\mathbb{L}}^T \boldsymbol{y}/|y|^2.
\end{aligned}$$

The vectors $\boldsymbol{n}(x, t)$ and $\boldsymbol{N}(y)$ are connected by the relation

$$\boldsymbol{n}(x, t)|_{x = e_{r,\mathbf{0}}(y,t)} = \frac{\widehat{\mathbb{L}}^T(r, \mathbf{0})\boldsymbol{N}(y)}{|\widehat{\mathbb{L}}^T(r, \mathbf{0})\boldsymbol{N}(y)|}\Big|_{S_{R_0}}.$$

In addition, we observe that $H(e_{r,\mathbf{0}}, t) + \frac{2}{R_0} = \mathcal{B}_0 r + l_5$, $\mathcal{B}_0 r$ being the first variation of $H + 2/R_0$ with respect to r and l_5 being a nonlinear remainder. By $\boldsymbol{n}_s$ we denote the normal to the surface

$$\Gamma_{t,s} = \{x \in \mathbb{R}^3 | x = y + sr(y, t)\boldsymbol{N}(y), \quad y \in S_{R_0}\}, \quad s \in (0, 1).$$

Remark 12.1.1 *Equation $z = y + r(y, t)\boldsymbol{N}(y)$ defines the surface Γ_t shifted by the vector $-\boldsymbol{h}(t)$.*

Finally, we recall that the condition $|\Omega_t^+| = 4\pi R_0^3/3$ and the fact that the barycenter of Ω_t^+ is placed at the origin $\{y = 0\}$ can be expressed in terms of the function r as follows:

$$\int_{S_{R_0}} \big((R_0 + r)^3 - R_0^3\big)\,\mathrm{d}S = 0, \quad \int_{S_{R_0}} y_j\big((R_0 + r)^4 - R_0^4\big)\,\mathrm{d}S = 0, \quad j = 1, 2, 3. \qquad (12.1.7)$$

Theorem 12.1.1 (Global L_2-Solvability for the Problem with Surface Tension) *Let* $\Sigma \in W_2^{3/2+l}$, $\boldsymbol{u}_0 \in \boldsymbol{W}_2^{1+l}(B)$, *and* $r_0 \in W_2^{2+l}(S_{R_0})$ *for some* $l \in (1/2, 1)$. *We assume also that compatibility conditions, as well as smallness conditions hold:*

$$\nabla \cdot \boldsymbol{u}_0 = l_2(\boldsymbol{u}_0, r_0),\ [\mu^{\pm}\Pi_0\mathbb{S}(\boldsymbol{u}_0)\boldsymbol{N}]|_{S_{R_0}} = \boldsymbol{l}_3(\boldsymbol{u}_0, r_0),\ [\boldsymbol{u}_0]|_{S_{R_0}} = 0,\ \boldsymbol{u}_0|_{\Sigma} = 0, \tag{12.1.8}$$

$$\|\boldsymbol{u}_0\|_{\boldsymbol{W}_2^{1+l}(B)} + \|r_0\|_{W_2^{2+l}(S_{R_0})} \leqslant \varepsilon. \tag{12.1.9}$$

In addition, we suppose that $\boldsymbol{f}$ *has finite norms* $\|\mathrm{e}^{bt}\boldsymbol{f}\|_{W_2^{l,l/2}(Q_\infty)}$, $\sup_{\tau>0}\|\mathcal{D}_x^{\boldsymbol{i}}\boldsymbol{f}\|_{Q_{\tau,\tau+T_0}}$, *where* $Q_\infty = \Omega \times (0,\infty)$, $Q_{\tau,\tau+T_0} = \Omega \times (\tau, \tau + T_0)$, $T_0 > 2$ *is an appropriate fixed number, and*

$$\|e^{bt}\boldsymbol{f}\|_{W_2^{l,l/2}(Q_\infty)} \leqslant \varepsilon, \quad b > 0, \quad \sup_{\tau>0} \|\mathcal{D}_x^{\boldsymbol{i}}\boldsymbol{f}\|_{Q_{\tau,\tau+T_0}} \leqslant \varepsilon, \quad |\boldsymbol{i}| = 1, 2. \tag{12.1.10}$$

Then problem (12.1.3) *has a unique solution* $(\boldsymbol{u}, q, r)$ *and it satisfies the inequality*

$$\begin{aligned} &\|\mathrm{e}^{at}\boldsymbol{u}\|_{W_2^{2+l,1+l/2}(D_\infty)} + \|\mathrm{e}^{at}\nabla q\|_{W_2^{l,l/2}(D_\infty)} + \|\mathrm{e}^{at}q\|_{W_2^{0,l/2}(D_\infty)} \\ &\quad + \|\mathrm{e}^{at}r\|_{W_2^{5/2+l,5/4+l/2}(G_\infty)} + \|\mathrm{e}^{at}\mathcal{D}_t r\|_{W_2^{3/2+l,3/4+l/2}(G_\infty)} \\ &\leqslant c_1(\varepsilon)\Big\{\|\mathrm{e}^{at}\boldsymbol{f}\|_{W_2^{l,l/2}(Q_\infty)} + \|\boldsymbol{u}_0\|_{\boldsymbol{W}_2^{1+l}(B)} + \|r_0\|_{W_2^{2+l}(S_{R_0})}\Big\} \end{aligned} \tag{12.1.11}$$

with some $a < b$, $D_\infty^{\pm} = B^{\pm} \times (0, \infty)$, $D_\infty = D_\infty^+ \cup D_\infty^-$, $G_\infty = S_{R_0} \times (0, \infty)$.

We recall that similar results in the Hölder spaces have been obtained in Chap. 7 (see Theorem 7.1.2).

Theorem 12.1.1 guarantees solution stability understood in the sense that velocity vector field differs a little from zero as well as pressure function does a little from a step function for small initial data and mass forces. In addition, the limit interface is a sphere $S_{R_0}(h_\infty)$ of the radius R_0; however, the center h_∞ of the limit sphere may be displaced slightly with respect to the origin, the barycenter of Ω_0^+, for arbitrarily small initial velocity and mass forces. This displacement will be evaluated by inequality (12.3.23) at the end of Sect. 12.3. There will be also given an estimate from below of the initial distance between the outer boundary and fluid interface sufficient for preventing the intersection of the surfaces in the future.

The proof of Theorem 12.1.1 consists of several steps. It is based on an exponential energy inequality for a solution of a linear problem, which implies exponential decay of a global solution to the problem in time.

12.2 Global Solvability of a Linear Problem

Along with (12.1.5), we consider the linear problem

$$
\begin{aligned}
&\mathcal{D}_t \boldsymbol{v} - \nu^{\pm}\nabla^2 \boldsymbol{v} + \frac{1}{\rho^{\pm}}\nabla p = \boldsymbol{f}, \qquad \nabla\cdot\boldsymbol{v} = f \ \text{ in } B^{\pm},\ t>0,\\
&\boldsymbol{v}(y,0) = \boldsymbol{v}_0(y) \ \text{ in } B \equiv B^+ \cup B^-, \qquad r(y,0) = r_0(y) \ \text{ on } S_{R_0},\\
&[\boldsymbol{v}]|_{S_{R_0}} = 0, \quad [\mu^{\pm}\Pi_0\mathbb{S}(\boldsymbol{v})\boldsymbol{N}]|_{S_{R_0}} = \boldsymbol{b}, \quad \boldsymbol{v}|_{\Sigma} = 0, \qquad (12.2.1)\\
&[\boldsymbol{N}\cdot\mathbb{T}(\boldsymbol{v},p)\boldsymbol{N}]|_{S_{R_0}} - \sigma\mathcal{B}_0 r|_{S_{R_0}} = b,\\
&\mathcal{D}_t r - \Big(\boldsymbol{v}\cdot\boldsymbol{N} - \frac{\boldsymbol{N}}{|B^+|}\cdot\int_{B^+}\boldsymbol{v}(y',t)\,\mathrm{d}y'\Big)\Big|_{S_{R_0}} = g, \quad \int_{B^+} p(y,t)\,\mathrm{d}y = 0,\ t>0.
\end{aligned}
$$

We introduce the notation: $D^{\pm}_{\Sigma T} = B^{\pm}\times(0,T)$, $D_{\Sigma T} = \cup D^{\pm}_{\Sigma T}$, $G_T = S_{R_0}\times(0,T)$ as well as the norm

$$
|u|^{(s+l,l/2)}_{G_T} = \|u\|_{W_2^{s+l,0}(G_T)} + \|u\|_{W_2^{l/2}\left(0,T;W_2^s(S_{R_0})\right)}.
$$

Theorem 12.2.1 (Local Solvability of the Linear Problem) *Let* $\Sigma\in W_2^{\frac{3}{2}+l}$, $r_0\in W_2^{2+l}(S_{R_0})$ *with* $l\in(1/2,1)$. *For arbitrary* $\boldsymbol{f}\in\boldsymbol{W}_2^{l,l/2}(D_{\Sigma T})$, $f\in W_2^{1+l,0}(D_{\Sigma T})$, $f=\nabla\cdot\boldsymbol{F}$, $\boldsymbol{F}\in\boldsymbol{W}_2^{0,1+\frac{l}{2}}(D_{\Sigma T})$, $[F_{\boldsymbol{N}}]|_{S_{R_0}}=0$, $\boldsymbol{v}_0\in W_2^{1+l}(B)$, $\boldsymbol{b}\in\boldsymbol{W}_2^{l+\frac{1}{2},\frac{l}{2}+\frac{1}{4}}(G_T)$, $\boldsymbol{N}\cdot\boldsymbol{b}=0$, $b\in W_2^{l+\frac{1}{2},0}(G_T)\cap W_2^{l/2}\big(0,T;W_2^{1/2}(S_{R_0})\big)$, $g\in W_2^{3/2+l,3/4+l/2}(G_T)$ *with* $T<\infty$ *satisfying compatibility conditions*

$$
\nabla\cdot\boldsymbol{v}_0 = f|_{t=0}, \quad [\mu^{\pm}\Pi_0\mathbb{S}(\boldsymbol{v}_0)\boldsymbol{N}]|_{S_{R_0}} = \boldsymbol{b}|_{t=0}, \quad [\boldsymbol{v}_0]|_{S_{R_0}} = 0, \quad \boldsymbol{v}_0|_{\Sigma} = 0, \qquad (12.2.2)
$$

problem (12.2.1) *has a unique solution* $(\boldsymbol{v},p,r)$: $\boldsymbol{v}\in\boldsymbol{W}_2^{2+l,1+\frac{l}{2}}(D_{\Sigma T})$, $\nabla p\in\boldsymbol{W}_2^{l,\frac{l}{2}}(D_{\Sigma T})$, $r(\cdot,t)\in W_2^{2+l}(S_{R_0})$ *for any* $t\in(0,T)$, *and*

$$
\begin{aligned}
&\|\boldsymbol{v}\|_{\boldsymbol{W}_2^{2+l,1+l/2}(D_{\Sigma T})} + \|\nabla p\|_{\boldsymbol{W}_2^{l,l/2}(D_{\Sigma T})} + \|p\|_{W_2^{0,l/2}(D_{\Sigma T})} + \|r\|_{W_2^{5/2+l,5/4+l/2}(G_T)}\\
&\quad + \|\mathcal{D}_t r\|_{W_2^{3/2+l,3/4+l/2}(G_T)} \leqslant c(T)\Big\{\|\boldsymbol{f}\|_{\boldsymbol{W}_2^{l,l/2}(D_{\Sigma T})} + \|f\|_{W_2^{1+l,0}(D_{\Sigma T})}\\
&\qquad + \|\boldsymbol{F}\|_{W_2^{0,1+l/2}(D_{\Sigma T})} + \|\boldsymbol{b}\|_{\boldsymbol{W}_2^{l+1/2,l/2+1/4}(G_T)} + |b|^{(1/2+l,l/2)}_{G_T}\\
&\qquad + \|g\|_{W_2^{3/2+l,3/4+l/2}(G_T)} + \|\boldsymbol{v}_0\|_{\boldsymbol{W}_2^{1+l}(B)} + \|r_0\|_{W_2^{2+l}(S_{R_0})}\Big\}. \qquad (12.2.3)
\end{aligned}
$$

Remark 12.2.1 *For any function* $\rho \in W_2^{1,1}(G_T)$, *it follows from trace theorem that*

$$\|\rho(\cdot,t)\|_{W_2^{1/2}(S_{R_0})} \leqslant c\Big\{\|\rho\|_{W_2^{1,0}(G_T)} + \|\mathcal{D}_t\rho\|_{G_T}\Big\},\ t\in[0,T],$$

which implies the inequality

$$\|r(\cdot,t)\|_{W_2^{2+l}(S_{R_0})} \leqslant c\Big\{\|r\|_{W_2^{5/2+l,0}(G_T)} + \|\mathcal{D}_t r\|_{W_2^{3/2+l,0}(G_T)}\Big\}.$$

This means that Theorem 12.2.1 *yields* $\Gamma_t \in W_2^{2+l}$ *for all* $t\in[0,T]$.

Proof Let r_1 be a function satisfying the conditions

$$r_1(y,0) = r_0(y),$$
$$\mathcal{D}_t r_1(y,0) = g(y,0) + \Big(\boldsymbol{v}_0(y)\cdot\boldsymbol{N}(y) - \frac{\boldsymbol{N}(y)}{|B^+|}\cdot\int\limits_{B+}\boldsymbol{v}_0(y')\,\mathrm{d}y'\Big) \equiv r_0'(y)$$

and the inequalities

$$\begin{aligned} &|r_1|_{G_T}^{(\frac{5}{2}+l,\frac{l}{2})} + \|\mathcal{D}_t r_1\|_{W_2^{\frac{3}{2}+l,\frac{3}{4}+\frac{l}{2}}(G_T)} \\ &\quad\leqslant c\Big\{\|r_1\|_{W_2^{\frac{5}{2}+l,\frac{5}{4}+\frac{l}{2}}(G_T)} + \|\mathcal{D}_t r_1\|_{W_2^{\frac{3}{2}+l,\frac{3}{4}+\frac{l}{2}}(G_T)}\Big\} \\ &\quad\leqslant c\Big\{\|r_0\|_{W_2^{2+l}(S_{R_0})} + \|r_0'\|_{W_2^{l+1/2}(S_{R_0})}\Big\}. \end{aligned} \tag{12.2.4}$$

Such r_1 exists due to Proposition 4.1 in [91] and trace theorem for the Sobolev–Slobodetskiĭ spaces.

Since $\mathcal{B}_0\boldsymbol{N} = 0$, in view of the fact that $\boldsymbol{N}$ is the eigenvector of the operator $\Delta_{S_{R_0}}$ with the eigenvalue $-\frac{2}{R_0^2}$, we can write

$$\begin{aligned} \mathcal{B}_0 r(y,t) &= \mathcal{B}_0 r_1(y,t) + \int_0^t \mathcal{B}_0\Big(\mathcal{D}_\tau\big(r(y,\tau) - r_1(y,\tau)\big)\Big)\,\mathrm{d}\tau \\ &= \mathcal{B}_0 r_1(y,t) + \int_0^t \mathcal{B}_0\Big(g(y,\tau) + \boldsymbol{v}(y,\tau)\cdot\boldsymbol{N}(y) - \mathcal{D}_\tau r_1(y,\tau)\Big)\,\mathrm{d}\tau. \end{aligned}$$

Consequently, system (12.2.1) can be transformed to the form:

$$\begin{aligned} &\mathcal{D}_t\boldsymbol{v} - \nu^\pm\nabla^2\boldsymbol{v} + \frac{1}{\rho^\pm}\nabla p = \boldsymbol{f}, \qquad \nabla\cdot\boldsymbol{v} = f \ \text{ in } B^\pm,\ \ t>0, \\ &\quad \boldsymbol{v}(y,0) = \boldsymbol{v}_0(y) \ \text{ in } B, \\ &[\boldsymbol{v}]|_{S_{R_0}} = 0, \quad [\mu^\pm\Pi_0\mathbb{S}(\boldsymbol{v})\boldsymbol{N}]\big|_{S_{R_0}} = \boldsymbol{b}, \quad \boldsymbol{v}|_\Sigma = 0, \end{aligned} \tag{12.2.5}$$

$$[\boldsymbol{N} \cdot \mathbb{T}(\boldsymbol{v}, p)\boldsymbol{N}]\big|_{S_{R_0}} - \sigma \boldsymbol{N} \cdot \mathcal{B}_0 \int_0^t \boldsymbol{v}|_{S_{R_0}} \,\mathrm{d}\tau = b' + \sigma \int_0^t B' \,\mathrm{d}\tau$$
$$+ 2\sigma \int_0^t \nabla_S \boldsymbol{v} : \nabla_S \boldsymbol{N} \,\mathrm{d}\tau \quad \text{on } S_{R_0},$$

$$\int_{B^+} p(y,t)\,\mathrm{d}y = 0, \quad t > 0,$$

where $b' = b + \sigma \mathcal{B}_0 r_1$, $B' = \mathcal{B}_0(g - \mathcal{D}_t r_1)$, ∇_S is surface gradient on S_{R_0}; $\mathbb{S} : \mathbb{T} \equiv S_{ij} T_{ij}$. Such problems have been investigated in Chap. 10, where, in particular, the solvability of (12.2.5) without the additional term in the last boundary condition and the estimate of its solution (Theorem 10.5.2)

$$\|\boldsymbol{v}\|_{W_2^{2+l,l+1/2}(D_{\Sigma T})} + \|\nabla p\|_{\boldsymbol{W}_2^{l,l/2}(D_{\Sigma T})} + \|p\|_{W_2^{0,l/2}(D_{\Sigma T})} \leqslant c(T)\Big\{\|\boldsymbol{f}\|_{\boldsymbol{W}_2^{l,l/2}(D_{\Sigma T})}$$
$$+ \|f\|_{W_2^{1+l,0}(D_{\Sigma T})} + \|\boldsymbol{F}\|_{\boldsymbol{W}_2^{0,1+l/2}(D_{\Sigma T})} + \|\boldsymbol{b}\|_{\boldsymbol{W}_2^{l+1/2,l/2+1/4}(G_T)}$$
$$+ |b'|_{G_T}^{(1/2+l,l/2)} + \|B'\|_{W_2^{l-1/2,l/2-1/4}(G_T)} + \|\boldsymbol{v}_0\|_{\boldsymbol{W}_2^{1+l}(B)}\Big\} \tag{12.2.6}$$

have been established. Inequality (12.2.6) together with (12.2.4) implies estimate (12.2.3) because the term $2\sigma \int_0^t \nabla_S \boldsymbol{v}(y,t) : \nabla_S \boldsymbol{N}(y)\,\mathrm{d}\tau$ is of lower order and has no essential influence on the final result. □

Now we consider problem (12.2.1) with $\boldsymbol{f} = 0$, $f = 0$, $\boldsymbol{b} = 0$, $b = 0$, $g = 0$ and with $r_0(y)$ satisfying orthogonality conditions

$$\int_{S_{R_0}} r_0(y)\,\mathrm{d}S = 0, \quad \int_{S_{R_0}} r_0(y) y_j \,\mathrm{d}S = 0, \quad j = 1,2,3, \tag{12.2.7}$$

obtained by linearization of (12.1.7). Since

$$\int_{S_{R_0}} \mathcal{D}_t r(y,t)\,\mathrm{d}S = \int_{S_{R_0}} \boldsymbol{v} \cdot \boldsymbol{N}\,\mathrm{d}S = \int_{B^+} \nabla \cdot \boldsymbol{v}(y,t)\,\mathrm{d}y = 0,$$

$$\int_{S_{R_0}} \mathcal{D}_t r(y,t) y_j \,\mathrm{d}S = \int_{S_{R_0}} y_j \Big\{\boldsymbol{v} \cdot \boldsymbol{N} - \frac{\boldsymbol{N}}{|B^+|} \cdot \int_{B^+} \boldsymbol{v}(y',t)\,\mathrm{d}y'\Big\}\,\mathrm{d}S$$
$$= \int_{B^+} \nabla \cdot \big(y_j \boldsymbol{v}(y,t)\big)\,\mathrm{d}y - \int_{B^+} \boldsymbol{v}\,\mathrm{d}y' \cdot \int_{B^+} \frac{\nabla y_j}{|B^+|}\,\mathrm{d}y$$
$$= \int_{B^+} v_j(y,t)\,\mathrm{d}y - \int_{B^+} v_j(y,t)\,\mathrm{d}y' \frac{|B^+|}{|B^+|} = 0,$$

conditions (12.2.7) are also satisfied for $r(y,t)$, $t>0$:

$$\int_{S_{R_0}} r(y,t)\,\mathrm{d}S=0, \quad \int_{S_{R_0}} r(y,t)y_j\,\mathrm{d}S=0, \quad j=1,2,3. \tag{12.2.8}$$

Theorem 12.2.2 (Global Solvability of the Linear Homogeneous Problem) *Problem* (12.2.1) *with* $\boldsymbol{f}=0$, $f=0$, $\boldsymbol{b}=0$, $b=0$, $g=0$ *and with* $\boldsymbol{v}_0\in W_2^{1+l}(B)$, $r_0\in W_2^{2+l}(S_{R_0})$, $l\in(1/2,1)$, *satisfying compatibility conditions* (12.2.2), *i.e.,*

$$\nabla\cdot\boldsymbol{v}_0=0, \quad [\mu^{\pm}\Pi_0\mathbb{S}(\boldsymbol{v}_0)\boldsymbol{N}]|_{S_{R_0}}=0, \quad [\boldsymbol{v}_0]|_{S_{R_0}}=0, \quad \boldsymbol{v}_0|_{\Sigma}=0, \tag{12.2.9}$$

and orthogonality conditions (12.2.7) *has a unique solution* $(\boldsymbol{v},p,r)$ *such that* $\boldsymbol{v}\in\boldsymbol{W}_2^{2+l,1+l/2}(D_{\Sigma\infty})$, $\nabla p\in\boldsymbol{W}_2^{l,l/2}(D_{\Sigma\infty})$, $r(\cdot,t)\in W_2^{2+l}(S_{R_0})$ *for any* $t\in(0,\infty)$. *This solution is subjected to the inequality*

$$\begin{aligned}&\|\mathrm{e}^{\beta t}\boldsymbol{v}\|^2_{\boldsymbol{W}_2^{2+l,1+\frac{l}{2}}(D_{\Sigma\infty})}+\|\mathrm{e}^{\beta t}\nabla p\|^2_{\boldsymbol{W}_2^{l,\frac{l}{2}}(D_{\Sigma\infty})}+\|\mathrm{e}^{\beta t}p\|^2_{W_2^{0,\frac{l}{2}}(D_{\Sigma\infty})}+\|\mathrm{e}^{\beta t}r\|^2_{W_2^{\frac{5}{2}+l,\frac{5}{4}+\frac{l}{2}}(G_\infty)}\\&\quad+\|\mathrm{e}^{\beta t}\mathcal{D}_t r\|^2_{W_2^{\frac{3}{2}+l,\frac{3}{4}+\frac{l}{2}}(G_\infty)}\leqslant c\big\{\|\boldsymbol{v}_0\|^2_{\boldsymbol{W}_2^{1+l}(B)}+\|r_0\|^2_{W_2^{2+l}(S_{R_0})}\big\}\end{aligned} \tag{12.2.10}$$

with certain $\beta>0$.

We outline the proof of (12.2.10). At first, weighted L_2-estimates of $\boldsymbol{v}$ and r are obtained.

Proposition 12.2.1 *A solution of* (12.2.1), (12.2.7) *with* $\boldsymbol{f}=0$, $f=0$, $\boldsymbol{b}=0$, $b=0$, $g=0$ *satisfies the inequality*

$$\|e^{\beta_1 t}\boldsymbol{v}(\cdot,t)\|^2_B+\|e^{\beta_1 t}r(\cdot,t)\|^2_{W_2^1(S_{R_0})}\leqslant c\big\{\|\boldsymbol{v}_0\|^2_B+\|r_0\|^2_{W_2^1(S_{R_0})}\big\}, \quad t>0, \tag{12.2.11}$$

where $\beta_1>0$, c *is independent of* t.

Proof Inequality (12.2.11) is obtained in the same way as inequality (7.2.9) in Proposition 7.2.1 and even easier because the triple $(\boldsymbol{v},p,r)$ solves a linear problem. The proof is based on energy relation

$$\frac{1}{2}\frac{\mathrm{d}}{\mathrm{d}t}\|\sqrt{\rho^{\pm}}\boldsymbol{v}\|^2_B-\sigma\int_{S_{R_0}}\boldsymbol{v}\cdot\boldsymbol{N}\mathcal{B}_0 r\,\mathrm{d}S+\frac{1}{2}\|\sqrt{\mu^{\pm}}\mathbb{S}(\boldsymbol{v})\|^2_B=0, \tag{12.2.12}$$

which, in view of the last boundary condition in (12.2.1) and the self-adjointness of the operator $\mathcal{B}_0$, implies that

$$\frac{1}{2}\frac{\mathrm{d}}{\mathrm{d}t}\Big(\|\sqrt{\rho^{\pm}}\boldsymbol{v}\|^2_B-\sigma\int_{S_{R_0}}r\mathcal{B}_0 r\,\mathrm{d}S\Big)+\frac{1}{2}\|\sqrt{\mu^{\pm}}\mathbb{S}(\boldsymbol{v})\|^2_B=0.$$

Similarly to (12.2.12), one can deduce the equality

$$\frac{\mathrm{d}}{\mathrm{d}t}\int_B \rho^{\pm}\boldsymbol{v}\cdot\boldsymbol{W}\,\mathrm{d}x-\int_B \rho^{\pm}\boldsymbol{v}\cdot\mathcal{D}_t\boldsymbol{W}\,\mathrm{d}x+\int_B\frac{\mu^{\pm}}{2}\mathbb{S}(\boldsymbol{v}):\mathbb{S}(\boldsymbol{W})\,\mathrm{d}x-\sigma\int_{S_{R_0}} r\mathcal{B}_0 r\,\mathrm{d}S=0, \tag{12.2.13}$$

where $\boldsymbol{W}$ is an auxiliary vector field satisfying the relations

$$\nabla\cdot\boldsymbol{W}(x,t)=0 \text{ in } B,\quad \boldsymbol{W}\cdot\boldsymbol{N}|_{S_{R_0}}=r,\quad [\boldsymbol{W}]|_{S_{R_0}}=0,\quad \boldsymbol{W}|_{\Sigma}=0,$$

$$\|\boldsymbol{W}\|_{\boldsymbol{W}_2^1(B)}\leqslant c\|r\|_{W_2^{1/2}(S_{R_0})},$$

$$\|\mathcal{D}_t\boldsymbol{W}\|_B\leqslant c\|\mathcal{D}_t r\|_{S_{R_0}}\leqslant c\big\{\|\boldsymbol{v}\cdot\boldsymbol{N}\|_{S_{R_0}}+\|\boldsymbol{v}\|_B\big\}.$$

We multiply (12.2.13) by a small $\gamma>0$ and add it to (12.2.12). Taking account of the fact that the form $-\int_{S_{R_0}} r\mathcal{B}_0 r\,\mathrm{d}S=\int_{S_{R_0}}(|\nabla_S r|^2-2R_0^{-2}r^2)\,\mathrm{d}S$ is positive definite if r is subject to conditions (12.2.8) [78] and making use of (12.2.4) and of the Korn inequality for $\boldsymbol{v}$, we show that the so-called generalized energy [58]

$$\mathcal{E}(t)=\frac{1}{2}\|\sqrt{\rho^{\pm}}\boldsymbol{v}\|_B^2-\sigma\int_{S_{R_0}} r\mathcal{B}_0 r\,\mathrm{d}S+\gamma\int_B\rho^{\pm}\boldsymbol{v}\cdot\boldsymbol{W}\,\mathrm{d}x$$

satisfies the estimate

$$\mathcal{D}_t\mathcal{E}(t)+2\beta_1\mathcal{E}(t)\leqslant 0,$$

where $\beta_1=\mathit{const}>0$. If γ sufficiently small, $\mathcal{E}$ is estimated from above and from below by the sum of the norms $c(\|\boldsymbol{v}\|_B^2+\|r\|^2_{W_2^1(S_{R_0})})$. Therefore, we arrive at (12.2.11). □

For obtaining bounds for higher-order norms of the solution similar to inequality (12.2.11), we invoke a local-in-time estimate of the solution. Keeping in mind forthcoming arguments, we assume that $T>2$.

Proposition 12.2.2 *Let $T>2$. The solution of problem* (12.2.1), (12.2.7) *with $\boldsymbol{f}=0$, $f=0$, $\boldsymbol{b}=0$, $b=0$, $g=0$ is subject to the inequality*

$$\begin{aligned}\|\boldsymbol{v}\|_{\boldsymbol{W}_2^{2+l,1+\frac{l}{2}}(D_{\Sigma t_0-1,t_0})}&+\|\nabla p\|_{\boldsymbol{W}_2^{l,\frac{l}{2}}(D_{\Sigma t_0-1,t_0})}+\|p\|_{W_2^{0,\frac{l}{2}}(D_{\Sigma t_0-1,t_0})}\\&+\|r\|_{W_2^{\frac{5}{2}+l,\frac{5}{4}+\frac{l}{2}}(G_{t_0-1,t_0})}+\|\mathcal{D}_t r\|_{W_2^{\frac{3}{2}+l,\frac{3}{4}+\frac{l}{2}}(G_{t_0-1,t_0})}\\&\leqslant c(\|\boldsymbol{v}\|_{Q_{t_0-3/2,t_0}}+\|r\|_{G_{t_0-3/2,t_0}}),\end{aligned} \tag{12.2.14}$$

where $2<t_0\leqslant T$, $D_{\Sigma t_1,t_2}=\cup B^{\pm}\times(t_1,t_2)$, $Q_{t_1,t_2}=\Omega\times(t_1,t_2)$, $G_{t_1,t_2}=S_{R_0}\times(t_1,t_2)$.

Proof We fix $t_0 \in (2,T)$ and multiply (12.2.14) by the cutoff function $\zeta_\lambda(t)$, smooth, monotone, equal to zero for $t \leqslant t_0 - 2 + \lambda/2$ and to one for $t \geqslant t_0 - 2 + \lambda$, where $\lambda \in (0,1]$, and such that $\dot\zeta_\lambda(t) \equiv \frac{d\zeta_\lambda(t)}{dt}$ and $\ddot\zeta_\lambda(t)$ satisfy the inequalities

$$\sup_{t\in\mathbb{R}} \left|\dot\zeta_\lambda(t)\right| \leqslant c\lambda^{-1}, \quad \sup_{t\in\mathbb{R}} |\ddot\zeta_\lambda(t)| \leqslant c\lambda^{-2}.$$

Then for $\boldsymbol{v}_\lambda = \boldsymbol{v}\zeta_\lambda$, $p_\lambda = p\zeta_\lambda$, $r_\lambda = r\zeta_\lambda$ we obtain

$$\begin{aligned}
&\mathcal{D}_t\boldsymbol{v}_\lambda - \nu^\pm\nabla^2\boldsymbol{v}_\lambda + \frac{1}{\rho^\pm}\nabla p_\lambda = \boldsymbol{v}\dot\zeta_\lambda, \quad \nabla\cdot\boldsymbol{v}_\lambda = 0 \text{ in } B^\pm, \ t>0,\\
&\qquad\qquad \boldsymbol{v}_\lambda(y,0) = 0 \text{ in } B^\pm, \quad r_\lambda(y,0) = 0 \text{ on } S_{R_0},\\
&[\boldsymbol{v}_\lambda]|_{S_{R_0}} = 0, \quad [\mu^\pm\Pi_0\mathbb{S}(\boldsymbol{v}_\lambda)\boldsymbol{N}]\big|_{S_{R_0}} = 0, \quad \boldsymbol{v}_\lambda|_\Sigma = 0,\\
&[\boldsymbol{N}\cdot\mathbb{T}(\boldsymbol{v}_\lambda,p_\lambda)\boldsymbol{N}]\big|_{S_{R_0}} - \sigma\mathcal{B}_0 r_\lambda|_{S_{R_0}} = 0, \quad \int_{B^+} p_\lambda(y,t)\,dy = 0,\\
&\mathcal{D}_t r_\lambda - \Big(\boldsymbol{v}_\lambda\cdot\boldsymbol{N} - \frac{\boldsymbol{N}}{|B^+|}\cdot\int_{B^+}\boldsymbol{v}_\lambda(y',t)\,dy'\Big)\Big|_{S_{R_0}} = r\dot\zeta_\lambda(t).
\end{aligned} \tag{12.2.15}$$

By Theorem 12.2.1 applied to system (12.2.15), estimate (12.2.3) holds for $\boldsymbol{v}_\lambda$, p_λ, r_λ which implies that

$$\begin{aligned}
&\|\boldsymbol{v}\|_{\boldsymbol{W}_2^{2+l,1+l/2}(D_{\Sigma t_1+\lambda,t_0})} + \|\nabla p\|_{\boldsymbol{W}_2^{l,l/2}(D_{\Sigma t_1+\lambda,t_0})} + \|p\|_{W_2^{0,l/2}(D_{\Sigma t_1+\lambda,t_0})}\\
&+ \|r\|_{W_2^{5/2+l,5/4+l/2}(G_{t_1+\lambda,t_0})} + \|\mathcal{D}_t r\|_{W_2^{3/2+l,3/4+l/2}(G_{t_1+\lambda,t_0})}\\
&\leqslant c\lambda^{-2}\Big\{\|\boldsymbol{v}\|_{\boldsymbol{W}_2^{l,l/2}(D_{\Sigma t_1+\lambda/2,t_0})} + \|r\|_{W_2^{3/2+l,3/4+l/2}(G_{t_1+\lambda/2,t_0})}\Big\},
\end{aligned} \tag{12.2.16}$$

where $t_1 = t_0 - 2$.

Now, we employ interpolation inequalities

$$\begin{aligned}
&\|\boldsymbol{v}\|_{\boldsymbol{W}_2^{l,l/2}(D_{\Sigma t_1+\lambda/2,t_0})} \leqslant \varkappa^2\|\boldsymbol{v}\|_{\boldsymbol{W}_2^{2+l,1+l/2}(D_{\Sigma t_1+\lambda/2,t_0})} + c\varkappa^{-l}\|\boldsymbol{v}\|_{Q_{t_1+\lambda/2,t_0}},\\
&\|r\|_{W_2^{3/2+l,0}(G_{t_1+\lambda/2})} \leqslant \varkappa^2\|r\|_{W_2^{5/2+l,0}(G_{t_1+\lambda/2,t_0})} + c\varkappa^{-3-2l}\|r\|_{G_{t_1+\lambda/2,t_0}},\\
&\|r\|_{W_2^{0,3/4+l/2}(G_{t_1+\lambda/2})} \leqslant \varkappa^2\|\mathcal{D}_t r\|_{W_2^{0,3/4+l/2}(G_{t_1+\lambda/2,t_0})} + c\varkappa^{-3/2-l}\|r\|_{G_{t_1+\lambda/2,t_0}},
\end{aligned}$$

which lead to

$$\Psi(\lambda) \leqslant c_1\varkappa^2\lambda^{-2}\Psi(\lambda/2) + c_2\varkappa^{-m}\lambda^{-2}K,$$

where $\Psi(\lambda)$ denotes the left-hand side of (12.2.16), $K = \|\boldsymbol{v}\|_{Q_{t_1,t_0}} + \|r\|_{G_{t_1,t_0}}$, $m = 3+2l$. Setting $\varkappa = \delta\lambda \leqslant 1$ with $\delta > 0$ and multiplying the above inequality by λ^{m+2}, we obtain

$$\Psi(\lambda)\lambda^{m+2} \leqslant c_1\delta^2 2^{m+2}\Psi(\lambda/2)(\lambda/2)^{m+2} + c_2\delta^{-m}K.$$

If now one takes δ so small that $c_1\delta^2 2^{m+2} < 1/2$, one can deduce by iterations that

$$\Psi(\lambda)\lambda^{m+2} \leqslant c_2\delta^{-m}(K + 2^{-1}K + 2^{-2}K + \ldots) \leqslant \frac{c_2\delta^{-m}}{1-1/2}K \leqslant 2c_2\delta^{-m}K.$$

For $\lambda = 1$, this inequality is equivalent to (12.2.14). □

Proof of Theorem 12.2.2 By Theorem 12.2.1 and Proposition 12.2.2, we have

$$\begin{aligned} e^{2\beta(T-j)}\Big\{ &\|\boldsymbol{v}\|^2_{\boldsymbol{W}_2^{2+l,1+l/2}(D_{\Sigma T-j-1,T-j})} + \|\nabla p\|^2_{\boldsymbol{W}_2^{l,l/2}(D_{\Sigma T-j-1,T-j})} \\ &+ \|p\|^2_{W_2^{0,l/2}(D_{\Sigma T-j-1,T-j})} + \|r\|^2_{W_2^{5/2+l,5/4+l/2}\left(G_{T-j-1,T-j}\right)} \\ &+ \|\mathcal{D}_t r\|^2_{W_2^{3/2+l,3/4+l/2}\left(G_{T-j-1,T-j}\right)}\Big\} \\ \leqslant c e^{2\beta(T-j)}\Big\{ &\|\boldsymbol{v}\|^2_{Q_{T-j-2,T-j}} + \|r\|^2_{G_{T-j-2,T-j}}\Big\}, \quad j = 0, 1, \ldots, [T]-2. \end{aligned} \tag{12.2.17}$$

Taking the sum of (12.2.17) from $j = 0$ to $j = [T]$–2, we obtain the inequality which implies that

$$Y^2_{T-[T]+1,T}(e^{\beta t}\boldsymbol{v}, e^{\beta t}p, e^{\beta t}r) \leqslant c\int_{T-[T]}^{T} e^{2\beta t}\Big(\|\boldsymbol{v}(\cdot,t)\|^2_\Omega + \|r(\cdot,t)\|^2_{S_{R_0}}\Big)\,dt, \tag{12.2.18}$$

where

$$\begin{aligned} Y_{t_1,t_2}(\boldsymbol{u},q,r) \equiv &\|\boldsymbol{u}\|_{\boldsymbol{W}_2^{2+l,1+l/2}(D_{\Sigma t_1,t_2})} + \|\nabla q\|_{\boldsymbol{W}_2^{l,l/2}(D_{\Sigma t_1,t_2})} + \|q\|_{W_2^{0,l/2}(D_{\Sigma t_1,t_2})} \\ &+ \|r\|_{W_2^{5/2+l,5/4+l/2}(G_{t_1,t_2})} + \|\mathcal{D}_t r\|_{W_2^{3/2+l,3/4+l/2}(G_{t_1,t_2})}. \end{aligned} \tag{12.2.19}$$

By adding the estimate

$$Y^2_{0,2}(\boldsymbol{v},p,r) \leqslant c\Big\{\|\boldsymbol{v}_0\|^2_{\boldsymbol{W}_2^{1+l}(B)} + \|r_0\|^2_{W_2^{2+l}(S_{R_0})}\Big\}$$

to (12.2.18), choosing $\beta < \beta_1$ and making use of (12.2.11), we arrive at an inequality, equivalent to (12.2.10). □

12.3 The Nonlinear Problem

We start with the construction of a solution to problem (12.1.3) in a finite time interval $(0, T_0)$ with T_0 to be fixed later.

Theorem 12.3.1 (Local Solvability of the Nonlinear Problem with Surface Tension) *Let $T_0 < \infty$ and let compatibility conditions* (12.1.8) *in Theorem* 12.1.1 *be satisfied. Then there exists a value $\varepsilon(T_0) \ll 1$ such that for small data:*

$$\|\boldsymbol{u}_0\|_{\boldsymbol{W}_2^{1+l}(B)} + \|r_0\|_{W_2^{2+l}(S_{R_0})} + \|\boldsymbol{f}\|_{\boldsymbol{W}_2^{l,l/2}(Q_{T_0})} \leqslant \varepsilon, \quad \|\nabla \boldsymbol{f}\|_{\boldsymbol{W}_2^{l,l/2}(Q_{T_0})} \leqslant \varepsilon, \tag{12.3.1}$$

problem (12.1.3) *has a unique solution $(\boldsymbol{u}, q, r)$ in the interval $(0, T_0)$ and the inequalities*

$$Y_{0,T_0}(\boldsymbol{u}, q, r) \leqslant c\Big\{N(\boldsymbol{u}_0, r_0) + \|\boldsymbol{f}\|_{\boldsymbol{W}_2^{l,l/2}(Q_{T_0})}\Big\}, \tag{12.3.2}$$

$$N(\boldsymbol{u}(\cdot, T_0), r(\cdot, T_0)) \leqslant \theta N(\boldsymbol{u}_0, r_0) + c\|\boldsymbol{f}\|_{\boldsymbol{W}_2^{l,l/2}(Q_{T_0})} \tag{12.3.3}$$

hold, where $\theta < 1$, Y_{0,T_0} is calculated by (12.2.19), *and*

$$N(\boldsymbol{w}, \rho) \equiv \|\boldsymbol{w}\|_{\boldsymbol{W}_2^{1+l}(B)} + \|\rho\|_{W_2^{2+l}(S_{R_0})}.$$

The proof of Theorem 12.3.1 relies on Theorem 12.2.1 and on the following estimates of the nonlinear terms.

Proposition 12.3.1 *If*

$$\|r(\cdot, t)\|_{W_2^{3/2+l}(S_{R_0})} + \|\mathcal{D}_t r(\cdot, t)\|_{W_2^{1/2+l}(S_{R_0})} + \|\boldsymbol{u}(\cdot, t)\|_B \leqslant \delta, \quad t \leqslant T, \tag{12.3.4}$$

where δ is a certain small positive number, then nonlinear terms (12.1.6) *and $\widehat{\boldsymbol{f}}(y, t) \equiv \boldsymbol{f}\big(e_{r,h}(y, t), t\big)$ are subjected to the inequalities*

$$\begin{aligned} Z(\boldsymbol{u}, q, r) \equiv \|\boldsymbol{l}_1(\boldsymbol{u}, r)\|_{\boldsymbol{W}_2^{l,l/2}(D_{\Sigma T})} + \|l_2(\boldsymbol{u}, r)\|_{W_2^{1+l,0}(D_{\Sigma T})} + \|\boldsymbol{L}(\boldsymbol{u}, r)\|_{W_2^{0,1+l/2}(Q_T)} \\ + \|\boldsymbol{l}_3(\boldsymbol{u}, r)\|_{W_2^{1/2+l,1/4+l/2}(G_T)} + |l_4(\boldsymbol{u}, r)|_{G_T}^{(1/2+l,l/2)} + |l_5(r)|_{G_T}^{(1/2+l,l/2)} \\ + \|l_6(\boldsymbol{u}, r)\|_{W_2^{3/2+l,3/4+l/2}(G_T)} \leqslant c Y^2(\boldsymbol{u}, q, r), \end{aligned} \tag{12.3.5}$$

$$\|\widehat{\boldsymbol{f}}\|_{\boldsymbol{W}_2^{l,l/2}(Q_T)} \leqslant c\Big\{\|\boldsymbol{f}\|_{\boldsymbol{W}_2^{l,l/2}(Q_T)} + \|\nabla \boldsymbol{f}\|_{Q_T} \sup_{t<T} \big(\|\mathcal{D}_t r(\cdot, t)\|_{W_2^{l+1/2}(S_{R_0})} + \|\boldsymbol{u}(\cdot, t)\|_B\big)\Big\}. \tag{12.3.6}$$

If $(\boldsymbol{u}, r)$ and $(\boldsymbol{u}', r')$ satisfy (12.3.4), *then*

$$Z(\boldsymbol{u}-\boldsymbol{u}',q-q',r-r') \leqslant c\delta Y(\boldsymbol{u}-\boldsymbol{u}',q-q',r-r'),$$
$$\|\widehat{\boldsymbol{f}}-\widehat{\boldsymbol{f}}'\|_{\boldsymbol{W}_2^{l,l/2}(Q_T)} \leqslant c\delta Y(\boldsymbol{u}-\boldsymbol{u}',q-q',r-r'), \tag{12.3.7}$$

where $\widehat{\boldsymbol{f}}' = \boldsymbol{f}(e_{r',\boldsymbol{h}'}(y,t),t)$, $\boldsymbol{h}' = |B^+|^{-1}\int_0^t\int_{B+}\boldsymbol{u}'(y,\tau)L'(y,\tau)\,\mathrm{d}y\,\mathrm{d}\tau$, L' *is the Jacobian of the transformation* $e_{r',\boldsymbol{h}'}$.

Proof Inequality (12.3.5) is established in the same way as a similar estimate in [59]. And inequality

$$\|\widehat{\boldsymbol{f}}\|_{Q_T} = \|\boldsymbol{f}(e_{r,\boldsymbol{h}}(y,t),t)\|_{Q_T} \leqslant c\|\boldsymbol{f}\|_{Q_T}$$

is obtained by the passage to the Eulerian coordinates $x = e_{r,\boldsymbol{h}}(y,t)$ under the integral sign and using the boundedness of the Jacobian L which follows from (12.3.4). The estimates

$$\int_0^T \mathrm{d}t \int_\Omega \mathrm{d}y \int_\Omega \frac{|\boldsymbol{f}(e_{r,\boldsymbol{h}}(y,t),t)-\boldsymbol{f}(e_{r,\boldsymbol{h}}(z,t),t)|^2}{|y-z|^{3+2l}}\,\mathrm{d}z$$
$$\leqslant c\int_0^T \mathrm{d}t \int_\Omega \mathrm{d}x \int_\Omega \frac{|\boldsymbol{f}(x,t)-\boldsymbol{f}(x',t)|^2}{|x-x'|^{3+2l}}\,\mathrm{d}x',$$
$$\int_0^T \mathrm{d}t \int_0^t \mathrm{d}\tau \int_\Omega \frac{|\boldsymbol{f}(e_{r,\boldsymbol{h}}(y,t),t)-\boldsymbol{f}(e_{r,\boldsymbol{h}}(y,t),t-\tau)|^2}{\tau^{1+l}}\,\mathrm{d}y$$
$$\leqslant c\int_0^T \mathrm{d}t \int_0^t \mathrm{d}\tau \int_\Omega \frac{|\boldsymbol{f}(x,t)-\boldsymbol{f}(x,t-\tau)|^2}{\tau^{1+l}}\,\mathrm{d}x$$

are proved in the same manner (for small δ). Finally, assuming that the vector-function $\boldsymbol{f}$ is extended outside Ω with preservation of class and making use of the relation

$$\boldsymbol{f}\big(e_{r,\boldsymbol{h}}(y,t),t\big) - \boldsymbol{f}\big(e_{r,\boldsymbol{h}}(y,t-\tau),t\big) =$$
$$= \int_0^1 \nabla\boldsymbol{f}\Big(e_{r,\boldsymbol{h}}(y,t) - \lambda\int_0^\tau (\boldsymbol{N}^*(y)\mathcal{D}_t r^*(y,t-\tau') + \dot{\boldsymbol{h}}(t)\chi(y))\,\mathrm{d}\tau',t\Big)\,\mathrm{d}\lambda$$
$$\times \int_0^\tau \Big(\boldsymbol{N}^*(y)\mathcal{D}_t r^*(y,t-\tau') + \dot{\boldsymbol{h}}(t)\chi(y)\Big)\,\mathrm{d}\tau',$$

we obtain, in view of (12.1.1), the inequality

$$\int_0^T \mathrm{d}t \int_0^t \frac{\mathrm{d}\tau}{\tau^{1+l}} \int_\Omega \big|\boldsymbol{f}(e_{r,\boldsymbol{h}}(y,t),t) - \boldsymbol{f}(e_{r,\boldsymbol{h}}(y,t-\tau),t)\big|^2\,\mathrm{d}y$$
$$\leqslant c(T)\|\nabla\boldsymbol{f}\|^2_{Q_T}\Big(\sup_{Q_T}|\mathcal{D}_t r^*(y,t)| + \sup_{t<T}\|\boldsymbol{u}(\cdot,t)\|_\Omega\Big)^2$$

$$\leqslant c(T)\|\nabla \boldsymbol{f}\|^2_{Q_T}\Big\{\sup_{t<T}\|\mathcal{D}_t r(\cdot,t)\|^2_{W_2^{l+1/2}(S_{R_0})}+\sup_{t<T}\|\boldsymbol{u}(\cdot,t)\|^2_{\Omega}\Big\}.$$

Inequality (12.3.7) is proved by applying the above estimates to the difference $\boldsymbol{f}(e_{r,\boldsymbol{h}},t)-\boldsymbol{f}(e_{r',\boldsymbol{h}'},t)=\int_0^1\nabla\boldsymbol{f}\Big(e_{r',\boldsymbol{h}'}+\lambda\big(\boldsymbol{N}^*(y,t)(r-r')+(\boldsymbol{h}-\boldsymbol{h}')\chi(y)\big),t\Big)\,\mathrm{d}\lambda\big(\boldsymbol{N}^*(y,t)(r-r')+(\boldsymbol{h}-\boldsymbol{h}')\chi(y)\big)$. □

Proof of Theorem 12.3.1 We go back to problem (12.1.3). The solution is sought in the form

$$\boldsymbol{u}=\boldsymbol{u}'+\boldsymbol{u}'',\quad q=q'+q'',\quad r=r'+r'',$$

where $(\boldsymbol{u}',q',r')$ and $(\boldsymbol{u}'',q'',r'')$ are solutions of the problems

$$\begin{aligned}
&\mathcal{D}_t\boldsymbol{u}'-\nu^\pm\nabla^2\boldsymbol{u}'+\frac{1}{\rho^\pm}\nabla q'=0,\qquad \nabla\cdot\boldsymbol{u}'=0\quad\text{in } B^\pm,\ t>0,\\
&\boldsymbol{u}'(y,0)=\boldsymbol{u}_0'(y)\ \text{ in } B,\quad r'(y,0)=r_0'(y)\ \text{ on } S_{R_0},\\
&[\boldsymbol{u}']|_{S_{R_0}}=0,\quad [\mu^\pm\Pi_0\mathbb{S}(\boldsymbol{u}')\boldsymbol{N}]\big|_{S_{R_0}}=0,\qquad \boldsymbol{u}'|_\Sigma=0,\\
&[-q'+\mu^\pm\boldsymbol{N}\cdot\mathbb{S}(\boldsymbol{u}')\boldsymbol{N}]\big|_{S_{R_0}}-\sigma\mathcal{B}_0 r'=0,\\
&\mathcal{D}_t r'-\Big(\boldsymbol{u}'-\frac{1}{|B^+|}\int_{B^+}\boldsymbol{u}'\,\mathrm{d}y\Big)\cdot\boldsymbol{N}=0\ \text{ on } S_{R_0},\quad \int_{B^+}q'(y,t)\,\mathrm{d}y=0,
\end{aligned}\tag{12.3.8}$$

and

$$\begin{aligned}
&\mathcal{D}_t\boldsymbol{u}''-\nu^\pm\nabla^2\boldsymbol{u}''+\frac{1}{\rho^\pm}\nabla q''=\boldsymbol{l}_1(\boldsymbol{u},q,r)+\widehat{\boldsymbol{f}}(y,t),\\
&\qquad\nabla\cdot\boldsymbol{u}''=l_2(\boldsymbol{u},r)\ \text{ in } B^\pm,\ t>0,\\
&\boldsymbol{u}''(y,0)=\boldsymbol{u}_0''(y)\ \text{ in } B,\quad r''(y,0)=r_0''(y)\ \text{ on } S_{R_0},\\
&[\boldsymbol{u}'']\big|_{S_{R_0}}=0,\quad [\mu^\pm\Pi_0\mathbb{S}(\boldsymbol{u}'')\boldsymbol{N}]\big|_{S_{R_0}}=\boldsymbol{l}_3(\boldsymbol{u},r),\qquad \boldsymbol{u}''|_\Sigma=0,\\
&[-q''+\mu^\pm\boldsymbol{N}\cdot\mathbb{S}(\boldsymbol{u}'')\boldsymbol{N}]\big|_{S_{R_0}}-\sigma\mathcal{B}_0 r''=l_4(\boldsymbol{u},r)+\sigma l_5(r),\ \int_{B^+}q''(y,t)\,\mathrm{d}y=0,\\
&\mathcal{D}_t r''-\Big(\boldsymbol{u}''-\frac{1}{|B^+|}\int_{B^+}\boldsymbol{u}''\,\mathrm{d}y\Big)\cdot\boldsymbol{N}=l_6(\boldsymbol{u},r)\ \text{ on } S_{R_0},
\end{aligned}\tag{12.3.9}$$

respectively; here the nonlinear terms l_i are given by relations (12.1.6).

We define the initial data $(\boldsymbol{u}_0'', r_0'')$ from the relations

$$\int_{S_{R_0}} r_0''(y)\,\mathrm{d}S = \int_{S_{R_0}} \Big(r_0 - \frac{\varphi(y,r_0)}{3R_0^2}\Big)\,\mathrm{d}S,$$

$$\int_{S_{R_0}} r_0''(y) y_j\,\mathrm{d}S = \int_{S_{R_0}} \Big(r_0 y_j - \frac{\psi_j(y,r_0)}{4R_0^3}\Big)\,\mathrm{d}S, \quad j=1,2,3,$$

$$[\boldsymbol{u}_0'']|_{S_{R_0}} = 0, \quad [\mu^{\pm}\Pi_0 \mathbb{S}(\boldsymbol{u}_0'')\boldsymbol{N}]\big|_{S_{R_0}} = \boldsymbol{l}_3(\boldsymbol{u}_0, r_0),$$

$$\nabla\cdot \boldsymbol{u}_0'' = l_2(\boldsymbol{u}_0, r_0) \text{ in } B, \quad \boldsymbol{u}_0'' = 0 \text{ on } \Sigma,$$

where $\varphi(y,r) = (R_0+r)^3 - R_0^3$, $\psi_j(y,r) = y_j\big((R_0+r)^4 - R_0^4\big)$ (see (12.1.7)), and the inequality

$$\|\boldsymbol{u}_0''\|_{\boldsymbol{W}_2^{1+l}(B)} + \|r_0''\|_{W_2^{2+l}(S_{R_0})} \leqslant c\varepsilon\big\{\|\boldsymbol{u}_0\|_{\boldsymbol{W}_2^{1+l}(B)} + \|r_0\|_{W_2^{2+l}(S_{R_0})}\big\}. \tag{12.3.10}$$

The function r_0'' can be defined as follows (cf. [90]):

$$r_0''(y) = \frac{I\boldsymbol{N}(y)\cdot \boldsymbol{y}}{3|B^+|} + \frac{\boldsymbol{I}\cdot \boldsymbol{N}(y)}{|B^+|},$$

where

$$I = -\int_{S_{R_0}} \frac{3r_0^2R_0 + r_0^3}{3R_0^2}\,\mathrm{d}S, \quad I_j = -\int_{S_{R_0}} \frac{y_i(6r_0^2R_0^2 + 4r_0^3R_0 + r_0^4)}{4R_0^3}\,\mathrm{d}S, \quad j=1,2,3.$$

And the vector field $\boldsymbol{u}_0''$ can be taken in the form: $\boldsymbol{u}_0'' = \boldsymbol{u}_1 + \boldsymbol{u}_2$, where

$$\nabla\cdot \boldsymbol{u}_1(y) = l_2(\boldsymbol{u}_0, r_0) \equiv \nabla\cdot(\mathbb{I} - \widehat{\mathbb{L}})\boldsymbol{u}_0 = (\mathbb{I} - \widehat{\mathbb{L}}^T)\nabla\cdot \boldsymbol{u}_0, \quad [\boldsymbol{u}_1]|_{S_{R_0}} = 0, \quad \boldsymbol{u}_1|_{\Sigma} = 0,$$

and $\boldsymbol{u}_2 = 0$ in B^-, $\boldsymbol{u}_2 = rot\boldsymbol{\Phi}(y)$ in B^+, moreover,

$$\boldsymbol{\Phi}|_{S_{R_0}} = \frac{\partial \boldsymbol{\Phi}}{\partial \boldsymbol{N}}\Big|_{S_{R_0}} = 0,$$

$$\mu^+\frac{\partial^2\boldsymbol{\Phi}}{\partial \boldsymbol{N}^2}\Big|_{S_{R_0}} = \Big(\boldsymbol{l}_3(\boldsymbol{u}_0, r_0) - [\mu^{\pm}\Pi_0\mathbb{S}(\boldsymbol{u}_1)\boldsymbol{N}]\big|_{S_{R_0}}\Big)\times \boldsymbol{N}\big|_{S_{R_0}}.$$

Thus, $\nabla\cdot \boldsymbol{u}_2 = 0$ and $\boldsymbol{u}_2|_{S_{R_0}} = 0$.

Since $[\boldsymbol{u}_0]|_{S_{R_0}} = 0$ and $[\widehat{\mathbb{L}}^T\boldsymbol{N}]|_{S_{R_0}} = 0$, compatibility condition $[\boldsymbol{N}\cdot(\mathbb{I} - \widehat{\mathbb{L}})\boldsymbol{u}_0]|_{S_{R_0}} = 0$ sufficient for constructing $\boldsymbol{u}_1$ is fulfilled. It remains to verify that $[\mu^{\pm}\Pi_0\mathbb{S}(\boldsymbol{u}_0'')\boldsymbol{N}]\big|_{S_{R_0}} = \boldsymbol{l}_3(\boldsymbol{u}_0, r_0)$. Indeed, it can be shown (see [90]) that

$$[\mu^{\pm}\Pi_0\mathbb{S}(\boldsymbol{u}_2)\boldsymbol{N}]\big|_{S_{R_0}} = \mu^+\Pi_0\mathbb{S}(\boldsymbol{u}_2^+)\boldsymbol{N}\big|_{S_{R_0}} = \mu^+\frac{\partial \boldsymbol{u}_2^+}{\partial \boldsymbol{N}}\Big|_{S_{R_0}} = \mu^+\boldsymbol{N}\times\frac{\partial^2\boldsymbol{\Phi}}{\partial \boldsymbol{N}^2}\Big|_{S_{R_0}}$$
$$= \boldsymbol{l}_3(\boldsymbol{u}_0, r_0) - [\mu^{\pm}\Pi_0\mathbb{S}(\boldsymbol{u}_1)\boldsymbol{N}]\Big|_{S_{R_0}}.$$

Hence, $\boldsymbol{u}_0'' = \boldsymbol{u}_1 + \boldsymbol{u}_2$ exists and satisfies (12.3.10) together with r_0''.

The initial functions $\boldsymbol{u}_0' = \boldsymbol{u}_0 - \boldsymbol{u}_0''$, $r_0' = r_0 - r_0''$ satisfy orthogonality conditions (12.2.7) and compatibility ones (12.2.9). Consequently, by Theorem 12.2.2, problem (12.3.8) is solvable on an infinite time interval and

$$\begin{aligned}
&\|e^{\beta t}\boldsymbol{u}'\|_{\boldsymbol{W}_2^{2+l,1+l/2}(D_{\Sigma T})} + \|e^{\beta t}\nabla q'\|_{\boldsymbol{W}_2^{l,l/2}(D_{\Sigma T})} + \|e^{\beta t}q'\|_{W_2^{0,l/2}(D_{\Sigma T})}\\
&\quad + \|e^{\beta t}r'\|_{W_2^{5/2+l,5/4+l/2}(G_T)} + \|e^{\beta t}\mathcal{D}_t r'\|_{W_2^{3/2+l,3/4+l/2}(G_T)}\\
&\quad \leqslant c_0\Big\{\|\boldsymbol{u}_0'\|_{\boldsymbol{W}_2^{1+l}(B)} + \|r_0'\|_{W_2^{2+l}(S_{R_0})}\Big\}, \quad \forall T \leqslant \infty,
\end{aligned} \tag{12.3.11}$$

whence, according to Remark 12.2.1, it follows that

$$\|\boldsymbol{u}'(\cdot,T)\|^2_{\boldsymbol{W}_2^{1+l}(B)} + \|r'(\cdot,T)\|^2_{W_2^{2+l}(S_{R_0})} \leqslant c_1 e^{-2\beta T}\Big\{\|\boldsymbol{u}_0'\|^2_{\boldsymbol{W}_2^{1+l}(B)} + \|r_0'\|^2_{W_2^{2+l}(S_{R_0})}\Big\}. \tag{12.3.12}$$

So, to get (12.3.3), we fix $T = T_0$ such that

$$c_1 e^{-\beta T_0} \leqslant \frac{\theta}{2} < \frac{1}{2}.$$

As for problem (12.3.9), it is solvable in the case of sufficiently small $\varepsilon(T_0)$ in (12.3.1). Indeed, a solution can be constructed by successive approximations according to the scheme:

$$\begin{aligned}
&\mathcal{D}_t\boldsymbol{u}_{m+1}'' - \nu^{\pm}\nabla^2\boldsymbol{u}_{m+1}'' + \frac{1}{\rho^{\pm}}\nabla q_{m+1}'' = \boldsymbol{l}_1(\boldsymbol{u}_m, q_m, r_m) + \widehat{\boldsymbol{f}}_m(y,t),\\
&\qquad \nabla\cdot\boldsymbol{u}_{m+1}'' = l_2(\boldsymbol{u}_m, r_m) \quad \text{in } B^{\pm},\ t>0,\\
&\int_{B^+} q_{m+1}''(y,t)\,\mathrm{d}y = 0,\\
&\boldsymbol{u}_{m+1}''(y,0) = \boldsymbol{u}_0''(y) \quad \text{in } B, \quad r_{m+1}''(y,0) = r_0''(y) \ \text{ on } S_{R_0},\\
&[\boldsymbol{u}_{m+1}'']\big|_{S_{R_0}} = 0, \quad [\mu^{\pm}\Pi_0\mathbb{S}(\boldsymbol{u}_{m+1}'')\boldsymbol{N}]\big|_{S_{R_0}} = \boldsymbol{l}_3(\boldsymbol{u}_m, r_m), \quad \boldsymbol{u}_{m+1}''|_{\Sigma} = 0,\\
&[-q_{m+1}'' + \mu^{\pm}\boldsymbol{N}\cdot\mathbb{S}(\boldsymbol{u}_{m+1}'')\boldsymbol{N}]\big|_{S_{R_0}} - \sigma\mathcal{B}_0 r_{m+1}'' = l_4(\boldsymbol{u}_m, r_m) + \sigma l_5(r_m),\\
&\mathcal{D}_t r_{m+1}'' - \Big(\boldsymbol{u}_{m+1}'' - \frac{1}{|B^+|}\int_{B^+}\boldsymbol{u}_{m+1}''\,\mathrm{d}y\Big)\cdot\boldsymbol{N} = l_6(\boldsymbol{u}_m, r_m) \quad \text{on } S_{R_0},
\end{aligned} \tag{12.3.13}$$

where $m = 1, 2, \ldots$, $\boldsymbol{u}_m = \boldsymbol{u}' + \boldsymbol{u}''_m$, $q_m = q' + q''_m$, $r_m = r' + r''_m$, $\widehat{\boldsymbol{f}}_m = \boldsymbol{f}\big(e_{r_m,\boldsymbol{h}_m}(y,t),t\big)$, $\boldsymbol{h}_m = |B^+|^{-1}\int_0^t\int_{B^+}\boldsymbol{u}_m(y,\tau)L_m(y,\tau)\,\mathrm{d}y\,\mathrm{d}\tau$, $L_m = L|_{\boldsymbol{u}=\boldsymbol{u}_m}$. For $m = 0$, we set $q''_0 = 0$, while we define $\boldsymbol{u}''_0(y,t)$ and $r''_0(y,t)$ as functions satisfying the initial conditions $\boldsymbol{u}''_0(y,0) = \boldsymbol{u}''_0(y)$, $r''_0(y,0) = r''_0(y)$, $\mathcal{D}_t r''_0(y,0) = 0$ $\big(\boldsymbol{u}''_0(y), r''_0(y)$ are constructed above$\big)$ and the inequalities

$$\begin{aligned}
\|\boldsymbol{u}''_0\|_{\boldsymbol{W}_2^{2+l,1+l/2}(D_{\Sigma T_0})} &+ \|r''_0\|_{W_2^{5/2+l,5/4+l/2}(G_{T_0})} + \|\mathcal{D}_t r''_0\|_{W_2^{3/2+l,3/4+l/2}(G_{T_0})} \\
&\leqslant c\Big\{\|\boldsymbol{u}''_0\|_{\boldsymbol{W}_2^{1+l}(B)} + \|r''_0\|_{W_2^{2+l}(S_{R_0})}\Big\} \\
&\leqslant c_1\varepsilon\Big\{\|\boldsymbol{u}_0\|_{\boldsymbol{W}_2^{1+l}(B)} + \|r_0\|_{W_2^{2+l}(S_{R_0})}\Big\}. \qquad (12.3.14)
\end{aligned}$$

Such $\boldsymbol{u}''_0, r''_0$ exist in view of inverse trace theorem and (12.3.10).

If the functions $\boldsymbol{u}''_m, q''_m, r''_m$ are known, then $\boldsymbol{u}''_{m+1}, q''_{m+1}, r''_{m+1}$ can be found by Theorem 12.2.1 as a solution to problem (12.3.13). By virtue of (12.2.3), (12.3.5) and (12.3.14),

$$Y''_{m+1} \equiv Y(\boldsymbol{u}''_{m+1}, q''_{m+1}, r''_{m+1}) \leqslant c(T_0)\Big\{\|\widehat{\boldsymbol{f}}_m\|_{\boldsymbol{W}_2^{l,l/2}(Q_{T_0})} + \varepsilon N_0 + Y''^2_m\Big\}, \qquad (12.3.15)$$

where $N_0 = \|\boldsymbol{u}_0\|_{\boldsymbol{W}_2^{1+l}(B)} + \|r_0\|_{W_2^{2+l}(S_{R_0})}$,

$$\begin{aligned}
Y(\boldsymbol{u}, q, r) \equiv &\|\boldsymbol{u}\|_{\boldsymbol{W}_2^{2+l,1+l/2}(D_{\Sigma T_0})} + \|\nabla q\|_{\boldsymbol{W}_2^{l,l/2}(D_{\Sigma T_0})} + \|q\|_{W_2^{0,l/2}(D_{\Sigma T_0})} \\
&+ \|r\|_{W_2^{5/2+l,5/4+l/2}(G_{T_0})} + \|\mathcal{D}_t r\|_{W_2^{3/2+l,3/4+l/2}(G_{T_0})},
\end{aligned}$$

$\widehat{\boldsymbol{f}}_m = \boldsymbol{f}\big(e_{r_m,\boldsymbol{h}_m}(y,t),t\big)$; $\quad e_{r_m,\boldsymbol{h}_m}(y,t) = y + r^*_m(y,t)\boldsymbol{N}^*(y) + \chi(y)\boldsymbol{h}_m(t)$, moreover,

$$\begin{aligned}
Y_m \equiv &Y(\boldsymbol{u}_m, q_m, r_m), \qquad Y' \equiv Y(\boldsymbol{u}', q', r'), \\
N^{(m)}(T_0) \equiv &\|\boldsymbol{u}_m(\cdot, T_0)\|_{\boldsymbol{W}_2^{1+l}(B)} + \|r_m(\cdot, T_0)\|_{W_2^{2+l}(S_{R_0})}, \quad m \geqslant 1.
\end{aligned}$$

We show by induction that (12.3.15) yields estimates for the norms Y''_m and Y_m uniform on m.

To this end, let us assume that $\boldsymbol{u}_m, r_m$ satisfy (12.3.4) with δ so small that, in view of (12.3.6), (12.3.11),

$$\begin{aligned}
\|\widehat{\boldsymbol{f}}_m\|_{\boldsymbol{W}_2^{l,l/2}(Q_{T_0})} &\leqslant c_{\boldsymbol{f}}\|\boldsymbol{f}\|_{\boldsymbol{W}_2^{l,l/2}(Q_{T_0})} + c'_{\boldsymbol{f}}\|\nabla\boldsymbol{f}\|_{Q_{T_0}} Y(\boldsymbol{u}_m, 0, r_m) \\
&\leqslant c_{\boldsymbol{f}}\|\boldsymbol{f}\|_{\boldsymbol{W}_2^{l,l/2}(Q_{T_0})} + c'_{\boldsymbol{f}}\|\nabla\boldsymbol{f}\|_{Q_{T_0}}\big(c_0 N_0 + Y''_m\big) \\
&\leqslant c_{\boldsymbol{f}}\|\boldsymbol{f}\|_{\boldsymbol{W}_2^{l,l/2}(Q_{T_0})} + c_2\varepsilon N_0 + c'_{\boldsymbol{f}}\varepsilon Y''_m.
\end{aligned}$$

In addition, let

$$Y''_m \leqslant 2c(T_0)\Big(c_{\boldsymbol{f}}\|\boldsymbol{f}\|_{\boldsymbol{W}_2^{l,l/2}(Q_{T_0})} + c_2\varepsilon N_0\Big). \tag{12.3.16}$$

Then, on one hand, by (12.3.11),

$$Y_m \leqslant Y' + Y''_m \leqslant cN_0 + 2c(T_0)\Big(c_{\boldsymbol{f}}\|\boldsymbol{f}\|_{\boldsymbol{W}_2^{l,l/2}(Q_{T_0})} + c_2\varepsilon N_0\Big) \leqslant c\varepsilon, \tag{12.3.17}$$

which guarantees the smallness of δ, and, on the other hand, according to estimates (12.3.15), (12.3.16),

$$Y''_{m+1} \leqslant c(T_0)\Big(c_{\boldsymbol{f}}\|\boldsymbol{f}\|_{\boldsymbol{W}_2^{l,l/2}(Q_{T_0})} + c_2\varepsilon N_0\Big)\Big\{1 + 2c'_{\boldsymbol{f}}\varepsilon c(T_0) + \varepsilon/c_2$$
$$+ 4c^2(T_0)\big(c_{\boldsymbol{f}}\|\boldsymbol{f}\|_{\boldsymbol{W}_2^{l,l/2}(Q_{T_0})} + c_2\varepsilon N_0\big)\Big\} \leqslant 2c(T_0)\big\{c_{\boldsymbol{f}}\|\boldsymbol{f}\|_{\boldsymbol{W}_2^{l,l/2}(Q_{T_0})} + c_2\varepsilon N_0\big\},$$

provided that

$$2c'_{\boldsymbol{f}}\varepsilon c(T_0) + \varepsilon/c_2 + 4c^2(T_0)\big(c_{\boldsymbol{f}}\|\boldsymbol{f}\|_{\boldsymbol{W}_2^{l,l/2}(Q_{T_0})} + c_2\varepsilon N_0\big) \leqslant 1.$$

By virtue of (12.3.14), inequality (12.3.16) holds for $m = 0$, hence, it is satisfied for all m. Moreover, from (12.3.12) and (12.3.16) it follows that

$$N^{(m)}(T_0) \leqslant c\Big(\mathrm{e}^{-\beta T_0} + c'(T_0)\varepsilon\Big)N_0 + c''(T_0)\|\boldsymbol{f}\|_{\boldsymbol{W}_2^{l,l/2}(Q_{T_0})} \leqslant \theta N_0 + c''\|\boldsymbol{f}\|_{\boldsymbol{W}_2^{l,l/2}(Q_{T_0})}, \tag{12.3.18}$$

if

$$c\Big(\mathrm{e}^{-\beta T_0} + c'(T_0)\varepsilon\Big) \leqslant \theta.$$

The convergence of $(\boldsymbol{u}''_m, q''_m, r''_m)$ to a solution of (12.3.9) follows from inequality (12.2.3) and (12.3.7).

Letting $m \to \infty$ in (12.3.17), (12.3.18), we arrive at (12.3.2), (12.3.3). □

Now we can complete the proof of Theorem 12.1.1.

Proof of Theorem 12.1.1 We extend a local solution of (12.1.3) guaranteed by Theorem 12.3.1 into the interval $t > 0$ step by step: first to the interval $(T_0, 2T_0)$, then to $(2T_0, 3T_0)$ and so forth. Let us suppose the solution is already found for $t < kT_0$. Then it can be defined for $t \in (kT_0, (k+1)T_0)$ as a solution to the problem with the initial conditions $\boldsymbol{u}(y, kT_0) = \boldsymbol{u}(y, kT_0 - 0) \equiv \boldsymbol{u}_k(y)$, $r(y, kT_0) = r(y, kT_0 - 0) \equiv r_k(y)$.

We write transformation (12.1.2) for $t > kT_0$ as follows:

$$x = y + \boldsymbol{h}(kT_0)\chi(y) + \boldsymbol{k}(t, k)\chi(y) + \boldsymbol{N}^*(y)r^*(y, t), \tag{12.3.19}$$

where $\boldsymbol{h}(kT_0)$ is already found and $\boldsymbol{k}(t,k) = \boldsymbol{h}(t) - \boldsymbol{h}(kT_0)$. The elements of the Jacobi matrix of this transformation are given by the formula

$$\mathbb{L}_{ij} = \Big\{\delta_{ij} + \big(h_i(kT_0) + k_i(t,k)\big)\frac{\partial\chi(y)}{\partial y_j} + \frac{\partial\big(N_i^* r^*(y)\big)}{\partial y_j}\Big\}_{i,j=1}^3.$$

Proposition 12.3.1 can be reformulated in the following way.

Proposition 12.3.2 *Let $k \in \mathbb{N}$. If inequality* (12.3.4) *holds for $t > kT_0$ and $|\boldsymbol{h}(kT_0)| \leqslant \delta$, then*

$$Z_k(\boldsymbol{u}, q, r) \leqslant c\Big\{\delta Y_k(\boldsymbol{u}, q, r) + Y_k^2(\boldsymbol{u}, q, r)\Big\},$$

where Z_k and Y_k are norms (12.3.5) *and* (12.2.19), *respectively, computed for $t \in (kT_0, (k+1)T_0)$. In addition, $\hat{\boldsymbol{f}}$ satisfies inequalities* (12.3.6) *and* (12.3.7) *on this time interval as well.*

Let us consider the case $k = 1$. From (12.3.2) and (12.3.3) it follows that

$$N_1 \equiv N(\boldsymbol{u}_1, r_1) \leqslant C\varepsilon.$$

Hence, by replacing ε with $C^{-1}\varepsilon$, we see that this problem is solvable on the interval $(T_0, 2T_0)$ and the estimates

$$Y_1^2(\boldsymbol{u}, q, r) \leqslant c\Big\{N_1^2 + \|\boldsymbol{f}\|^2_{\boldsymbol{W}_2^{l,l/2}(Q_{T_0,2T_0})}\Big\},$$

$$N_2^2 \leqslant \theta^2 N_1^2 + c\|\boldsymbol{f}\|^2_{\boldsymbol{W}_2^{l,l/2}(Q_{T_0})} \leqslant C\varepsilon^2,$$

are satisfied, where

$$N_k = N(\boldsymbol{u}_k, r_k).$$

The constants in these estimates need not coincide with the constants in (12.3.2), (12.3.3), because of the presence of the extra term with $\boldsymbol{h}(T_0)$ in (12.3.19), but, as will be shown below, the differences between these constants are of order δ for all $k > 0$.

If the solution is found for $t < kT_0$ and the inequalities

$$N_j^2 \leqslant \theta^2 N_{j-1}^2 + c\|\boldsymbol{f}\|^2_{W_2^{l,l/2}(Q_{(j-1)T_0,jT_0})}, \qquad \theta < 1, \tag{12.3.20}$$

$$Y_j^2 \leqslant c\Big\{N_{j-1}^2 + \|\boldsymbol{f}\|^2_{W_2^{l,l/2}(Q_{(j-1)T_0,jT_0})}\Big\}, \quad j = 1, \ldots, k-1,$$

are proved, then

$$N_j^2 \leqslant \ldots \leqslant \theta^{2j} N_0^2 + c\sum_{i=0}^{j-1} \theta^{2(j-1-i)} \|\boldsymbol{f}\|^2_{W_2^{l,l/2}(Q_{iT_0,(i+1)T_0})} \leqslant c\theta^{2(j-1)}\varepsilon^2 \tag{12.3.21}$$

with the constants c independent of j (we have used inequalities (12.1.10) for $\boldsymbol{f}$). Since $\theta^j \to 0$ for $j \to \infty$, the right-hand side of (12.3.21) is less than ε^2 for $j \geqslant j_0$, and the replacement of ε with $C^{-1}\varepsilon$ can be made only a finite number of times.

The estimate of $\boldsymbol{h}(jT_0)$ can be obtained at every step. Let $\theta_1 > \theta$ ($\theta_1 = \mathrm{e}^{-aT_0}$, $a < b$). We take the sum of (12.3.20) multiplied by θ_1^{-2j}. This leads to

$$\sum_{j=0}^{k} \theta_1^{-2j} N_j^2 \leqslant N_0^2 + \frac{\theta^2}{\theta_1^2} \sum_{j=1}^{k} \theta_1^{-2j+2} N_{j-1}^2 + c\sum_{j=1}^{k} \theta_1^{-2j} \|\boldsymbol{f}\|^2_{W_2^{l,l/2}(Q_{(j-1)T_0,jT_0})}$$

and

$$\sum_{j=0}^{k} \theta_1^{-2j} N_j^2 \leqslant \frac{\theta_1^2}{\theta_1^2-\theta^2} N_0^2 + \frac{c\theta_1^2}{\theta_1^2-\theta^2} \sum_{j=1}^{k} \theta_1^{-2j} \|\boldsymbol{f}\|^2_{W_2^{l,l/2}(Q_{(j-1)T_0,jT_0})}.$$

Hence, by the embedding theorem,

$$|\boldsymbol{h}(kT_0)| = \frac{3}{4\pi R_0^3} \Big| \int_0^{kT_0} \int_{\Omega_t^+} \boldsymbol{v}(\cdot,t)\,\mathrm{d}x\,\mathrm{d}t \Big| \leqslant c\sqrt{T_0} \Big(\sum_{j=0}^{k-1} \theta_1^{-2j} \int_{jT_0}^{(j+1)T_0} \|\boldsymbol{u}(\cdot,t)\|^2_{W_2^{l+1}(B+)}\,\mathrm{d}t \Big)^{\frac{1}{2}}$$

$$\leqslant c\Big(N_0^2 + \sum_{j=0}^{k-1} \theta_1^{-2j} \|\boldsymbol{f}\|^2_{W_2^{l,l/2}(Q_{jT_0,(j+1)T_0})} \Big)^{\frac{1}{2}} \leqslant c\varepsilon \tag{12.3.22}$$

with the constants c independent of k. Finally, by passing to the limit as $k \to \infty$ in

$$\sum_{j=0}^{k} \theta_1^{-2j} Y_j^2(\boldsymbol{u}, q, r) \leqslant c\Big\{ N_0^2 + \sum_{j=0}^{k} \theta_1^{-2j} \|\boldsymbol{f}\|^2_{W_2^{l,l/2}(Q_{jT_0,(j+1)T_0})} \Big\},$$

we obtain an inequality equivalent to (12.1.11). In addition, the passage to the limit in (12.3.22) allows us to estimate the limiting position $h(\infty)$ of the inner drop barycenter:

$$|h(\infty)| \leqslant c_2 \Big\{ \|\mathrm{e}^{at}\boldsymbol{f}\|_{W_2^{l,l/2}(Q_\infty)} + \|\boldsymbol{u}_0\|_{\boldsymbol{W}_2^{1+l}(B)} + \|r_0\|_{W_2^{2+l}(S_{R_0})} \Big\} \leqslant 2c_2\varepsilon. \tag{12.3.23}$$

From embedding theorems, it follows that

$$\max_{G_\infty} |r| \leqslant c_1\Big\{\|e^{at}\boldsymbol{f}\|_{W_2^{l,l/2}(Q_\infty)} + \|\boldsymbol{u}_0\|_{\boldsymbol{W}_2^{1+l}(B)} + \|r_0\|_{W_2^{2+l}(S_{R_0})}\Big\}.$$

It is clear that if $2(c_1 + c_2)\varepsilon$ is less than the initial distance between the surfaces Γ_t and Σ, the intersection of these surfaces will be never possible for any $t > 0$. □

We show that one can construct a solution to problem (12.1.3) under less restrictive assumptions on $\boldsymbol{f}$.

We introduce the norms

$$|||\boldsymbol{u}, q, r||| = \sum_{j=0}^{\infty} Y_j(\boldsymbol{u}, q, r), \quad |||\boldsymbol{f}||| = \sum_{j=0}^{\infty} \|\boldsymbol{f}\|_{W_2^{l,l/2}(Q_{jT_0,(j+1)T_0})}, \tag{12.3.24}$$

$$|||\boldsymbol{f}|||_\eta = \sum_{j=0}^{\infty} \eta_j^{-1}\|\boldsymbol{f}\|_{W_2^{l,l/2}(Q_{jT_0,(j+1)T_0})},$$

where $\eta = \{\eta_j\}_0^\infty \in (0,1)$ and $\eta_j \to 0$ for $j \to \infty$.

Theorem 12.3.2 *Let* $\boldsymbol{u}_0 \in W_2^{l+1}(B)$, $r_0 \in W_2^{2+l}(S_{R_0})$, *and let* $\boldsymbol{f}$ *have finite norms* (12.3.24). *Assume that compatibility conditions* (12.1.8), *as well as smallness conditions* (12.1.9) *and the inequalities*

$$\sup_{\tau>0} \|\mathcal{D}_x^{\boldsymbol{i}}\boldsymbol{f}\|_{Q_{\tau,\tau+T_0}} \leqslant \varepsilon, \quad |\boldsymbol{i}| = 1, 2, \quad |||\boldsymbol{f}||| + |||\boldsymbol{f}|||_\eta \leqslant \varepsilon$$

are satisfied. Then there exists a solution of (12.1.3) *defined for* $t > 0$ *and satisfying the estimate*

$$|||\boldsymbol{u}, q, r||| \leqslant c\Big\{\|\boldsymbol{u}_0\|_{W_2^{l+1}(B)} + \|r_0\|_{W_2^{2+l}(S_{R_0})} + |||\boldsymbol{f}|||\Big\}. \tag{12.3.25}$$

Proof We follow the arguments in the proof of Theorem 12.1.1 presented above. The inequalities

$$N_j \leqslant \theta N_{j-1} + c\|\boldsymbol{f}\|_{W_2^{l,l/2}(Q_{(j-1)T_0,jT_0})}, \quad Y_j \leqslant c\big(N_{j-1} + c\|\boldsymbol{f}\|_{W_2^{l,l/2}(Q_{jT_0,(j+1)T_0})}\big),$$

equivalent to (12.3.20), imply that

$$N_j \leqslant \theta^j N_0 + c\sum_{i=0}^{j-1} \theta^{j-1-i}\|\boldsymbol{f}\|_{W_2^{l,l/2}(Q_{iT_0,(i+1)T_0})} \leqslant \theta^j N_0 + c\varkappa_j|||\boldsymbol{f}|||_\eta,$$

where $\varkappa_j = \max_{i \leqslant (j-1)} \theta^{j-1-i}\eta_i \leqslant \max(\theta^{[(j-1)/2]}\eta_{[(j-1)/2]}) \to 0$ as $j \to \infty$, $[k]$ means the integral part of k. Consequently, the solution of (12.1.3) is extendable to the whole half-axis $t > 0$ (if $h(kT_0)$ is small); moreover, we have

$$\sum_{j=1}^{k} N_j \leqslant \frac{1}{1-\theta}\Big\{N_0 + c\sum_{i=0}^{k} \|\boldsymbol{f}\|_{W_2^{l,l/2}(Q_{iT_0,(i+1)T_0})}\Big\}$$

and

$$\sum_{j=0}^{k} Y_j(\boldsymbol{u}, q, r) \leqslant c\Big\{N_0 + \sum_{i=0}^{k-1} \|\boldsymbol{f}\|_{W_2^{l,l/2}(Q_{iT_0,(i+1)T_0})}\Big\}.$$

Using this inequality, we estimate $h(kT_0)$: $|h(kT_0)| \leqslant c\varepsilon$. Letting $k \to \infty$, we arrive at (12.3.25). □

Conclusions

We summarize our research. Let us formulate the results presented in the book for the problem on the motion of two incompressible fluids with an unknown interface in the complete setting. So, we have proved:

(1) The existence of a local-in-time unique solution to the problem in the Hölder spaces with power-law weight at infinity in $\mathbb{R}^3$, while time interval on which the solution exists depends on the data of the problem;
(2) Global-in-time solvability in the ordinary Hölder spaces of the problem with nonnegative coefficient of surface tension on fluid interface in a bounded domain for small initial data;
(3) Local unique solvability in the Hölder spaces of the problem on thermocapillary convection for a drop in a liquid medium;
(4) The existence of a unique solution to the two-phase problem in the Oberbek–Boussinesq approximation on a sufficiently small time interval for arbitrary data and on an infinite time interval for small data;
(5) The existence of a global solution in the Sobolev–Slobodetskiĭ spaces to the problem with positive surface tension coefficient in a closed vessel for sufficiently small initial velocities, small mass forces decreasing at infinity and initial interface close to the sphere and not touching the walls of the vessel, the interface being smoother by 1/2 in comparison with the outer boundary;
(6) Global L_2-solvability of the problem without taking surface tension into account. (Problem data should be small but the interface may have any shape; its initial and further regularity coincides with the smoothness of a solid boundary.)

We conclude that the estimates obtained for a solution of the problem governing the motion of a two-phase fluid in a bounded domain and with the surface tension guarantee the stability of a solution over time, i.e., the

© The Author(s), under exclusive license to Springer Nature Switzerland AG 2021

I. V. Denisova, V. A. Solonnikov, *Motion of a Drop in an Incompressible Fluid*, Advances in Mathematical Fluid Mechanics,
https://doi.org/10.1007/978-3-030-70053-9

velocity of the fluid decays with time, the pressure tends to a time-dependent step function, and the interface does to a sphere centered at the barycenter of the internal fluid; moreover, this center shifts in the general case from the initial position, no matter how small the initial data of the problem are.

In the absence of surface tension, a similar result has been obtained for an arbitrary smooth initial interface. If the data are small enough, this surface changes slowly and its limiting shape is close to the initial one.

References

1. Abels H., On general solutions of two-phase flows for viscous incompressible fluids. Interfaces Free Bound. **9**(1), 31–65 (2007)
2. R.A. Adams, *Sobolev Spaces* (Academic Press, New York, San Francisco, London, 1975), 270 p.
3. M.S. Agranovich, M.I. Vishik, Elliptic problems with a parameter and parabolic problems of general type. Uspekhi Mat. Nauk **19**(117) , 53–161 (1964) (English: Russ. Math. Surv. **19**, 53–157 (1964))
4. G. Allain, Small-time existence for the Navier–Stokes equations with a free surface. Appl. Math. Optim. **16**(1), 37–50 (1987)
5. L.K. Antanovskiĭ, B.K. Kopbosynov, Nonstationary thermocapillary drift of a drop of viscous liquid. J. Appl. Mech. Tech. Phys. **27**, 208–213 (1986). https://doi.org/10.1007/BF00914730
6. J.T. Beale, The initial value problem for the Navier–Stokes equation with a free boundary. Commun. Pure Appl. Math. **34**(3), 359–392 (1981)
7. J.T. Beale, Large-time regularity of viscous surface waves. Arch. Ration. Mech. Anal. **84**(4), 307–352 (1984)
8. J. Bemelmans, Liquid drop in a viscous fluid under the influence of gravity and surface tension. Manuscripta Math. **36**(1), 105–123 (1981)
9. O.V. Besov, V.P. Il'in, S.M. Nikolskiĭ, *Integral Representation of Functions and Theorems of Imbedding* (Nauka, Moscow, 1975), 480 p.
10. G.I. Bizhanova, V.A. Solonnikov,On free boundary problems for second-order parabolic equations. Algebra Analis **12**(6), 98–139 (2000) (in Russian) (English transl. in St. Petersburg Math. J. **12**(6), 949–981) (2001)
11. V.O. Bytev, Unsteady motion of rotating ring of viscous incompressible liquid with free boundary. PMTF **3**, 82–88 (1970) (in Russian) (English transl. in J. Appl. Mech. Tech. Phys. **11**, 432–438 (1970). https://doi.org/10.1007/BF00908073
12. I.V. Denisova, Investigation of the problem of droplet motion in a liquid medium. Preprint LOMI, R-9-89, Leningrad: LOMI AN SSSR, 1989, 16 pp. (in Russian)
13. I.V. Denisova, The motion of a drop in a flow of a liquid. Dinamika Sploshn. Sredy SOAN SSSR **93/94**, 32–37 (1989) (in Russian)

© The Author(s), under exclusive license to Springer Nature Switzerland AG 2021

I. V. Denisova, V. A. Solonnikov, *Motion of a Drop in an Incompressible Fluid*, Advances in Mathematical Fluid Mechanics,
https://doi.org/10.1007/978-3-030-70053-9

14. I.V. Denisova, Problem on unsteady motion of a drop in a viscous incompressible flow. Thesis for the Degree of Kandidat in Phys.-Math. Sci., Leningrad: LOMI AN SSSR (1989) (manual) 134 p. (in Russian)

15. I.V. Denisova, A priori estimates of the solution of a linear time dependent problem connected with the motion of a drop in a fluid medium. Trudy Mat. Inst. Steklov. **188**, 3–21 (1990) (in Russian) (English transl. in Proc. Steklov Inst. Math. **3**, 1–24) (1991)

16. I.V. Denisova, Solvability in Hölder spaces of a linear problem concerning the motion of two fluids separated by a closed surface. Algebra Analiz **5**(4), 122–148 (1993) (in Russian) (English transl. in St. Petersburg Math. J. **5**(4), 765–787 (1994)

17. I.V. Denisova, Problem of the motion of two viscous incompressible fluids separated by a closed free interface. Acta Appl. Math. **37**, 31–40 (1994)

18. I.V. Denisova, Classical solvability of the problem describing the evolution of a drop in a liquid medium, in *Navier–Stokes Equations and Related Nonlinear Problems* ed. by A. Sequeira (Plenum Press, New York, 1995), pp. 191–199

19. I.V. Denisova, On the problem of thermocapillary convection for two incompressible fluids separated by a closed interface. Progr. Nonlin. Differ. Equ. Their Appl. **61**, 45–64 (2005)

20. I.V. Denisova, Model problem connected with the motion of two incompressible fluids. Adv. Math. Sci. Appl. **17**(1), 195–223 (2007)

21. I.V. Denisova, Global solvability of a problem on two fluid motion without surface tension. Zap. Nauchn. Sem. S.-Peterburg. Otdel. Mat. Inst. Steklov. (POMI) **348**, 19–39 (2007). English transl. in J. Math. Sci. **152**(5), 625–637 (2008)

22. I.V. Denisova, Motion of two viscous immiscible fluids. Thesis for the degree of Doctor of Physics and Mathematics, St. Petersburg State University, St. Petersburg (2012) 333 p. (in Russian)

23. I.V. Denisova, Global L_2-solvability of a problem governing two-phase fluid motion without surface tension. Port. Math. **71**(1), 1–24 (2014)

24. I.V. Denisova, Global classical solvability of an interface problem on the motion of two fluids. RIMS Kokyuroku Ser. Kyoto Univ. **1875**, 84–108 (2014)

25. I.V. Denisova, Global solvability of the problem on two-phase capillary fluid motion in the Oberbeck–Boussinesq approximation, in *Mathematical Fluid Dynamics, Present and Future* ed. by Yu. Suzuki, Yo. Shibata. Springer Proc. Math. Stat., vol. 183 (2016), pp. 49–70

26. I.V. Denisova, Š. Nečasová, The Oberbeck–Boussinesq Approximation for the Motion of Two Incompressible Fluids, Zap. Nauchn. Sem. S.-Peterburg. Otdel. Mat. Inst. Steklov. (POMI) **362**, 92–119 (2008) (English transl. in J. Math. Sci. **159**(4), (2009) 436–451)

27. I.V. Denisova, V.A. Solonnikov, Solvability of the linearized problem on the motion of a drop in a liquid flow. Zap. Nauchn. Sem. Leningrad. Otdel. Mat. Inst. Steklov. (LOMI). **171**, 53–65 (1989) (in Russian) (English transl. in J. Soviet Math. **56**(2), 2309–2316) (1991)

28. I.V. Denisova, V.A. Solonnikov, Solvability in Hölder spaces for a model initial boundary–value problem generated by a problem on the motion of two fluids. Zap. Nauchn. Sem. Leningrad. Otdel. Mat. Inst. Steklov. (LOMI) **188**, 5–44 (1991) (in Russian) (English Transl. J. Math. Sci. **70**(3), 1717–1746) (1994)

29. I.V. Denisova, V.A. Solonnikov, Classical solvability of the problem on the motion of two viscous incompressible fluids. Algebra Anal. **7**(5), 101–142 (1995) (in Russian) (English transl. in St.Petersburg Math. J. **7**(5), 755–786 (1996))

30. I.V. Denisova, V.A. Solonnikov, Global solvability of a problem governing the motion of two incompressible capillary fluids. Zap. Nauchn. Sem. S.-Peterburg. Otdel. Mat. Inst. Steklov. (POMI) **397**, 20–52 (2011) (in Russian) (English Transl. J. Math. Sci. **185**(5), 668–686 (2012))

31. I.V. Denisova, V.A. Solonnikov, L_2-theory for a two-phase incompressible fluid with taking the surface tension into account, Preprint POMI, 12/2017. St. Petersburg, 2017, 29 pp. (in Russian). http://www.pdmi.ras.ru/preprint/2017/17-12.html

32. I.V. Denisova, V.A. Solonnikov, L_2-theory for two incompressible fluids separated by a free interface. Topol. Methods Nonlinear Anal. **52**, 213–238 (2018). https://doi.org/10.12775/TMNA.2018.019

33. Yo. Giga, Sh. Takahashi, On global weak solutions of the nonstationary two-phase Stokes flow. SIAM J. Math. Anal. **25**, 876–893 (1994)

34. K.K. Golovkin, Certain conditions for the smoothness of a function of several variables and estimates of convolution operators. Dokl. Akad. Nauk SSSR **139**(3), 524–527 (1961) (in Russian) (English transl. in Soviet Math. Doklady Acad. Nauk SSSR **139–141**, 949–953 (1961))

35. K.K. Golovkin, On equivalent norms in fractional spaces. Tr. Mat. Inst., Akad. Nauk SSSR, **66**, 364–383 (1962) (in Russian) (English Transl. Proc. Steklov Inst. Math. AMS Transl. Ser 2, **81**, 257–280 (1969))

36. K.K. Golovkin, V.A. Solonnikov, Estimates of convolution operators. Zap. Nauchn. Semin. Leningr., Otd. Mat. Inst. Steklova **7**, 6–86 (1968) (in Russian) (English transl. in Semin. Math., V.A. Steklov Math. Inst., Leningrad, vol. 7 (1968), pp. 1–36)

37. N.M. Günther, La théorie du potentiel et ses applications aux problèmes fondamentaux de la physique mathématique, Paris, Gauthler–Villars, 1934, 303 pp.

38. J. Hadamard, Mouvement permanent lent d'une sphère liquide et visqueuse dans un liquide visqueux. Compt. rend. Acad. sd. **152**(25), 1735–1738 (1911)

39. D.D. Joseph, Yu.Y. Renardy, *Fundamentals of Two-Fluids Dynamics*, Part I. Math. Theory and Appl. (Springer, Berlin, 1993)

40. M. Köhne, Ja. Prüss, M. Wilke, Qualitative behaviour of solutions for the two-phase Navier–Stokes equations with surface tension. Math. Ann. **356**(2), 737–792 (2013)

41. G. Korn, T. Korn, *Mathematical Handbook (for Scientists and Engineers)*, 2nd edn. (McGraw-Hill Book Company, New York 1968), 833 p.

42. O.A. Ladyzhenskaya, *The Mathematical Theory of Viscous Incompressible Flow*. (Nauka, Moscow, 1961), 288 p. (English transl. Gordon and Breach, Science Publishers, New York-London-Paris (1969))

43. O.A. Ladyzhenskaya, *Boundary Value Problems of Mathematical Physics* (Nauka, Moscow, 1973), 408 p. (English transl., Springer, Berlin, 1985)

44. O.A. Ladyzhenskaya, On multiplicators in Hölder spaces with nonhomogeneous metrics. Zap. Nauchn. Sem. S.-Peterburg. Otdel. Mat. Inst. Steklov. (POMI) **271**, 156–174 (2000) (in Russian) (English transl. in J. Math. Sci. **115**(6), 2792–2802 (2003))

45. O.A. Ladyzhenskaya, V.A. Solonnikov, Some problems of vector analysis and generalized formulations of boundary value problems for the Navier–Stokes equations, Zap. Nauchn. Sem. Leningrad. Otdel. Mat. Inst. Steklov. (LOMI) **59**, 81–116 (1976) (in Russian) (English transl. in J. Soviet Math. **10**(2) 257–286 (1978))

46. O.A. Ladyzhenskaya, N.N. Ural'tseva, *Linear and Quasilinear Equations of Elliptic Type* (Nauka, Moscow, 1964), 540 p. (in Russian) (English transl. in Academic Press, New York, 1968, 495 pp.; French. transl., Dunod, Paris, 1969)

47. O.A. Ladyzhenskaya, V.A. Solonnikov, N.N. Ural'tseva, *Linear and Quasilinear Equations of Parabolic Type* (Nauka, Moscow, 1967) (English Transl. Math. Monogr., vol 23, Amer. Math. Soc., Providence, 1968), 2nd edn., 1988, 648 pp.)

48. M.V. Lagunova, V.A. Solonnikov, Nonstationary problem of thermocapillary convection, Leningrad Branch of Steklov Mathematical Institute (LOMI). Preprint E-13-89, Leningrad 1989, 28 p.

49. L.D. Landau, E.M. Lifshitz, *Course of Theoretical Physics*, vol. 6 (Nauka, Moscow, 1986), 736 p. (English Transl. Landau and Lifshitz: Fluid Mechanics, Elsevier, Amsterdam, 2013) ISBN: 978-1-483-16104-4

50. O.M. Lavrentieva, *The Motion of a Viscous Rotating Ring Incompressible Fluid.* Dep. in VINITI, 27.11.84, No 7562-84, Moscow, 1984 (in Russian)

51. P.D. Lax, A.N. Milgram, Parabolic equations. Ann. Math. Stud. **33**, 167–189 (1954)

52. A. Lunardi, Maximal space regularity in nonhomogeneous initial boundary-value parabolic problem. Num. Funct. Anal. Optim. **10**, 323–349 (1989)

53. I.Sh. Moghilevskiĭ, V.A. Solonnikov, Solvability of a noncoercive initial boundary-value problem for the Stokes system in Hölder classes of functions. Z. Anal. Anwend. **8**(4), 329–347 (1989) (in Russian)

54. I.Sh. Mogilevskiĭ, V.A. Solonnikov, *On the Solvability of an Evolution Free Boundary Problem for the Navier–Stokes Equations in Hölder Spaces of functions.* Mathematical Problems Relating to Navier–Stokes Equations, Ser. on Advances in Math. Appl. Sci., ed. by G.P. Galdi, vol. 11 (World Sci. Publ., Singapore, 1992), pp. 105–181

55. A. Nouri, F. Poupaud, Y. Demay, *An Existence Theorem for the Multifluid Stokes Problem.* Prepubl. Math. No. 357, Univ. de Nice–Sophia–Antipolis, 1993

56. A. Nouri, F. Poupaud., An existence theorem for the multifluid Navier–Stokes problem. J. Differ. Equ. **123**, 71–88 (1995)

57. F.K.G. Odqvist, Uber die Randwetaufgaben der Hydrodynamik zäher Flüssigkeiten. Math. Z. **32**(3), 329–375 (1930)

58. M. Padula, On the exponential stability of the rest state of a viscous compressible fluid. J. Math. Fluid Mech. **1**, 62–77 (1999)

59. M. Padula, V.A. Solonnikov, On the local solvability of free boundary problem for the Navier–Stokes equations. Probl. Mat. Anal. **50**, 87–112 (2010) (J. Math. Sci. **170**(4), 522–553)

60. Ja. Prüss, G. Simonett, On the two-phase Navier–Stokes equations with surface tension. Interfaces Free Bound. **12**(3), 311–345 (2010)

61. Ja. Prüss, G. Simonett, in *Analytic solutions for the two-phase Navier–Stokes equations with surface tension and gravity.* Parabolic Problems, Progr. Nonlin. Diff. Eq. and Their Appl., ed. by Escher J. et al., vol. 80 (2011), pp. 507–540

62. Ja. Prüss, G. Simonett, *Moving Interfaces and Quasilinear Parabolic Evolution Equations.* Monographs in Mathematics, vol. 105 (Birkhäuser, Basel, 2016)

63. V.V. Pukhnachov, *Motion of a Viscous Fluid with Free Boundaries.* Textbook, Novosibirsk University, Novosibirsk, 1989, 96 p. (in Russian)

64. V.V. Pukhnachov, Thermocapillary convection under low gravity, in *Fluid Dynamics Transactions*, vol. 14 (Warszawa: PWN, 1989), pp. 145–204

65. V.Ya. Rivkind, The stationary motion of a weakly deformed drop in the flow of a viscous fluid. Zap. Nauchn. Sem. Leningrad. Otdel. Mat. Inst. Steklov. (LOMI) **69**, 157–170 (1977) (in Russian) (English transl. in J. Soviet Math. **10** (1), 110–119 (1978))

66. V.Ya. Rivkind, Stationary motion of a viscous drop taking into account its deformation. Zap. Nauchn. Sem. Leningrad. Otdel. Mat. Inst. Steklov. (LOMI) **84**, 220–243 (1979) (in Russian) (English transl. in J. Soviet. Math. **21**(3), 405–420 (1983))

67. V.Ya. Rivkind, A priori estimates and the method of successive approximations for solution of the problem of movement of a drop. Trudy Mat. Inst. Steklov. **159**, 150–166 (1983) (in Russian) (English transl. in Proc. Steklov Inst. Math. **159**, 155–172 (1984))

68. V.Ya. Rivkind, N.B. Friedman, On the Navier–Stokes equations with discontinuous coefficients. Zap. Nauchn. Sem. Leningrad. Otdel. Mat. Inst. Steklov. (LOMI) **38**, 137–148 (1973) (in Russian) (English transl. in J. Soviet Math. **8**(4), 456–464 (1977))

69. Ya.A. Roitberg, Z.G. Sheftel', *Nonlocal boundary-value problems for elliptic equations and systems.* Sib. Matem. J. **XIII**(1), 165–181 (1972) (in Russian) (English transl. in Sib. Math. J. **13**, 118–129 (1972) https://doi.org/10.1007/BF00967646

70. W. Rybczynski, *Über die fortschreitende Bewegung einer flüssigen Kugel in einem zähen Medium.* Bull. Int. Acad., Sci. Cracovia, Cl. Sci. Math. Nat., Ser. A (1911), pp. 40–44

71. J. Schauder, Potentialtheoretische Untersuchungen. Math. Z. **33**, 602–640 (1931)

72. Yo. Shibata, S. Shimizu, Maximal $L_p - L_q$-regularity for the two-phase Stokes equations. Model problems. J. Differ. Equ. **251**, 373–419 (2011)

73. S. Shimizu, Local solvability of free boundary problems for the two-phase Navier–Stokes equations with surface tension in the whole space. Progr. Nonlin. Diff. Eq. Their Appl. **80**, 647 686 (2011)

74. V.A. Solonnikov, A priori estimates for some boundary value problems. Dokl. Akad. Nauk SSSR **138**(4), 781–784 (1961) (in Russian)

75. V.A. Solonnikov, A priori estimates of a solution to a second-order equation of parabolic type, in *Boundary Value Problems of Mathematical Physics. Part 1, Collection of Articles.* Trudy Mat. Inst. Steklov., vol. 70, Moscow–Leningrad (1964), pp. 133–212 (in Russian)

76. V.A. Solonnikov, Estimates of the solution of the non-stationary Navier–Stokes system. Zap. Nauchn. Sem. Leningrad. Otdel. Mat. Inst. Steklov. (LOMI) **38**, 153–231 (1973) (English transl. in J. Soviet Math. **8**(4), 467–529 (1977))

77. V.A. Solonnikov, Estimates of the solution of an initial-boundary value problem for a linear nonstationary Navier–Stokes system, Zap. Nauchn. Sem. Leningrad. Otdel. Mat. Inst. Steklov. (LOMI) **59**, 178–254 (1976) (in Russian) (English transl. in J. Soviet Math. **10**(2), 336–393 (1978))

78. V.A. Solonnikov, On non-stationary motion of a finite liquid mass bounded by a free surface. Zap. Nauchn. Sem. S.-Peterburg. Otdel. Mat. Inst. Steklov. (POMI) **152**, 137–157 (1986) (in Russian) (English transl. in J. Soviet Math. **40**(5), 672–686 (1988))

79. V.A. Solonnikov, On the evolution of an isolated volume of a viscous incompressible capillary fluid for large values of time. Vestn. LSU, Ser. 1 **3**(15), 49–55 (1987) (in Russian) (English transl. in Vestn. Leningr. Univ., Math. **20**(3), 52–58 (1987; Zbl 0654.76029))

80. V.A. Solonnikov, On the transient motion of an isolated volume of viscous incompressible fluid, Izv. Akad. Nauk SSSR, Ser. Mat. **51**(5), 1065–1087 (1987) (in Russian) (English transl. in Math. USSR–Izv. **31**(2), 381–405 (1988))

81. V.A. Solonnikov, On non-stationary motion of a finite isolated mass of self-gravitating fluid. Algebra i Analiz. **1**(1), 207–249 (1989) (in Russian) (English transl. in Leningrad Math. J. **1**(1), 227 276 (1990))

82. V.A. Solonnikov, On an initial-boundary value problem for the Stokes systems arising in the study of a problem with a free boundary. Trudy Mat. Inst. Steklov. **188**, 150–188 (1990) (in Russian) (English transl. in Proc. Steklov Inst. Math. **3**, 191–239 (1991))

83. V.A. Solonnikov, Solvability of the problem of evolution of a viscous incompressible fluid bounded by a free surface on a finite time interval. Algebra i Analiz **3**(1), 222–257 (1991) (in Russian) (English transl. in St. Petersburg Math. J. **3**(1), 189–220 (1992))

84. V.A. Solonnikov, On a steady motion of a drop in an infinite liquid medium. Zap. Nauchn. Sem. S.-Peterburg. Otdel. Mat. Inst. Steklov. (POMI) **233**, 233–254 (1996) (English transl. in J. Math. Sci. **93**(5), 784–799 (1999))

85. V.A. Solonnikov, Estimates of solutions of the second initial boundary–value problem for the Stokes system in the spaces of functions with Hölder continuous derivatives with respect to the spacial variables. Zap. Nauchn. Sem. S.-Peterburg. Otdel. Mat. Inst. Steklov. (POMI) **259**, 254–279 (1999) (English transl. in J. Math. Sci. **109**(5), 1997–2017 (2002))

86. V.A. Solonnikov, Initial boundary–value problem for generalized Stokes equations in the half-space. Zap. Nauchn. Sem. S.-Peterburg. Otdel. Mat. Inst. Steklov. (POMI) **271**, 224–275 (2000) (in Russian) (English transl. J. Math. Sci. **115**(6) (2003))

87. V.A. Solonnikov, Generalized energy estimates in a free boundary problem for a viscous incompressible fluid. Zap. Nauchn. Sem. S.-Peterburg. Otdel. Mat. Inst. Steklov. (POMI) **282**, 216–243 (2001) (English transl. in J. Math. Sci. **120**(5), 1766–1783 (2004))

88. V.A. Solonnikov, Lectures on evolution free boundary problems: classical solutions. Lect. Notes Math. **1812**, 123–175 (2003)

89. V.A. Solonnikov, On the stability of a uniformly rotating viscous incompressible self-gravitating liquid. Zap. Nauchn. Sem. S.-Peterburg. Otdel. Mat. Inst. Steklov. (POMI) **348**, 165–208 (2007) (English. transl. in J. Math. Sci. **152**, 713–740 (2008). https://doi.org/10.1007/s10958-008-9090-7

90. V.A. Solonnikov, On problem of stability of equilibrium figures of uniformly rotating viscous incompressible liquid, in *Instability in Models Connected with Fluid Flows. II.* Int. Math. Ser., ed. by C. Bardos, A. Fursikov, vol. 7 (Springer, New York, 2008), pp. 189–254

91. V.A. Solonnikov, On the linear problem arising in the study of a free boundary problem for the Navier–Stokes equations. Algebra i Analiz **22**(6), 235–269 (2010) (English transl. in St. Petersburg Math. J. **22**(6), 1023–1049 (2011)).

92. V.A. Solonnikov, L_p-theory of the problem of motion of two incompressible capillary fluids in a container. Probl. Mat. Anal. **75**, 93–152 (2014) (English. transl. in J. Math. Sci. **198**(6), 761–827 (2014))

93. V.A. Solonnikov, I.V. Denisova, Classical Well-Posedness of Free Boundary Problems in Viscous Incompressible Fluid Mechanics, in *Handbook of Mathematical Analysis in Mechanics of Viscous Fluids I* (Springer, Berlin, 2017), 1–86. https://doi.org/10.1007/978-3-319-10151-4_27-2

94. V.A. Solonnikov, V.E. Shchadilov, A certain boundary value problem for the stationary system of Navier–Stokes equations. Trudy Mat. Inst. Steklov. **125**, 196–210 (1973) (English. transl. in Proc. Steklov Inst. Math. **125**, 186–199 (1973))

95. Sh. Takahashi, On global weak solutions of the nonstationary two-phase Navier–Stokes flow. Adv. Math. Sci. Appl. **5**, 321–342 (1995)

96. N. Tanaka, Global existence of two phase nonhomogeneous viscous incompressible fluid flow. Commun. Partial Differ. Equ. **18**(1 and 2), 41–81 (1993)

97. N. Tanaka, Two-phase free boundary problem for viscous incompressible thermocapillary convection. Jpn. J. Mech. **21**, 1–41 (1995)

98. J.A. Thorpe, *Elementary Topics in Differential Geometry* (Springer, New York-Heidelberg-Berlin, 1978)

Printed in the United States
by Baker & Taylor Publisher Services